TETRAHEDRON ORGANIC CHEMISTRY SERIES
Series Editors: J E Baldwin, FRS & R M Williams

VOLUME 18

Ligand Coupling Reactions

With

Heteroatomic Compounds

Related Pergamon Titles of Interest

BOOKS

Tetrahedron Organic Chemistry Series:
CARRUTHERS: Cycloaddition Reactions in Organic Synthesis
DEROME: Modern NMR Techniques for Chemistry Research
GAWLEY & AUBÉ: Principles of Asymmetric Synthesis
HASSNER & STUMER: Organic Syntheses Based on Named Reactions
and Un-named Reactions
McKILLOP: Advanced Problems in Organic Reaction Mechanisms
OBRECHT: Solid Supported Combinatorial and Parallel Synthesis of Small
Molecular-Weight Compound Libraries
PERLMUTTER: Conjugate Addition Reactions in Organic Synthesis
SESSLER & WEGHORN: Expanded, Contracted & Isomeric Porphyrins
SIMPKINS: Sulphones in Organic Synthesis
TANG & LEVY: Chemistry of *C*-Glycosides
WILLIAMS: Synthesis of Optically Active Alpha-Amino Acids 2nd Edition*
WONG & WHITESIDES: Enzymes in Synthetic Organic Chemistry

JOURNALS

BIOORGANIC & MEDICINAL CHEMISTRY
BIOORGANIC & MEDICINAL CHEMISTRY LETTERS
TETRAHEDRON
TETRAHEDRON: ASYMMETRY
TETRAHEDRON LETTERS

*Full details of all Elsevier Science publications/free specimen copy of any Elsevier Science
journal are available on request from your nearest Elsevier Science office*

* In Preparation

Ligand Coupling Reactions With Heteroatomic Compounds

CNRS - Universités d'Aix-Marseille 1 et 3, Faculté des Sciences Saint Jérôme
Marseille, France

PERGAMON

An Imprint of Elsevier Science

ELSEVIER SCIENCE Ltd
The Boulevard, Langford Lane
Kidlington, Oxford OX5 1GB, UK

Library of Congress Cataloging-in-Publication Data
A catalog record from the Library of Congress has been applied for.

British Library Cataloguing in Publication Data
A catalogue record from the British Library has been applied for.

First Edition 1998

ISBN 0 08 042794 4 Hardcover
ISBN 0 08 042793 6 Flexicover

To Dominique and Adrien

Contents

FOREWORD

The whole field of ligand coupling has only emerged in the last decade as a basis for new synthetic transformations. As Professor Finet shows in this comprehensive survey, the recent clarification of reaction mechanisms of ligand coupling process around heteroatom centres, now provides an understanding of these reactions which are certain to permit their application in organic synthesis, thereby achieving transformations which are quite difficult to achieve by other methods. This book will be of interest to all chemists engaged in organic synthesis both in academia and industry.

J E Baldwin, FRS

Preface.

The concept of the role of ligand coupling in Organic Synthesis is due (in alphabetical order) to Oae and to Trost. These authors made seminal contributions to the field. The early work on iodonium compounds, also of seminal importance, by Beringer was confused by the belief, shared by many, that such reactions must be radical in character. The early work of McEwen on antimony compounds also does not distinguish between radical coupling and ligand coupling.

More recent work on bismuth, lead and iodonium compounds has clarified the importance of ligand coupling over radical coupling.

The field has now attained a maturity and universality where a comprehensive presentation of ligand coupling in Organic Synthesis in a single monograph is urgently needed.

This book by Dr. Jean-Pierre Finet, Directeur de Recherche au C.N.R.S. at Marseille, provides in a thorough and scholarly way, a balanced coverage of the whole field. It should receive a warm reception from synthetic and mechanistic chemists.

D.H.R. Barton

D.H.R. Barton

Dow Distinguished Professor of Chemical Invention

This book owes much to Professor Sir Derek Barton, first and foremost, for his many seminal contributions to the chemistry of ligand coupling reactions with various heteroatomic elements, and second for giving me the opportunity to work in this field and to bring my own contribution to a better understanding of the importance of this mechanistic concept in heteroatomic chemistry. It is with a very great sadness for me that Professor Barton passed away just a few weeks before this book, that he had initiated, was published. I would like to dedicate it to his memory as a tribute to the perenniality of his enlightening spirit.

This book would not have been possible without the support of the "Centre National de la Recherche Scientifique" which gave me the bread and butter together with the necessary freedom for bringing this enterprise to completion.

I am also very thankful to all my colleagues for their help, fruitful discussions and various contributions. I would particularly mention Professors Dervilla M.X. Donnelly, Kin-ya Akiba, Victor A. Dodonov, Shigeru Oae, John T. Pinhey and Paul Tordo, as well as Doctors Christiane Bernard-Henriet, Alexei Fedorov, Patrick J. Guiry, Alexei V. Gushchin, Pascal Suisse de Sainte Claire and Mr Sébastien Combes. I would also like to thank all the collaborators with whom I have been associated over the years of my involvement in this field. Their contributions helped me to shape up my understanding of this concept.

This book constitutes an attempt to cover as comprehensively as possible the various facets of the reactions of organic chemistry occurring by the ligand coupling mechanism. Considering the number of heteroelements which are covered in this monograph, I am highly conscious of the difficulties to reach this goal of complete coverage. Therefore, I am aware that the work of some contributors to the field will unfortunately have been overlooked or not been given its due. I do apologise in advance and comments are welcome.

This book will hopefully make organic chemists more familiar with the chemistry of the different elements reacting by this still unsufficiently recognized mechanism and will foster new developments in this promising field.

March 1998

Abbreviations

An	4-Methoxyphenyl
Bipy	Bipyridine
BPR	Berry pseudorotation
BTMG	*N-tert*-butyl-*N',N',N'',N''*-tetramethylguanidine
Bz	Benzoyl
Bzl	Benzyl
DMAP	4-Dimethylaminopyridine
DME	1,2-Dimethoxyethane
DMF	*N,N*-dimethylformamide
DMSO	Dimethylsulfoxide
DPE	1,1-Diphenylethylene
HMPA	Hexamethylphosphorotriamide
HTIB	Hydroxytosyloxyiodobenzene
L.C.	Ligand coupling
LDA	Lithium diisopropylamide
L.E.	Ligand exchange
Ms	Methanesulfonyl
Ox	Oxidation
Ψ	Pseudorotation
Phen	Phenanthroline
Py	Pyridyl
RT	room temperature
S.E.T.	single electron transfer
SP	square pyramid
TBP	trigonal bipyramid
TBDMS	*tert*-Butyldimethylsilyl
Tf	Trifluoromethanesulfonyl
THF	Tetrahydrofuran
THP	Tetrahydropyranyl
TMG	*N,N,N',N'*-tetramethylguanidine
TMS	Trimethylsilyl
Tol	Methylphenyl
TR	Turnstile rotation
Ts	4-Toluenesulfonyl

Chapter 1

Introduction

1.1 DEFINITIONS

The word "**ligand**" was first introduced by Stock in 1917 in a discussion on the similarities between carbon and silicon chemistry.[1] This term has been mostly used by chemists to describe atoms or groups attached to a central atom in coordination compounds or organometallic compounds.[2] It has been used only infrequently by organic chemists, although Fieser employed it to describe the substituents of a phosphorus compound.[3] In the past decade, it has gained wider acceptance in organic chemistry, in agreement with its original definition as given by Stock:[1]

> "Zur Verhütung von Mißverständnissen sind die Bedeutung einiger hier gebrauchter Worte erklärt. Affinität ist der Ausdruck für die Festigkeit, mit welcher ein Element andere Elemente oder Radicale (allgemein: »Liganden« [ligare, binden]; die Einführung eines - bisher fehlenden - Wortes für diesen, wohl ohne weiteres verständlichen Begriff vereinfacht die Ausdruckweise) bindet. - Valenz bedeutet die Krafteinheit, die einen einwertigen Liganden binden kann; positive Valenzen binden negative, negative Valenzen positive Liganden. - Wertigkeit ist die Zahl der Valenzen, die ein Atom betätigt; Höchstwertigkeit die Höchstzahl der bei einem Element beobachteten Valenzen."

and translated by Brock[2] as:

> "To prevent misunderstandings the meaning of several words used here must be explained. Affinity is the expression for the firmness with which one element binds other elements or radicals (generally: "Ligands" (*ligare* [Latin], to bind); the introduction of a word hitherto lacking simplifies the manner of expression for this immediately clear concept).- Valence (*Valenz*) means the unit of force which can bind a univalent ligand; positive valencies bind negative ligands, negative valencies bind positive ligands.- Atomicity (Wertigkeit) is the number of valencies which an atom manifests; the highest atomicity is the highest number of valencies observed for an element."

The term "**reductive elimination**" is widely used in organometallic chemistry, particularly in organotransition metal chemistry, where it is generally associated with the reverse "**oxidative addition**".[4] However, in organic chemistry, reductive elimination is a term generally associated with various reactions dealing with the regio- and stereo-controlled synthesis of alkenes from 1,2-disubstituted alkanes, as well as the union of two fragments or the creation of a ring.[5] This is, for example, the case of the reductive elimination of 1,2 or 1,4-dihalides or halohydrins, of oxiranes and thiiranes, the Julia alkenylation or the reductive elimination of β-substituted nitroalkanes.[5] Ramberg-Bäcklund reaction is another variety of reductive elimination of an episulfone.[6] This reaction has proven very efficient for the synthesis of large rings, such as the core structure of neocarzinostatin.[7]

The term "**extrusion reaction**" defines reactions in which an atom or group of atoms bonded to two other atoms is lost. Among the most frequently lost atoms or group of atoms are: azo, carbonyl, sulfonyl, thio, carbon dioxide. The two atoms formerly bonded to the lost group can become directly bonded together: this is the restricted definition of March.[8]

$$A \diagup \overset{B}{\diagdown} C \longrightarrow A-C \ + \ B$$

If the fragmentation of the two atoms from the group, to which they were bonded, takes place in a concerted process resulting in the formation of a conjugated system, this is a cheletropic extrusion. Thus, the two possibilities can be found in the case of the loss of a SO_2 group: SO_2 extrusion from an episulfone, as in the Ramberg-Bäcklund reaction, is an extrusion reaction, but SO_2 extrusion from a cyclic sulfolene is a cheletropic reaction.[9]

$$A \diagup \overset{B}{\diagdown} C \longrightarrow A \diagdown \diagup C \ + \ B$$

The term "**cheletropic reaction**" defines reactions in which two σ bonds to a single atom are made or broken concertedly.[10] A widely studied example, theoretically as well as experimentally, is the cheletropic elimination of sulfur dioxide from a cyclic sulfolene to generate stereochemically pure dienes or polyenes.[11]

1.2 LIGAND COUPLING

The term "**ligand coupling**" in hypervalent compounds was used by Oae to describe an extrusion of two ligands, with a pair of electrons, from a central atom which undergoes a return to its more stable valency.[12]

$$Nu^\ominus \ + \ L_nM \diagup \overset{X}{\diagdown} Y \longrightarrow Y^\ominus \ + \ L_nM \diagup \overset{X}{\diagdown}{}_{Nu} \longrightarrow NuX \ + \ L_nM$$

According to Oae, to be a ligand coupling the process requires a cohesive interaction between an axial and an equatorial ligand by overlap of orbitals of both axial and equatorial ligands.[12] If the coupling reaction between the axial and the equatorial ligands proceeds concertedly, the configuration of both ligands will be retained in the final product.

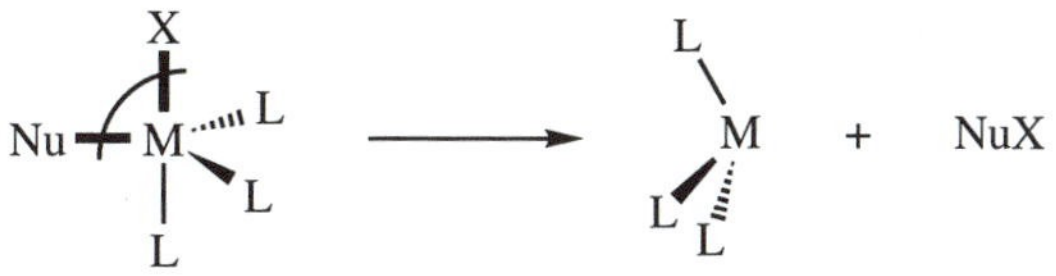

denotes coupling of the two ligands on M atom

In chapter 2, we shall see that this stereochemical conception is too narrow to accommodate all the systems which belong to the group of ligand coupling reactions. These can be defined as reactions involving the intramolecular coupling of two groups bonded to a highervalent heteroatomic element occurring in a concerted manner. As stated by Whitesides, "the most valuable single type of information to have in characterizing the mechanism of a reaction that makes or breaks bonds at a tetrahedral carbon atom is the stereochemistry of the transformation at that carbon".[13] In the case of the ligand coupling mechanism, formation of C-C, C-H or C-X bonds must proceed with retention of configuration at carbon. The same retention of configuration is observed in the case of ligand coupling reactions with stereochemically defined vinyl systems.

This concept of ligand coupling can also be applied to a variety of organometallic reactions, as in the following example:[14,15]

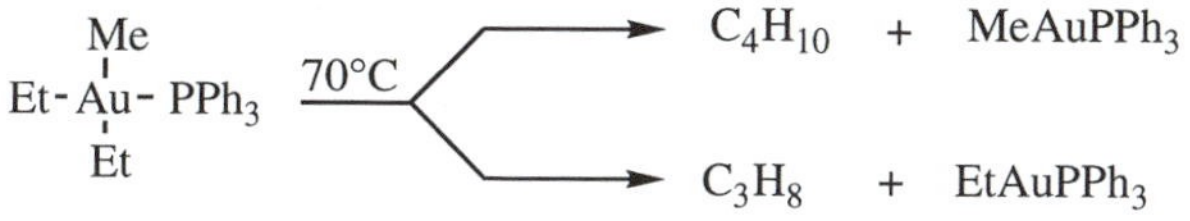

ref. 15

The ligand coupling mechanism has also been suggested to occur during a number of oxidation reactions involving metal oxides, such as for example:[16]

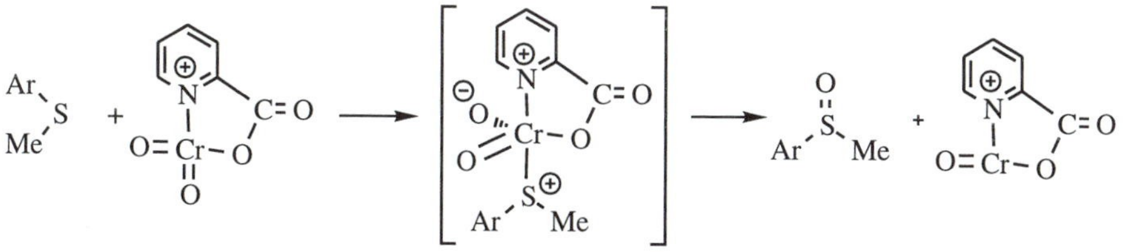

In this book, the ligand coupling reactions which are surveyed are multistep sequences involving the formation of an hypervalent intermediate generated by one of the two following types: either substitution of a σ-ligand by the nucleophile or addition of the nucleophilic substrate on a π-ligand:

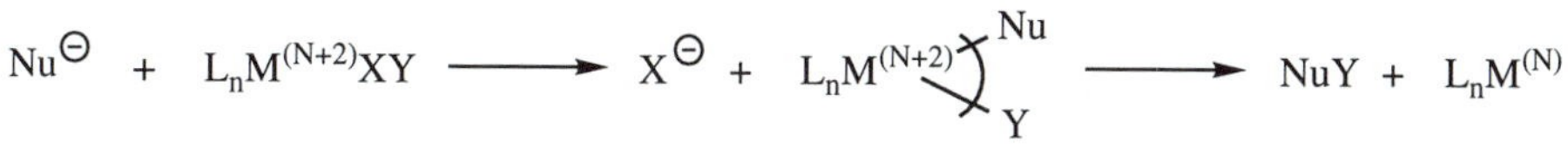

or

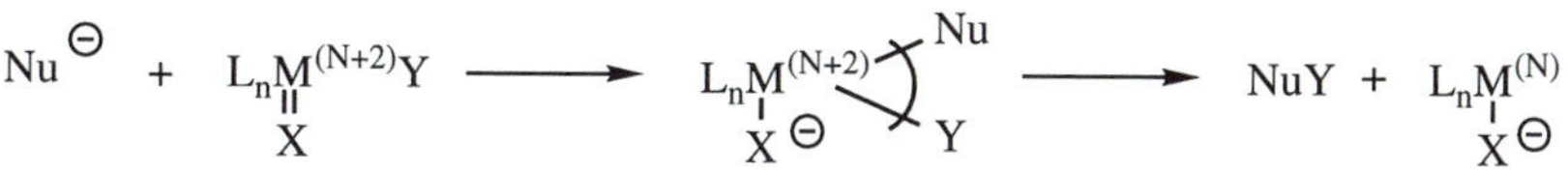

The compounds involved in these two types of reactions which will be surveyed in the following chapters are the higher-valent derivatives of the main-group p-block elements which include:

- "hypervalent organo-nonmetallic" molecules of group V-VIII, as defined by Musher[17-19]

- "main-group organometallic" compounds in their higher valences, which afford, after the ligand coupling step, an inorganic by-product which is stable in the lower oxidation state [e.g. organolead Pb(IV) to inorganic Pb(II) salts[20] or organothallium Tl(III) to inorganic Tl(I) salts].

Although they are not ligand coupling reactions occurring on the main-group element derivative, but due to their synthetic interest, I shall also describe some copper-catalysed reactions of main-group higher-valent compounds. However, it will exclude the reductive elimination reactions which are frequently observed in the case of organotransition metal derivatives. With these metals, the reductive elimination step can take place either on the stoichiometric highervalent organometal derivative, or as part of the oxidative addition-reductive elimination catalytic system. This is featured by cobalt, iron, rhodium, palladium, platinum, nickel, copper among others.[4,21]

A classification of hypervalent organo-nonmetallic derivatives, which have an electron-rich multicenter bonding system, has been devised by Perkins *et al.* in 1980.[22] This system, (the N-X-L system), describes the valence electron count and the coordination number. For an atom (X), the overall bonding system is described first by the number of valence shell electrons (N) formally directly associated with the atom (X), and by the number of ligands (L) directly bonded to it. A number of examples is shown below.

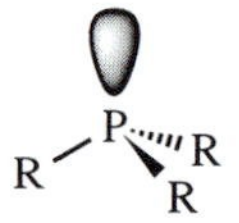

8-P-3 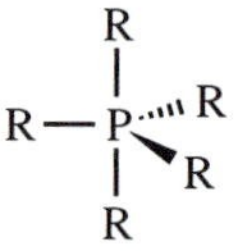   10-P-5 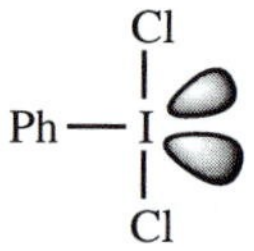   10-I-3

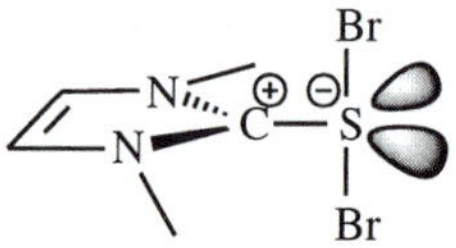

10-S-3 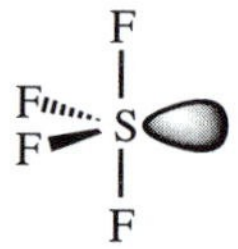   10-S-4    12-S-6

1.3 PSEUDOROTATION

Pentacoordinate model structures have been used in the interpretation of a wide variety of reactions occurring on phosphorus derivatives.[23] These pentacoordinate compounds can exist as two basic structures, trigonal bipyramid (TBP) and square pyramid (SP), which are generally non rigid and undergo facile stereomutation. The ligand rearrangement involved in the stereomutation is generally called pseudorotation, and two mechanisms have been proposed. In 1960, Berry suggested that the stereomutation in group V compounds takes place via a TBP-SP-TBP pathway, which results in an exchange between the apical and equatorial ligands through bond bending.[24] In the equilibrium TBP (D_{3h}) conformation, a molecule ML_5 has three equatorial bonds (e) and two apical bonds (a). In the first step of the Berry pseudorotation mechanism, the TBP structure evolves into a tetragonal pyramid (C_{4v}). During the deformation, one equatorial bond becomes the apical bond of the square pyramidal structure and the other two equatorial and two apical bonds move concertedly to form the basal bonds of the square pyramid. In the second part of the process, the two formerly equatorial and the two formerly apical bonds exchange their positions in such a way that eventually the two former apical bonds become equatorial in the new conformation. (Scheme 1.1)

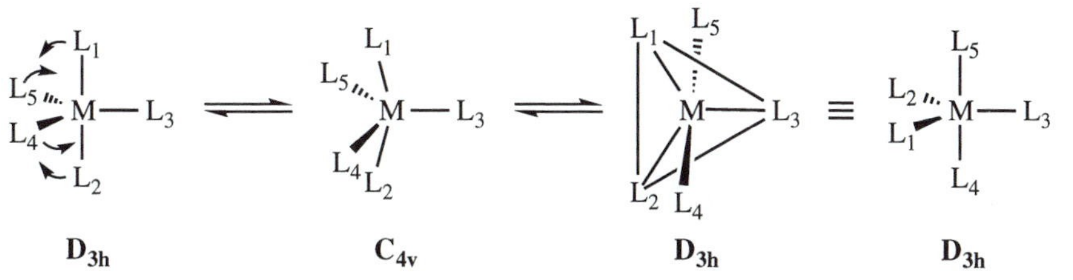

Scheme 1.1: The Berry pseudorotation

An alternative mechanism, the turnstile rotation mechanism (TR), was later proposed by Ugi *et al.* to explain these intramolecular rearrangements, particularly when the pentavalent structure incorporates ring systems.[25] The turnstile mechanism consists of three consecutive steps. The first step is a deformation of the (D_{3h}) conformation, in which there is an angle compression between L_4 and L_5, so that the angle $\theta(L_4\text{-}M\text{-}L_5)$ closes from 120° down to 90°. The three groups L_2, L_4 and L_5 form with M a regular trigonal pyramid. Simultaneously, the other two groups L_1 and L_3 undergo an angle deviation $\theta_1 = ca.$ 9° towards the L_2 group, the angle $\theta(L_1\text{-}M\text{-}L_3)$ remaining constant equal to 90°. In the new conformation, the C_3 symmetry axis of the trigonal pyramid corresponds to the C_2 symmetry axis of the isosceles triangle formed by L_1, L_3 and M. (Scheme 1.2.1)

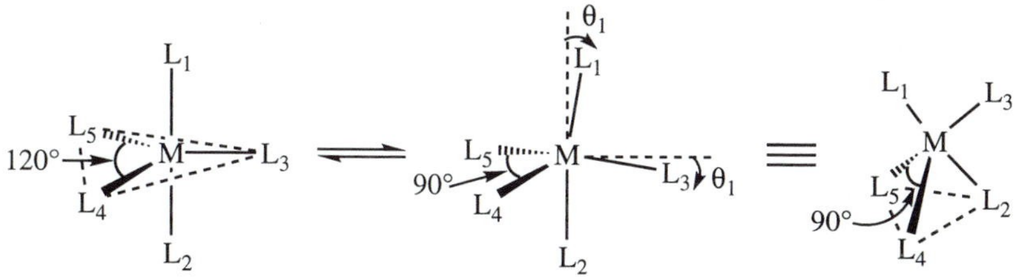

Scheme 1.2.1: The turnstile rotation mechanism: first step

The second step corresponds to a torsion of the two units by 60°, so that the new arrangement is equivalent to the arrangement reached at the end of the first step. The second step is represented by a topside view in Scheme 1.2.2.

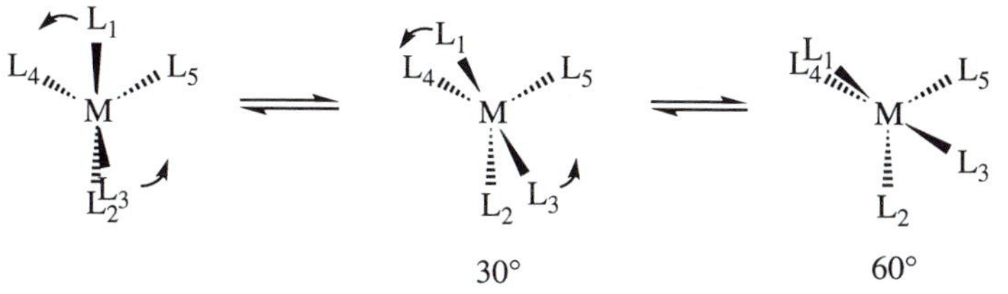

Scheme 1.2.2: The turnstile rotation mechanism: second step

The third step is the reverse path of the first step, the two groups L_3 and L_1 undergoing an angle deviation $\theta_{-1} = ca.$ $9°$. At the same time, there is an angle decompression between L_2 and L_5, so that the angle $\theta(L_2$-M-$L_5)$ opens up from $90°$ to $120°$. The overall moves restore the stable new TBP structure. (Scheme 1.2.3)

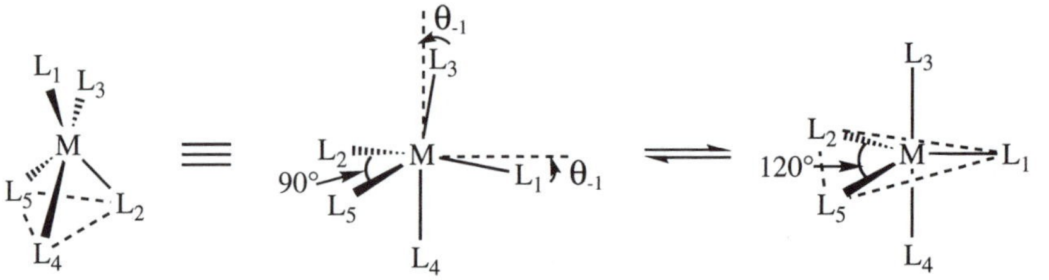

Scheme 1.2.3: The turnstile rotation mechanism: third step

Although the Berry pseudorotation mechanism (BPR) is the most frequently used to interpret the experimental results, it has been shown that these two mechanisms can be topologically equivalent.[26] The Berry pseudorotation can be viewed as a $(1 + 4)$ process, as the equatorial group L_5 remains in equatorial position in the new TBP structure. The turnstile mechanism can be viewed as a $(2 + 3)$ process, in which one equatorial and one apical groups interchange their positions, and in parallel, two equatorial and the other apical group also interchange their positions. However, in the latter three groups, one equatorial group eventually remains in the equatorial plane. This overall turnstile mechanism results in the interchange between two apical and two equatorial groups, as in the Berry pseudorotation. In the following scheme (Scheme 1.3), the interchanges between the two extreme configurations occurring in the two mechanisms are summarized. For clarity sake, the equatorial ligands are marked e_1, e_2 and e_3, and the apical ligands are considered equivalent.

Berry Pseudorotation: a (1+4) process

(aa ee)   (ee aa)

Turnstile Rotation: a (2+3) process

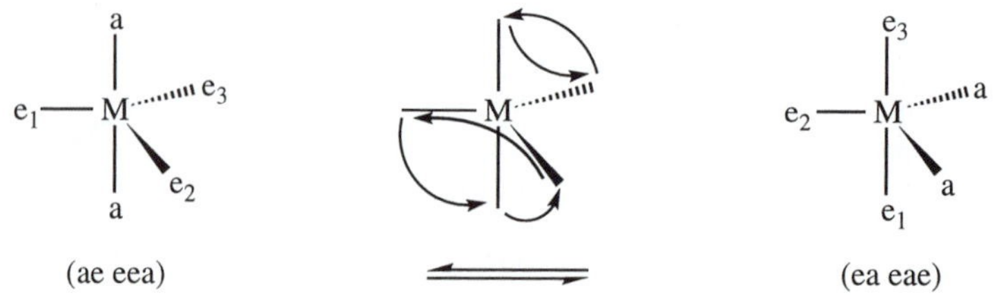

(ae eea)   (ea eae)

Scheme 1.3: Topological equivalence of Berry pseudorotation and turnstile rotation mechanisms

In the following chapters of this book, when a pseudorotation mechanism is involved in the overall sequence of reactions, it will be denoted in the reaction schemes using the formalism shown below, according to the convention introduced by Oae. This formalism does not bear any significance on the nature of the pathway followed by the topological transformation.

1.4 ORGANIZATION OF THE BOOK

In chapter 2, the general aspects of the ligand coupling mechanism are discussed: historical development, theoretical and experimental evidences. The following chapters will review the ligand coupling reactions which have been reported for a number of heteroatoms belonging to the main-group elements, and more specifically those for which a range of reactions of synthetic interest in organic chemistry has been described. We shall study the reactions involving organic compounds of hypervalent sulfur (Ch. 3), phosphorus (Ch. 4), iodine (Ch. 5) and bismuth (Ch. 6). In the case of lead compounds, we shall review the ligand coupling reactions of lead tetraacetate, as an acetate group is introduced in the substrate, and the second part of the chapter will be devoted to the chemistry of the highervalent organolead triacetates (Ch. 7). In the last chapter (Ch. 8), we shall review the chemistry of the elements which, from the point of view of the range of ligand coupling reactions applicable in organic synthesis, appears as narrower. This is particularly the case of tellurium, selenium and, to a lesser degree, of antimony and thallium.

This book is intended to give to the synthetic chemist a useful introduction to the various facets of the chemistry which can be performed by reactions explained by this still insufficiently recognized mechanism in the chemistry of main-group p-block elements.

1.5 REFERENCES

1. Stock, A. *Ber. Dtsch. Chem. Ges.* **1917**, *50,* 170-182.
2. Brock, W.H.; Jensen, K.A.; Jørgensen, C.K.; Kauffman, G.B. *Polyhedron* **1983**, *2,* 1-7.
3. Fieser, L.F.; Fieser, M. *Advanced Organic Chemistry*; Reinhold: New York, **1961**; p. 484. where the description of "a central phosphorus atom surrounded by complexing phenyl groups as ligands" is mentionned.
4. Collman, J.P.; Hegedus, L.S.; Norton, J.R.; Finke, R.G. *Principles and Applications of Organotransition Metal Chemistry*; University Science Books: Mill Valley, **1987**.
5. Kocienski, P. Reductive Elimination, Vicinal Deoxygenation and Vicinal Desilylation, in *Comprehensive Organic Synthesis*; Trost, B.M.; Fleming, I., Eds.; Pergamon Press: Oxford, **1991**; Vol. 6, Ch. 5.2, pp. 975-1010.
6. For a recent review: Simpkins, N. *Sulphones in Organic Synthesis*; Pergamon Press: Oxford, **1994**; Ch. 7, pp. 273-284.
7. Wender, P.A.; Harmata, M.; Jeffrey, D.; Mukai, C.; Suffert, J. *Tetrahedron Lett.* **1988**, *29,* 909-912.
8. March, J. *Advanced Organic Chemistry*, 4th ed.; J. Wiley & Sons: New York, **1992**; pp. 1045-1050.
9. Guziec, F.S.Jr; Sanfilippo, L.J. *Tetrahedron* **1988**, *44,* 6241-6285.

10.   Woodward, R.B.; Hoffmann, R. *The Conservation of Orbital Symmetry*; Academic Press: New York, **1970**; pp. 152-163.

11.   Mock, W.L. *J. Am. Chem. Soc.* **1969**, *91*, 5682-5684; *ibid.* **1975**, *97*, 3666-3672; *ibid.* **1975**, *97*, 3673-3680.

12.   **a)** Oae, S. *Croatica Chem. Acta* **1986**, *59,* 129-151. **b)** Oae, S.; Uchida, Y. *Acc. Chem. Res.* **1991**, *24*, 202-208. **c)** Oae, S. *Pure Appl. Chem.* **1996**, *68,* 805-812.

13.   Bock, P.L.; Boschetto, D.J.; Rasmussen, J.R.; Demers, J.P.; Whitesides, G.M. *J. Am. Chem. Soc.* **1974**, *96*, 2814-2825.

14.   Negoro, T.; Oae, S. *Reviews on Heteroatom Chemistry* **1995**, *13*, 235-272.

15.   Komiya, S.; Ozaki, S.; Endo, I.; Inoue, K.; Kasuga, N.; Ishizaki, Y. *J. Organomet. Chem.* **1992**, *433*, 337-351.

16.   Srinivasan, C.; Rajagopal, S.; Chellamani, A. *J. Chem. Soc., Perkin Trans. 2* **1990**, 1839-1843.

17.   Musher, J.I. *Angew. Chem., Int. Ed. Engl.* **1969**, *8*, 54-68.

18.   Martin, J.C. *Science* **1983**, *221*, 509-514.

19.   For a discussion on the definition of hypervalent compounds, see: **a)** Schleyer, P.v.R. *Chem. Eng. News* **1984**, *62*, (May, 28), 4. **b)** Martin, J.C. *Chem. Eng. News* **1984**, *62*, (May, 28), 4 and 70. **c)** Harcourt, R.D. *Chem. Eng. News* **1985**, *63*, (January, 21), 3-4.

20.   **a)** Kaupp, M.; Schleyer, P.v.R. *J. Am. Chem. Soc.* **1993**, *115*, 1061-1073. **b)** Kolis, J.W. *Chemtracts-Inorg. Chem.* **1993**, *5*, 62-65.

21.   Crabtree, R.H. *The Organometallic Chemistry of the Transition Metals*; 2nd ed.; J. Wiley & Sons: New York, **1994**.

22.   Perkins, C.W.; Martin, J.C.; Arduengo, A.J.; Lau, W.; Alegria, A.; Kochi, J.K. *J. Am. Chem. Soc.* **1980**, *102*, 7753-7759.

23.   Holmes, R.R. *Pentacoordinated Phosphorus*; ACS Monographs N° 175 and 176; American Chemical Society: Washington, DC, **1980**; Vol. I and II.

24.   Berry, R.S. *J. Chem. Phys.* **1960**, *32*, 933-938.

25.   Ugi, I.; Marquarding, D.; Klusacek, H.; Gokel, G.; Gillespie, P. *Angew. Chem., Int. Ed. Engl.* **1970**, *9,* 703-730; Ugi, I.; Marquarding, D.; Klusacek, H.; Gillespie, P.; Ramirez, F. *Acc. Chem. Res.* **1971**, *4,* 288-296.

26.   Kutzelnigg, W.; Wasilewski, J. *J. Am. Chem. Soc.* **1982**, *104,* 953-960. Wang, P.; Agrafiotis, D.K.; Streitwieser, A.; Schleyer, P.v.R. *J. Chem. Soc., Chem. Commun.* **1990**, 201-203.

Chapter 2

The Ligand Coupling Mechanism

2.1 HISTORICAL BACKGROUND

The idea of a ligand coupling mechanism occurring in the course of fragmentation reactions of heteroatomic derivatives came out slowly from experimental observations which could not be accommodated by the classical S_N mechanisms. The decomposition of a tetravalent sulfurane intermediate was invoked by Franzen *et al.*[1] in the reaction of triarylsulfonium salts with organolithium reagents, and by Wiegand and Mc Ewen[2] and by Oae[3] to explain the thermal decomposition of triarylsulfonium halides. To rationalize the experimental results of the reactions of hypervalent sulfur derivatives, Trost[4-6] brilliantly defined the general characteristics of the mechanism. Although some theoretical studies were performed in the mid 1970's,[7] it is not before the next decade that the body of experimental evidences became large enough to allow a relatively clear understanding of the ligand coupling mechanism. The term itself, *"ligand coupling"* started around 1985 to be used by Oae[8,9] to describe a particular type of reductive elimination taking place with the hypervalent derivatives of main-group elements. Its occurrence was clearly claimed mostly in the case of sulfur and phosphorus compounds by Oae and his group.

In spite of the similarities between the experimental results observed in the chemistry of tetravalent organosulfur compounds and the chemistry of trivalent organoiodane compounds, it took a long time before ligand coupling was considered an important mechanism in the hypervalent iodine chemistry. This may be ascribed to the fact that in the 1930's, Sandin had observed that the decomposition of diaryliodonium salts in the presence of pyridine led to ring *C*-arylated products by a free radical pathway.[10] Following this initial observation, Beringer and his coworkers,[11-14] as well as most of the chemists subsequently involved in the chemistry of iodonium compounds,[15] used to consider that the overall aromatic nucleophilic substitution reactions were best explained by the exclusive occurrence of free radical species. These radical species may derive from the decomposition of a 9-I-2 iodane intermediate, itself a product of electron transfer between the iodonium salt and the substrate.[16,17] In the mid 1980's, Barton *et al.* clearly observed the dichotomy in the overall mechanistic picture of the reactivity of iodonium salts.[18] They concluded that, in nucleophilic aromatic substitution reactions of diaryliodonium salts, two competing mechanisms are taking place: a free radical component affording arene products, among others, and a ligand coupling mechanism leading to the products of nucleophilic aromatic substitution. They also showed that this competition occurs in a number of other nucleophilic aromatic substitution systems. The importance of these observations was only slowly acknowledged. Indeed, even recently, the free radical pathway still remained considered as the most valuable alternative in iodonium chemistry.[19] The ligand coupling mechanism was later claimed by Moriarty,[20-22] to be implied in a number of reactions of hypervalent iodine compounds in general and by Grushin to rationalise the experimental results of the reactions of diaryliodonium salts in particular.[23] It was sometimes evoked, with care, as for example by Kurosawa *et al.* who suggested the *"occurrence of a rapid, in-cage radical coupling and/or a non-radical, S_NAr-type pathway"*, to explain the results of the reaction of arylthallium (III) derivatives with nitronate salts.[24] In the last decade, it has gained wider acceptance and is now associated with a number of reactions of organic chemistry involving the presence of organic derivatives of main-group p-block elements.[25-40]

In this chapter, we will focus on the ligand coupling mechanism itself from two levels: the theoretical point of view and the experimental observations which have shed some light on the intricacies of this still poorly understood mechanism. The question of whether a reaction takes place via a ligand coupling mechanism or through radical species will be dealt with in the following individual chapters.

2.2 MECHANISTIC ASPECTS

In the reaction of hypervalent or highervalent reagents with nucleophilic substrates, different pathways can be invoked to explain the outcome of the reaction, being compatible with the stereochemical observations. In a number of cases in the sulfur chemistry, the reactions of alkyl pyridyl sulfoxides with nucleophiles lead to nucleophilic-type substitution products with retention of configuration. A classical S_N2 mechanism is therefore excluded. Two main possibilities can be considered. In the first one (Scheme 2.1, path A), the nucleophile reacts with the electrophilic heteroatomic center to yield a covalent intermediate which undergoes in a second stage a reductive elimination to afford the coupled product together with a by-product derived from the heteroatom, which has experienced a reduction of its oxidation level from (n+2) to (n). In the second one, the frontside S_N2 mechanism (Scheme 2.1, path B), the nucleophile interacts directly with the M-X bond, which is going to be cleaved, by an approach on the nodal surface of the σ* orbital of M-X.[41-43]

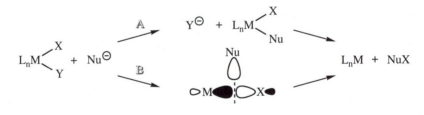

Scheme 2.1

Usually disfavoured in classical S_N2 reactions, the frontside attack can be facilitated by the lower electronegativity of the hypervalent atom and the reduction of its oxidation level. For some specific systems, such a frontside mechanism can be operating.[44] However, in a number of hypervalent elements, classical S_N2 reactions were reported to occur on the heteroatom, affording a new hypervalent derivative. This intermediate can be stable enough to be isolated or, on the contrary, can sometimes remain undetected. In a second stage, a "reductive elimination" step, in the organometallic meaning, takes place on this intermediate to yield the coupling product together with the heteroatom-containing by-product. This "reductive elimination" step can be conceived to occur by different possibilities. These can be heterolytic, homolytic or concerted. In the heterolytic pathway (Scheme 2.2, path 1), a pair of ions will be generated either free or in a cage of solvent. In the homolytic pathway (Scheme 2.2, paths 2 and 3), either one of the two fragments which will constitute the coupled product can leave as a free radical and then recombine with the second fragment still bound to the heteroatomic containing radical fragment. The last pathway (Scheme 2.2, path 4) is a concerted one, which can involve more or less polar transition states, depending on the nature of the substituents of the intermediate: this is the "**ligand coupling**" mechanism, represented by an arc of a circle between the bonds undergoing the coupling.

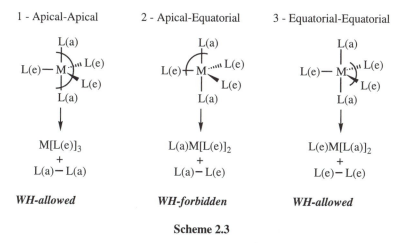

$$\begin{array}{ccc}
 & \overset{1}{\longrightarrow} & \left[L_nMA^{\oplus}\ B^{\ominus} \right] \\
 & \overset{2}{\longrightarrow} & \left[L_nMA^{\bullet}\ B^{\bullet} \right] \\
L_nM\underset{B}{\overset{A}{\diagdown}} & \overset{3}{\longrightarrow} & \left[L_nMB^{\bullet}\ A^{\bullet} \right] \\
 & \overset{4}{\longrightarrow} & L_nM{-}A \\
 & & B
\end{array} \quad L_nM + AB$$

Scheme 2.2

2.2.1 Ligand coupling: theoretical models

A number of theoretical studies on hypervalent compounds have been performed. They focus on the structure, stability and interconversion between the different possible conformers. The studies related to the chemical evolution of these species into a pair of two new species are more limited. Two types of approaches have been reported on the elimination of two ligands from a hypervalent compound: the orbital symmetry conservation and the *ab initio* calculations. These approaches always deal with simplistic model compounds in which the heteroatom bears generally hydrogen, fluorine or chlorine ligands.

2.2.1.1 The orbital symmetry conservation model

Analysis of the fragmentation reaction of PR_5 from the point of view of the conservation of orbital symmetry led Hoffmann *et al.* to suggest a simple answer.[7] The least-motion mode of concerted departure from the stable D_{3h} structure, coupling between an apical group and an equatorial group, is symmetry-forbidden, or **WH (Woodward-Hoffmann) forbidden**. On the other hand, both equatorial-equatorial and apical-apical fragmentations are symmetry-allowed, or **WH (Woodward-Hoffmann) allowed** (Scheme 2.3).

1 - Apical-Apical	2 - Apical-Equatorial	3 - Equatorial-Equatorial

$$\begin{array}{ccc}
L(a) & L(a) & L(a) \\
L(e){-}M{\cdots}L(e) & L(e){-}M{\cdots}L(e) & L(e){-}M{\cdots}L(e) \\
L(a) & L(a) & L(a) \\
\downarrow & \downarrow & \downarrow \\
M[L(e)]_3 & L(a)M[L(e)]_2 & L(e)M[L(a)]_2 \\
+ & + & + \\
L(a){-}L(a) & L(a){-}L(e) & L(e){-}L(e) \\
\textit{WH-allowed} & \textit{WH-forbidden} & \textit{WH-allowed}
\end{array}$$

Scheme 2.3

In this system, the elimination reaction $PR_5 \longrightarrow PR_3 + R_2$ was studied on the reaction path from the more stable pentavalent D_{3h} bipyramid trigonal structure to the products PR_3 with a C_{3v} structure. The apical-apical fragmentation starts from the D_{3h} structure and evolves directly into PR_3 possessing a planar D_{3h} structure and eventually into the more stable pyramidal C_{3v} structure. Although this process appears sterically unlikely, it can be seen as a continuation of the Berry pseudorotation, in which the two apical ligands move to the equatorial positions. In the case of the equatorial-equatorial fragmentation, the first formed intermediate PR_3 presents a T-shaped C_{2v} structure, which eventually evolves into the C_{3v} structure. (Scheme 2.4)

Apical-apical fragmentation:

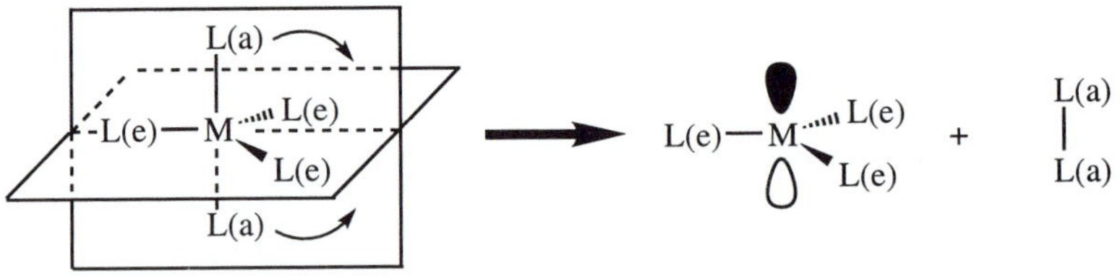

L(a) ⟶ L(e)—M⋯L(e), L(e), L(a) ⟶ L(e)—M⋯L(e), L(e) + L(a)—L(a)

Equatorial-equatorial fragmentation:

L(a), L(e)—M⋯L(e), L(e), L(a) ⟶ L(e)—M, L(a) + L(e)—L(e)

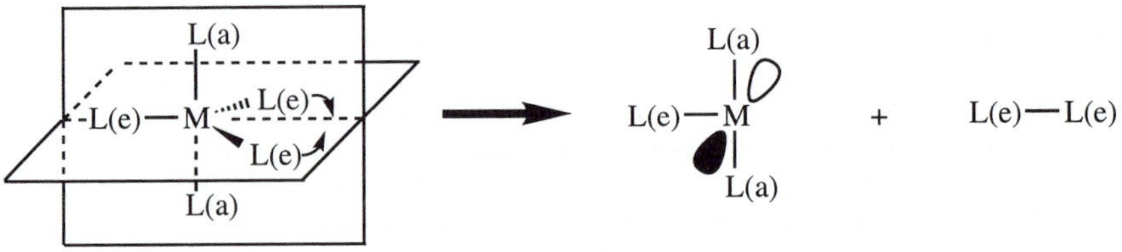

Scheme 2.4

The very fast Berry[45] and turnstile[46,47] pseudorotation processes can lead to the presence of a number of stereoisomers. The Berry process occurs through a transition state possessing the C_{4v} square-pyramidal geometry with an energy barrier of *ca.* 2-3 kcal/mol in the case of phosphorus derivatives.[48,49] (Scheme 2.5)

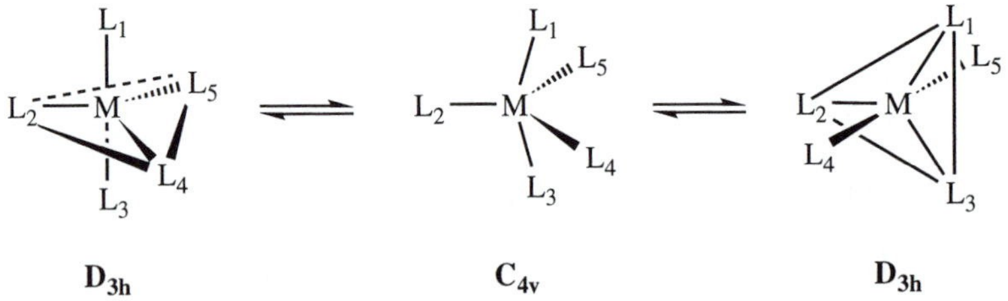

L_1, L_2—M, L_5, L_3, L_4 ⇌ L_1, L_2—M, L_5, L_3, L_4 ⇌ L_1, L_2—M, L_5, L_4, L_3

D_{3h} C_{4v} D_{3h}

Scheme 2.5: The Berry pseudorotation

In the C_{4v} square-pyramidal transition state, the fragmentation reaction of PR_5 can also take place to form PR_3 and R_2. A similar orbital symmetry conservation study led to the general conclusion that, from the three possible coupling processes, only two were symmetry-allowed: L_1-L_3 and L_2-L_3. As the permutation processes between the five ligands borne by the pentasubstituted central heteroatom can afford twenty different combinations for a trigonal bipyramid (TBP),[50,51] the stability will be governed by a combination of steric and electronic factors, the knowledge of which is important for the prediction of the outcome of the fragmentation process in the case of more substituted substrates.

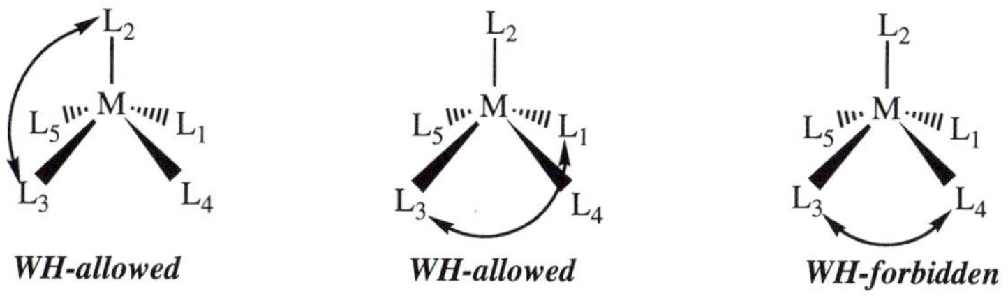

WH-allowed   WH-allowed   WH-forbidden

Scheme 2.6

According to the Woodward-Hoffmann rules, the suprafacial apical-equatorial fragmentation cannot occur by a concerted pathway. It will require either a zwitterionic or a biradical transition state in order to take place.[52]

2.2.1.2 *Ab initio* calculation models

Although not observed, the phosphorane PH_5 has been one of the most extensively studied hypervalent compound. The *ab initio* calculations on PH_5 have focused on the electronic structure, the equilibrium geometry, the barriers to Berry and turnstile pseudoratotation, and the thermodynamic stability of PH_5 compared to $PH_3 + H_2$.[53-59] A number of *ab initio* calculations were also reported on the influence of the nature of the substituents on the stabilities of the phosphoranes.[60-63] This led to the concept of "equatophilicity" or "equatoriphilicity", according to which less electronegative ligands or ligands forming covalent bonds to phosphorus prefer the equatorial position.[64-66] The barrier height of the elimination reaction of H_2 from PH_5 has been much less studied.[48,58,67] In the case of the heavier analogues of PH_5, the reports are very scarce: one or two on each element, AsH_5,[49] SbH_5[49,68] and BiH_5.[49,69] Moreover, the periodic trends were compared in the case of the XH_5 species with X = P, As, Sb and Bi.[49] In the series of heteroatomic compounds bearing four substituents, the *ab initio* calculations focused also mostly on the structure and thermodynamic stabilities.[58,70-76] Very few reports were concerned with the elimination reaction to form $XH_2 + H_2$ from XH_4.[77,78]

2.2.1.2.1 The XR_5 system

In the first reported studies, Kutzelnigg and Wasilewski performed their calculations at two levels of sophistication.[48] At the intermediate level (SCF with polarisation functions), they found two saddle points, one corresponding to the Woodward-Hoffmann allowed eq-eq concerted process and the other to a non-least motion variant of a WH-forbidden process, leading to an ion pair $PH_4^+H^-$ with an extremely flat region of the potential energy surface between the two saddle points. At a higher level of sophistication (CEPA method, including the electron correlation), only the concerted transition state was obtained. The transition state of the concerted path was claimed to have a C_{2v} symmetry. However, with an optimized structure for PH_5 under C_{2v} symmetry, Reed and von Schleyer found that at 6-31G* the reaction has two imaginary frequencies. But with the C_s symmetry leading to a slightly deformed geometry, only one imaginary frequency was found.[67] The presence of catalytic amounts of acids was suggested to considerably lower the barrier for decomposition of PH_5.[48]

In their recently reported work, Moc and Morokuma performed *ab initio* MO calculations using effective core potentials (ECP) on central atoms.[49] The calculations were done on a series of hypervalent XH_5 hydrides, with X = P, As, Sb and Bi. They showed that, in the Berry pseudorotation, the XH_5 (D_{3h}) structures are local minima, and the XH_5 (C_{4v}) structures are transition states, with pseudorotation barriers of about 2 kcal/mol.[49] (Scheme 2.7 and Table 2.1)

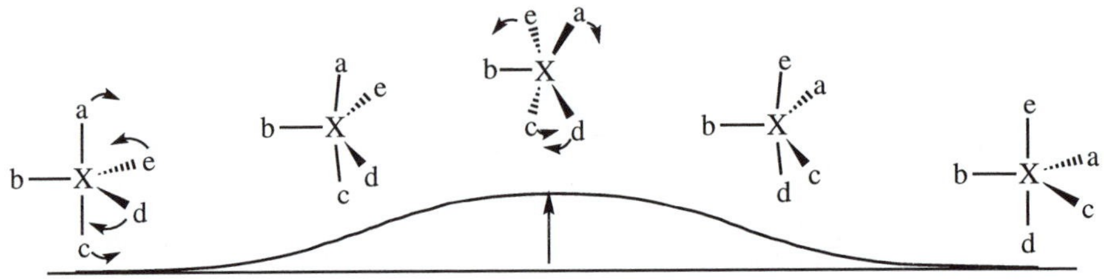

Scheme 2.7: *ab initio* model of the Berry pseudorotation

Table 2.1: Energy barriers (in kcal/mol) to Berry pseudorotation for XH_5.[49]

species	MP4/ECP	MP2/ECP
PH_5	1.9	2.0
AsH_5	2.1	2.2
SbH_5	2.3	2.2
BiH_5	1.9	

The periodic trend in the thermodynamic stabilities of XH_5 relative to $XH_3 + H_2$ was also studied.[49] The XH_5 (D_{3h}) structures are thermodynamically unstable compared to the systems XH_3 (C_{3v}) + H_2 by over 40 kcal/mol: 44-45 kcal/mol for PH_5 to 73-75 kcal/mol for BiH_5 with an irregular trend. (Scheme 2.8 and Table 2.2)

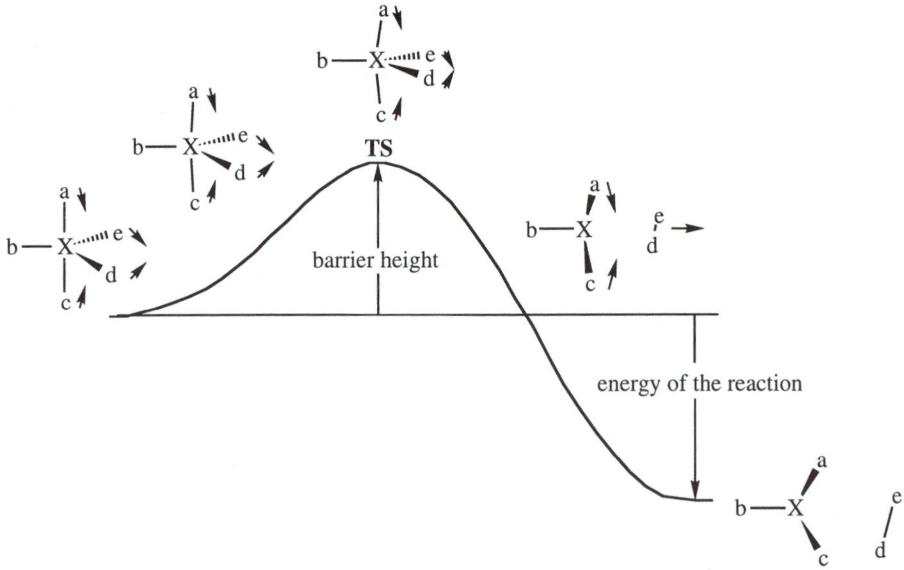

Scheme 2.8: *ab initio* model of a ligand coupling process

Table 2.2: Energies (in kcal/mol) of the reactions $XH_3 + H_2 \longrightarrow XH_5$.[49]

reaction	RHF/ECP	MP4/ECP
$PH_3 + H_2 \longrightarrow PH_5$	44.8	45.3
$AsH_3 + H_2 \longrightarrow AsH_5$	55.2	54.6
$SbH_3 + H_2 \longrightarrow SbH_5$	50.0	50.2
$BiH_3 + H_2 \longrightarrow BiH_5$	75.4	73.0

As the barrier heights to H_2 loss are relatively high and similar for all species (30-34 kcal/mol), all the XH_5 species are kinetically stable.[49] However, the isolation of XH_5 based on $XH_3 + H_2$ reaction seemed unrealistic due to the large barrier to overcome: 75 to 105 kcal/mol. (Scheme 2.8 and Table 2.3)

Table 2.3: Barrier heights (in kcal/mol) for the H_2 elimination reactions from XH_5.[49]

reaction	RHF/ECP	MP2/ECP	MP4/ECP	MP4/ECP/ZPE
$PH_5 \longrightarrow PH_3 + H_2$	45.5	32.6	32.6	30.2
$AsH_5 \longrightarrow AsH_3 + H_2$	42.8	33.3	32.9	30.9
$SbH_5 \longrightarrow SbH_3 + H_2$	47.9	38.1	36.8	34.9
$BiH_5 \longrightarrow BiH_3 + H_2$	38.5	34.0	33.5	31.5

The activation barrier for the H_2 elimination by an equatorial-equatorial process for PH_5 [30.2 kcal/mol for the MP4/ECP/ZPE (ZPE = zero-point energy)],[49] correlate well with the previously reported values after appropriate corrections: 30.7 kcal/mol (at MP4/6.31++G**+ZPE)[67] and 33.6 (after correction of the original value of 36 kcal/mol, obtained by CEPA with polarization functions, for appropriate ZPE values).[48] A good correlation can be noted between these values and those reported for the ligand coupling of SbH_5 (- 49.7 kcal/mol), the barrier height (37.8 kcal/mol), as well as for the Berry pseudorotation (2.3 kcal/mol), which were calculated by MP2/DZdp/MP2/DZdp.[68] Similarly, the large endothermicity of the $BiH_3 + H_2 \longrightarrow BiH_5$ reaction of 73 kcal/mol correlates well with the value obtained by Schwerdtfeger of *ca.* 79 kcal/mol, based on the quadratic configuration interaction (QCI) relativistic pseudopotential calculation.[69]

The optimized transition states for the elimination of H_2 from XH_5 (D_{3h}) possessed C_s symmetry. In the case of X = P, As and Sb, elimination occurs by an equatorial-equatorial process. In the case of X = Bi, elimination occurs by a zwitterionic apical-equatorial process, this being related to the high exothermicity (- 73 kcal/mol) of the reaction BiH_5 (D_{3h}) $\longrightarrow$ BiH_3 (C_{3v}) + H_2, instead of - 45 to - 50 kcal/mol for PH_5, AsH_5 and SbH_5.[49]

2.2.1.2.2 The XR₄ system

Theoretical studies of the XR_4 system (X = S, Se or Te) have been performed either by MO analysis[79] or by *ab initio* calculations.[58,70-76] They have mostly focused on the bonding, structure and thermodynamic stabilities of some derivatives of these compounds, particularly XH_4, XF_4 and XMe_4. However, the transition structures for H_2 eliminations from the XH_4 system have been studied only by two groups.[77,78] Configuration interaction studies on the reaction $SH_4 \longrightarrow SH_2 + H_2$ led Yoshioka *et al.* to conclude that there was a decomposition pathway going through a C_{2v} transition state. The *ab initio* activation energy was predicted to be 42.5 kcal/mol. An energy difference of 72.4 kcal/mol between SH_4 (C_{4v}) and the $SH_2 + H_2$ products was predicted at the TZ+P CI level with corrections for unlinked clusters.[77] However, in their more recent study Moc *et al.* found that, for XH_4 (X = S, Se and Te), the transition states of C_{2v} symmetry, although symmetry-allowed, are disfavoured over highly polar transition structures which have no symmetry and which do not correspond to a least-motion pathway.[78] The most favourable transition structure is highly polarized with a C_1 symmetry. This can be considered as a trigonal bipyramid, in which one apical ligand and one equatorial ligand are coupled, the third equatorial position being occupied by the lone pair of X. (Scheme 2.9) The energy barriers to the dissociation reactions were predicted to be 16 kcal/mol for SH_4 and SeH_4, and 23 kcal/mol for TeH_4. The MP4 energy differences between XH_4 (C_{4v}) and $XH_2 + H_2$ were found to be 79.8 kcal/mol for SH_4, 73.4 kcal/mol for SeH_4 and 56.2 kcal/mol for TeH_4. Based on simple qualitative arguments, it was considered that electron-donating groups will be expected to facilitate the elimination process by stabilization of the "cationic" fragment in the C_1 transition state and electron-acceptors will slow down the reactions. Moreover, this model should be applicable to more complex products with organic substituents.[78]

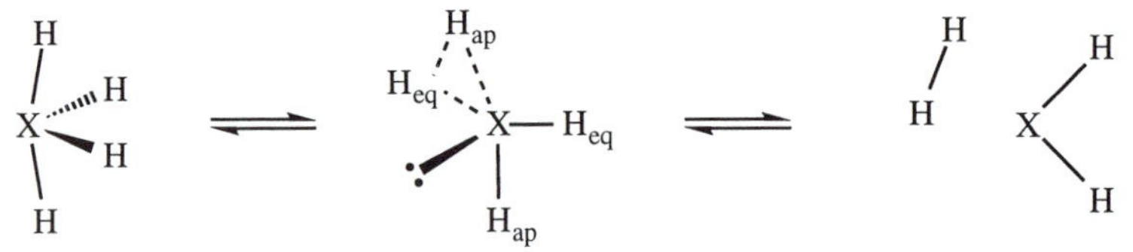

C_{4v} C_1 C_{2v}

Scheme 2.9

In the case of the tetramethylchalcogens XMe_4, the barriers to Berry pseudorotation were estimated to be about 4.8, 3.6 and 0.7 kcal/mol respectively for the S, Se and Te compounds. All XMe_4 compounds are thermodynamically unstable compared to XMe_2 + ethane. The estimated binding energies at both SCF and MP2 levels of theory are about 84 kcal/mol for SMe_4, 72 kcal/mol for $SeMe_4$ and 54 kcal/mol for $TeMe_4$.[76]

2.2.1.2.3 The VXR_n system with V= vinyl group

Bimolecular substitution at the vinyl carbon have been explained by either an Ad_N-E mechanism, leading to retention of configuration, or by a S_N2-Vin mechanism leading to inversion of configuration.[80] *Ab initio* calculations at the 3-21G*//3-21G*, 6-31G*//3-21G* or 6-311G**//3-21G* levels gave significant indications on the relevance of the symmetry of the LUMO of the electrophile and the stereochemical outcome of the reaction.[81] In general, LUMO of π symmetry corresponds to Ad_N-E mechanism, and LUMO of σ symmetry corresponds to S_N2-Vin mechanism. However, vinyliodonium compounds failed to give the correspondence. In the case of the non-activated vinyl-iodonium salts, the σ symmetry corresponding to the S_N2-Vin mechanism with inversion of configuration is supported by the experimental observation of inversion in the reaction of β-alkylvinyl-iodonium tetrafluoroborates with halides, although no mechanism was suggested.[82] The same σ symmetry was predicted for the β-sulfonylvinyliodonium analogue, but in this case the experimental observation is in contradiction, as reaction of β-sulfonylvinyliodonium tetrafluoroborates with halides led to products with retention of configuration. A ligand coupling mechanism was postulated as a possibility in this case.[83]

2.2.2 Ligand coupling: experimental studies

The ligand coupling reaction is an important reaction in the chemistry of hypervalent compounds. A large number of synthetic applications show the growing interest in these reactions. However the experimental study of the intimate mechanism of ligand coupling has been poorly developed.

2.2.2.1 Types of experimental ligand coupling systems

The survey of the literature of all the ligand coupling processes which have been described with the various heteroatomic main-group elements shows that they can be classified in two general categories, themselves subdivided into subclasses:

- **type A** : *ipso-ipso* coupling between two σ bonded ligands   TYPE **LC**
In the ligand coupling involving two ligands linked to the heteroatom by a σ bond, the process can be more or less synchronous. In the case of a reaction between two similar ligands or between ligands of fairly similar polarity, the reaction will be relatively synchronous. It could therefore be considered as an homocoupling for which the abbreviation " **LCH** " could be used.

$$L_nM \overset{A}{\underset{A}{\big\langle}} \longrightarrow L_nM + A\!-\!A \quad \text{or} \quad L_nM \overset{A}{\underset{B}{\big\langle}} \longrightarrow L_nM + A\!-\!B$$

A typical example of this type is the thermal decomposition of tetraaryltelluranes.[84,85]

$$\left[Me\!-\!\langle\bigcirc\rangle\!- \right]_4 Te \xrightarrow{\Delta} Me\!-\!\langle\bigcirc\rangle\!-\!\langle\bigcirc\rangle\!-\!Me \; + \; Me\!-\!\langle\bigcirc\rangle\!-\!Te\!-\!\langle\bigcirc\rangle\!-\!Me$$

On the other hand, when the two ligands are of sufficiently different polarity, the reaction is analogous to an internal nucleophilic substitution. The reaction will be mostly non-synchronous with a polar transition state. For this case, the abbreviation " **LCN** " could be suggested. A large number of examples were discovered by Oae *et al.* in their extensive studies of the ligand coupling of sulfoxide derivatives.[23,24]

$$\underset{R^1}{\overset{Ar}{>}}S\!=\!O \; + \; R^2M \longrightarrow \underset{R^1}{\overset{Ar_{\prime\prime\prime\prime}}{\underset{R^2}{\big|}}}\overset{OM}{\underset{\big|}{S}}\!-\!: \longrightarrow Ar\!-\!R^2$$

The type A ligand coupling processes (*ipso-ipso*), whether of the synchronous **LCH** type or of the polar **LCN** type, are the most frequently occurring mode of coupling. They are involved in a large number of reactions in the chemistry of sulfur, selenium, tellurium, phosphorus, antimony and iodine derivatives.

- **type B** : *ipso*-allyl coupling between one σ bonded ligand and one allylic atom   TYPE **LC'**
By analogy with the S_N2 and S_N2' mechanisms, the second class of ligand coupling reactions involving an allylic system can be named LC'. The relative polarity of the two ligands allows the distinction between two subtypes of LC' coupling reactions.

$$L_nM \overset{A}{\underset{Z\!-\!\!}{\big\langle}}\!\diagdown\!\diagup\!/ \longrightarrow L_nM \; + \; Z\!\diagup\!\diagdown\!\diagup\!^A$$

The nature of the ligand A will govern the philicity of the process. If the ligand A, originally present in the organoelement derivative, plays the role of a cationic equivalent, the ligand containing the allylic moiety will play the nucleophilic role. The substitution of ligand A will therefore correspond to an internal nucleophilic substitution and this process could be named: " **LC$_{N'}$** ".

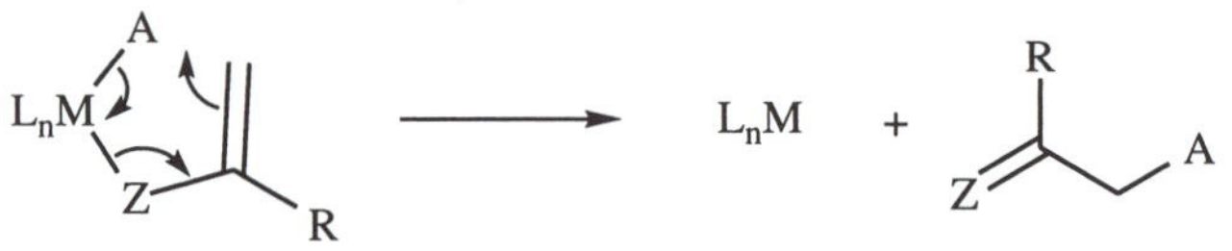

In the opposite sense of polarity of the reactants, ligand A plays the nucleophilic role and the ligand containing the allylic moiety will behave as the electrophile. The substitution of ligand A will therefore correspond to an electrophilic substitution and this process could be named: " **LC$_{E'}$** ".

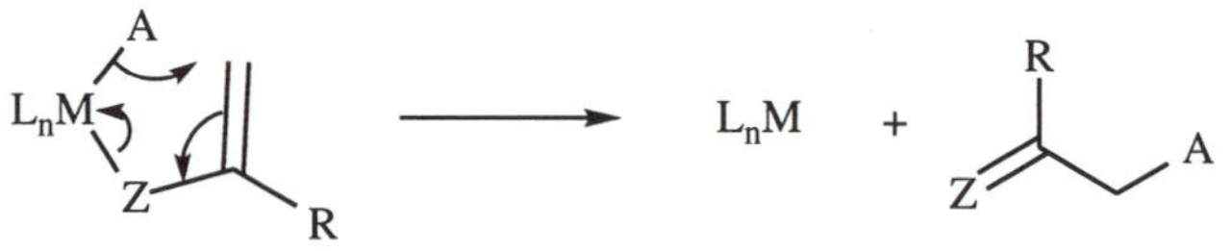

The type B (*ipso*-allyl) coupling is less frequently observed. It happens mostly in the reactions of enolic compounds with iodine, bismuth and lead derivatives. It is also found in the reactions of nitronate salts with organobismuth, organolead and organothallium compounds. This ligand coupling mechanism involves an unsymmetrical transition state. The organic fragment originally present in the organoelement derivative behaves as the cationic equivalent component: this is a " **LC$_{N'}$** " process. An example of this type of ligand coupling is the arylation of ketones by reaction with pentavalent organobismuth reagents.[86]

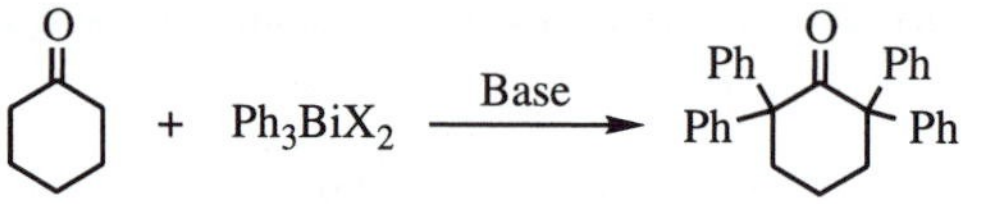

The elementary phenylation reaction with pentavalent organobismuth reagents is itself a two-steps sequence which proceeds first by the formation of a covalent pentavalent substrate-bismuth intermediate. In the second step, this intermediate then undergoes a ligand coupling process.[35,36]

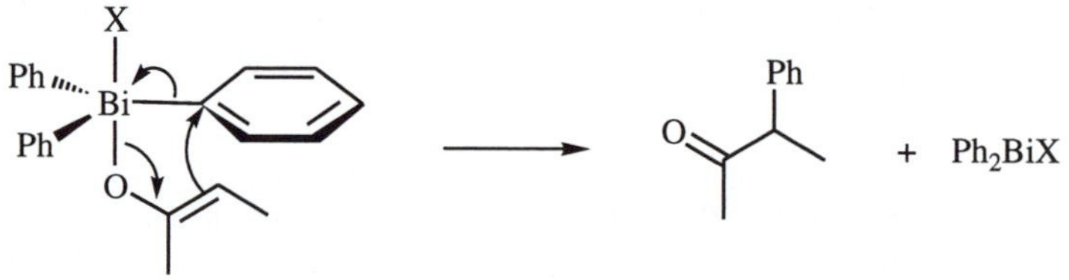

A typical example of the " **LC$_{E'}$** " type is the reaction of lead tetraacetate with enol-type substrates such as, among others, phenols and ketones. In these reactions, the acetoxy ligand plays the role of the nucleophilic component.[87]

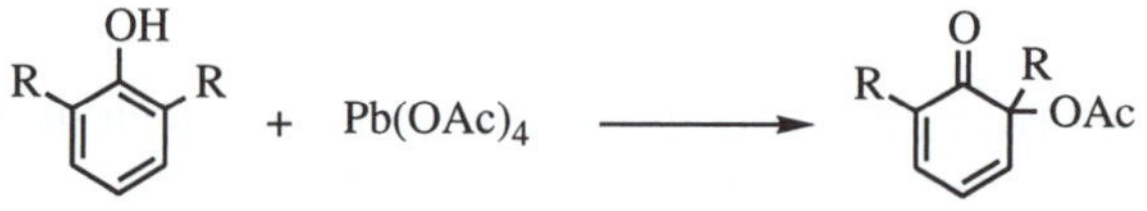

As X-ray studies of various acetoxylead derivatives have shown that the acetoxy ligand behaves as a bidentate ligand,[88,89] the transition state of the oxidation reactions can be considered to involve a structure between a five and a seven-coordinate transition state:

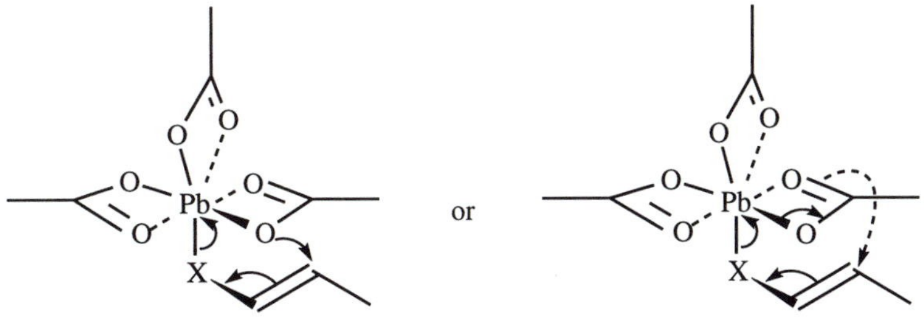

The experimental studies on the mechanism of the ligand coupling reaction have shown that the pathway going from the reactants to the coupled products can be rationalized either by the orbital symmetry conservation model or by the *ab initio* model. In the orbital symmetry conservation model, the coupling step can take place either by apical-apical bonding interaction or by equatorial-equatorial bonding interaction. On the other hand, in the *ab initio* model, the coupling step takes place by apical-equatorial interaction *via* a polar transition state. The experimental study of the mechanism of the ligand coupling reaction presents a number of difficulties and specific requirements. The evolution of the system must be concerted, but not necessarily synchronous,[90] and it must take place irreversibly. An important problem is associated with the presence of very fast Berry pseudorotation, which is common for most hypervalent derivatives of iodine, phosphorus, selenium, sulfur and tellurium. Moreover, the large diversity of substituents carried by the main-group element, especially the presence of heteroatoms which can also influence the reaction path, limits the generalization of mechanistic studies dealing with the ligand coupling mechanism itself. Unshared pair of electrons on the central atom may also affect the course of the reaction. These different conditions are difficult to be all satisfied in the same system, and therefore the number of experimental studies dealing with the ligand coupling mechanism do not compare with the wealth of the synthetic applications which are now known.

2.2.2.2 The orbital symmetry conservation model

This model has been used to rationalize the experimental data obtained in the case of ligand coupling reactions implying organotellurium, organoantimony, organoiodine and organobismuth compounds.

2.2.2.2.1 Organotellurium compounds

Thermolysis of tetraaryltellurane derivatives has been reported to afford the corresponding biaryl and the diaryltellurium compounds, together with small amounts of the arenes.[84,91,92] In 1977, Barton *et al.* showed that, in different solvents or in the absence of solvent, thermal decomposition of tetraaryltellurane derivatives afforded nearly quantitatively the diaryltellurides and biaryl derivatives. Such was the case in benzene, toluene as well as in triethylsilane or in the presence of an easily polymerizing solvent such as styrene. When the thermolysis reaction was performed in *tert*-BuSH, only the di-*tert*-butyldisulfide was obtained quantitatively, without any biaryl compounds. As addition of a powerful hydrogen atom transfer reagent, 1,4-dihydrobenzene, did not alter significantly the ratio of products, the intermediacy of free radicals was excluded and the thermal decomposition was suggested to occur by a concerted pathway.[84]

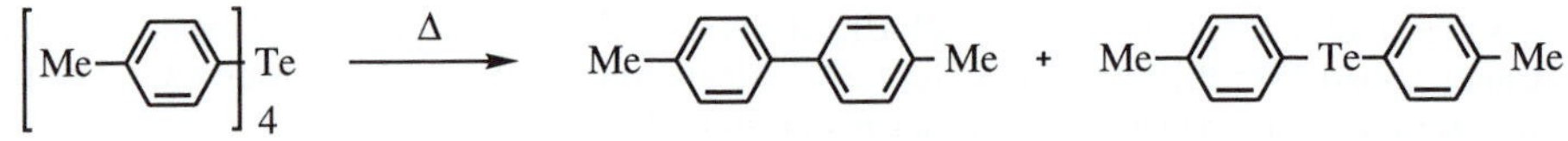

$$Ph_4Te + 2\ t\text{-BuSH} \longrightarrow 2\ PhH + Ph_2Te(S\text{-}t\text{-}Bu)_2 \longrightarrow (t\text{-}Bu\text{-}S)_2 + Ph_2Te$$

The thermal decomposition was explained by the orbital symmetry conservation model, using either the apical-apical (a)-(a) coupling or the equatorial-equatorial (e)-(e) coupling.[85]

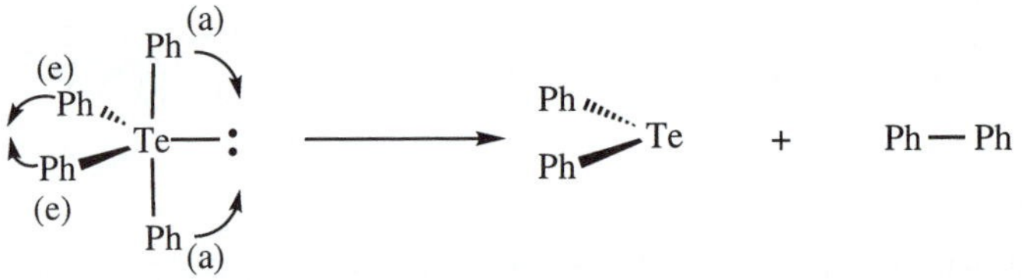

By contrast, decomposition of the stereochemically more rigid bis-(2,2'-biphenylylene)tellurium requires higher temperature and leads essentially to products derived from the biradical intermediate.[93]

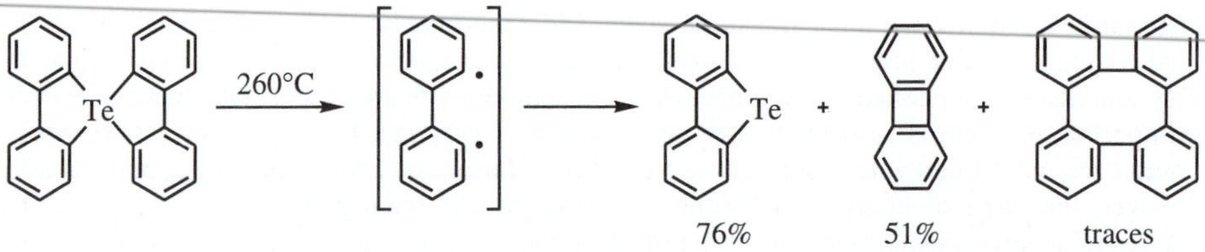

Scheme 2.10

2.2.2.2.2 Organoantimony compounds

Thermolysis of pentaphenylantimony at 225°C for 3 hours was reported by Mc Ewen *et al.* in 1968 to afford only triphenylantimony and biphenyl in nearly quantitative yields, by a direct intramolecular process.[94]

$$Ph_5Sb \xrightarrow[\text{3 h / sealed tube}]{225°C\ /\ C_6H_6} Ph_3Sb + Ph\text{-}Ph$$
$$\qquad\qquad\qquad\qquad 89\% \quad 90\%$$

However, it is only recently that Akiba *et al.* have carefully studied the mechanism of the ligand coupling reaction of various pentavalent organoantimony derivatives.[40,95] Thermal decomposition of triarylbis(phenylethynyl)antimony derivatives without solvent led to mixtures of bis(phenylacetylene) and aryl(phenylacetylene). Whatever the nature of the aryl group (electron-rich or electron-poor), the corresponding biaryl was never detected. It was suggested that the ligand coupling reaction took place by apical-apical coupling, as the yield of the aryl(phenylacetylene) product increased in parallel to the increasing electron-withdrawing ability of the aryl group. This is consistent with the presence of a higher amount of the conformer possessing an apical aryl ligand when this aryl group is substituted by increasingly more electron-withdrawing substituents.[95]

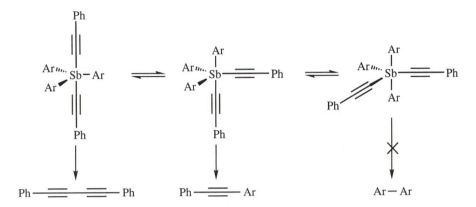

Scheme 2.11: Topological transformations during the thermolysis of triarylbis(phenylethynyl)antimony compounds

Table 2.4: Ratio of products of the thermolysis of triarylbis(phenylethynyl)antimony compounds[40]

	Ratio			
Ar	Ph$-\!\!\equiv\!\!-\!\!\equiv\!\!-$Ph	Ar$-\!\!\equiv\!\!-$Ph	Ar-Ar	Overall yield (%)
4-MeC$_6$H$_4$	76	24	0	93
Ph	66	34	0	99
4-ClC$_6$H$_4$	50	50	0	64

In the case of unsymmetrical pentaarylantimony derivatives Ar$_n$SbTol$_{5-n}$ in which Ar = 4-CF$_3$C$_6$H$_4$ and Tol = 4-MeC$_6$H$_4$, thermolysis of their solutions in benzene at 220°C led to ligand coupling products with a greater selectivity for Ar-Ar or Ar-Tol coupling rather than for Tol-Tol (< 10% for n ≥ 1).[40] These observations were explained by a preferential coupling between the two apical ligands, in spite of the presence of a very fast Berry pseudorotation, leading to topological transformations affording mixtures of all the conformers in various proportions. However, a difficulty in the interpretation of these results arose from the occurrence of equilibration by ligand exchange between unsymmetrical pentaarylantimony compounds, already observed in benzene at 60°C. This ligand exchange reaction took place in spite of the absence of either unshared electron pairs or halogen atoms.[40]

Table 2.5: Selectivity of the ligand coupling reactions of pentaarylantimony[40]

$$\text{Tol}_n\text{SbAr}_{5-n} \xrightarrow{\ 220°C\ /\ C_6H_6\ } \text{Tol-Tol}\ +\ \text{Tol-Ar}\ +\ \text{Ar-Ar}$$

	Tol-Tol	Tol-Ar	Ar-Ar
TolSbAr$_4$	0	100	-
Tol$_2$SbAr$_3$	10	2	88
Tol$_3$SbAr$_2$	2	98	0
Tol$_4$SbAr	-	21	79

Tol = 4-MeC$_6$H$_4$ and Ar = 4-CF$_3$C$_6$H$_4$

Flash vacuum thermolysis was more significant, as the bitolyl product, Tol-Tol, was not detected from any of the reactions with mixed compounds such as Ar_4TolSb or Ar_3Tol_2Sb. Even the tritolyl derivative, Ar_2Tol_3Sb, for which a significant amount of conformer with two tolyl groups in apical position could be expected, failed to lead to the formation of Tol-Tol. Ligand coupling was suggested to take place only by coupling between the two apical ligands on two particular stereoisomers in equilibrium by very fast Berry pseudorotation.

Table 2.6: Ligand coupling reactions of pentaarylantimony by Flash Vacuum Thermolysis[40]

$$Tol_nSbAr_{5-n} \xrightarrow{\text{F.V.T.}} \text{Tol-Ar} + \text{Ar-Ar}$$

	Tol-Ar	Ar-Ar
$TolSbAr_4$	64	36
Tol_2SbAr_3	42	58
Tol_3SbAr_2	24	76

$Tol = 4\text{-}MeC_6H_4$ and $Ar = 4\text{-}CF_3C_6H_4$

2.2.2.2.3 Organoiodine compounds

A number of mechanistic pathways have been claimed in order to explain the various facets of the chemistry of hypervalent iodine compounds. Apart from the oxidation reactions, the uses of hypervalent iodine compounds in synthesis started with the independent studies of Beringer and Neilands on iodonium salts.[96] Different proposals were made to explain the range of products: formation of tricoordinate iodine compounds, formation of benzyne, reductive decomposition, or nucleophilic substitution. Two types of mechanisms emerged as playing key roles: a polar pathway and a free radical pathway. A delicate balance between these two extremes was observed to be strongly dependent upon a number of factors, such as the nature of the nucleophile, the nature of the counterion and the solvent. Moreover, the two mechanisms may be occurring simultaneously. For example, in the reaction of enolates with diaryliodonium salts, the *C*-arylation products are formed by a ligand coupling process, and the competing free radical component leads to various reduction products, such as arenes.[18] It is only recently that the ligand coupling mechanism was clearly recognized to play a major role and to be able to explain a number of mechanistic specificities. First suggested by Budylin *et al.* in 1981,[97] the ligand coupling mechanism was widely used by Moriarty and Vaid in their review in 1990.[11] The pieces of the mechanistic puzzle were put together by Grushin in 1992.[23,98] The ligand coupling mechanism is now more and more taken into account in the chemistry of hypervalent iodine derivatives.

In his account, Moriarty suggested that reactions of nucleophiles with diaryliodonium salts led to a tricoordinate iodine derivative which subsequently decomposed to coupled products by either one of two mechanisms.[21] In the first one (path a), a direct ligand coupling was suggested to occur on the iodane itself. In the second mechanism (path b), the iodane collapsed to a pair of radicals which then evolved to the products. (Scheme 2.12)

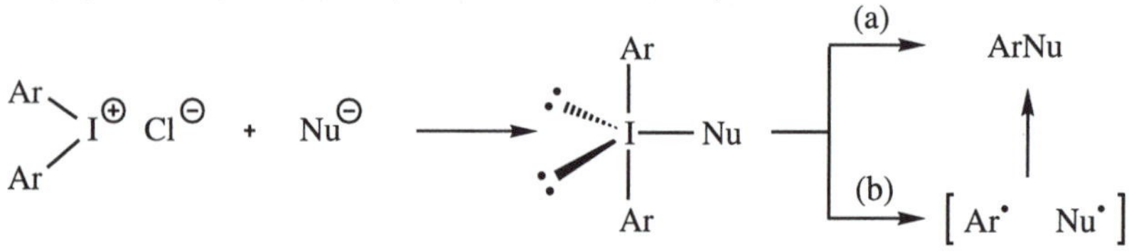

Scheme 2.12

Grushin *et al.* proposed a unified mechanism for the reaction of nucleophiles with diaryliodonium salts. The first step is the formation of a 10-I-3 hypervalent iodine intermediate.[23,98] These 10-I-3 iodane derivatives present a trigonal bipyramidal structure in the crystalline state, assuming the two unshared electron pairs behave as phantom ligands.[99] In solution, trigonal bipyramidal iodane compounds undergo topological transformations by Berry pseudorotation or by turnstile rotation.[100] The interaction between any two of the three ligands coplanar with the iodine atom in the C_{4v} structure is symmetry allowed. The nucleophile will prefer to bind to the larger group as a greater decrease of the steric strain will result. This explains the observed *ortho*-effect.[101-103] When no significant steric difference exists between the two substituents, the nucleophile will bind to the more electron-deficient ligand, as this will induce a larger decrease of the positive charge on the iodine atom. When strongly electron-withdrawing groups, such as nitro or cyano, are present in *ortho* or *para* positions, the *ipso* attack under classical S_NAr mechanism, with formation of Meisenheimer complexes, can become a competitive pathway.

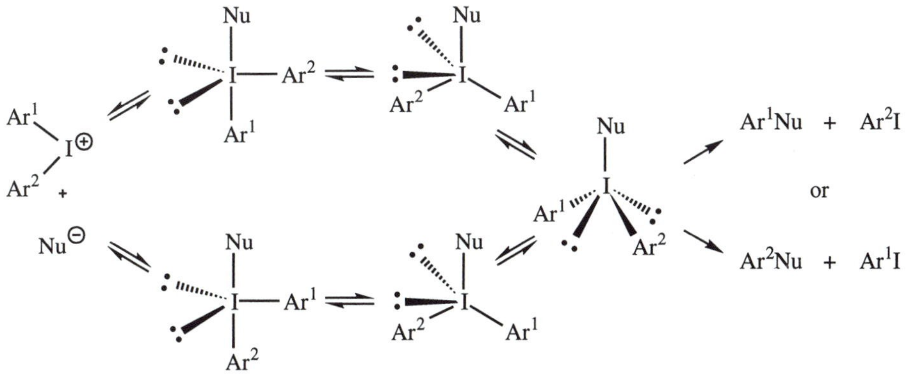

Scheme 2.13

The influence of the nature of the nucleophile on the magnitude of the *ortho*-effect[102] is also related to a steric effect on the 10-I-3 intermediate. Larger nucleophiles cause a larger steric effect, and this will result in a lower regioselectivity. The difference between NO_2^- and N_3^- anions is also explained by this effect, as azide ion can only coordinate by terminal nitrogen atoms, whereas nitrite ion can coordinate either by the terminal oxygen atoms or by the central nitrogen atom, resulting in a greater steric strain.

The divergent outcome between the reactions of diaryliodonium salts with charged nucleophiles and with neutral nucleophiles was also explained by this model. In principle, the 10-I-3 intermediate, which is formed in the first step of the overall process, can decompose by two routes. When intramolecular rotation is possible, ligand coupling takes place easily. The second possibility is the homolytic cleavage of the iodine-nucleophile bond, leading to a pair of radicals formed by one electron reduction of the iodonium cation.

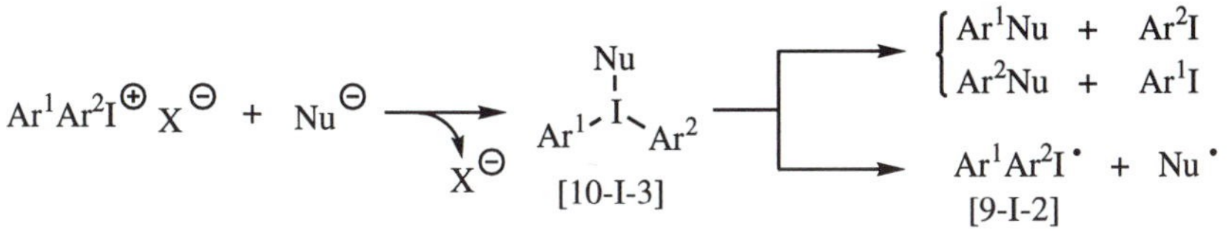

Scheme 2.14

The regiospecificity of the radical reactions of diaryliodonium salts was clearly demonstrated by Tanner *et al.* who showed that the 9-I-2 intermediate itself can follow two possible pathways for radical decomposition.[16] The difference between the energy of the two transition states depends on the electronic effects of the substituents. The substituent allowing a greater delocalization of the negative charge will correspond to the pathway with the lower energy (E_2 in scheme 2.15), and will leave as the free aryl radical. Moreover, the chain character of the radical pathway was supported by the action of 1,1-diphenylethylene which altered the ratio PhH to Ph-Ph in the reaction of diphenyliodonium tetrafluoroborate with potassium cyanide in a dioxane-water mixture.[104] In this reaction, the radical reductive cleavage is the main pathway, whereas the ligand coupling is the predominant pathway in the same system using nitrite, azide or thiocyanate ions as nucleophiles.[104]

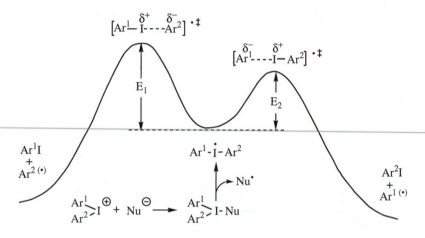

Scheme 2.15

Depending on the nature of the nucleophile, the 10-I-3 intermediate is either neutral or charged. Reaction of a neutral nucleophile with the iodonium salt leads to a charged 10-I-3 intermediate, and reaction of a charged nucleophile, by contrast, leads to a neutral intermediate. When the 10-I-3 intermediate is charged, the major part of the positive charge is located on the atom which has shared its lone pair with the iodine atom. Electrostatic interactions between the unshared pair of electrons of the iodine atom and the atom bearing the positive charge results in an increase of the energy barrier to rotation. The intermediate then cannot reach the favourable transition state for ligand coupling, and therefore follows the homolytic pathway (Scheme 2.16). On the other hand, when the 10-I-3 intermediate is neutral, no electrostatic interaction impedes its free rotation and the optimal configuration can be easily reached to favour the ligand coupling pathway (Scheme 2.17).

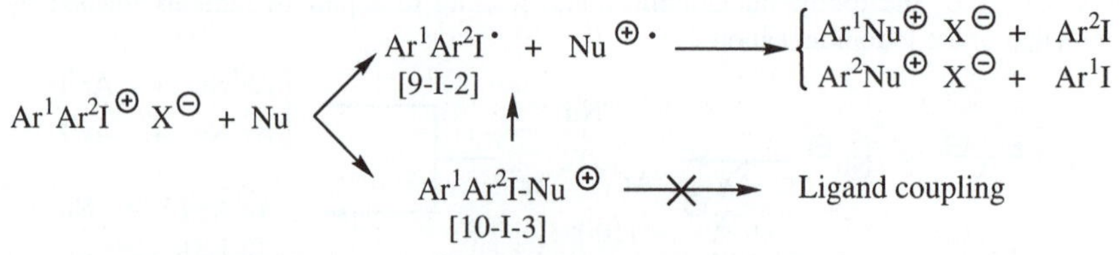

$$Ar^1Ar^2I^{\oplus} \ X^{\ominus} + Nu \left\langle \begin{array}{l} Ar^1Ar^2I^{\bullet} \ + \ Nu^{\oplus \bullet} \longrightarrow \left\{ \begin{array}{l} Ar^1Nu^{\oplus} \ X^{\ominus} + \ Ar^2I \\ Ar^2Nu^{\oplus} \ X^{\ominus} + \ Ar^1I \end{array} \right. \\ [9\text{-}I\text{-}2] \\ \\ \uparrow \\ Ar^1Ar^2I\text{-}Nu^{\oplus} \longrightarrow\!\!\!\times\!\!\!\longrightarrow \ \text{Ligand coupling} \\ [10\text{-}I\text{-}3] \end{array} \right.$$

Scheme 2.16: Reaction of a diaryliodonium salt with a neutral nucleophile

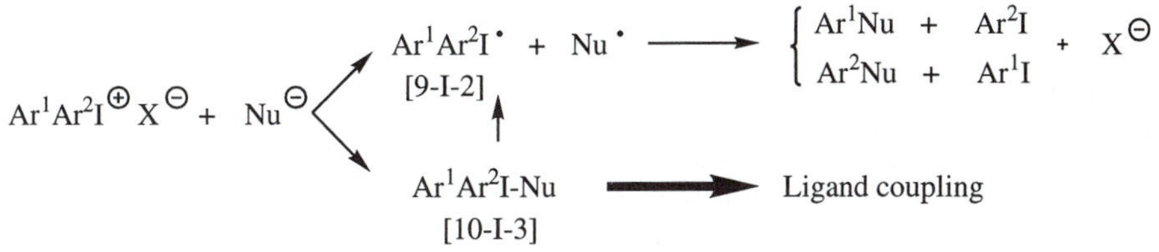

Scheme 2.17: Reaction of a diaryliodonium salt with a charged nucleophile

The phenomenon of inhibition of the rotation also explains the occurrence of a radical pathway in the reaction of the rigid dibenzo[*b,d*]iodolium cation with charged nucleophiles which otherwise react by ligand coupling mechanism with stereochemically labile diaryliodonium cations. In the case of the dibenzo[*b,d*]iodolium system, rearrangement of the trigonal bipyramid of the 10-I-3 intermediate to the C_{4v} structure is forbidden by the rigidity of the five-membered ring, only the homolytic path is possible.[105-107]

Thermal decomposition of 10*H*-dibenzo[b,e]iodinium chloride:[106]

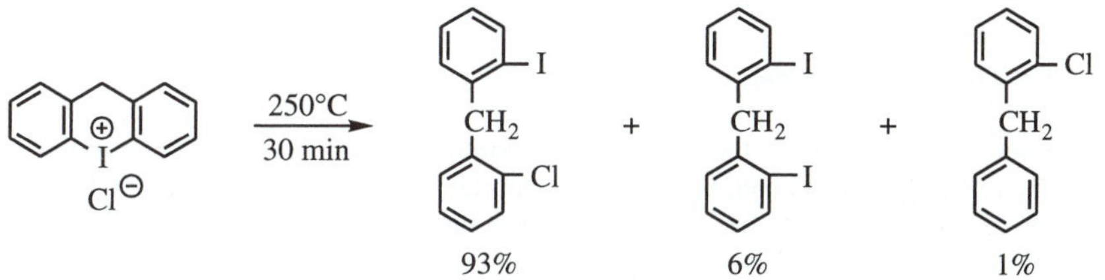

Thermal decomposition of dibenzo[b,e]iodolium chloride:[106]

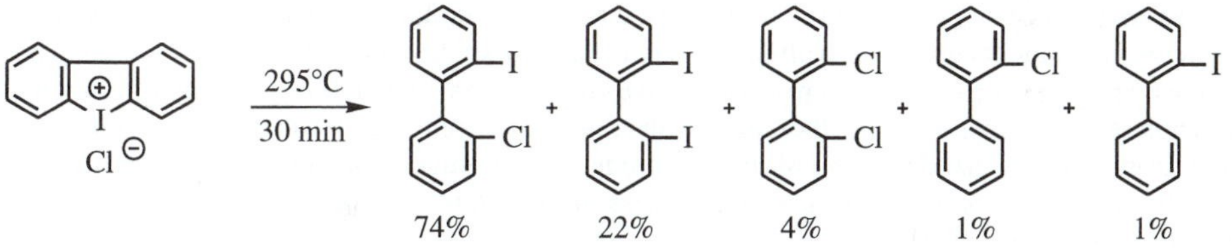

2.2.2.2.4 Organobismuth compounds

The chemistry of pentavalent organobismuth compounds involving arylation reactions of a variety of nucleophiles has been explained by the occurrence of a ligand coupling mechanism. As most of these reactions concern ambident allylic systems, the mechanistic studies were limited to the influence of the nature of the substituents: relative migratory aptitude in the case of mixed pentavalent organobismuth derivatives[108] or regiochemistry of the arylation, *O*- *vs* *C*-arylation, as a function of the nature of the substituents present on the phenolic ring.[109]

However, a reaction can be reasonably explained by the orbital symmetry conservation model: the thermal decomposition of (*p*-nitrophenoxy)triphenylbismuth derivatives.[110,111] Indeed, coupling between the two apical ligands is very easy in the decomposition of (*p*-nitrophenoxy) tetraphenyl-bismuth which affords a high yield of the ligand coupling product, 4-nitrodiphenyl ether.[112] But when one phenyl group is replaced by a more electron-withdrawing ligand, then completely different mixtures are observed.[108] With a chlorine atom, topological transformations give rise to conformations in which equatorial-equatorial interaction can occur, and this is reflected by the formation of a significant amount of biphenyl. However, when the second apical ligand is a strongly electron-withdrawing group, for example a trifluoroacetoxy ligand, the conformation becomes more locked with the three phenyl ligands in the equatorial positions. Equatorial-equatorial interaction is again the only possible pathway, and indeed the reaction affords a good yield of biphenyl. (Table 2.7)

Table 2.7: Thermal decomposition of 4-nitrophenoxytriphenylbismuth derivatives

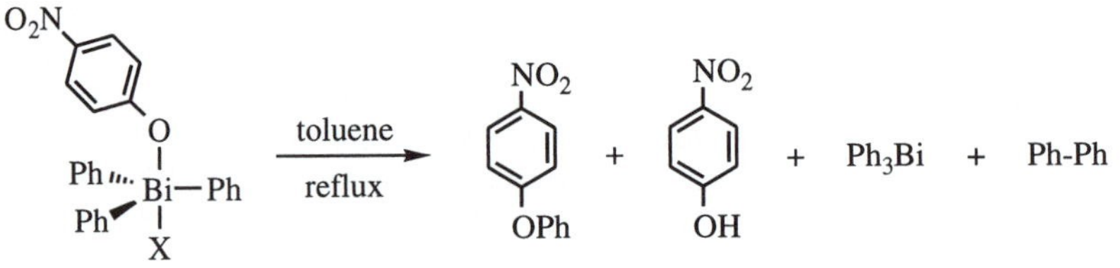

X = Ph	95%	-	79%	-	ref. 112
X = Cl	2%	19%	17%	27%	ref. 108
X = OCOCF$_3$	-	34%	4%	69%	ref. 108

2.2.2.3 The *ab initio* model

The *ab initio* model involves the least-motion coupling interaction *via* a highly polar transition state, in which an apical ligand couples with an equatorial ligand. This theoretical model is supported by a number of experimental studies, which have been performed on hypervalent sulfur compounds. Its application has been extended by analogy to the rationalization of the experimental data obtained in the chemistry of organophosphorus and organolead derivatives. However, in these cases, the experimental studies did not address the ligand coupling mechanism itself, but the general reaction mechanism involving the multistep sequence.

The reaction of organometallic compounds, such as organolithium and Grignard reagents, with triarylsulfonium salts and diarylsulfoxides has been well studied in part because of the intriguing mechanistic questions which were raised. Five major types of reaction pathways involving the organometallic compound and the sulfur substrate have been suggested to play a more or less significant role (mechanisms a-e). Moreover, products of decomposition of the sulfur substrate itself can also sometimes be met (mechanism f). These mechanisms are represented below in the case of triarylsulfonium salts.[3,5,113]

a - Aromatic nucleophilic substitution (S$_N$Ar):

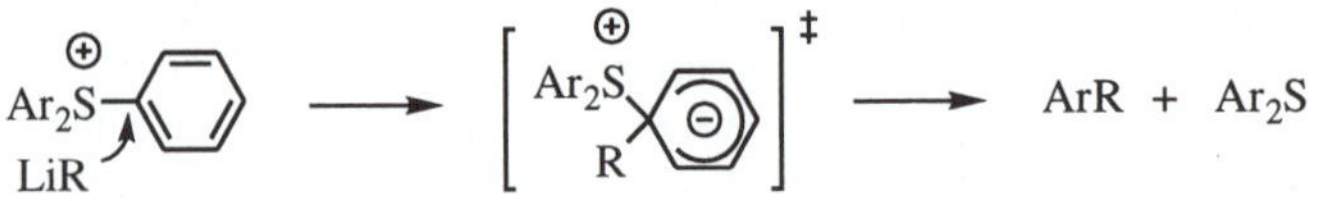

b - Benzyne mechanism:

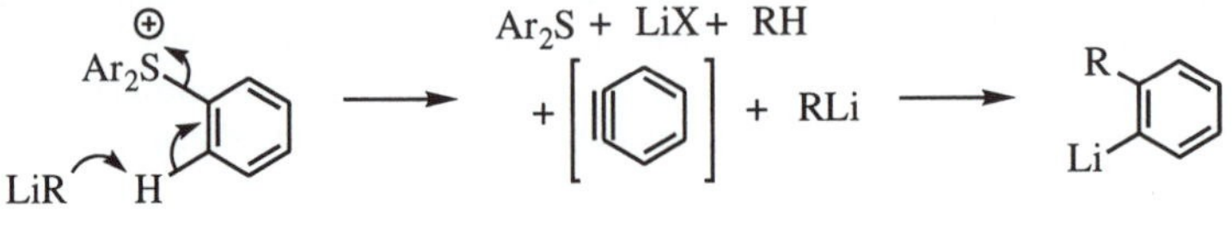

c - Single electron transfer mechanism (S.E.T.):

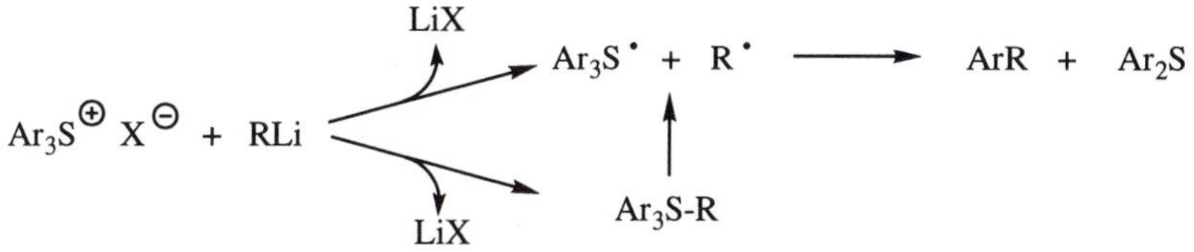

$$Ar_3S^{\oplus} X^{\ominus} + RLi$$

$$\xrightarrow{\quad LiX \quad} Ar_3S^{\bullet} + R^{\bullet} \longrightarrow ArR + Ar_2S$$

$$\uparrow$$

$$Ar_3S\text{-}R$$

$$\xrightarrow{\quad LiX \quad}$$

d - Nucleophilic attack on sulfur leading to a tetravalent sulfur intermediate undergoing ligand coupling:

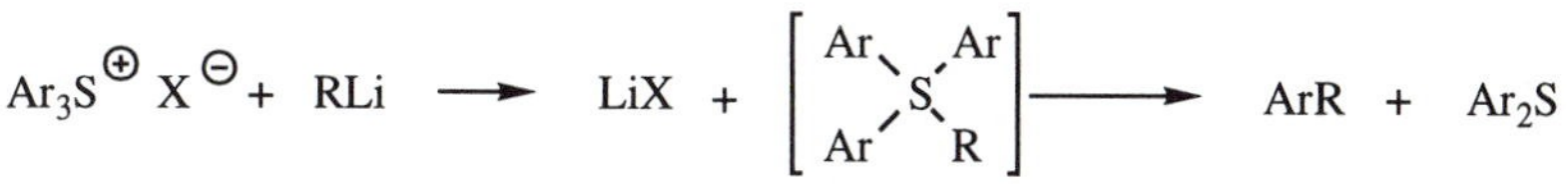

$$Ar_3S^{\oplus} X^{\ominus} + RLi \longrightarrow LiX + \left[\begin{array}{c} Ar \quad Ar \\ S \\ Ar \quad R \end{array} \right] \longrightarrow ArR + Ar_2S$$

e - Ligand exchange by S_N2 substitution on sulfur:

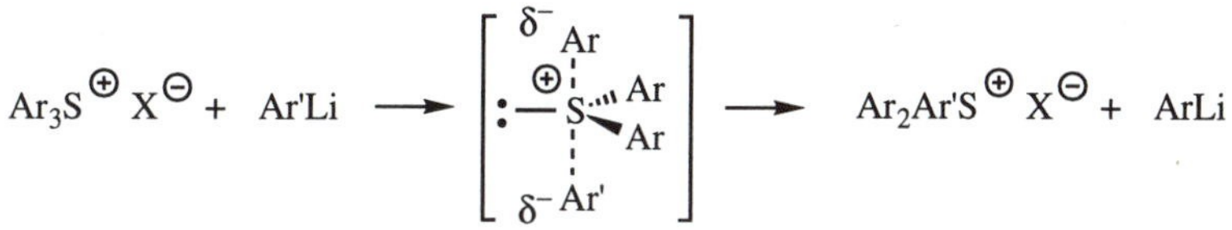

$$Ar_3S^{\oplus} X^{\ominus} + Ar'Li \longrightarrow \left[\begin{array}{c} \delta^- Ar \\ \oplus \\ S \cdots Ar \\ Ar \\ \delta^- Ar' \end{array} \right] \longrightarrow Ar_2Ar'S^{\oplus} X^{\ominus} + ArLi$$

f - Intramolecular decomposition of the reagent itself:

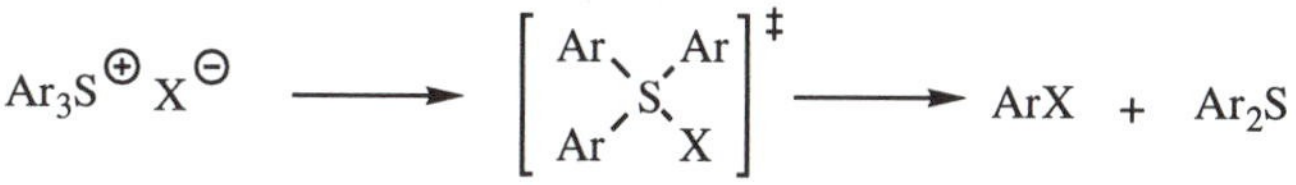

$$Ar_3S^{\oplus} X^{\ominus} \longrightarrow \left[\begin{array}{c} Ar \quad Ar \\ S \\ Ar \quad X \end{array} \right]^{\ddagger} \longrightarrow ArX + Ar_2S$$

The experimental studies on the chemistry of sulfur compounds have led Oae to develop the concept of ligand coupling,[8,9] in which two stereochemical elements are implied: the intramolecular coupling of two groups is concerted with extrusion of the reduced heteroatom, and the stereochemistry of the reagents is maintained throughout the whole process to eventually afford products with retention of configuration. The importance of these facts was deduced from the mechanistic studies performed on sulfonium derivatives which dealt mostly with the stereochemical evolution of the intermediate. The retention of stereochemistry was more extensively studied in the case of the reactions of sulfoxide derivatives.

2.2.2.3.1 Mechanistic studies with sulfonium compounds

The reaction of triarylsulfonium salts or diarylsulfoxides with organolithium or Grignard reagents affords diarylsulfide and biaryl coupling products. The formation of sulfuranes with four carbon-sulfur bonds has been postulated for a long time. In 1971, Sheppard reported the detection of tetrakis(pentafluorophenyl)sulfurane by ^{19}F NMR at low temperature, as an unstable intermediate in the reactions of pentafluorophenyllithium with fluorosulfurane derivatives which afforded decafluorobiphenyl and bis(pentafluorophenyl)sulfide.[114]

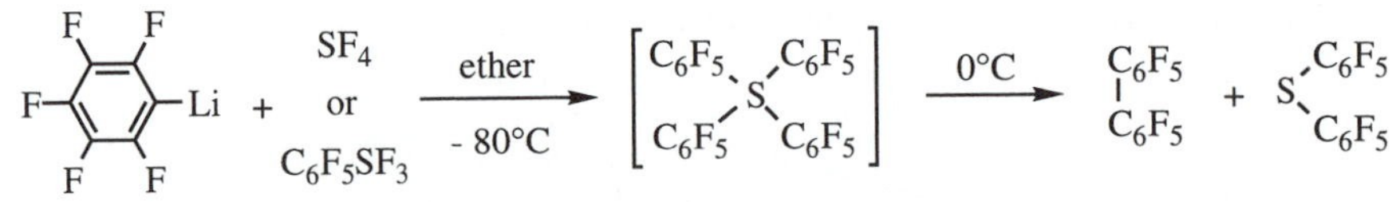

It is only recently that the formation of tetraphenylsulfurane in the reaction of triphenylsulfonium or diphenylsulfoxide with phenyllithium was demonstrated by low temperature NMR experiments.[115] Suggested by Trost[5] to be the most promising candidate as a stable sulfurane, the spirosulfurane, bis (2,2'-biphenylylene)sulfurane, was isolated recently.[116] Its structure was determined by X-ray crystallography to be a slightly distorted pseudo-trigonal-bipyramid, with two apical S-C bonds, two equatorial S-C bonds and the lone pair as the third equatorial ligand.

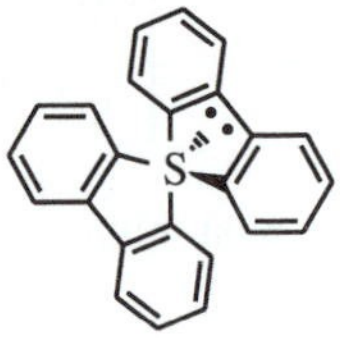

In their first report,[4] Trost *et al.* studied the reaction of organometallic reagents with triarylsulfonium salts at -78°C in THF, which led to mixtures of diarylsulfide and biaryl by *ipso-ipso* coupling. At about the same time, Khim and Oae reported a slightly different outcome for these reactions.[3] However, as their experiments have been performed in ether under reflux, competitive side reactions such as occurrence of an arynic pathway and ligand exchange took a significant weight in the overall reaction. The molecular flexibility of the postulated intermediate tetraarylsulfurane led Trost to suggest for the ligand coupling a transition state such as:

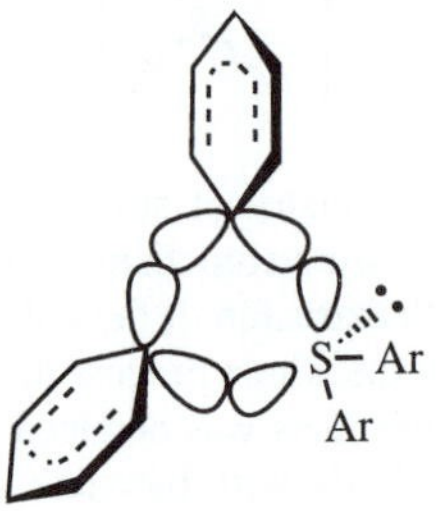

The coupling reaction involves overlap of the π-systems of two of the ligands with concomitant cleavage of the carbon-sulfur bonds. To determine the stereochemistry of the coupling, they subsequently studied the reaction of the more rigid S-aryldibenzothiophenium salts with organolithium reagents, which was suggested to form the labile diaryl bis(2,2'-biphenylylene) sulfurane.[5,6] The formation of a compound of this type, diphenyl bis(2,2'-biphenylylene)sulfurane ($Ar^1 = Ar^2 = Ph$), was detected only in 1995 by low temperature NMR studies.[117]

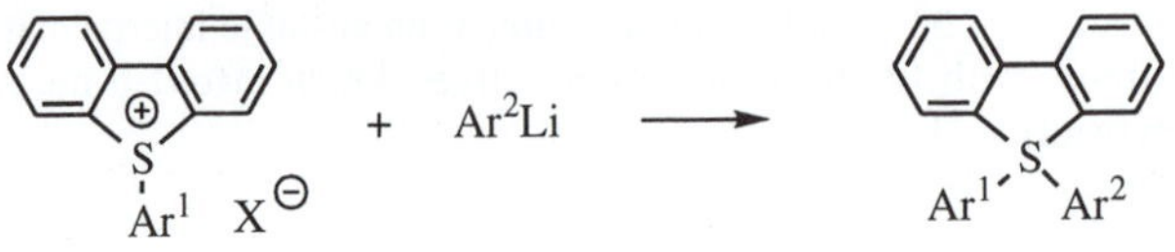

To achieve a bonding interaction between the π-systems of two ligands, two modes of coupling were considered not to require a too important distortion. Due to the trigonal bipyramid structure with the five-membered ring in equatorial-apical orientation, the two interactions $A = e_1\text{-}a_1$ and $B = e_1\text{-}a_2$ were discarded. On the other hand, the least-motion most favoured interactions were considered to be $C = a_1\text{-}e_2$, $D = a_2\text{-}e_2$ and, to a lesser degree, $E = e_1\text{-}e_2$.[5]

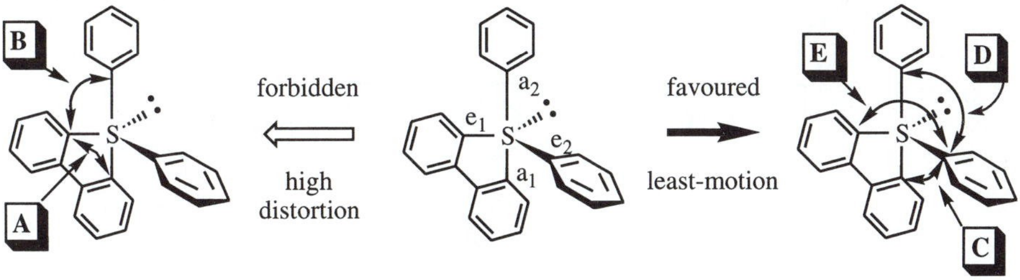

Scheme 2.18

Due to the inductive electron-withdrawing effect of a benzene substituent, the biphenylyl group is more activated towards coupling than a phenyl group. Thus, the $a_1\text{-}e_2$ is favoured over the $a_2\text{-}e_2$ coupling, and, in the case when $Ar^1 = Ar^2 = Ph$, only the phenyl-biphenylyl coupling was obtained (Scheme 2.19, pathway $\mathbb{A}$). The same mode of coupling was observed for phenyl ligands substituted with electron-donating groups.

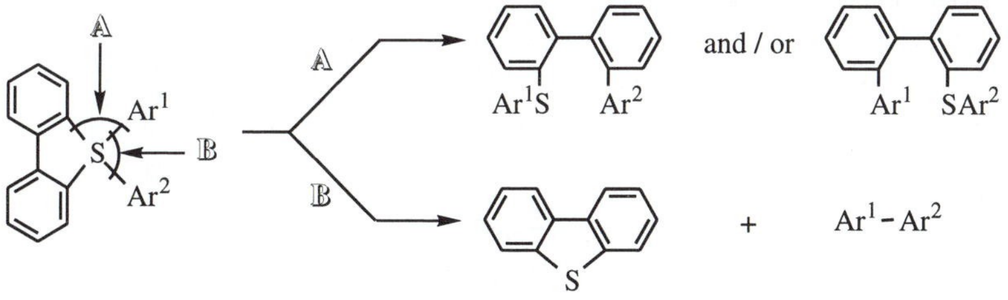

Scheme 2.19

On the other hand, when the phenyl ligands were substituted with electron-withdrawing groups, the two modes of coupling (pathways $\mathbb{A}$ and $\mathbb{B}$) were found: $a_1\text{-}e_2$ and $a_2\text{-}e_2$. Similar trends were observed in the case when the two aryl groups, Ar_1 and Ar_2, were different: coupling takes place preferentially between the most electron-depleted groups. The fact that the addition of the aryllithium derived from the more or the less electron-poor group did not interfere with the eventual ratios of coupled products shows that, in this system, topological interconversion between conformers is faster than ligand coupling.[4-6]

By contrast, a stereochemical effect was observed in the reaction of S-aryldibenzothiophenium salts with vinyllithium reagents. Assuming that the entering nucleophilic group occupies the apical position, the kinetic product has the vinyl group in the apical position. It can undergo only vinyl-aryl or aryl-biphenylyl coupling. If the aryl ligand is a phenyl, unsubstituted or substituted by electron-withdrawing groups, the rapid rate of coupling leads mostly to substituted styrene and dibenzothiophene with a small amount of aryl-biphenylyl coupling. If the aryl ligand is substituted by electron-donating groups, the vinyl-aryl coupling is so slowed that topological isomerisation takes place efficiently. Therefore, the vinyl-biphenylyl coupling now takes place predominantly or exclusively.[5,6]

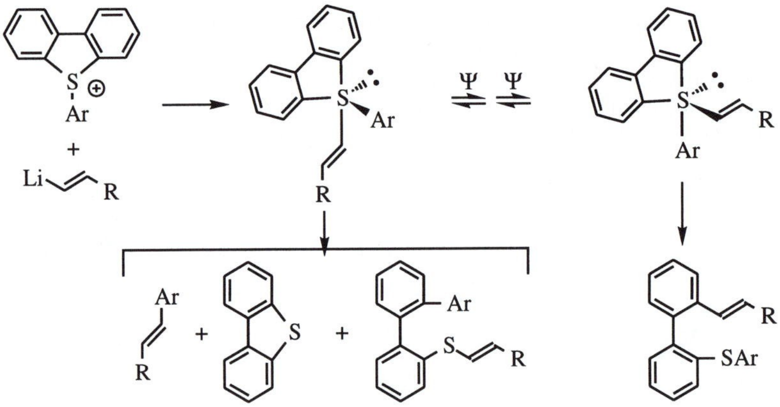

Scheme 2.20

In a series of papers,[113,118-122] Hori and Shimizu *et al.* have studied a wider scope of substrates and their results were in good general agreement with the reports of Trost and Oae. When the dibenzothiophene ring was replaced by a more flexible 9,9-dimethylthioxanthene system, the similar general reactivity patterns were found. However, ligand exchange reactions also took place, as the reaction conditions were similar to those reported by Oae[3]: the reactions were in general performed in ether at room temperature or under reflux.[119]

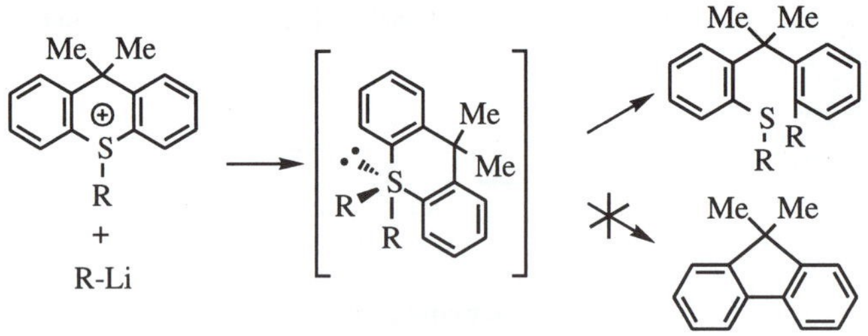

Scheme 2.21

With the even more flexible 10,11-dihydrodibenzo[*b,f*]thiepinium systems, reaction of different aryllithium reagents used in large excess led to the various types of coupling, with a slight predominance for the formation of dihydrophenanthrene (47-66%), resulting from coupling along pathway B. This system appears to behave as the tetraarylsulfurane derivatives, which do not contain cyclic rings and lead to statistical mixtures of coupling products.[120]

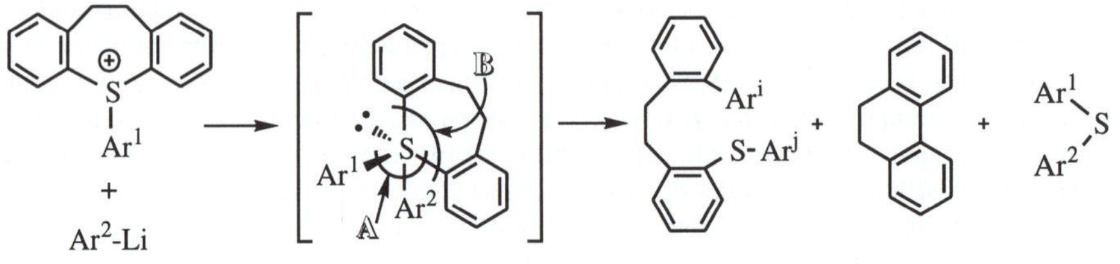

Scheme 2.22

2.2.2.3.2 Mechanistic studies with sulfoxide compounds

The extensive work of Oae and his group on the reactions of sulfoxides, in particular with organometallic compounds, led to the definition of the stereochemical criteria for ligand coupling.[8,9] The reaction of sulfoxides with organomagnesium or organolithium reagents leads to a variety of products depending on the nature of the sulfoxide, the organometallic reagent and the reaction conditions. The reaction of diarylsulfoxides with aryl Grignard or aryllithium has been shown to lead to biaryl formation by a sequence of transition steps involving the formation of a hydroxysulfurane. The coupling itself results from a major pathway by a benzyne intermediate, the ligand coupling pathway constituting only a minor pathway.[123]

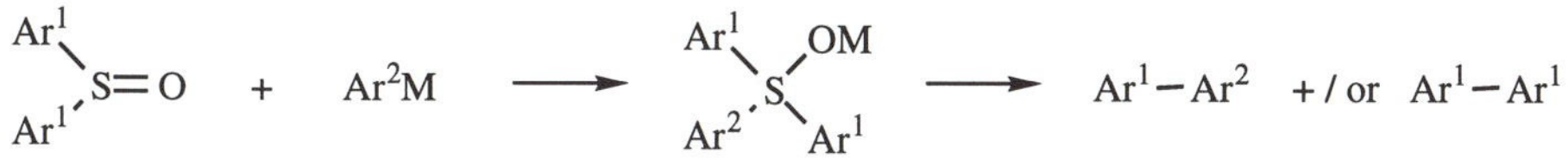

Sulfoxides containing a methylsulfinyl group react with organomagnesium halides to give sulfides.[124,125] The reaction was suggested to involve deprotonation, formation of a methylydene-sulfonium intermediate and addition of the alkyl Grignard on the α-carbon.[125]

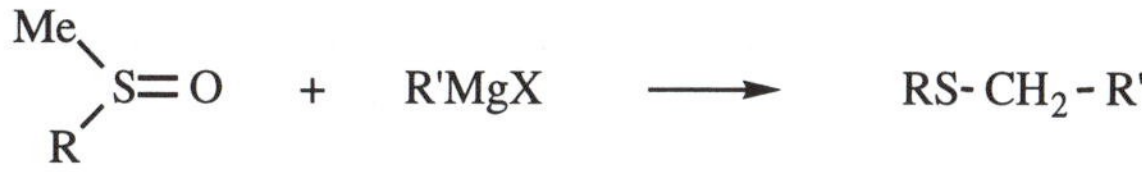

In a third type of behaviour, sulfoxides bearing an electron-poor aryl or heteroaryl group react with organomagnesium or organolithium reagents to afford ligand coupling products.[27-29]

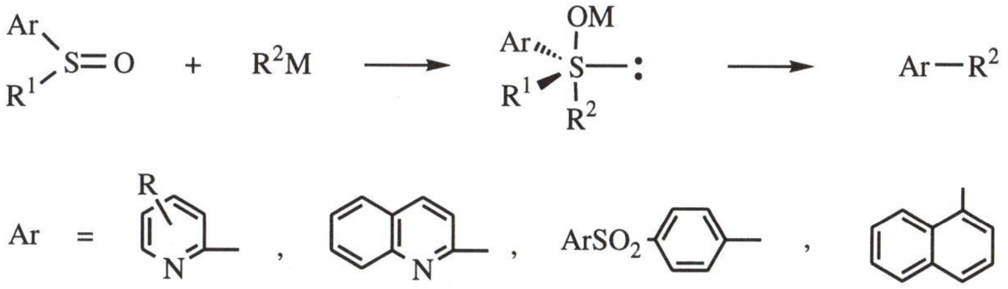

The sulfoxides prone to ligand coupling possess an aryl group such as a 2-pyridyl substituted or not, quinolinyl, arylsulfonyl or naphthyl. The organic fragment of the Grignard or organolithium reagents can be aryl, heteroaryl, vinyl, benzyl, allyl, but also *sec-* and *tert-*alkyl groups.

The reaction of a tricoordinate sulfoxide with nucleophiles usually leads to a tetrasubstituted pentacoordinate σ-sulfurane, in which the pair of unshared electrons plays the role of the fifth ligand. The attack of the nucleophile takes place so that, in the first formed putative intermediate, the incoming nucleophile occupies the apical position. Depending on the nature of the substituting groups, this intermediate can evolve under different pathways, which can take place alone, in combination or in competition. The major pathways which can be followed by the intermediate sulfurane are ligand exchange, topological transformations and ligand coupling. The stereochemical outcome of the reactions may thus become difficult to interpret.

a - **Ligand exchange**:

A pure ligand exchange process leads to products with inversion of configuration, as in a typical S_N2 reaction. However, if topological transformations take place, the final product can present retention of configuration. Ligand exchange is one of the most studied reaction in the case of hypervalent compounds. An example of ligand exchange with inversion of configuration is the oxygen exchange reaction of sulfoxides in acetic anhydride. The rate of racemisation was twice the rate of oxygen exchange, $k_{rac} = 2\ k_{ex}$.[126]

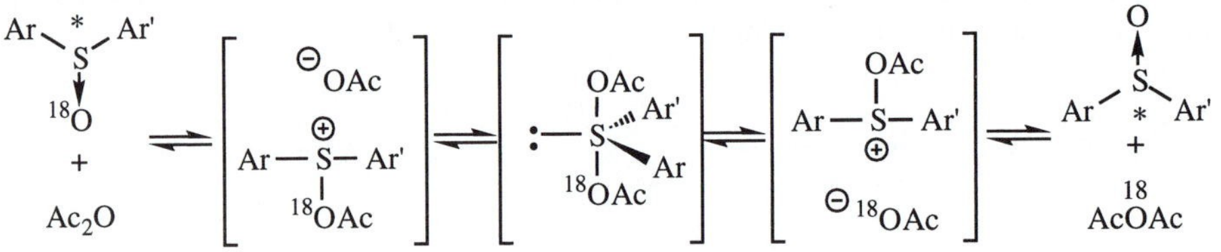

Scheme 2.23

Reaction of alkyllithium compounds with optically active arylmethylsulfoxides leads to the corresponding dialkylsulfoxides which present inversion of configuration.[127]

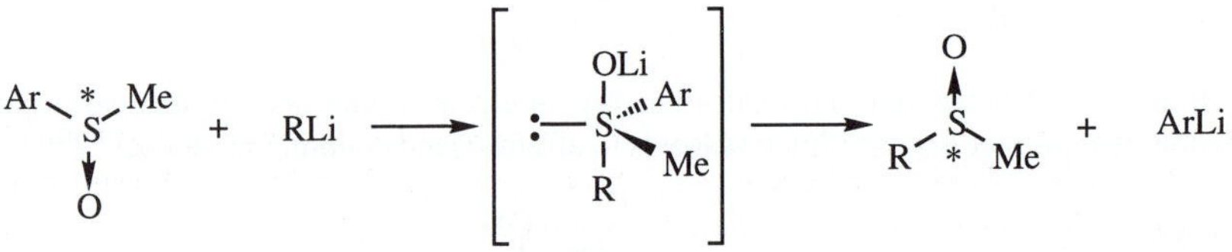

If the intermediate sulfurane undergoes pseudorotation, the ligand exchange can proceed with retention of configuration. A substantial oxygen exchange took place when methyl p-tolylsulfoxide was heated at 150°C in dimethylsulfoxide. In this reaction, the exchange was easier with the p-methyl than with the p-chloro derivative. Moreover, very little racemisation was observed, $k_{rac} \ll k_{ex}$. Assuming that the entering and leaving groups approach or leave the sulfoxide along the apical position, the retention of configuration results from pseudorotation.[128]

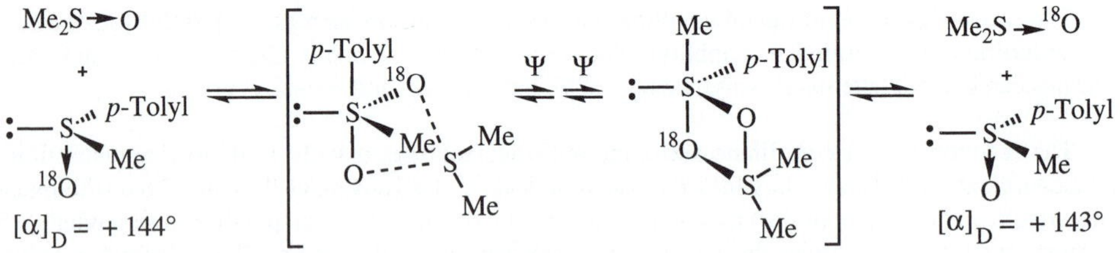

Scheme 2.24

When a sulfoxide, bearing a combination of ligands other than pyridyl and benzyl, is treated with Grignard reagents, ligand exchange and ligand coupling take place independently or concurrently.[129,130] For example, when methyl-2-pyridylsulfoxide is treated with a Grignard reagent which does not give an intermediate prone to ligand coupling, ligand exchange takes place to lead to

2-pyridylmagnesium halide and a new sulfoxide. In a second step, the newly formed pyridyl Grignard reagent reacts with methyl-2-pyridylsulfoxide giving a dipyridyl sulfurane, which affords eventually 2,2'-bipyridine as a ligand coupling product.

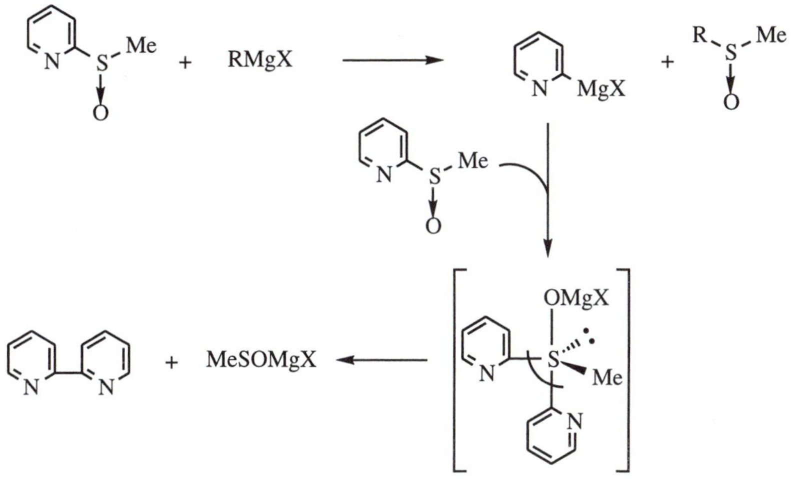

Scheme 2.25

The ease of ligand exchange appeared to be controlled essentially by the apicophilicity of the exchanged ligand.[18] But the steric hindrance can play a very significant role, as shown by the reaction of alkylpyridylsulfoxides with phenylmagnesium bromide.[129]

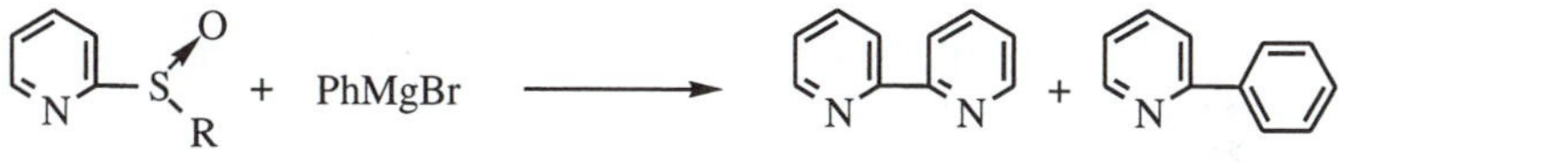

R =		
Me	68%	9%
Et	56%	4%
i-Pr	59%	6%
t-Bu	0%	85%

b - **Topological transformations**:

Berry pseudorotation (BPR) or turnstile rotation (TR) mechanisms can lead to products with retention of configuration or with racemization. When a topological transformation takes place at a competitive rate with the ligand coupling process, mixtures of products can be obtained, as for example in the reaction of 2-pyridylsulfoxides with Grignard reagents. The attack of the nucleophile on the sulfoxide leads to a first sulfurane intermediate, in which the incoming nucleophile occupies an apical position. When pseudorotation can occur, the 2-pyridyl group always remain in the equatorial position, while a benzyl group occupies an apical position. Ligand coupling will result from the interaction between the equatorial 2-pyridyl group and the apical benzyl group.

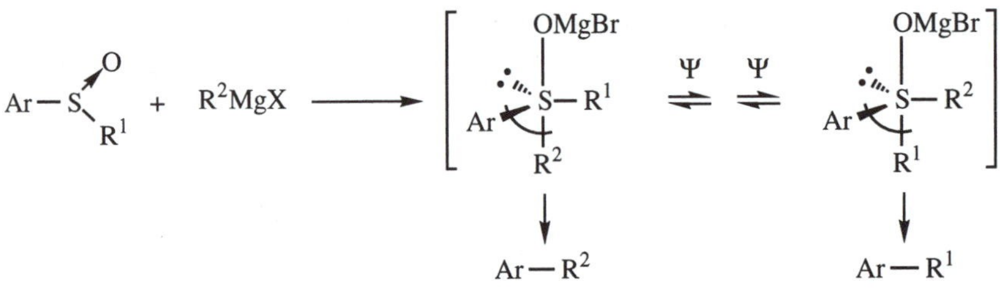

Scheme 2.26

In the reaction of benzyl-d_2 2-pyridylsulfoxide with unlabelled benzylmagnesium chloride, the coupling product was predominantly the unlabelled 2-benzylpyridine.[131] The rate of pseudorotation appears much smaller than the rate of coupling. However, increase of the reaction temperature alters slightly the ratio between the two products.

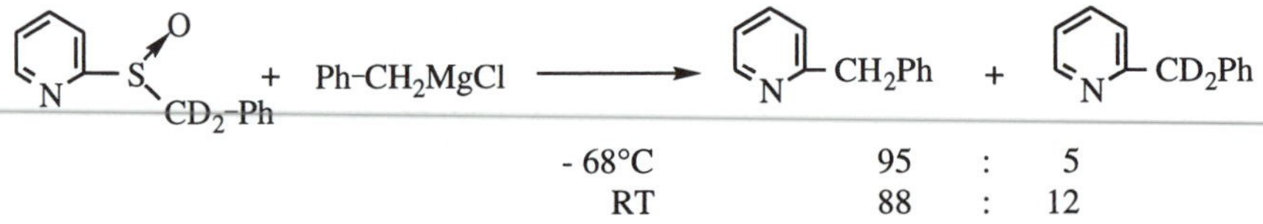

- 68°C	95 :	5
RT	88 :	12

The effect of temperature increase on the relative ratio between ligand coupling products is more or less intense, but always present. (Table 2.8)

Table 2.8: Influence of the reaction temperature on the ratio of products[132]

Ar	R^1	R^2	$ArR^1 : ArR^2$ at RT	$ArR^1 : ArR^2$ at 50°C
2-pyridyl	4-Me-C$_6$H$_4$-CH$_2$	C$_6$H$_5$-CH$_2$	16 : 84	20 : 80
2-pyridyl	C$_6$H$_5$-CH$_2$	4-Me-C$_6$H$_4$-CH$_2$	44 : 56	52 : 48
PhSO$_2$C$_6$H$_4$	4-Me-C$_6$H$_4$-CH$_2$	C$_6$H$_5$-CH$_2$	20 : 80	19 : 81
PhSO$_2$C$_6$H$_4$	C$_6$H$_5$-CH$_2$	4-Me-C$_6$H$_4$-CH$_2$	77 : 23	46 : 54

The electronic nature of the ligands plays an important role on the extent of the pseudorotation phenomenon. When there is a competition between two substituted benzyl groups, the more electron-poor benzyl group will preferentially occupy the apical position, favouring the coupling process.[132] When the incoming nucleophile bears the most electron-withdrawing group, ligand coupling will occur more rapidly than pseudorotation. However, when the incoming nucleophile is more electron-rich, pseudorotation will take a significant weight in the competition between the two pathways, so that the more electron-poor ligand will tend to occupy eventually the apical position. (Table 2.9)

Table 2.9: Competition between benzyl and *p*-chlorobenzyl ligands[132]

R[1]	R[2]	ArR[1] : ArR[2] at RT
C_6H_5-CH_2	4-Cl-C_6H_4-CH_2	9 : 91
4-Cl-C_6H_4-CH_2	C_6H_5-CH_2	65 : 35

The steric size of the two ligands does not generally play a very important role in the outcome of the reaction. For example, in the competition between *para*-tolyl and *ortho*, *para*-dimethylbenzyl groups, the steric size of the ligands did not play a noticeable role in the case of the 2-pyridylsulfoxide system. In going from tolyl to the xylyl incoming nucleophile, the ligand coupling products involving the incoming group increased only from 56% to 59% (Tables 2.8 and 2.10). However, in the case of the 4-(phenylsulfonyl)phenylsulfoxide system, the steric size of the ligands plays a more significant role, although not determinant. When the competition takes place between benzyl and tolyl ligands, the favoured coupling product is derived from the benzyl group whatever the incoming group [80% or 77% of 1-benzyl 4-(phenylsulfonyl)benzene]. On the other hand, when the competition happens between benzyl and the 2,4-dimethylbenzyl groups, the effect of the pseudorotation is more pronounced. Indeed, when the 2,4-dimethylbenzyl group is the incoming nucleophile, the ratios observed at room temperature (47 : 53) are nearly identical to those obtained at 50°C, when the incoming nucleophile is the 4-tolyl group (46 : 54).

Table 2.10: Competition between benzyl and *ortho*, *para*-dimethylbenzyl groups[132]

Ar	R[1]	R[2]	ArR[1] : ArR[2] at RT
2-pyridyl	2,4-$Me_2C_6H_3$-CH_2	C_6H_5-CH_2	17 : 83
2-pyridyl	C_6H_5-CH_2	2,4-$Me_2C_6H_3$-CH_2	41 : 59
$PhSO_2C_6H_4$	2,4-$Me_2C_6H_3$-CH_2	C_6H_5-CH_2	12 : 88
$PhSO_2C_6H_4$	C_6H_5-CH_2	2,4-$Me_2C_6H_3$-CH_2	47 : 53

c - **Ligand coupling**:

As the ligand coupling process results from an intramolecular interaction between two organic fragments which are linked to the heteroatom by a σ-bond, the original configuration of the two fragments will not be affected during the reaction. The ligand coupling process must afford the products with retention of configuration.[8,9] This stereochemical aspect of the ligand coupling mechanism was demonstrated by the observation of retention of configuration in a variety of reactions involving sulfoxides bearing sp^3 as well as sp^2 carbon ligands directly linked to the sulfur atom. These ligands are either optically active groups or geometric isomers of vinylic groups.

α - Ligands with optically active sp³ carbons:

As the 2-pyridyl group is not chiral, the conservation of the stereochemical integrity of the second ligand of the sulfoxide, upon reaction with a Grignard reagent, was observed for the first time in the system shown in scheme 2.27. Optically pure S-(-) or R-(+) sulfides were oxidized to their corresponding sulfoxides, which were resolved into their pure diastereoisomers: S,S and S,R from the S sulfide and R,S and R,R from the R sulfide. Reaction of the pure sulfoxides or of the mixture of sulfoxides derived from the optically pure parent sulfides with methylmagnesium bromide led to the coupling products with retention of the original configuration. This was unequivocally confirmed by the X-ray study of the derived crystalline *N*-methylpyridynium salts.[133,134]

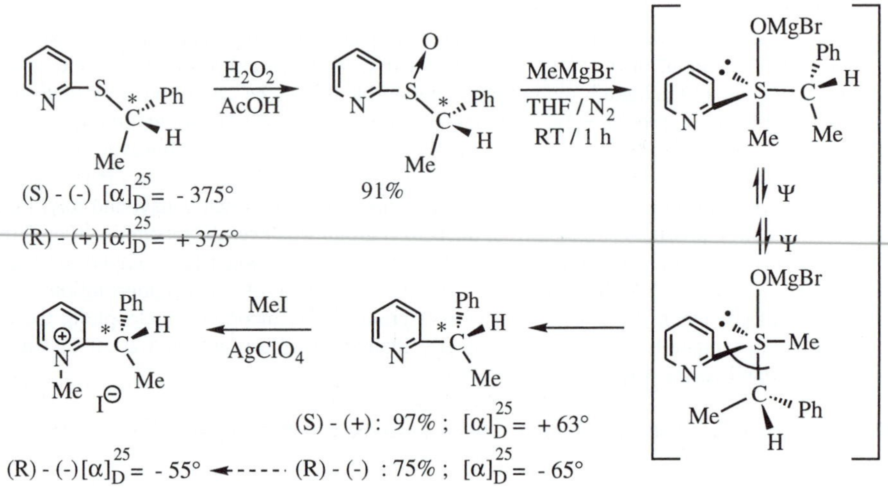

Scheme 2.27

Similar results were obtained with optically active (1R)-phenylethyl-2-quinolinyl (R)- and (S)-sulfoxides, which both led to (R)-2-(1-phenylethyl)quinoline.[135]

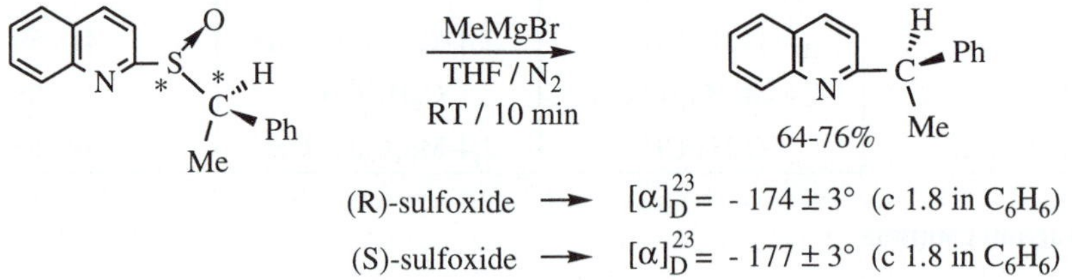

(R)-sulfoxide → $[\alpha]_D^{23} = -174 \pm 3°$ (c 1.8 in C_6H_6)

(S)-sulfoxide → $[\alpha]_D^{23} = -177 \pm 3°$ (c 1.8 in C_6H_6)

In these systems, the possibility of a chelation effect between the nitrogen atom of the 2-pyridyl group and the magnesium atom of the incoming Grignard reagent could affect the stereochemical outcome of the reaction. However, a similar complete retention of configuration was also observed in the case of the benzenesulfonylphenyl analogous system.[136]

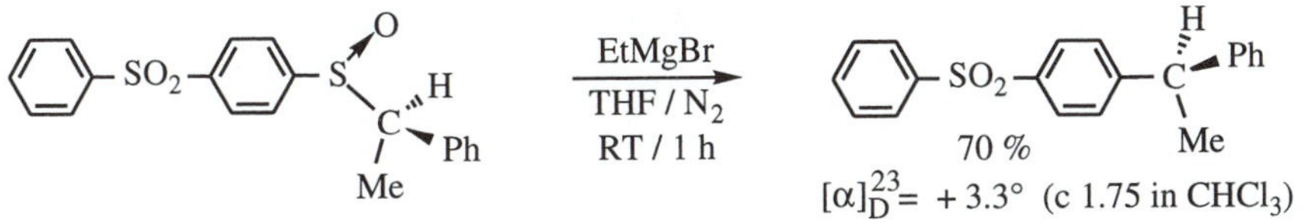

These results led Oae *et al.* to conclude that ligand coupling between the equatorially preferred 2-pyridyl or electron-withdrawing benzenesulfonylphenyl groups and the apicophilic 1-phenylethyl group proceeds concertedly.

β - Ligands with sp³ allylic carbons:

If ligand coupling is proceeding concertedly, then retention of configuration should also be observed with sp³ ligands prone to facile isomerisation or rearrangement, such as allylic groups. This postulate was indeed confirmed with crotyl and 1-methallyl groups, which retained the configuration of the starting sulfoxides in the ligand coupling product. The reaction of the crotyl sulfoxide with ethyl magnesium bromide afforded two products. The normal ligand coupling product was obtained in 49% yield. Diphenylsulfone isolated in 30% yield is a by-product formed by ligand exchange between the substrate and the ethyl Grignard reagent. In the case of the 1-methallylsulfoxide, the ligand coupling product was the only product.[137]

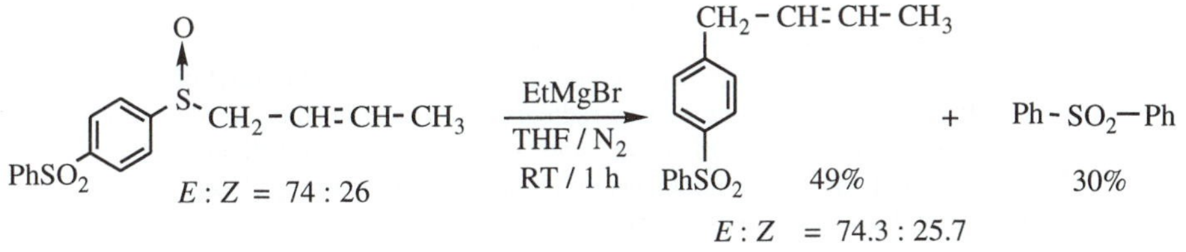

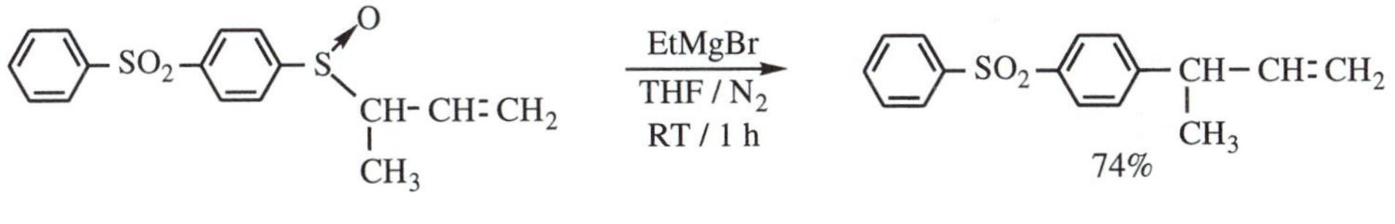

γ - Ligands with sp² vinylic carbons:

Reaction of arylvinylsulfoxides with Grignard reagents also led to coupling products showing retention of the geometry of the vinyl double bond, as well with the benzenesulfonylphenyl as with the 2-pyridylsulfoxide derivatives.[137] In the synthesis of α-stilbazoles, the yields of ligand coupling products were generally modest. In this case, the ligand coupling process was in competition with ligand exchange, resulting in a more complex mixture of products.[138]

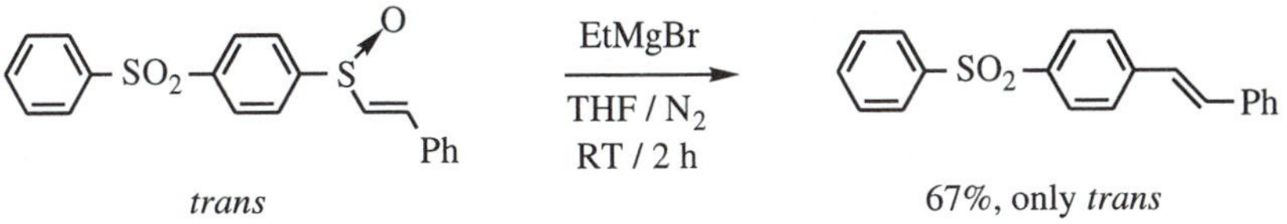

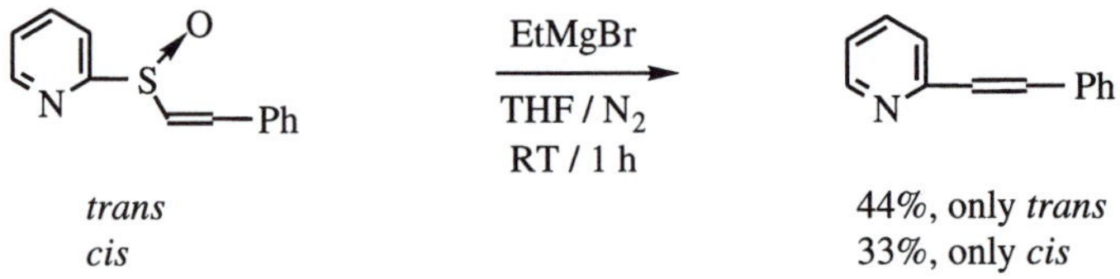
O
S
N
Ph
EtMgBr
THF / N2
RT / 1 h
N
Ph
trans
cis
44%, only trans
33%, only cis

2.2.3 Ligand coupling: concerted or radical species ?

The observation of a competition between the pathways followed by the 10-I-3 intermediates in the arylation reactions with diaryliodonium salts led Moriarty to consider whether the ligand coupling was taking place by a concerted mechanism (path a) or with the intervention of radical species (path b).[21] (Scheme 2.28)

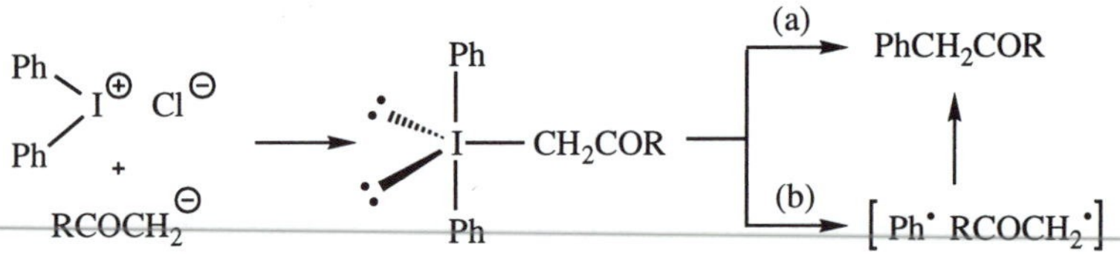
Ph
I
Cl
Ph
+
RCOCH2
Ph
I
CH2COR
Ph
(a)
PhCH2COR
(b)
[Ph• RCOCH2•]

Scheme 2.28

In a number of reactions, use of classical spin trapping agents led to the ESR observation of spin adducts, which were due to the presence of free radicals. For example, the presence of phenyl-*tert*-butylnitroxide was considered as a proof of the intermediacy of phenyl radicals in the reaction of diphenyliodonium salts with phenolate anions leading to arylphenyl ethers.[16] However, the qualitative observation of free radical species cannot be considered as a proof of a radical mechanism, as they can result from a competitive decomposition pathway. Two major experimental approaches were used to determine the real influence of free radical species: *a*) quantitative evaluation of the effect of the addition of free radical traps and *b*) the use of substrates containing internal traps.

2.2.3.1 Addition of free radical traps

In the decomposition of tetraaryltellurane derivatives, an hydrogen atom transfer reagent was used to exclude the occurrence of free radicals. Addition of 1,4-dihydrobenzene, an hydrogen atom transfer reagent, did not affect the relative yields of products of the thermal decomposition of tetra-*p*-tolyltellurane. The effective scavenging of tolyl radicals was demonstrated by the decomposition of *p*-tolylazotriphenylmethane, which under the same conditions did not give any bitolyl products. Similarly, styrene was used in this reaction and failed to induce any modification of the outcome of the reaction. Therefore, this ligand coupling reaction, a type **LC$_\mathbf{H}$** coupling, does not involve any free radical species.[84,85]

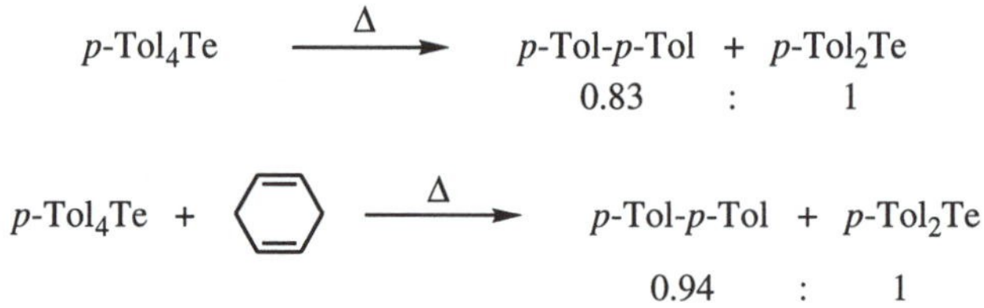
p-Tol4Te
Δ
p-Tol-p-Tol + p-Tol2Te
0.83 : 1
p-Tol4Te +
Δ
p-Tol-p-Tol + p-Tol2Te
0.94 : 1

$$p\text{-Tol}-N=N-CPh_3 \quad + \quad \bigcirc \quad \xrightarrow{\Delta} \quad \text{Toluene} \quad + \quad Ph_3CH$$

$$80\% \quad : \quad \text{quantitative}$$

An early report of the beneficial influence of 1,1-diphenylethylene (DPE) on the yields of alkylaryl ether obtained in the reaction of diaryliodonium salts with sodium alkoxides showed that radical chain reactions compete efficiently with the O-arylation reaction. By contrast, addition of diphenylpicrylhydrazyl, a stable free radical species, had no significant influence on the yields of products obtained in the absence of additives. In this case, the O-arylation reaction was considered to be a direct nucleophilic aromatic substitution reaction, without the involvement of any transient covalent intermediate.[139] (Table 2.11)

Table 2.11: Influence of additives on the reaction of diphenyliodonium tetrafluoroborate with sodium ethoxide[139]

$$Ph_2\overset{\oplus}{I} \ \overset{\ominus}{BF_4} \quad + \quad EtONa \quad \longrightarrow \quad PhH \ + \ PhOEt \ + \ PhI \ + \ Ph\text{-}Ph$$

Reaction conditions	PhH (%)	PhOEt (%)	PhI (%)	Ph-Ph (%)
Air	66	14	92	0.32
Argon	68	14	92	0.42
DPE (1 equiv.)	6-6.2	77-80	98-100	0.26-0.3
DPPH (0.05 equiv.)	53	26	92	0.26

DPE = 1,1-diphenylethylene; DPPH = *N,N*-Diphenylpicrylhydrazyl

A similar beneficial effect of 1,1-diphenylethylene on the yields of arylation products was later observed in the synthesis of benzonitrile derivatives by reaction of diaryliodonium salts with potassium cyanide[104] and in the reaction of diaryliodonium salts with the sodium salt of nitroalkanes.[17] In the latter case, the reaction was therefore considered to result from intermediate inner-sphere radicals.[17] Some years later, Barton *et al.* showed that 1,1-diphenylethylene acts as an efficient inhibitor of the radical chain process in the reaction of enolates with diaryliodonium salts. They concluded that the arylation products arose from a non-radical process.[18]

BTMG = *N-tert*-butyl-*N',N',N'',N''*-tetramethylguanidine

Similarly, a number of other arylation reaction systems were tested for the presence and study of the influence of free radicals. Addition of 1,1-diphenylethylene (DPE) in reactions of organobismuth, organolead and organosulfur reactions failed to affect the outcome of these reactions, which therefore were considered as occurring without the intervention of free radical species.[18,140,141]

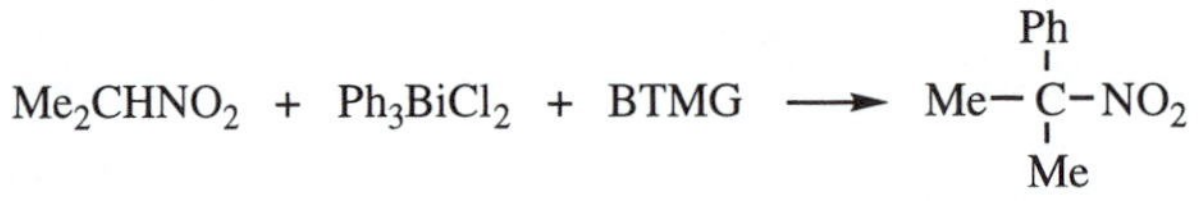

$$Me_2CHNO_2 + Ph_3BiCl_2 + BTMG \longrightarrow Me-\underset{\underset{Me}{|}}{\overset{\overset{Ph}{|}}{C}}-NO_2$$

No DPE	77%
With DPE	78%

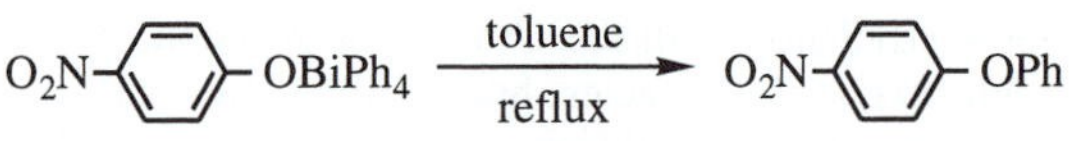

$$O_2N-\!\!\!\bigcirc\!\!\!-OBiPh_4 \xrightarrow[\text{reflux}]{\text{toluene}} O_2N-\!\!\!\bigcirc\!\!\!-OPh$$

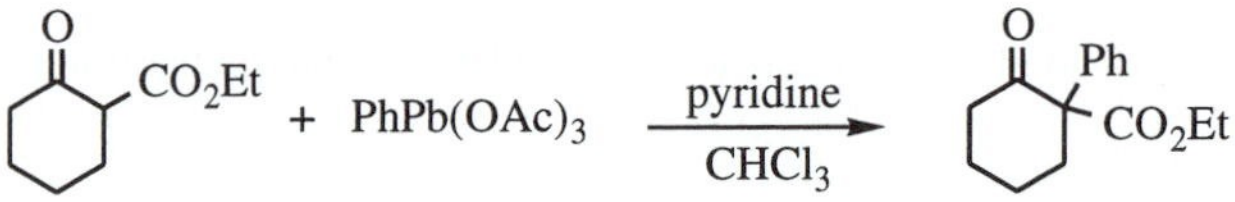

No DPE	95%
With DPE	90%

Cyclohexanone with CO$_2$Et + PhPb(OAc)$_3$ $\xrightarrow[\text{CHCl}_3]{\text{pyridine}}$ product

No DPE	78%
With DPE	76%

Coumarin (MeO, OH) + MeO-substituted aryl Pb(OAc)$_3$ (MeO, OMe) $\xrightarrow[\text{CHCl}_3]{\text{pyridine}}$ arylated coumarin product

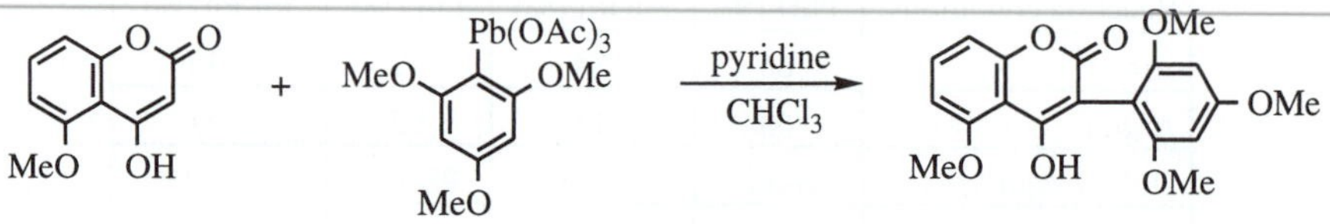

No DPE	94%
With DPE (2 equiv.)	90%
With DPE (10 equiv.)	87%

Nitrosobenzene was also used as a quantitative spin trap for phenyl free radicals, to study their involvement in arylation reactions with organobismuth reagents. Again, ESR observations indicated the formation of phenyl free radicals. However, quantitative trapping measurements showed that only a small amount of free radical derived adducts (2-3%) was formed. The yield of *C*-arylated products was not affected by the presence of this free radical trap.[18]

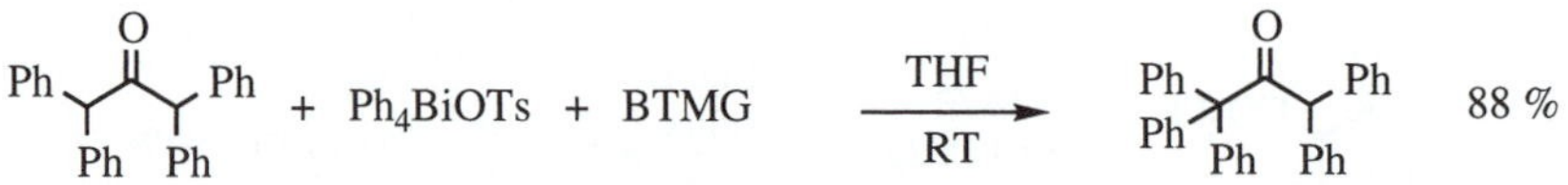

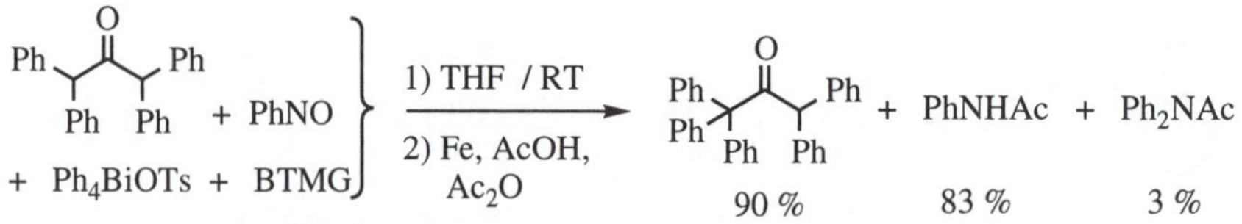

Ph–CO–Ph (Ph, Ph) + Ph$_4$BiOTs + BTMG $\xrightarrow[\text{RT}]{\text{THF}}$ Ph–CO–Ph (Ph, Ph, Ph, Ph) 88 %

Ph–CO–Ph (Ph, Ph) + PhNO + Ph$_4$BiOTs + BTMG $\xrightarrow[\text{2) Fe, AcOH, Ac}_2\text{O}]{\text{1) THF / RT}}$ Ph–CO–Ph (Ph, Ph, Ph, Ph) + PhNHAc + Ph$_2$NAc

90 %	83 %	3 %

2.2.3.2 Internal free radical traps

A different approach was employed by Pinhey *et al.* to test the presence of aryl radical species in aromatic substitution by aryllead triacetates.[142] An excellent system bearing an internal trap for free radical was developed by Beckwith *et al.* who observed that the thermal decomposition of *ortho*-allyloxyphenyldiazonium salts in the presence of sodium iodide afforded the corresponding benzofuran derivatives functionalised on the *C*-3 atom in high yields.[143] The use of an aryllead triacetate bearing this type of reactive group acting as an internal trap in the reaction of various nucleophiles failed to show the presence of products derived from aryl radicals.[142]

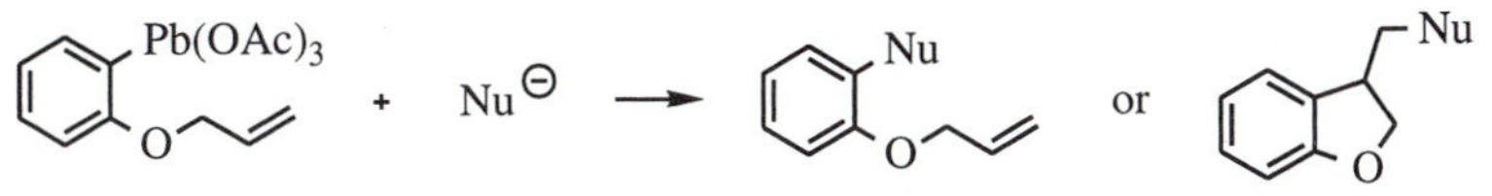

Indeed, reaction of *ortho*-allyloxyphenyllead triacetate with inorganic nucleophiles such as iodide and azide ions or with organic nucleophiles such as ethyl cyclopentanonecarboxylate, mesitol or the sodium salt of nitropropane always resulted in the exclusive formation of the *C*-aryl products. The radical derived products, 3-substituted dihydrobenzofurans, were not detected even in trace amounts.[142]

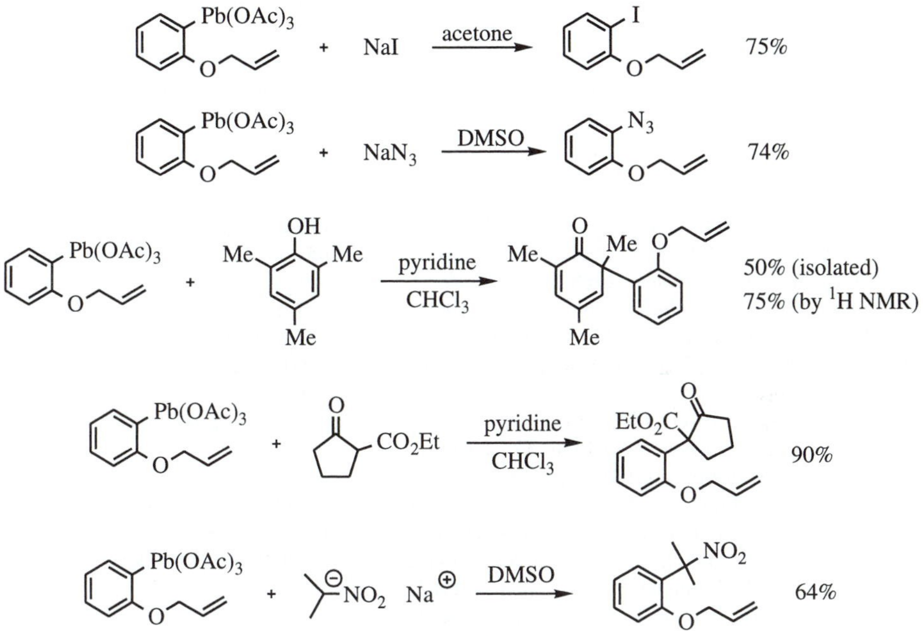

75%

74%

50% (isolated)
75% (by ^{1}H NMR)

90%

64%

The thermal decomposition of an (*ortho*-allyloxyphenyl)ketomethyllead diacetate did not afford the cyclized products resulting from aryl radical intermediates.[144]

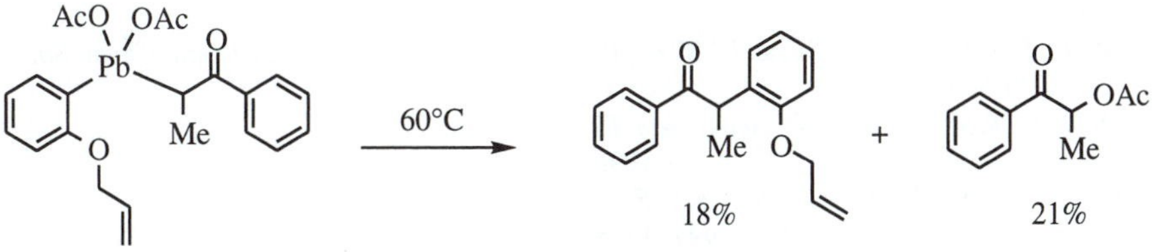

18%   21%

2.2.4 Conclusion

The studies of the mechanism of ligand coupling have led to the unequivocal conclusion of the existence of this mechanism. They have also shed some light on the stereochemical pathways involved in this mechanism. But a number of questions still remain unsolved or poorly understood. Due to the variety of systems which can undergo the ligand coupling reaction, the main difference remains between the two major modes of coupling: the symmetry allowed subtypes *apical-apical* or *equatorial-equatorial* or the symmetry forbidden *apical-equatorial*. However, this latter mode is the preferred one, when the transition state is highly polar involving zwitterionic transition states. A second difficulty is the frequent involvement of topological transformations and/or concurrent pathways, such as ligand exchange, which can bring some further complications to the complete mechanistic picture. Moreover, the pseudorotation phenomenon is very important in the ligand coupling reactions with sulfoxides, but the mode of stereomutation (Berry pseudorotation or turnstile rotation) is poorly understood. These uncertainties forbid generalization of the conclusions which can be drawn from the experimental results obtained with one heteroelement.

Nevertheless, the general characteristics of the ligand coupling mechanism are reproduced in all the systems: the coupling step is concerted with more or less polar transition states. It does not involve free radicals or ionic species and leads to retention of configuration. The involvement of this mechanism has been extended to explain some aspects of the chemistry of other elements, such as organolead or organophosphorus, for which very few or no stereochemical studies have been reported at this stage.

2.3 REFERENCES

1. Franzen, V.; Joschek, H.I.; Mertz, C. *Liebigs Ann. Chem.* **1962**, *654*, 82-91.
2. Wiegand, G.H.; Mc Ewen, W.E. *Tetrahedron Lett.* **1965**, 2639-2642. Wiegand, G.H.; Mc Ewen, W.E. *J. Org. Chem.* **1968**, *33*, 2671-2675.
3. Khim, Y.H.; Oae, S. *Bull. Chem. Soc. Jpn.* **1969**, *42*, 1968-1971.
4. Trost, B.M.; LaRochelle, R.W.; Atkins, R.C. *J. Am. Chem. Soc.* **1969**, *91*, 2175-2177.
5. LaRochelle, R.W.; Trost, B.M. *J. Am. Chem. Soc.* **1971**, *93*, 6077-6086.
6. Trost, B.M.; Arndt, H.C. *J. Am. Chem. Soc.* **1973**, *95*, 5288-5298.
7. Hoffmann, R.; Howell, J.M.; Muetterties, E.L. *J. Am. Chem. Soc.* **1972**, *94*, 3047-3058.
8. Oae, S. *Croatica Chem. Acta* **1986**, *59*, 129-151.
9. Oae, S. *Phosphorus Sulfur* **1986** *27*, 13-29.
10. Sandin, R.B.; Kulka, M.; McCready, R. *J. Am. Chem. Soc.* **1937**, *59*, 2014-2015.
11. Beringer, F.M.; Forgione, P.S.; Yudis, M.D. *Tetrahedron* **1960**, *8*, 49-63.
12. Beringer, F.M.; Galton, S.A.; Huang, S.J. *J. Am. Chem. Soc.* **1962**, *84*, 2819-2823.
13. Beringer, F.M.; Forgione, P.S. *Tetrahedron* **1963**, *19*, 739-748.
14. Beringer, F.M.; Falk, R.A. *J. Chem. Soc.* **1964**, 4442-4451.
15. Varvoglis, A. *The Organic Chemistry of Polycoordinated Iodine*; VCH: New York, **1992**.
16. Tanner, D.D.; Reed, D.W.; Setiloane, B.P. *J. Am. Chem. Soc.* **1982**, *104*, 3917-3923.
17. Singh, P.R.; Khanna, R.K. *Tetrahedron Lett.* **1982**, *23*, 5355-5358.
18. Barton, D.H.R.; Finet, J.-P.; Giannotti, C.; Halley, F. *J. Chem. Soc., Perkin Trans. 1* **1987**, 241-249.
19. Banks, J.T.; Garcia, H.; Miranda, M.A.; Pérez-Prieto, J.; Scaiano, J.C. *J. Am. Chem. Soc.* **1995**, *117*, 5049-5054.
20. Moriarty, R.M.; Prakash, O. *Acc. Chem. Res.* **1986**, *19*, 244-250.
21. Moriarty, R.M.; Vaid, R.K. *Synthesis* **1990**, 431-447.
22. Moriarty, R.M.; Vaid, R.K.; Koser, G.F. *Synlett* **1990**, 365-383.
23. Grushin, V.V. *Acc. Chem. Res.* **1992**, *25*, 529-536.

24. Kurosawa, H.; Okada, H.; Sato, M.; Hattori, T. *J. Organomet. Chem.* **1983**, *250*, 83-97.
25. Oae, S. *Reviews on Heteroatom Chemistry* **1988**, *1*, 304-335.
26. Oae, S.; Uchida, Y. *Reviews on Heteroatom Chemistry* **1989**, *2*, 76-91.
27. Oae, S.; Furukawa, N. in *Advances in Heterocyclic Chemistry*; Katritzky, A.R., Ed.; Academic Press: San Diego, **1990**; Vol. 48, pp. 1-63.
28. Oae, S. *Reviews on Heteroatom Chemistry* **1991**, *4*, 195-225.
29. Oae, S.; Uchida, Y. *Acc. Chem. Res.* **1991**, *24*, 202-208.
30. Oae, S. *Organic Sulfur Chemistry - Structure and Mechanism*; CRC Press: Boca Raton, **1991**; Ch. 5, pp. 183-201.
31. Negoro, T.; Oae, S. *Reviews on Heteroatom Chemistry* **1993**, *9*, 123-153.
32. Negoro, T.; Oae, S. *Reviews on Heteroatom Chemistry* **1995**, *13*, 235-272.
33. Oae, S. *Main Group Chemistry News* **1996**, *4*, 10-17.
34. Oae, S. *Pure Appl. Chem.* **1996**, *68*, 805-812.
35. Barton, D.H.R.; Finet, J.-P. *Pure Appl. Chem.* **1987**, *59*, 937.
36. Finet, J.-P. *Chem. Rev.* **1989**, *89*, 1487.
37. Barton, D.H.R. in *Heteroatom Chemistry*; Block, E., Ed.; VCH: New York, **1990**; pp. 95-104.
38. Barton, D.H.R.; Parekh, S.I. *Pure Appl. Chem.* **1993**, *65*, 603-610.
39. Pinhey, J.T. *Pure Appl. Chem.* **1996**, *68*, 819-824.
40. Akiba, K.-y. *Pure Appl. Chem.* **1996**, *68*, 837-842.
41. Bickelhaupt, F.M.; Ziegler, T.; Schleyer, P.v.R. *Organometallics* **1995**, *14*, 2288-2296.
42. Anh, N.T.; Minot, C. *J. Am. Chem. Soc.* **1980**, *102*, 103-107. Minot, C. *Nouv. J. Chim.* **1980**, *5*, 319-325.
43. Glukhovtsev, M.N.; Pross, A.; Schlegel, H.B.; Bach, R.D.; Radom, L. *J. Am. Chem. Soc.* **1996**, *118*, 11258-11264.
44. Umemoto, T. *Chem. Rev.* **1996**, *96*, 1757-1777.
45. Berry, R.S. *J. Chem. Phys.* **1960**, *32*, 933-938.
46. Ugi, I.; Marquarding, D.; Klusacek, H.; Gokel, G.; Gillespie, P. *Angew. Chem., Int. Ed. Engl.* **1970**, *9*, 703-730.
47. Ugi, I.; Marquarding, D.; Klusacek, H.; Gillespie, P.; Ramirez, F. *Acc. Chem. Res.* **1971**, *4*, 288-296.
48. Kutzelnigg, W.; Wasilewski, J. *J. Am. Chem. Soc.* **1982**, *104*, 953-960.
49. Moc, J.; Morokuma, K. *J. Am. Chem. Soc.* **1995**, *117*, 11790-11797.
50. Musher, J.I. *J. Am. Chem. Soc.* **1972**, *94*, 5662-5665.
51. Musher, J.I. *J. Chem. Ed.* **1974**, *51*, 94-97.
52. Woodward, R.B.; Hoffmann, R. *Angew. Chem., Int. Ed. Engl.* **1969**, *8*, 781-853.
53. Rauk, A.; Allen, L.C.; Mislow, K. *J. Am. Chem. Soc.* **1972**, *94*, 3035-3040.
54. Keil, F.; Kutzelnigg, W. *J. Am. Chem. Soc.* **1975**, *97*, 3623-3632.
55. Shih, S.-K.; Peyerimhoff, S.D.; Buenker, R.J. *J. Chem. Soc., Faraday Trans. 2* **1979**, *75*, 379-389.
56. Kutzelnigg, W.; Wallmeier, H.; Wasilewski, J. *Theor. Chim. Acta* **1979**, *51*, 261-273.
57. Trinquier, G.; Daudey, J.-P.; Caruana, G.; Madaule, Y. *J. Am. Chem. Soc.* **1984**, *106*, 4794-4799.
58. Ewig, C.S.; Van Wazer, J.R. *J. Am. Chem. Soc.* **1989**, *111*, 1552-1558.
59. Wang, P.; Agrafiotis, D.K.; Streitwieser, A.; Schleyer, P.v.R. *J. Chem. Soc., Chem. Commun.* **1990**, 201-203.
60. McDowell, R.S.; Streitwieser, A. Jr *J. Am. Chem. Soc.* **1985**, *107*, 5849-5855.
61. Michels, H.H.; Montgomery, J.A. Jr *J. Chem. Phys.* **1990**, *93*, 1805-1813.
62. Reed, A.E.; Schleyer, P.v.R. *J. Am. Chem. Soc.* **1990**, *112*, 1434-1445.
63. Wang, P.; Zhang, Y.; Glaser, R.; Reed, A.E.; Schleyer, P.v.R.; Streitwieser, A. *J. Am. Chem. Soc.* **1991**, *113*, 55-64.

64. Wasada, H.; Hirao, K. *J. Am. Chem. Soc.* **1992**, *114*, 16-27.
65. Wasada, H.; Hirao, K. *J. Am. Chem. Soc.* **1992**, *114*, 4444.
66. **a**) Morokuma, K.; Mathieu, S.; Dorigo, A.E. *The 1989 International Chemical Congress of Pacific Basin Societies* **1989**, phys590. **b**) Mathieu, S.; Morokuma, K. *Annu. Rev. Instit. Mol. Sci.* **1990**, 18.
67. Reed, A.E.; Schleyer, P.v.R. *Chem. Phys. Lett.* **1987**, *133*, 553-561.
68. Akiba, K.-y. *Pure Appl. Chem.* **1996**, *68*, 837-842.
69. Schwerdtfeger, P.; Heath, G.A.; Dolg, M.; Bennett, M.A. *J. Am. Chem. Soc.* **1992**, *114*, 7518-7527.
70. Koutecky, V.B.; Musher, J.I. *Theoret. Chim. Acta (Berlin)* **1974**, *33*, 227-238.
71. Schwenzer, G.M.; Schaefer, H.F.III *J. Am. Chem. Soc.* **1975**, *97*, 1393-1397.
72. Gleiter, R.; Veillard, A. *Chem. Phys. Lett.* **1976**, *37*, 33-39.
73. Eggers, M.D.; Livant, P.D.; McKee, M.L. *Theochem* **1989**, *55*, 69-89.
74. Fowler, J.E.; Schaefer, H.F.III *J. Am. Chem. Soc.* **1994**, *116*, 9596-9601.
75. Marsden, C.J.; Smart, B.A. *Aust. J. Chem.* **1994**, *47*, 1431-1440.
76. Marsden, C.J.; Smart, B.A. *Organometallics* **1995**, *14*, 5399-5409.
77. Yoshioka, Y.; Goddard, J.D.; Schaefer, H.F.III *J. Chem. Phys.* **1981**, *74*, 1855-1863.
78. Moc, J.; Dorigo, A.E.; Morokuma, K. *Chem. Phys. Lett.* **1993**, *204*, 65-72.
79. Chen, M.M.L.; Hoffmann, R. *J. Am. Chem. Soc.* **1976**, *98*, 1647-1653.
80. Rappoport, Z. *Acc. Chem. Res.* **1992**, *25*, 474-479.
81. Lucchini, V.; Modena, G.; Pasquato, L. *J. Am. Chem. Soc.* **1995**, *117*, 2297-2300.
82. Ochiai, M.; Oshima, K.; Masaki, Y. *J. Am. Chem. Soc.* **1991**, *113*, 7059-7061.
83. Ochiai, M.; Oshima, K.; Masaki, Y. *Tetrahedron Lett.* **1991**, *32*, 7711-7714.
84. Barton, D.H.R.; Glover, S.A. ; Ley, S.V. *J. Chem. Soc., Chem. Commun.* **1977**, 266-267.
85. Glover, S.A. *J. Chem. Soc., Perkin Trans. 1* **1980**, 1338-1344.
86. Barton, D.H.R.; Blazejewski, J.-C.; Charpiot, B.; Finet, J.-P.; Motherwell, W.B.; Papoula, M.T.B.; Stanforth, S.P. *J. Chem. Soc., Perkin Trans. 1* **1985**, 2667-2675.
87. Wessely, F.; Zbiral, E.; Sturm, H. *Chem. Ber.* **1960**, *93*, 2840-2851.
88. Schürmann, M.; Huber, F. *Acta Crystallogr., Sect. C* **1994**, *50*, 1710-1713.
89. Huber, F.; Preut, H.; Scholz, D.; Schürmann, M. *J. Organomet. Chem.* **1992**, *441*, 227-239.
90. Dewar, M.J.S. *J. Am. Chem. Soc.* **1984**, *106*, 209-219.
91. Wittig, G.; Fritz, H. *Liebigs Ann. Chem.* **1952**, *577*, 39-46.
92. Cohen, S.C.; Reddy, M.L.N.; Massey, A.G. *J. Organomet. Chem.* **1968**, *11*, 563-566.
93. Hellwinkel, D.; Fahrbach, S.I. *Liebigs Ann. Chem.* **1968**, *712*, 1-20.
94. Shen, K.-w.; McEwen, W.E.; Wolf, A.P. *J. Am. Chem. Soc.* **1969**, *91*, 1283-1288.
95. Akiba, K.-y.; Okinaka, T.; Nakatani, M.; Yamamoto, Y. *Tetrahedron Lett.* **1987**, *28*, 3367-3368.
96. Koser, G.F. *Halonium Ions* in *The Chemistry of Functional Groups, Supplement D*; Patai, S.; Rappoport, Z., Eds.; J. Wiley & Sons: New York, **1983**; Ch. 25, pp. 1265-1351.
97. Budylin, V.A.; Ermolenko, M.S.; Chugtai, F.A.; Kost, A.N. *Khim. Geterotsikl. Soedin.* **1981**, 1494-1496; *Chem. Abstr.* **1982**, *96*, 142617n.
98. Grushin, V.V.; Demkina, I.I.; Tolstaya, T.P. *J. Chem. Soc., Perkin Trans. 2* **1992**, 505-511.
99. Koser, G.F. *Hypervalent Halogen Compounds* in *The Chemistry of Functional Groups, Supplement D*; Patai, S.; Rappoport, Z., Eds.; J. Wiley & Sons: New York, **1983**; Ch. 18, pp. 721-811.
100. Reich, H.J.; Cooperman, C.S. *J. Am. Chem. Soc.* **1973**, *95*, 5077-5078.
101. Yamada, Y.; Okawara, M. *Bull. Chem. Soc. Jpn.* **1972**, *45*, 1860-1863.
102. Lancer, K.M.; Wiegand, G.H. *J. Org. Chem.* **1976**, *41*, 3360-3364.
103. Olah, G.A.; Sakakibara, T.; Asensio, G. *J. Org. Chem.* **1978**, *43*, 463-468.

104. Lubinkowski, J.J.; Gomez, M.; Calderon, J.L.; McEwen, W.E. *J. Org. Chem.* **1978**, *43*, 2432-2435.
105. Sato, T.; Shimada, S.; Shimizu, K.; Hata, K. *Bull. Chem. Soc. Jpn.* **1970**, *43*, 1918.
106. Sato, T.; Shimizu, K.; Moriya, H. *J. Chem. Soc., Perkin Trans. 1* **1974**, 1537-1539.
107. Kotali, E.; Varvoglis, A. *J. Chem. Soc., Perkin Trans. 1* **1987**, 2759-2763.
108. Barton, D.H.R.; Yadav-Bhatnagar, N.; Finet, J.-P.; Khamsi, J.; Motherwell, W.B.; Stanforth, S.P. *Tetrahedron* **1987**, *43*, 323-332.
109. Barton, D.H.R.; Bhatnagar, N.Y.; Finet, J.-P.; Motherwell, W.B. *Tetrahedron* **1986**, *42*, 3111-3122.
110. Barton, D.H.R.; Lester, D.J.; Motherwell, W.B.; Barros Papoula, M.T. *J. Chem. Soc., Chem. Commun.* **1980**, 246-247.
111. Barton, D.H.R.; Blazejewski, J.-C.; Charpiot, B.; Lester, D.J.; Motherwell, W.B.; Papoula, M.T.B. *J. Chem. Soc., Chem. Commun.* **1980**, 827-829.
112. Barton, D.H.R.; Bhatnagar, N.Y.; Blazejewski, J.-C.; Charpiot, B.; Finet, J.-P.; Lester, D.J.; Motherwell, W.B.; Papoula, M.T.B.; Stanforth, S.P. *J. Chem. Soc., Perkin Trans. 1* **1985**, 2657-2665.
113. Hori, M.; Kataoka, T.; Shimizu, H.; Miyagaki, M. *Chem. Pharm. Bull.* **1974**, *22*, 1711-1720.
114. Sheppard, W.A. *J. Am. Chem. Soc.* **1971**, *93*, 5597-5598.
115. Ogawa, S.; Matsunaga, Y.; Sato, S.; Erata, T.; Furukawa, N. *Tetrahedron Lett.* **1992**, *33*, 93-96.
116. Ogawa, S.; Matsunaga, Y.; Sato, S.; Iida, I.; Furukawa, N. *J. Chem. Soc., Chem. Commun.* **1992**, 1141-1142.
117. Sato, S.; Furukawa, N. *Tetrahedron Lett.* **1995**, *36*, 2803-2806.
118. Hori, M.; Kataoka, T.; Shimizu, H.; Miyagaki, M. *Chem. Lett.* **1972**, 515-520.
119. Hori, M.; Kataoka, T.; Shimizu, H.; Miyagaki, M. *Chem. Pharm. Bull.* **1974**, *22*, 2004-2013.
120. Hori, M.; Kataoka, T.; Shimizu, H.; Miyagaki, M.; Murase, M. *Chem. Pharm. Bull.* **1974**, *22*, 2014-2019.
121. Hori, M.; Kataoka, T.; Shimizu, H.; Miyagaki, M. *Chem. Pharm. Bull.* **1974**, *22*, 2020-2029.
122. Hori, M.; Kataoka, T.; Shimizu, H.; Miyagaki, M.; Takagi, T. *Chem. Pharm. Bull.* **1974**, *22*, 2030-2041.
123. Andersen, K.K.; Yeager, S.A.; Peynircioglu, N.B. *Tetrahedron Lett.* **1970**, 2485-2488. Ackerman, B.K.; Andersen, K.K.; Karup-Nielsen, I.; Peynircioglu, N.B.; Yeager, S.A. *J. Org. Chem.* **1974**, *39*, 964-968.
124. Oda, R.; Yamamoto, K. *J. Org. Chem.* **1961**, *26*, 4679-4681.
125. Nesmeyanov, A.N.; Kalyavin, V.A.; Reutov, O.A. *Zh. Org. Khim.* **1978**, *14*, 2465-2470, *Chem. Abstr.* **1979**, *90*, 120944r.
126. Oae, S.; Kise, M. *Tetrahedron Lett.* **1967**, 1409-1413.
127. Lockard, J.P.; Schroeck, C.W.; Johnson, C.R. *Synthesis* **1973**, 485-486.
128. Oae, S.; Yokoyama, M.; Kise, M.; Furukawa, N. *Tetrahedron Lett.* **1968**, 4131-4134.
129. Oae, S.; Kawai, T.; Furukawa, N. *Phosphorus Sulfur* **1987**, *34*, 123-132.
130. Furukawa, N.; Ogawa, S.; Matsumura, K.; Fujihara, H. *J. Org. Chem.* **1991**, *56*, 6341-6348.
131. Oae, S.; Takeda, T.; Kawai, T.; Furukawa, N. *Phosphorus Sulfur* **1987**, *34*, 133-137.
132. Wakabayashi, S.; Ishida, M.; Takeda, T.; Oae, S. *Tetrahedron Lett.* **1988**, *29*, 4441-4444.
133. Oae, S.; Kawai, T.; Furukawa, N. *Tetrahedron Lett.* **1984**, *25*, 69-72.
134. Oae, S.; Kawai, T.; Furukawa, N.; Iwasaki, F. *J. Chem. Soc., Perkin Trans. 2* **1987**, 405-411.
135. Uenishi, J.; Yamamoto, A.; Takeda, T.; Wakabayashi, S.; Oae, S. *Heteroatom Chem.* **1992**, *3*, 73-79.
136. Oae, S.; Takeda, T.; Wakabayashi, S.; Iwasaki, F.; Yamazaki, N.; Katsube, Y. *J. Chem. Soc., Perkin Trans. 2* **1990**, 273-276.
137. Oae, S.; Takeda, T.; Wakabayashi, S. *Tetrahedron Lett.* **1988**, *29*, 4445-4448.

138. Wakabayashi, S.; Kiyohara, Y.; Kameda, S.; Uenishi, J.; Oae, S. *Heteroatom Chem.* **1990**, *1,* 225-232.
139. Lubinkowski, J.J.; Knapczyk, J.W.; Calderon, J.L.; Petit, L.R.; McEwen, W.E. *J. Org. Chem.* **1975**, *40*, 3010-3014.
140. Barton, D.H.R.; Donnelly, D.M.X.; Finet, J.-P.; Guiry, P.J. *J. Chem. Soc., Perkin Trans. 1* **1992**, 1365-1375.
141. Barton, D.H.R.; Donnelly, D.M.X.; Guiry, P.J.; Finet, J.-P. *J. Chem. Soc., Perkin Trans. 1* **1994**, 2921-2926.
142. Morgan, J.; Pinhey, J.T. *J. Chem. Soc., Perkin Trans. 1* **1993**, 1673-1676.
143. Beckwith, A.L.J.; Meijs, G.F. *J. Org. Chem.* **1987**, *52*, 1922-1930. Beckwith, A.L.J.; Palacios, S.M. *J. Phys. Org. Chem.* **1991**, *4*, 404-412.
144. Morgan, J.; Buys, I.; Hambley, T.; Pinhey, J.T. *J. Chem. Soc., Perkin Trans. 1* **1993**, 1677-1681.

Chapter 3

Ligand Coupling Involving Organosulfur Compounds

The rich chemistry of organosulfur compounds involves a number of mechanisms allowing a wide range of synthetic transformations performed by a variety of pathways going through anionic, cationic, radical intermediates or concerted mechanisms. Although not so common, the ligand coupling mechanism has found a number of implications in the chemistry of organosulfur compounds, from a preparative point of view as well as from the mechanistic point of view.[1,2] Six types of organosulfur derivatives have been found to undergo ligand coupling more or less easily: the sulfonium salts, the sulfoxides, the sulfilimines, the sulfones, the polysulfanes and the oxathietanes.

3.1 SULFONIUM SALTS

The behaviour of sulfonium salts towards nucleophiles is strongly dependent upon the nature of the ligands present in the sulfonium salt. A large number of reactions involve either classical S_N2 substitutions or ylide formation. However, the ligand coupling mechanism is frequently observed in the reactions of triarylsulfonium salts and their analogues. Depending on the reaction conditions, competitive pathways such as free-radicals or occurrence of benzyne intermediates can be observed. The interaction of a nucleophile with a tricoordinate sulfonium salt affords a tetracoordinate σ-sulfurane. Its decomposition by the ligand coupling mechanism leads to products of a formal S_N2 substitution. (Scheme 3.1) Although the synthetic interest of these reactions is relatively limited, they have played an historically determinant role in the birth of the ligand coupling concept. Indeed, it is in the course of their studies on the reactivity of triarylsulfonium salts with organometallic reagents that Trost et al. described the general characteristics of this mechanism.[3]

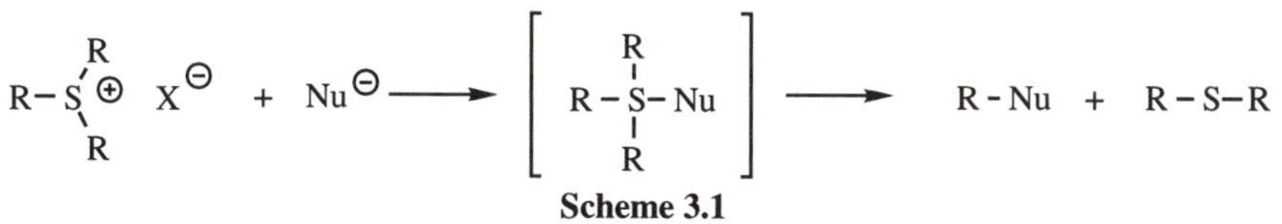

$$R-\overset{\overset{R}{|}}{\underset{\underset{R}{|}}{S}}{}^{\oplus}\ X^{\ominus}\ +\ Nu^{\ominus} \longrightarrow \left[R-\overset{\overset{R}{|}}{\underset{\underset{R}{|}}{S}}-Nu \right] \longrightarrow R-Nu\ +\ R-S-R$$

Scheme 3.1

3.1.1 Reactions of alkyldiarylsulfonium salts with organolithium reagents

Treatment of an alkyldiphenylsulfonium salt with aryl- or alkyl-lithium usually results in the generation of the highly reactive diphenylsulfoniumalkylide, which reacts with aldehydes and ketones to form oxiranes.[4,5] However, the generation of the sulfonium ylide can lead to side products resulting from ligand coupling reaction. For example, reaction of ethyldiphenylsulfonium tetrafluoroborate (**1**) with phenyllithium, followed by addition of cyclohexanone, led to the expected oxirane (**2**) among a mixture of products containing biphenyl (**4**) and ethylphenylsulfide (**5**). The formation of these products (4) and (5) can be explained as resulting from a ligand coupling taking place on an intermediate sulfurane, ethyltriphenylsulfurane, which is formed competitively during the generation of the ylid of (1).[6]

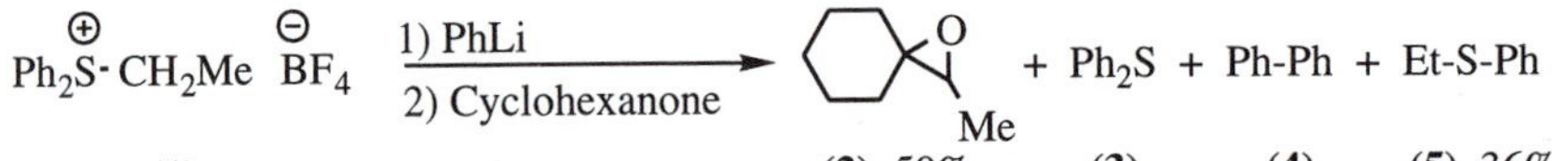

$$Ph_2\overset{\oplus}{S}\cdot CH_2Me\ \overset{\ominus}{BF_4} \quad \xrightarrow[\text{2) Cyclohexanone}]{\text{1) PhLi}} \quad \text{(2)} + Ph_2S + Ph\text{-}Ph + Et\text{-}S\text{-}Ph$$

(1)   (2), 50%   (3)   (4)   (5), 36%

Ligand coupling takes place in the course of the evolution of the products of the reaction of triphenylcyclopropenium bromide (6) with trimethylsulfonium methylide (7).[7,8] (Scheme 3.2)

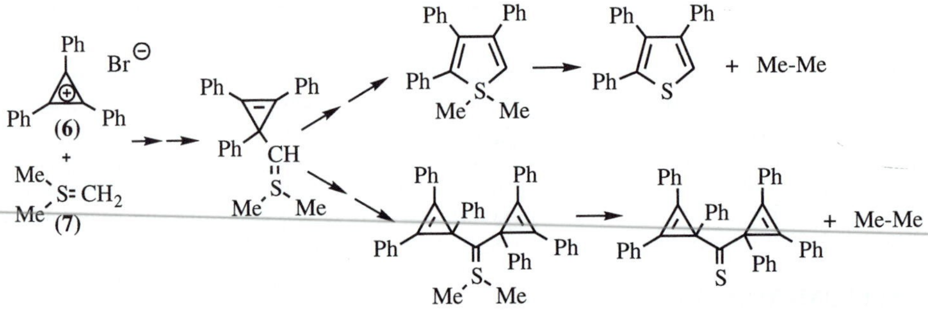

Scheme 3.2

3.1.2 Reactions of triarylsulfonium salts with organolithium reagents

In 1952, Wittig and Fritz reported that treatment of a triphenylsulfonium salt with phenyllithium afforded directly diphenylsulfide (3) and biphenyl (4).[9] They also showed that treatment of the triphenylsulfonium salt with trityl sodium led to tetraphenylmethane, although in poor yield. Later, in 1962, Sheppard described the reaction of phenyllithium with phenylsulfur trifluoride or with sulfur tetrafluoride at - 80°C, which afforded biphenyl and diphenylsulfide.[10] At the same time, Franzen *et al.* suggested that a tetracoordinate sulfurane was intermediately formed and collapsed to lead to the ligand coupling products.[11]

$$Ph_3\overset{\oplus}{S}\ \overset{\ominus}{Br} + PhLi \longrightarrow Ph\text{-}Ph + Ph_2S \qquad \text{ref. 9}$$

$$PhSF_3 + PhLi \longrightarrow Ph\text{-}Ph + Ph_2S$$
$$SF_4 + PhLi \longrightarrow Ph\text{-}Ph + Ph_2S \qquad \text{ref. 10}$$

A number of mechanisms, such as the occurrence of benzyne intermediates, single electron transfer leading to free radicals, nucleophilic aromatic substitution and ligand coupling could explain these reactions. The benzyne mechanism was easily excluded by consideration of the reaction of tri-*p*-tolylsulfonium tetrafluoroborate with *p*-tolyllithium.[12-14] The exclusive formation of the *p,p'*-disubstituted coupling product was totally incompatible with a benzyne mechanism, as 3,4-dehydrotoluene undergoes nucleophilic addition predominantly in the *meta* position.[15-17] Although the nucleophilic aromatic substitution was frequently claimed to explain the related reactions involving hypervalent diaryliodonium salts (Chapters 2 and 5), the ligand coupling mechanism was already suggested by Sheppard in 1962, who claimed that "*the biphenyl and diphenylsulfide no doubt arise from the decomposition of tetraphenylsulfur*".[10] Indirect evidence of the formation of the tetraaryl σ-sulfuranes as intermediates were reported by Trost *et al.*, who extensively studied the mechanism in a series of papers,[13,14,18] and also by Mislow[19] and Hori.[20-25] Strong support for this mechanism came in 1971, when Sheppard reported the detection of tetrakis(pentafluorophenyl)sulfurane (**8**) by ^{19}F NMR at low temperature, as an unstable intermediate in the reactions of pentafluorophenyllithium with perfluorosulfurane derivatives which afford decafluorobiphenyl (**9**) and bis(pentafluorophenyl) sulfide (**10**).[26,27]

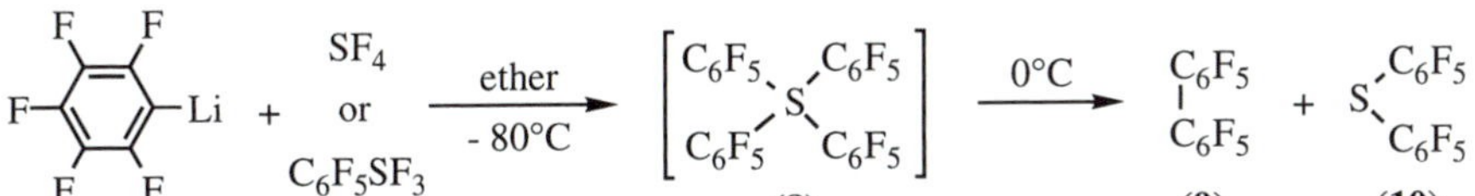

(8) **(9)** **(10)**

Scheme 3.3

It is only recently that the tetraphenylsulfurane (**11**) was detected by low temperature NMR experiments in the reaction of triphenylsulfonium salt with phenyllithium.[28,29] 2,2'-Biphenylylenediphenylsulfurane (**12**), prepared by a similar method, was also recently detected by low temperature NMR experiments.[30] As predicted by Trost,[14] the bis(2,2'-biphenylylene)sulfurane (**13**) was eventually isolated and its X-ray structure determined.[31]

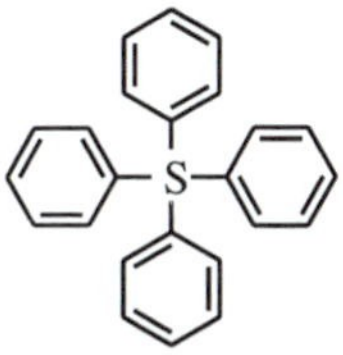

(11)

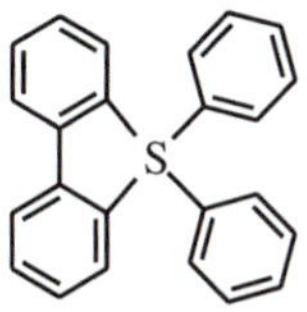

(12)

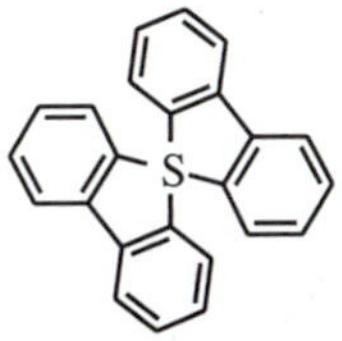

(13)

The synthetic utility of the reaction of triarylsulfonium salts with nucleophilic agents to form carbon-carbon bonds has not been much developed.[13,14,32,33] Formation of biaryls upon treatment of triarylsulfonium salts with aryllithium occurs in high yields. Although the reaction is more interesting for the preparation of symmetrical biaryls, it can also be applied to the case of unsymmetrical biaryls, preferably if one of the aryl groups bears an electron-withdrawing substituent. In case of a competition between ligands of different electronic activity, the group bearing the least electron-donating substituents is preferentially involved in the ligand coupling. The more electron-rich ligands usually remain bound to the sulfur atom.

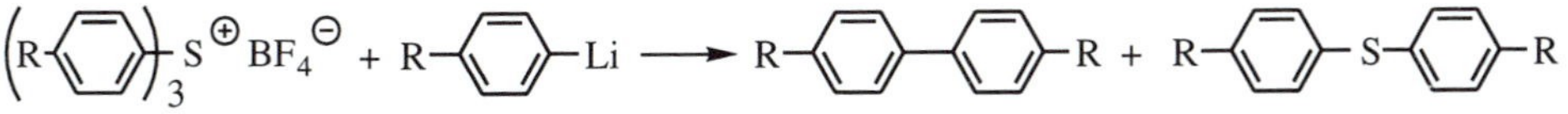

$$Tol_3S^{\oplus} BF_4^{\ominus} + Tol\text{-}Li \longrightarrow \underset{100\%}{Tol\text{-}Tol} + \underset{100\%}{Tol\text{-}S\text{-}Tol}$$

$$Tol_3S^{\oplus} BF_4^{\ominus} + Ph\text{-}Li \longrightarrow \underset{24\%}{Ph\text{-}Ph} + \underset{40\%}{Tol\text{-}Ph} + \underset{6\%}{Tol\text{-}Tol} + \underset{40\%}{Tol_2S} + \underset{3\%}{Tol\text{-}S\text{-}Ph}$$

$$Ph_3S^{\oplus} \; OSO_2CF_3^{\ominus} \; + \; Ph\text{-}Li \; \longrightarrow \; Ph\text{-}Ph \; + \; Ph\text{-}S\text{-}Ph \; + \; others$$
$$\qquad\qquad\qquad\qquad\qquad\qquad\quad 33\% \qquad\quad 33\%$$

A number of ligand coupling reactions involving the treatment of an heteroaryllithium reagent with triphenylsulfonium salts leading to phenyl-heteroaryl compounds [(14) - (17) for example] have also been reported.[32,33]

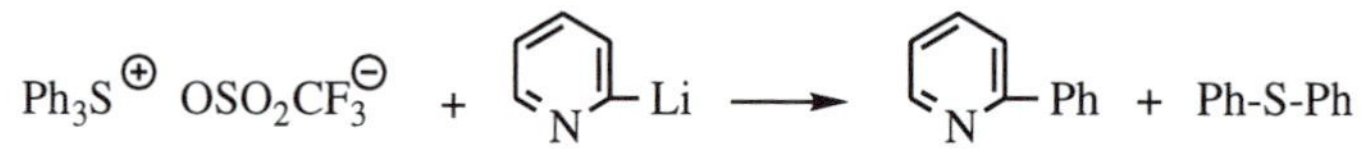

(14), 70%   73%

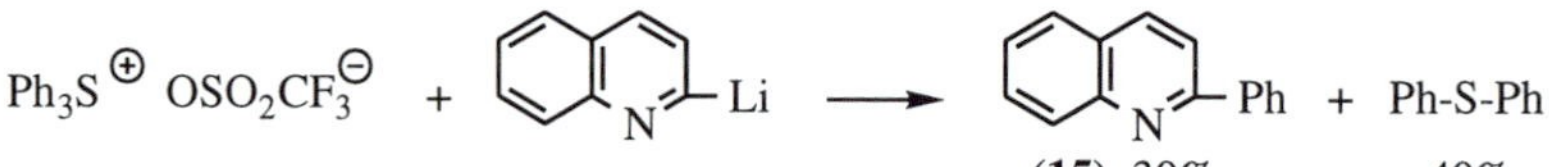

(15), 30%   40%

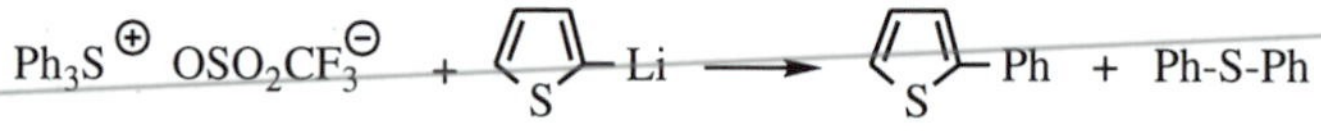

(16), 42%   40%

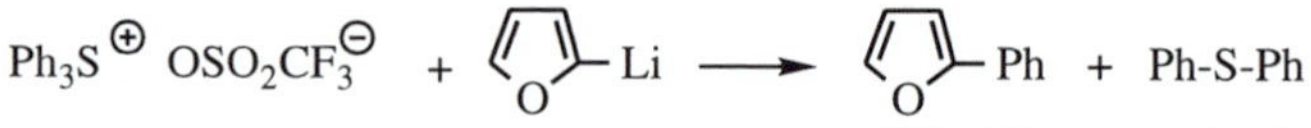

(17), 8%   38%

Reaction of the triarylsulfonium salts with vinyllithium or its derivatives led to good to high yields of the corresponding styrenes. With stereochemically pure substituted vinyllithium, the integrity of the stereochemistry was retained. *Cis-* and *trans-*propenyllithium (18) and (20) afforded the corresponding *cis-* and *trans-*propenylbenzene (19) and (21) respectively with no crossover.[13,14]

$$Ar_3S^{\oplus} \; BF_4^{\ominus} \; + \; \diagup\!\!\!\!\diagup Li \; \longrightarrow \; \diagup\!\!\!\!\diagup Ar \qquad 80\text{-}98\%$$

$$Ph_3S^{\oplus} \; BF_4^{\ominus} \; + \; Me\diagdown\!\!=\!\!\diagup Li \; \longrightarrow \; Me\diagdown\!\!=\!\!\diagup Ph \; + \; Ph\text{-}Ph$$
(18)   (19), 63%   (4), 38%

$$Ph_3S^{\oplus} \; BF_4^{\ominus} \; + \; Me\diagup\!\!\!\diagdown\!\!\diagup Li \; \longrightarrow \; Me\diagup\!\!\!\diagdown\!\!\diagup Ph \; + \; Ph\text{-}Ph$$
(20)   (21), 67%   (4), 6%

Allyllithium reacted with triphenylsulfonium salts to afford quantitatively allylbenzene (22). On the other hand, reaction with benzyllithium resulted in an intractable mixture. With saturated alkyllithium derivatives such as *n*-butyllithium or *tert*-butyllithium, only biphenyl (4) as ligand coupling product was isolated.[13,14] Moreover, the reaction with *n*-butyllithium gave a mixture of products, arising from ligand coupling taking place on the butyltriphenylsulfurane intermediate and from a competing ligand exchange reaction.[11,32-34]

$$Ph_3S^{\oplus} \; BF_4^{\ominus} \; + \; \diagup\!\!\!\diagdown\!\!\diagup Li \; \longrightarrow \; \diagup\!\!\!\diagdown\!\!\diagup Ph \qquad (22), 90\% \qquad\qquad \text{ref. 13,14}$$

$$Ph_3S^\oplus BF_4^\ominus + \textit{t}\text{-BuLi} \longrightarrow \text{Ph-Ph} \quad (\textbf{4}), 51\%$$

ref. 13,14

$$Ph_3S^\oplus OSO_2CF_3^\ominus + \textit{n}\text{-BuLi} \longrightarrow \text{Ph-Ph} + \text{Bu-S-Ph} + \text{Ph-S-Ph} + \text{others}$$

ref. 32,33

$$(\textbf{4}), 31\% \qquad 26\% \qquad (\textbf{3}), 14\%$$

A number of reports have dealt with the reaction of various nucleophilic agents with cyclic analogues of triarylsulfonium salts, such as *S*-aryldibenzothiophenium salts,[13,14,18,20-22,24] *S*-aryl-9,9-dimethylthioxanthenium salts,[22] *S*-aryl-10,11-dihydrodibenzo[*b,f*]thiepinium salts.[23] The reactivity patterns followed those observed with triarylsulfonium salts. When the two phenyl groups of the heterocyclic dibenzosulfonium system were unsymmetrically substituted, the entering nucleophile became mostly coupled to the more electron-depleted phenyl group, and the more electron-rich phenyl group remained bound to the sulfur atom.

3.1.3 Reactions of triarylsulfonium salts with heteroatomic nucleophiles

Depending on the nature of the counterion, triarylsulfonium salts are more or less stable. The most stable compounds have a non-nucleophilic counterion. On the other hand, the presence of a nucleophilic counterion such as a halide (Cl, Br or I) frequently results in the relatively facile decomposition by thermolysis to afford quantitatively the diarylsulfide and the corresponding aryl halide. The observation of the preferential relief of the steric strain in the case of phenyl-*p*-tolyl-(2,5-dimethylphenyl)sulfonium salts (**23**) was explained by the intramolecular decomposition of a first formed tetracoordinate intermediate.[35,36]

$$Ar_3S^\oplus X^\ominus \xrightarrow{\Delta} ArX + Ar\text{-}S\text{-}Ar$$

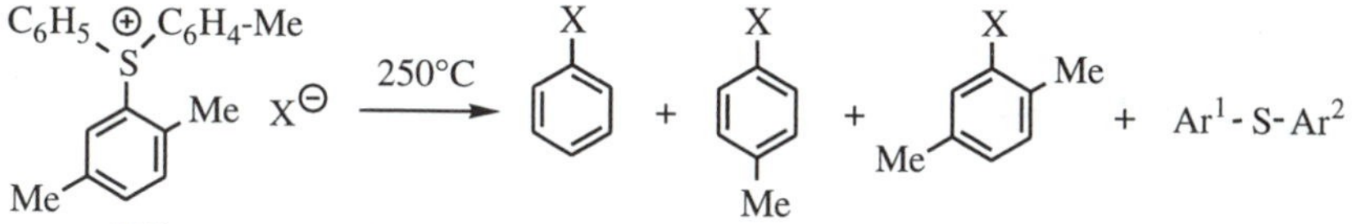

	X = Cl	1.5	1	5
	X = Br	1.9	1	11.7
	X = I	3.3	1	15.7

Such a ligand coupling mechanism may be involved in the site-specific one-step bromination of substituted benzenes by dimethylsulfoxide-hydrogen bromide.[37]

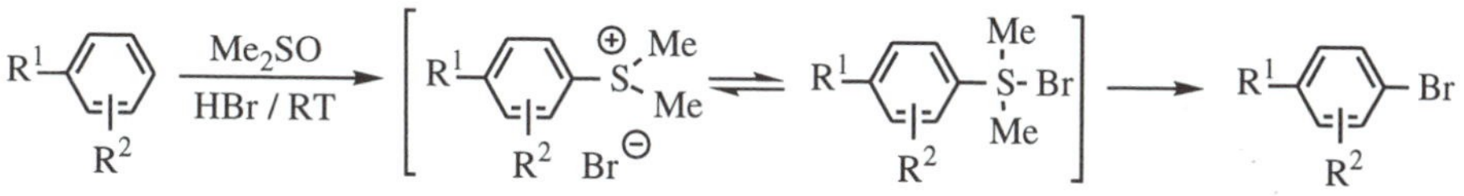

Scheme 3.4

The reaction of triarylsulfonium halides with sodium alkoxides at high temperatures afforded mixtures of alkyl aryl ethers (**24**), diarylsulfides, hydrocarbons and aldol resins or a ketone if the alkoxide is derived from a secondary alcohol.[25,36,38-41]

$$Ar_3S^{\oplus} X^{\ominus} + RO^{\ominus} \xrightarrow{\Delta} RO\text{-}Ar + Ar_2S + ArH + \text{Carbonyl Derivatives}$$

$$(24)$$

In the case of tertiary alcoholates, formation of the corresponding *tert*-alkyl aryl ether [(**24**), R = *tert*-alkyl] is observed. For example, the reaction of diphenyl-*p*-tolylsulfonium bromide with sodium *tert*-butoxide led to modest yields of *tert*-butyl phenyl ether and *tert*-butyl *p*-tolyl ether together with trace amounts of acetone.[36] The formation of the alkyl aryl ethers (**24**) and diarylsulfides occurs through a ligand coupling pathway and the other products, hydrocarbons and aldol resins or a ketone, derive from a competing free-radical chain reaction. Indeed, when the free radical trap 1,1-diphenyl-ethylene (DPE) was added to the reaction of diphenyl-*p*-tolylsulfonium iodide with sodium ethoxide at 80°C, the yields of the ethers increased from 22.1% for PhOEt and 0.8% for *p*-TolOEt in absence of DPE to 75.5% for PhOEt and 2.1% for *p*-TolOEt in the presence of the radical trap DPE.[41]

Triarylsulfonium halides reacted with sodium arylthiolates to afford mixtures of diarylsulfides. The reaction can be explained by a ligand coupling process taking place on a (arylthio)triarylsulfurane and is free of any competing free radical chain component as no diaryldisulfides were observed.[25,39]

$$Ar^1_3S^{\oplus} X^{\ominus} + Ar^2S^{\ominus} \longrightarrow Ar^1\text{-}S\text{-}Ar^2 + Ar^1\text{-}S\text{-}Ar^1$$

The reaction of triarylsulfonium halides with piperidine, used as solvent, in the presence of potassium amide at 100-110°C afforded a mixture of the derived *N*-arylpiperidine (**25**) in high yield, together with the diarylsulfide and/or arylthiol derived from the reagent by ligand exchange.[39]

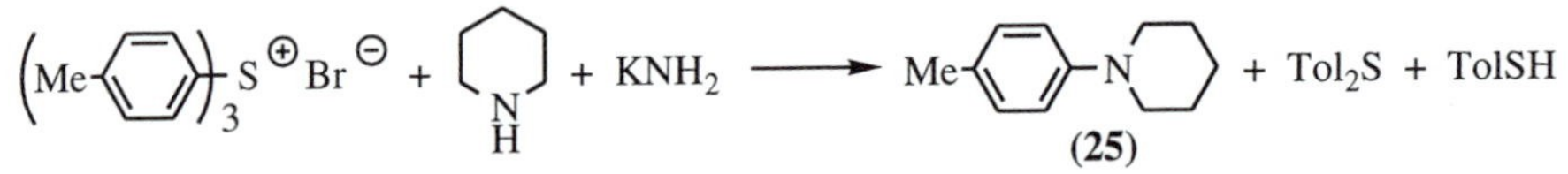

$$(25)$$

The reaction of the bis(2,2'-biphenylylene)sulfurane (**13**) with a range of alcohols and phenols led to the corresponding 2-alkoxy or 2-aryloxybiphenyls (**27**) in high yields.[42] The reaction was considered to proceed by proton-catalysed ring opening and formation of an oxysulfurane [10-S-4(C$_3$O)] (**26**), undergoing subsequently a ligand coupling process or an *ipso*-substitution. In the case of highly hindered phenols, such as 2,6-di-*tert*-butylphenol, a poor yield of the 2-aryloxybiphenyl [(**27**), R = 2,6-di-*tert*-butylphenyl] was obtained. The reaction afforded instead a range of products resulting from the occurrence of a single electron transfer mechanism.

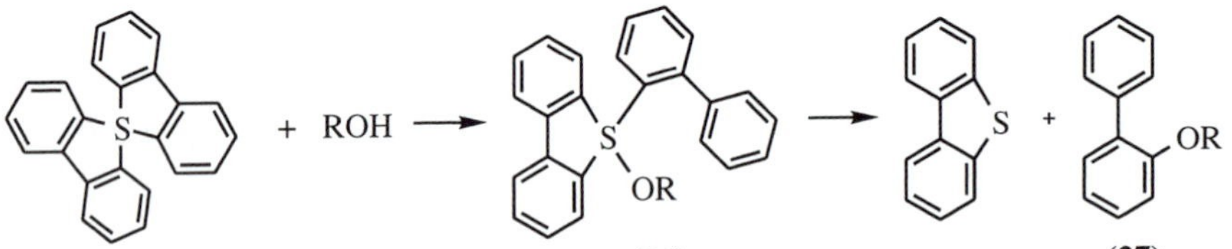

(**13**)   (**26**)   (**27**)

Scheme 3.5

Reaction of 1,2,3-triarylbenzo[*b*]thiophenium salts (**28**) with alkoxides resulted in the exclusive formation of the (Z)-alkoxysubstituted alkenes (**29**). The ring-opening reaction, proceeding with complete retention of configuration, was in favour of a ligand coupling mechanism.[43]

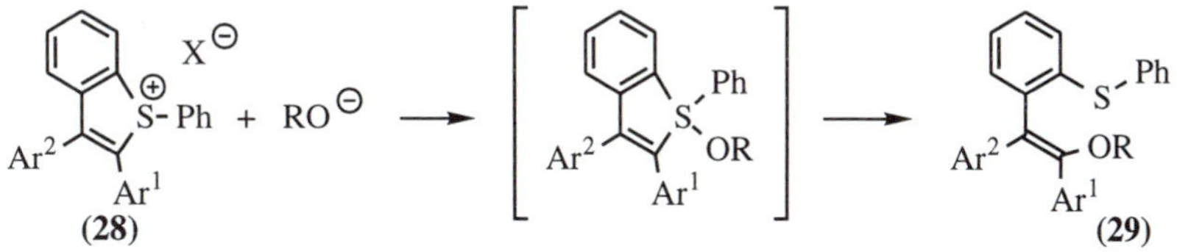

Scheme 3.6

The reaction of these same 1,2,3-triarylbenzo[*b*]thiophenium salts **(28)** with halide anions showed a different outcome. With iodide, only cleavage of the Ph-S bond to give **(30)** was observed. With other halides (X = Cl or Br), cleavage of the Ph-S bond was only favoured over the ring-opening reaction giving mixtures of **(30)** and **(31)**.[44]

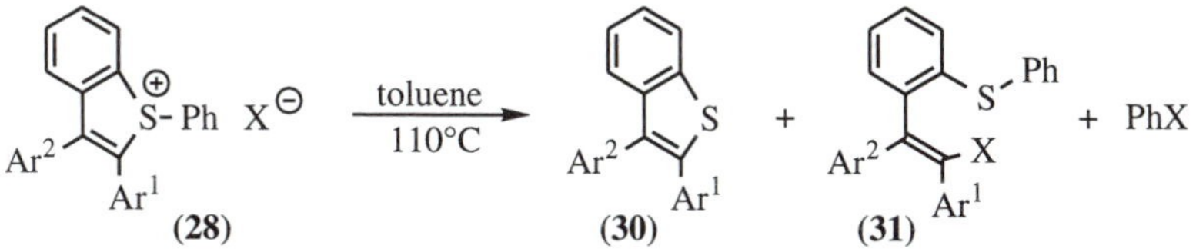

Ratio of products:	X =	Cl	51	:	49
		Br	62-82	:	38-18
		I	100	:	0

3.1.4 Reactions of allyl and benzylsulfonium salts

Treatment of allylsulfonium salts with organolithium reagents is a convenient way to generate vinylsulfonium ylides. However, reactions of allyldimethylsulfonium salts with bases lead exclusively to homoallyl methylsulfides *via* [3,2] sigmatropic rearrangements. On the other hand, generation of the diphenylsulfonium allylide **(33)** by reaction of allyldiphenylsulfonium tetrafluoroborate **(32)** with *n*-butyllithium is accompanied by the formation of volatile products, such as biphenyl **(4)** and *n*-butyl-phenylsulfide, which resulted from the decomposition of an intermediate tetracoordinate sulfurane. The formation of these side products can be avoided by use of the bulkier *tert*-butyllithium.[45,46]

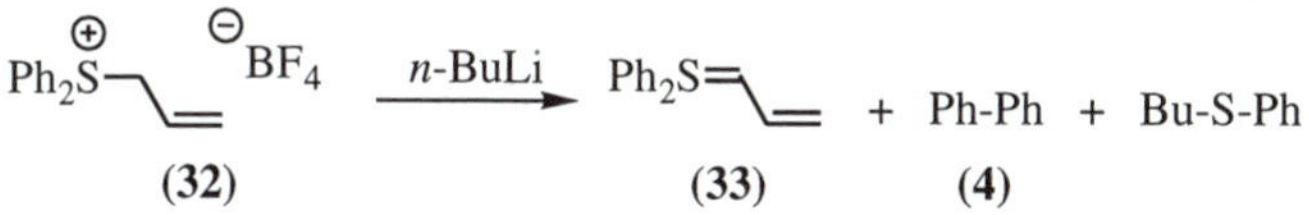

The mechanism of the thermal decomposition of benzyldialkylsulfonium salts as well as their behaviour towards nucleophiles has been the subject of long-standing investigations.[47-52] A range of mechanistic pathways have been invoked, such as S_N1, mixed S_N1/S_N2, concerted Hughes-Ingold S_N2, ion-dipole complex or preassociation-concerted mechanism. Young and Ruekberg, on their side, suggested that the formation of a sulfurane, followed by ligand coupling, offers a reasonable explanation to a range of reactions involving benzylsulfonium salts.[53] The ligand coupling mechanism, taking place on a covalent neutral sulfurane (with charged nucleophiles) or on a covalent charged sulfurane intermediate (with neutral nucleophiles) would therefore be an alternative to the closely related preassociation-concerted mechanism, which considers ion-pair formation instead of the covalently bonded sulfurane. Young *et al.* also considered that the ligand coupling mechanism can be extended to rationalize the unusual efficiency of methyltransferase enzymes.[53]

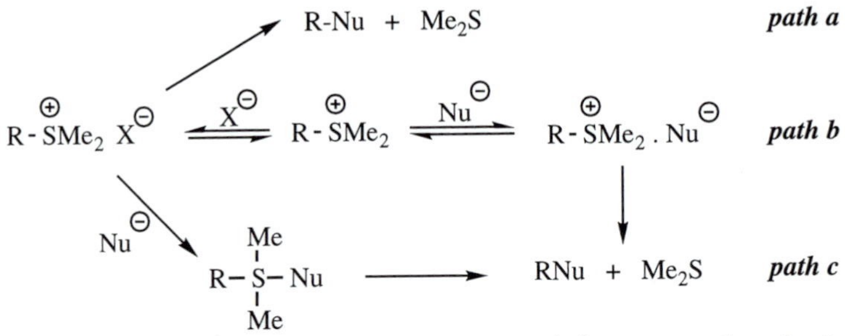

"path a" is a S_N2 mechanism, "path b" is a preassociation-concerted mechanism, "path c" is a ligand coupling mechanism

Scheme 3.7

3.1.5 Reactions of perfluorosulfonium salts

Due to the high electronegativities of fluorine (4.0) and R_f groups (CF_3: 3.45), electrophilic perfluoroalkylation is not as common a reaction as electrophilic alkylation of nucleophiles. However, adequately substituted hypervalent reagents have proved efficient perfluoroalkylating agents. Among them, a range of reagents possessing a sulfonium structure have been studied and used, particularly for trifluoromethylation.[54] The choice of the reagent can be made on the basis of the nucleophilicity of the substrate. Whereas acyclic trifluoromethyl diarylsulfonium salts [type (34)] are fairly unreactive,[55,56] a variety of dibenzothiophenium salts [(35) - (37)] have been used. Their reactivity order is highly dependent upon the nature of the substituents present on the dibenzoheterocyclic structure:

electron-donating < H < electron-withdrawing

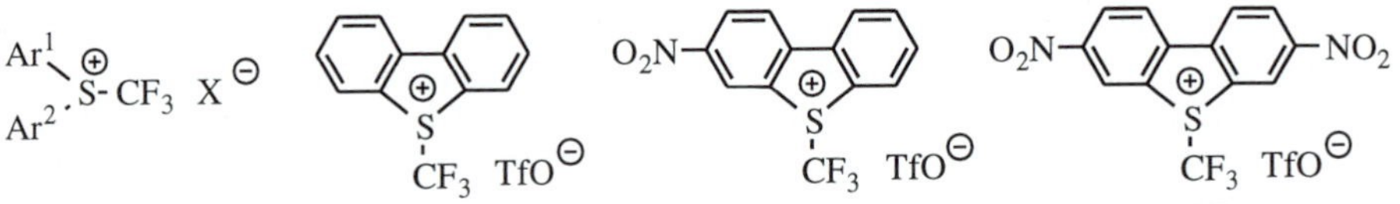

Water-soluble reagents were also developed. They showed a similar reactivity order: (38) < (39)[57]

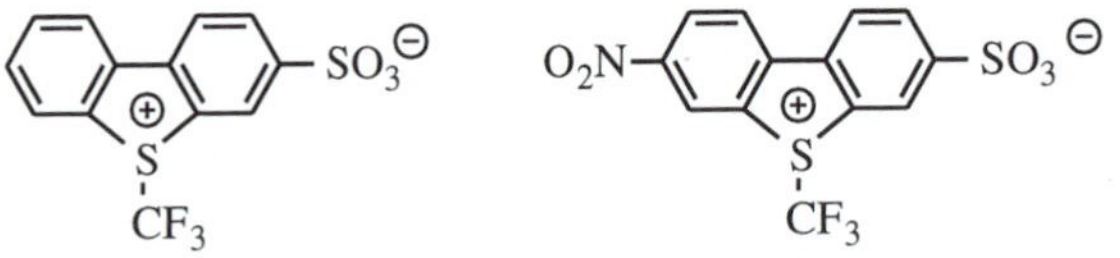

A wide range of nucleophilic substrates of different reactivity were trifluoromethylated with these reagents. The substrates include carbanions, activated aromatics, heteroaromatics, enol silyl ethers, enamines, phosphines, thiolate ions and iodide anions.[56-59] (Scheme 3.8) The least reactive substrates, such as triphenylphosphine, aniline and phenols, require the use of the most reactive dinitro derivative. Most of the reactions can be conveniently performed with the unsubstituted *S*-trifluoromethyl dibenzothiophenium salt (35). The least reactive sulfonium salts are the acyclic sulfonium compounds which reacted only with the sodium thiolates.[55,59]

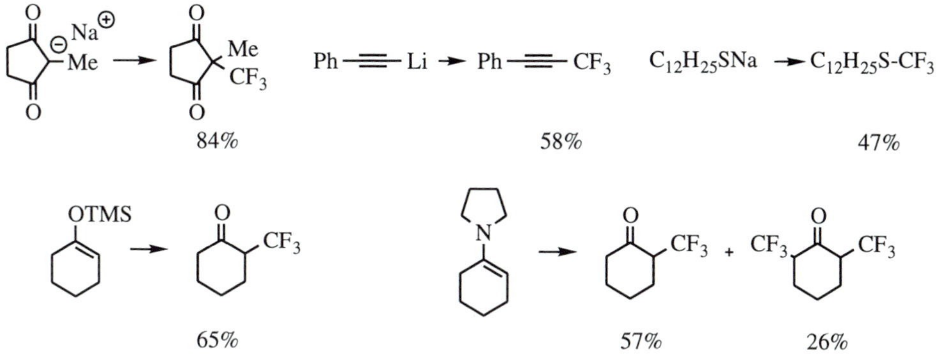

84%   58%   47%

65%   57%   26%

Scheme 3.8: Examples of trifluoromethylation reactions with salt (**35**)

Umemoto *et al.* suggested that the mechanism of these trifluoromethylation reactions occurs by a bimolecular ionic substitution mechanism, sometimes competing with a CF_3 free radical chain reaction.[59] This ionic mechanism was considered to be a side-on attack of the nucleophile to the S-CF_3 bond.[54,57,60] However, in view of the general similarities between the reactivity of these compounds and other types of ligand coupling reagents, the ligand coupling mechanism is a more likely alternative. This possibility was nevertheless excluded by Umemoto *et al.*, in part on kinetic arguments and in part because of the reactions of alkyl-, aryl-lithium and -magnesium halides which did not afford the corresponding trifluoromethyl derivatives. This may be explained by a competion between ligand coupling and ligand exchange, which could become the exclusive pathway in some instances.

The reaction of the lithium enolate of 2-methyl-1-indanone with the thiophenium salt (**35**) leading to the 2-trifluoromethyl derivative in 51% yield is an exception.[59] With all other *in situ* generated enolates of ketones, no trifluoromethylation was observed. To moderate the reactivity of the enolates, a boron Lewis acid (**40**) was added to form the boron complexes. This made a regio-, diastereo- and enantio-selective trifluoromethylation possible in good to high yields.[61]

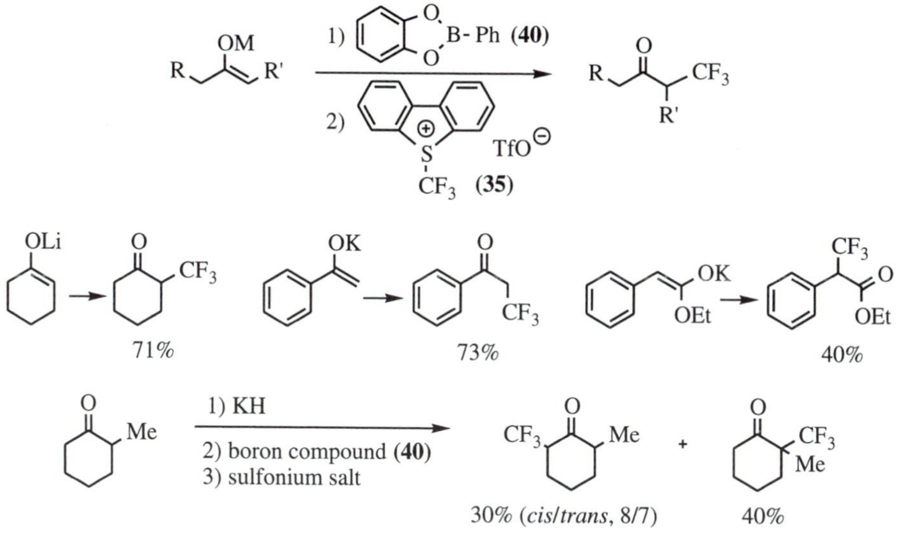

71%   73%   40%

30% (*cis/trans*, 8/7)   40%

The reaction is not limited to trifluoromethylation. *S*-(perfluoroalkyl)dibenzothiophenium reagents **(41)** can be prepared in the same manner as the trifluoromethyl compounds and they reacted similarly to the trifluoromethyl reagents.[61]

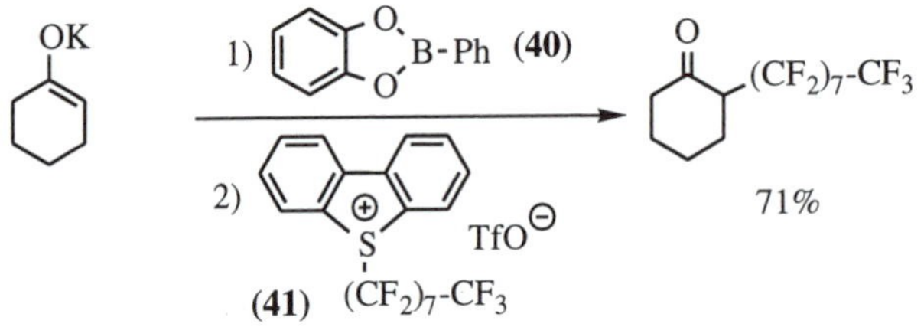

3.2 SULFOXIDES

Sulfoxides have found a wide variety of applications in organic synthesis, either as activating group or as reactive centers. As such, they have been successfully employed in a number of enantioselective synthesis.[62]

In a comparatively small number of systems, unsymmetrical sulfoxide derivatives undergo ligand coupling reactions upon treatment with nucleophilic agents. The largest group of these reactions implies the 2-pyridyl group as one of the ligands involved in the coupling reaction, while a limited number of other aromatic groups appeared able to undergo the ligand coupling reaction. Among the most frequently used nucleophilic reagents, organomagnesium or organolithium compounds react with sulfoxides to lead to a variety of products depending on the nature of the sulfoxide, the organometallic reagent and the reaction conditions. The reaction of diarylsulfoxides with aryl Grignard or aryllithium leads to biaryl formation by a sequence of transition steps involving the formation of a hydroxysulfurane. The coupling itself results from a major pathway going through a benzyne intermediate, the ligand coupling pathway constituting only a minor pathway.[63,64] The reaction of sulfoxides bearing an heteroaryl group or an electron-poor aryl group with organomagnesium or organolithium reagents leads to ligand coupling products.[65-67] The sulfoxides prone to ligand coupling possess an aryl group such as pyridyl, quinolinyl, arylsulfonyl or naphthyl, and also vinyl or alcynyl groups. The organic fragment of the Grignard or organolithium reagents can be benzyl, allyl, vinyl, aryl, heteroaryl, and also *sec-* and *tert*-alkyl groups.

The reaction of a sulfoxide with nucleophiles leads to the formation of a tetrasubstituted σ-sulfurane intermediate. This presents a trigonal bipyramidal structure in which the pair of unshared electrons plays the role of the fifth ligand. During the attack of the nucleophile on the sulfoxide, the incoming nucleophile occupies in the first place the apical position. Depending on the nature of the ligands present on the oxysulfurane, this intermediate can follow different pathways, which can take place alone, in combination or in competition, to afford the different combinations of ligand coupling products. (Scheme 3.9)

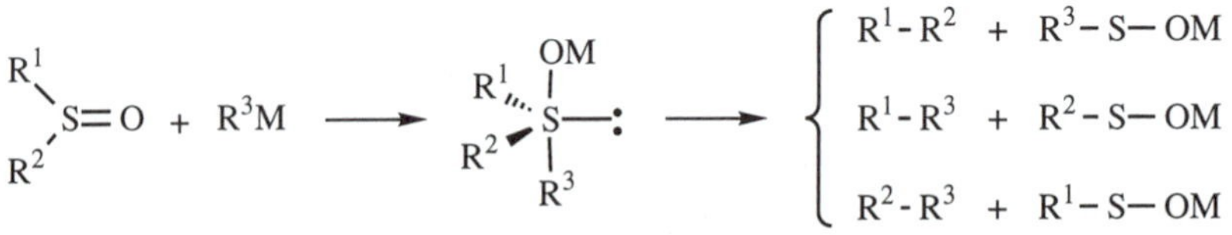

Scheme 3.9

3.2.1 Aryl-benzyl coupling

The reaction of arylbenzylsulfoxides with Grignard reagents led Oae *et al.* to the first experimental observation of the retention of configuration in ligand coupling reactions involving sp^3 carbon atoms. When the optically active S-(-) or R-(+) (1-phenylethyl)pyridylsulfides (**42**) and (**43**) were oxidised to their corresponding sulfoxides, and these sulfoxides treated with methylmagnesium bromide, the coupling products, S-(+) or R-(-) 2-(1-phenylethyl)pyridine (**44**) and (**45**), were isolated in good to high yields, with retention of the original configuration of the sp^3 carbon atom.[68,69]

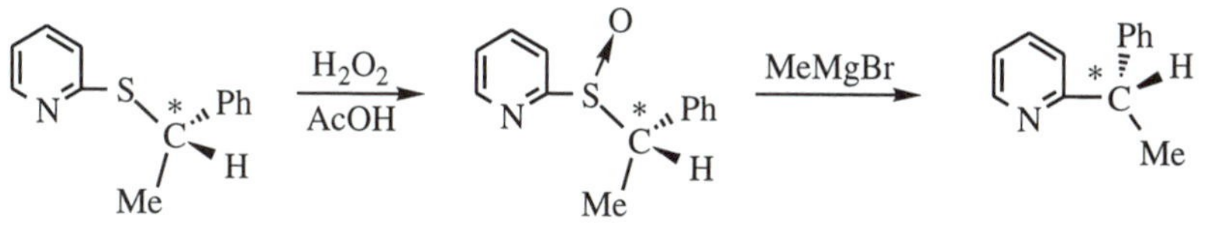

(**42**), (S)-(-)

$[\alpha]_D^{25} = -375°$

(**43**), (R)-(+)

$[\alpha]_D^{25} = +375°$

91%

(**44**), (S)-(+): 97%

$[\alpha]_D^{25} = +63°$

(**45**), (R)-(-): 75%

$[\alpha]_D^{25} = -65°$

Scheme 3.10

When there is a competition between two substituted benzyl groups, the electronic nature of the ligands plays an important role on the extent of the pseudorotation phenomenon. The more electron-poor benzyl group will preferentially occupy the apical position, favouring the coupling process.[70] When the incoming benzyl nucleophile is a less electron-rich group than the already present benzyl group, ligand coupling will occur more rapidly than pseudorotation. On the other hand, when the incoming nucleophile is more electron-rich than the already present benzyl group, pseudorotation will take a significant weight in the competition between the two processes, so that the more electron-poor ligand will tend to occupy more or less completely the apical position. This is partly verified for the tolyl/phenyl pair and more clearly verified for the 4-chlorophenyl/phenyl pair of competing benzyl ligands. (Table 3.1)

Table 3.1: Influence of the electronic nature of the ligands in the competition between two benzyl groups[70]

R^1	R^2	(**47**) : (**48**) at RT
4-Me-C_6H_4	C_6H_5	16 : 84
C_6H_5	4-Me-C_6H_4	44 : 56
C_6H_5	4-Cl-C_6H_4	9 : 91
4-Cl-C_6H_4	C_6H_5	65 : 35

The relative steric size of the two ligands in competition for ligand coupling does not generally play a very important role on the outcome of the reaction. For example, comparison of the ratios of products in the case of a competition between a benzyl ligand and a *para*-tolylmethyl or an (*ortho*, *para*-dimethylphenyl)methyl group does not show any significant difference. On going from the tolyl to the xylyl derivative as incoming nucleophile, the yields of the ligand coupling products (**48**) derived from the incoming group increased only from 56% to 59%. (Table 3.2)

Table 3.2: Influence of the steric size of the benzyl ligands on the ratio of products (**47**) and (**48**)[70]

R^1	R^2	(**47**) : (**48**) at RT
C$_6$H$_5$	4-Me-C$_6$H$_4$	44 : 56
C$_6$H$_5$	2,4-Me$_2$C$_6$H$_3$	41 : 59
4-Me-C$_6$H$_4$	C$_6$H$_5$	16 : 84
2,4-Me$_2$C$_6$H$_3$	C$_6$H$_5$	17 : 83

The reaction temperature plays a relatively minor role in modifying the rate of pseudorotation as well at low temperature (- 68°C)[71] as at higher temperatures (+ 50°C).[70]

Table 3.3: Influence of the reaction temperature on the ratio of products[70,71]

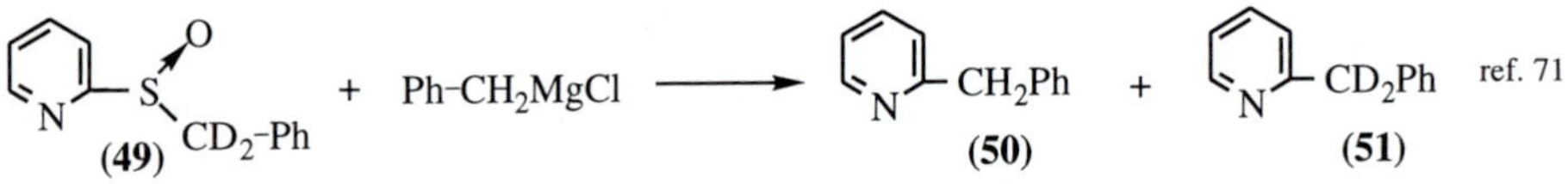

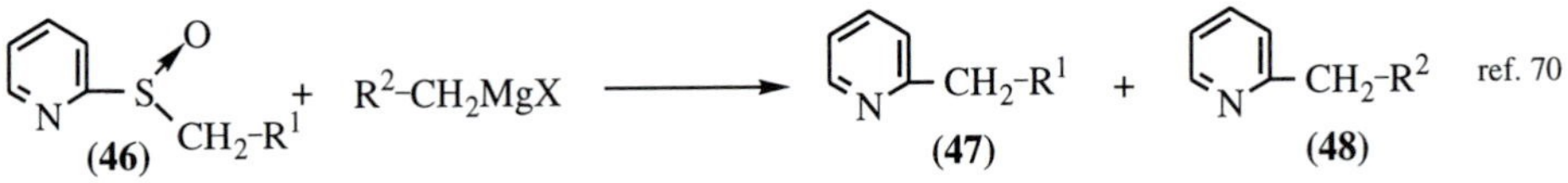

R^1	R^2	(**47**) : (**48**) at RT	(**47**) : (**48**) at 50°C
4-Me-C$_6$H$_4$	C$_6$H$_5$	16 : 84	20 : 80
C$_6$H$_5$	4-Me-C$_6$H$_4$	44 : 56	52 : 48

The oxysulfurane intermediate undergoing the ligand coupling reaction between the 2-pyridyl group and the benzyl group can be formed by action of other types of organometallic reagents. The benzylpyridine (**50**) can be formed by the reaction of benzylpyridylsulfoxide (**52**) with a benzyl Grignard reagent as well as with different Grignard reagents or organolithium compounds.

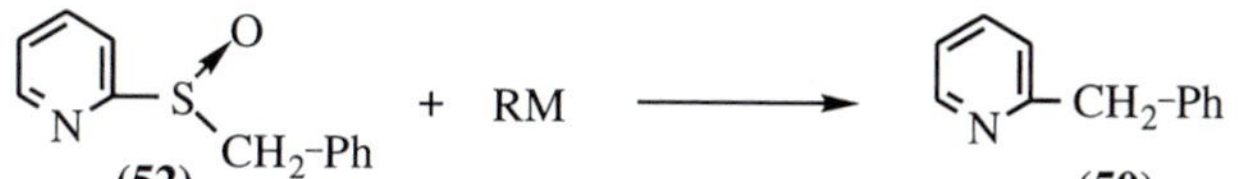

RM =	PhMgBr	98%
	MeMgBr	83%
	n-BuLi	46%

In the case of the reaction of the benzyl (4-pyridyl)sulfoxide (53) with phenylmagnesium bromide, a similar reaction of ligand coupling took place easily to lead to 4-benzylpyridine (54).[72]

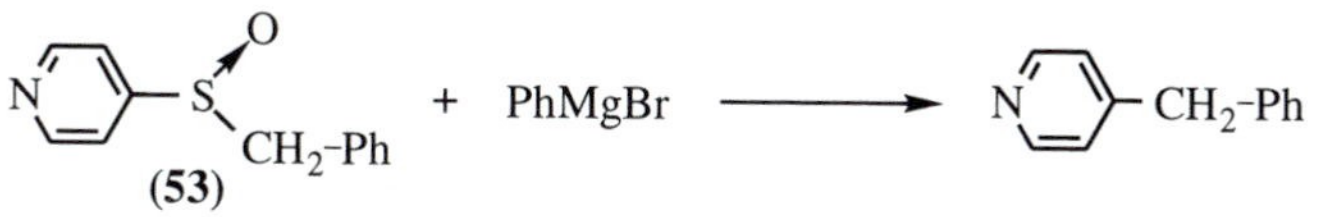

However, the similar reaction performed on the 3-pyridyl isomer (55) led to a completely different result. Ligand coupling did not occur, and the only isolated products resulted from a ligand exchange reaction occurring on the first formed oxysulfurane.[72]

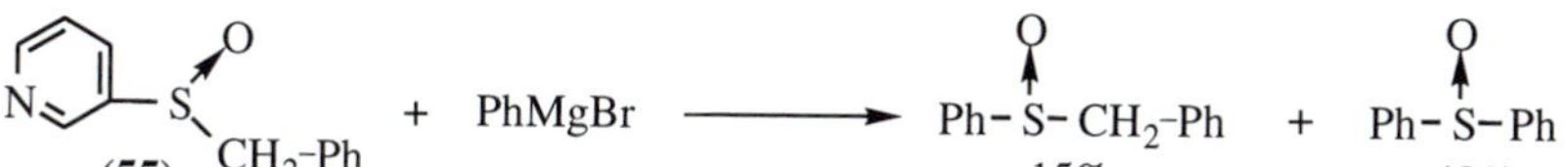

The benzyl group can also couple efficiently with other aromatic groups, such as 2-quinolinyl or 4-benzenesulfonylphenyl. The retention of configuration, originally found in the reaction of (1-phenylethyl) pyridylsulfoxides with organometallic reagents, was also observed in the case of optically active (1R)-phenylethyl 2-quinolinyl (R)- and (S)-sulfoxides (58), which both led to (R)-2-(1-phenylethyl) quinoline (59),[73] as well as in the case of the (R)- and (S)-(1-phenylethyl) benzenesulfonylphenyl analogues (60) leading to (61).[74]

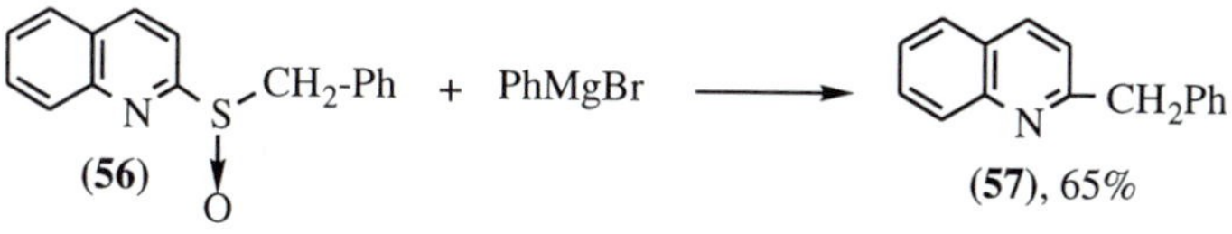

(56)   + PhMgBr ⟶   (57), 65%

(58)   MeMgBr ⟶   (59), 64-76%

$[\alpha]_D^{23} = -177 \pm 3°$ (c 1.8 in C$_6$H$_6$)

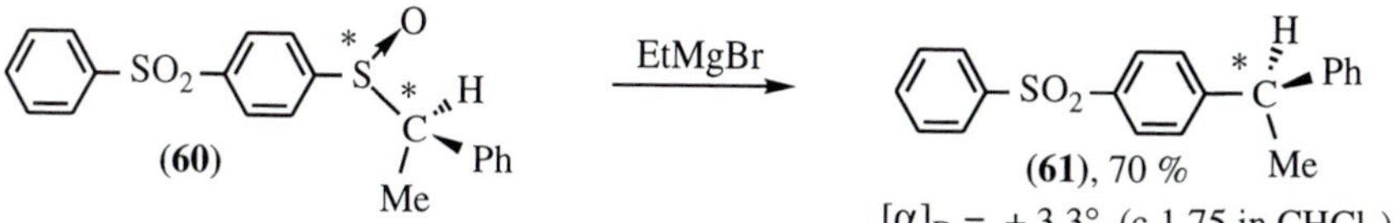

(60)   EtMgBr ⟶   (61), 70 %

$[\alpha]_D = +3.3°$ (c 1.75 in CHCl$_3$)

In the 4-(phenylsulfonyl)phenylsulfoxide system (62), the steric size of the substituted benzyl ligands plays a relatively significant role in the case of a competition between two benzyl groups. In the case of a competition between xylylmethyl against benzyl, the rate of ligand coupling is higher than the rate of pseudorotation, and the relative amount of ligand coupling product resulting from the interaction between the diphenylsulfonyl group and the incoming nucleophile is more important than in the case of the competition between 4-tolylmethyl and benzyl groups. The reaction involving a competition between two substituted benzyl ligands in the intermediate oxysulfurane showed a more important influence of the reaction temperature when the incoming nucleophile is a 4-tolyl group than when the incoming nucleophile is the phenyl group.[70]

Table 3.4: Ligand coupling reactions of 4-(phenylsulfonyl)phenylsulfoxide derivatives[70]

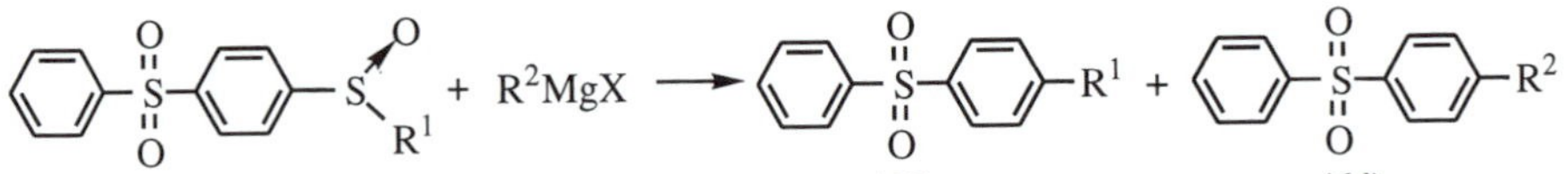

R¹	R²	(63) : (64) at RT	(63) : (64) at 50°C
C_6H_5-CH_2	4-Me-C_6H_4-CH_2	77 : 23	46 : 54
C_6H_5-CH_2	2,4-$Me_2C_6H_3$-CH_2	47 : 53	
4-Me-C_6H_4-CH_2	C_6H_5-CH_2	20 : 80	19 : 81
2,4-$Me_2C_6H_3$-CH_2	C_6H_5-CH_2	12 : 88	

When a 2-pyrimidylsulfoxide derivative was treated with a Grignard reagent, the outcome of the reaction was very dependent upon the nature of the incoming nucleophile. Unusually, the phenyl group appeared to be a better ligand coupling group than the benzyl group. The 2-benzylpyrimidine (66) was formed only in a poor yield in the reaction of the phenysulfoxide (65) with benzylmagnesium chloride, and was not formed in the reaction of the benzylsulfoxide (68) with phenylmagnesium bromide.[72,75]

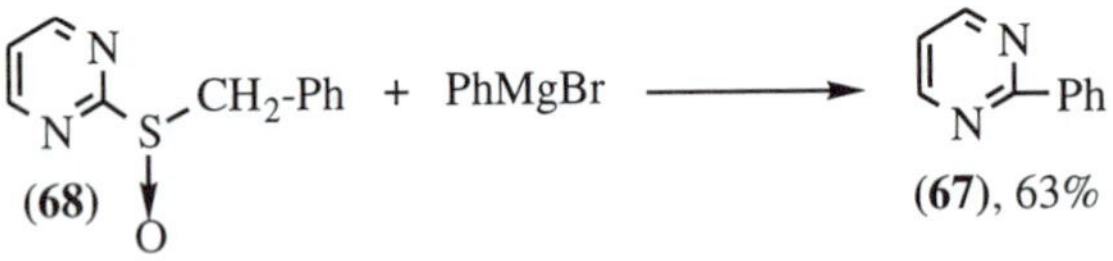

In the case of a competition reaction involving two aryl groups and a benzyl group, electronic and steric effects played a significant role.[72]

$$(50), 34 \quad : \quad (69), 66$$

$$(66), 14 \quad : \quad (69), 86$$

Although more electron-attracting than a 2-pyridyl group, the 2-benzothiazolyl ligand did not couple with the benzyl group. Ligand exchange was the only pathway followed by the first formed oxysulfurane. Secondarily, a new oxysulfurane possessing two benzothiazolyl ligands may be formed by action of the ligand exchange product, 2-benzothiazolylmagnesium halide, on the 2-benzothiazolylsulfoxide, eventually leading to 2,2'-bibenzothiazol **(71)**.[76]

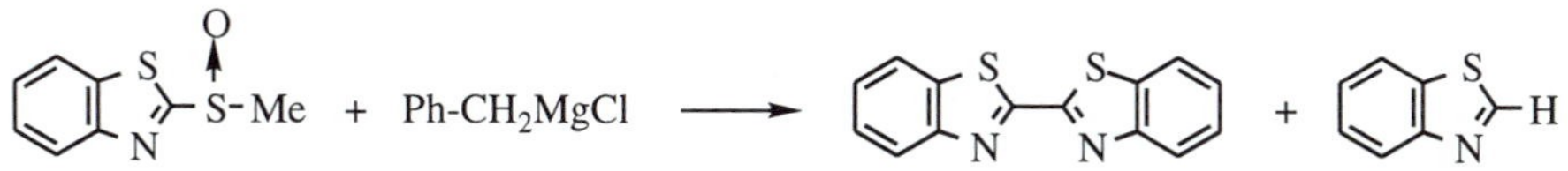

$$(70) \qquad\qquad (71), 6\% \qquad\qquad 65\%$$

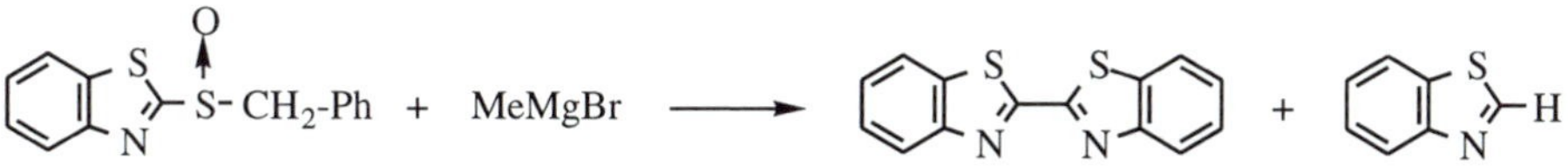

$$(72) \qquad\qquad (71), 39\% \qquad\qquad 40\%$$

Baker *et al.* have extended this ligand coupling reaction to the case of benzyl and substituted naphthylsulfoxide derivatives, which showed a somewhat different behaviour. Indeed, the reaction of phenylmagnesium bromide with benzylpyridylsulfoxide afforded a quantitative yield of 2-benzyl pyridine (98%).[70] By contrast, in the case of the benzylnaphthylsulfoxide **(73)**, the reaction with 4-tolylmagnesium bromide led only to the 1-(4-tolyl)naphthalene derivative **(76)** as ligand coupling product. Upon reaction with ethylmagnesium bromide, the benzylnaphthalene product was obtained but in poor yields (13-26%), the major product being the 1-ethylnaphthalene derivative **(77)** in 62-69% yields.[77,78]

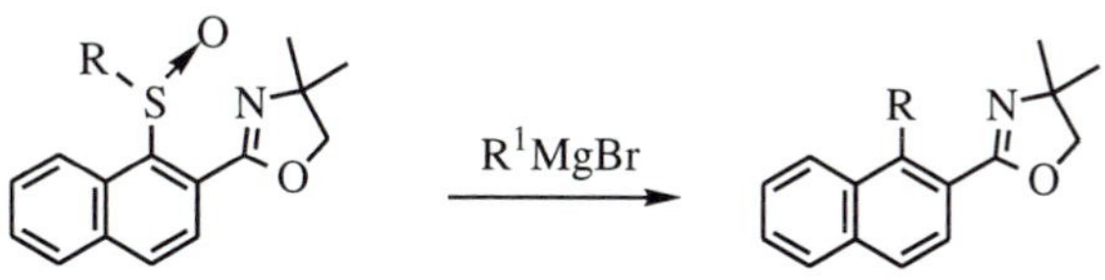

(73), R = PhCH$_2$
(74), R = ClC$_6$H$_4$CH$_2$

(75), R = H
(76), R = MeC$_6$H$_4$
(77), R = Et
(78), R = PhCH$_2$
(79), R = ClC$_6$H$_4$CH$_2$

(73) + 4-TolMgBr $\longrightarrow$ **(75)** (10%) + **(76)** (65%)

(73) + EtMgBr $\longrightarrow$ **(75)** (10%) + **(77)** (69%) + **(78)** (13%)

(74) + EtMgBr $\longrightarrow$ **(75)** (4%) + **(77)** (62%) + **(79)** (26%)

When optically active (2-alkoxycarbonyl)naphthylsulfoxide derivatives (80) and (81) were treated with (1-methylfluorenyl)lithium, high yields of coupling products, the 1-fluorenylnaphthalene-2-carboxylic acid derivatives (82) and (83), were obtained with a good enantioselectivity. In the case of the *tert*-butyl naphthylsulfoxide (80), the ester group of the product of ligand coupling was reduced with lithium aluminum hydride to afford the alcohol, which showed a 78% enantiomeric excess.[79]

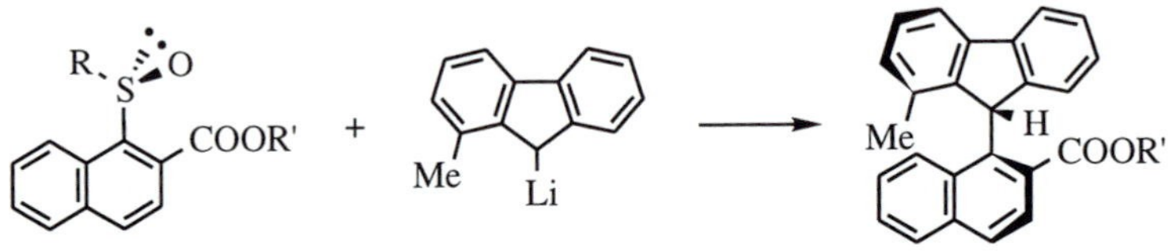

(80), R = *t*-Bu, R' = *i*-Pr   (82), quantitative yield, $[\alpha]_D = +96°$
(81), R = 4-Tol, R' = (1R)-menthyl   (83), 91%, 70% de

3.2.2 Aryl-allyl coupling

The 2-pyridyl group couples readily with allylic groups in the reaction of allylic pyridylsulfoxide (84) with Grignard reagents, such as phenylmagnesium bromide.[68,80]

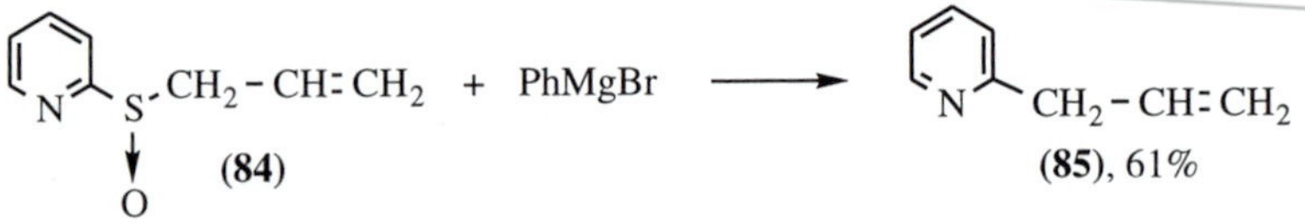

Reaction of methyl quinolinylsulfoxide (86) with allylmagnesium chloride or with 1-methylallyl-magnesium chloride led to the 2-allyl and 2-(1-methylallyl)quinoline, (87) and (88) respectively, contaminated with unseparable isomeric α,β-unsaturated derivatives, in high yields.[81]

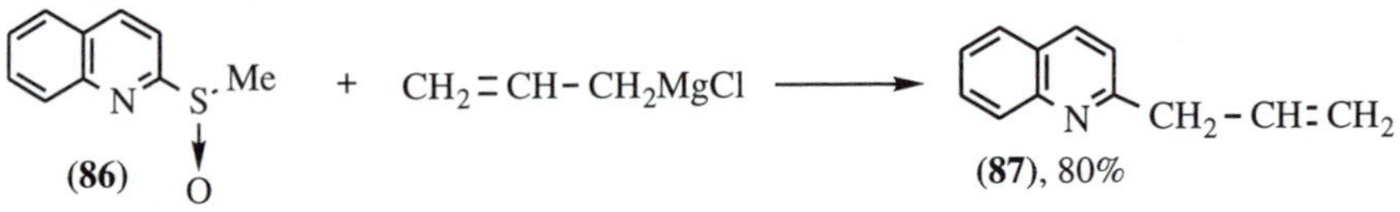

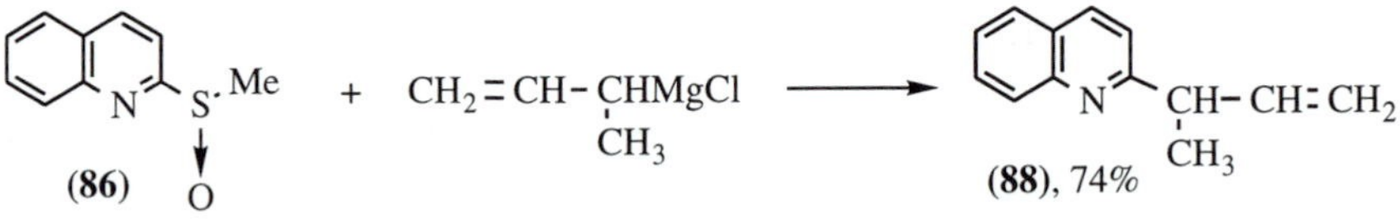

Retention of the stereochemistry of the starting sulfoxides in the ligand coupling product was clearly demonstrated in the case of the coupling of the 4-benzenesulfonylphenyl group with a crotyl group. In the reaction of the sulfoxide (89) with ethylmagnesium bromide, the coupling product with complete retention of the original stereochemistry (90) was accompanied by diphenylsulfone, formed through ligand exchange reaction.[82]

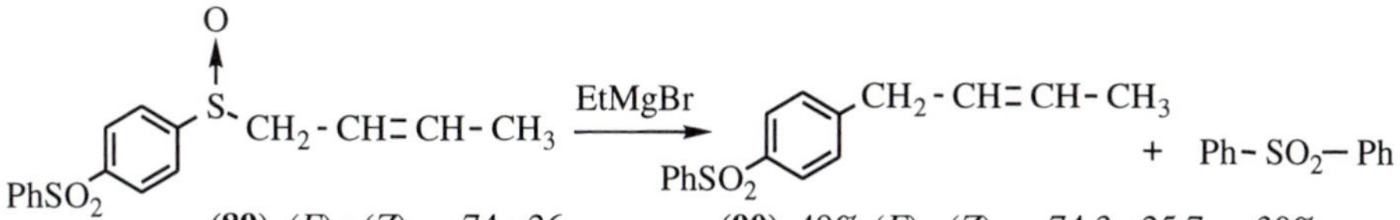

(89), $(E) : (Z) = 74 : 26$ **(90)**, 49% $(E) : (Z) = 74.3 : 25.7$ 30%

The reaction of (4-benzenesulfonylphenyl) benzylsulfoxide **(91)** with allylmagnesium halide afforded only the benzyl derived ligand coupling product **(69)**, without any allyl-aryl coupling product.[82]

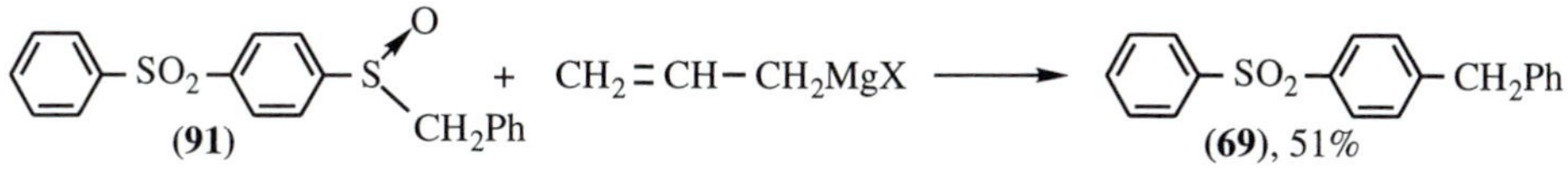

(91)   **(69)**, 51%

The reverse system, reaction of (4-benzenesulfonylphenyl) allylsulfoxide **(92)** with benzyl-magnesium halide, led to the same result.

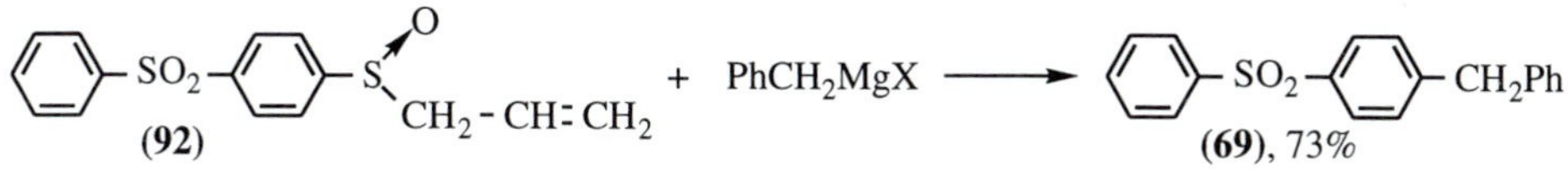

(92)   **(69)**, 73%

From these observations, the following order of reactivity or "migration ability" was deduced:

> Benzyl » Allyl › Phenyl group

In the case of the more sterically hindered 1-methylallyl derivative **(93)**, the normal ligand coupling product **(94)** was the only observed product and formed in high yield. The presence of diphenylsulfone was not observed.[82]

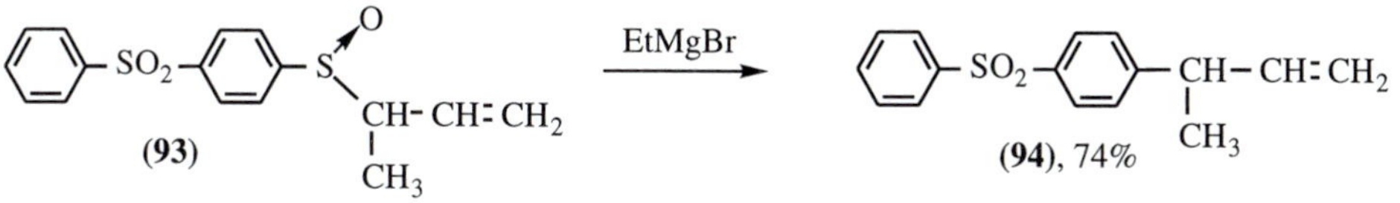

(93)   **(94)**, 74%

3.2.3 Aryl-vinyl coupling

Although less reactive than the benzyl and allyl groups, vinyl groups are also prone to ligand coupling reactions. The original geometric configuration of the vinylsulfoxides [**(95)** and **(97)**] is retained throughout the whole reaction to lead to the arylvinyl derivatives [**(96)** and **(98)** respectively] with the same geometry.[82]

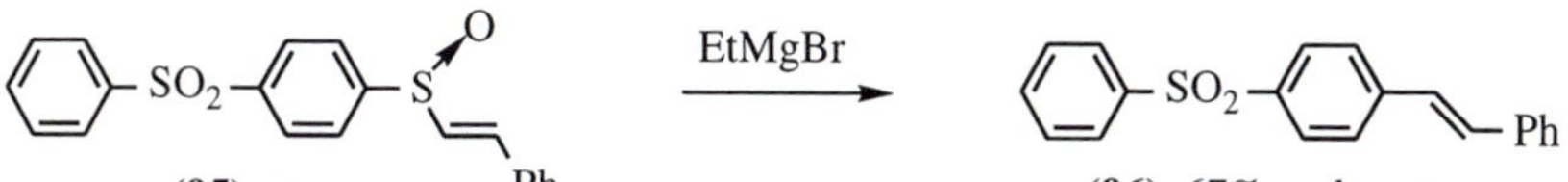

(95), *trans*   Ph   **(96)**, 67%, only *trans*

(97), *trans*　　　　　　　　　　　**(98)**, 44%, only *trans*

However, these reactions are frequently contaminated with side products resulting from ligand exchange and subsequent reactions. In the case of the pyridyl vinylsulfoxide **(99)**, the major product of ligand coupling coupling was in fact 2,2'-bipyridyl **(101)**, resulting from a first ligand exchange followed by reaction on a second molecule of the sulfoxide **(99)**, and finally ligand coupling occurring on an hydroxysulfurane bearing two 2-pyridyl ligands. [83]

(99), *cis*　　　　　　**(100)**, 33%, only *cis*　　**(101)**, 58%　　　**(102)**, 22%

These side reactions became predominant or the only pathway, when the styryl group contained an electron-rich aryl group, such as a 4-methoxyphenyl or a 4-dimethylaminophenyl group. However, this ligand coupling reaction was used in a synthesis of α-stilbazoles. The yields of ligand coupling products were generally relatively modest, as the ligand coupling process was in competition with ligand exchange, leading to a mixture of products.[83]

(103), *trans*　　　　**(104)**, 53%, only *trans*　　**(101)**, 30%　　　**(105)**, 14%

3.2.4 Coupling between two aryl or heteroaryl groups

The formation of a biaryl bond can be realized by the reaction of an arylsulfoxide with various organometallic reagents, such as organolithium or Grignard reagents. Depending on the nature of the ligands bound to the sulfoxide center and the nature of the organic fragment of the organometallic reagent, the products of the reaction can be derived either from a direct ligand coupling reaction or from more complex pathways involving ligand exchange and reaction of the newly formed organometallic species with the substrate, followed eventually by ligand coupling.

$$R^1_{}\!\!\diagdown S \blacktriangleright O \; + \; R^3M \longrightarrow R^1\text{-}R^2 \quad \text{and/or} \quad R^1\text{-}R^3$$
$$R^2\!\!\diagup$$

or

$$a\text{-} \quad R^1_{}\!\!\diagdown S \blacktriangleright O \; + \; R^3M \longrightarrow R^3_{}\!\!\diagdown S \blacktriangleright O \; + \; R^1M$$
$$R^2\!\!\diagup \qquad\qquad\qquad R^2\!\!\diagup$$

$$b\text{-} \quad R^1M \; + \; R^1_{}\!\!\diagdown S \blacktriangleright O \longrightarrow R^1\text{-}R^1$$
$$R^2\!\!\diagup$$

Two major types of reaction systems have been observed: coupling between two aryl groups (section 3.2.4.1) and coupling involving one or two heteroaryl groups (section 3.2.4.2).

3.2.4.1 Aryl-aryl coupling

The early observations of an aryl-aryl coupling were made during the studies of the reaction of a diarylsulfoxide with an organometallic reagent in the case of simple non-activated aromatic systems. Andersen *et al.* described the formation of biphenyl (**4**) in good yields upon treatment of diphenylsulfoxide (**106**) with phenyllithium.[64,84]

$$\begin{array}{c} \text{Ph} \\ \diagdown \\ \text{Ph} \end{array} \!\! S\!\!\rightarrow\!\! O \quad + \quad \underset{\text{(4-5 equiv.)}}{\text{PhLi}} \quad \xrightarrow{\text{Et}_2\text{O}} \quad \text{Ph-S-Ph} \quad + \quad \text{Ph-Ph}$$

$$\qquad\qquad \textbf{(106)} \qquad\qquad\qquad\qquad\qquad\qquad \textbf{(3)}, 87\% \qquad \textbf{(4)}, 65\%$$

As the formation of triphenylsulfonium bromide had already been reported to result from the reaction of diphenylsulfoxide with phenylmagnesium bromide,[85] the occurrence of an intermediate triphenylsulfonium species was suggested to take place in the reaction of diphenylsulfoxide with phenyllithium. The evolution of this intermediate could then follow the ligand coupling pathway, as observed by Trost *et al.* in the case of the reactions of triarylsulfonium salts (section 3.1.2). However, the reaction involving the reagents bearing 4-tolyl ligands, (**107**) and (**108**), led to significant amounts of the *m,p'*-bitolyl isomer (**111**) (26%).[63,64]

$$\textbf{(107)} \qquad\qquad \textbf{(108)} \qquad\qquad \textbf{(109)}, 66\% \qquad \textbf{(110)}, 31\% \qquad \textbf{(111)}, 26\%$$

The ligand coupling pathway was therefore considered to be a minor pathway. These results were explained by two different pathways for the evolution of the first formed tetracoordinated triarylsulfonium species, formed through addition of the lithium reagent on the sulfoxide group.

Step 1: Addition of the organolithium on the sulfoxide group

This species can break down to yield an aryne intermediate, which subsequently reacts with the organolithium reagent to lead to equivalent amounts of the *para-para* and *meta-para* isomers. (Scheme 3.11, step 2.A) In the second possibility, an heteroatom-containing anion is eliminated (LiO⁻ in the sulfoxide case) and the resulting triarylsulfonium cation reacts with the organolithium reagent to afford the tetraarylsulfurane. Ligand coupling then yields only the *para-para* coupling product. (Scheme 3.11, step 2.B) Moreover, a third possibility of evolution may arise in the case of reactions involving different ligands on the reagent and substrate: ligand exchange, which generates a new organometallic reagent which can be unreactive with the sulfoxide or react with it to afford a new type of intermediate sulfurane undergoing ligand coupling. (Scheme 3.11, step 2.C)

Step 2.A: Aryne mechanism

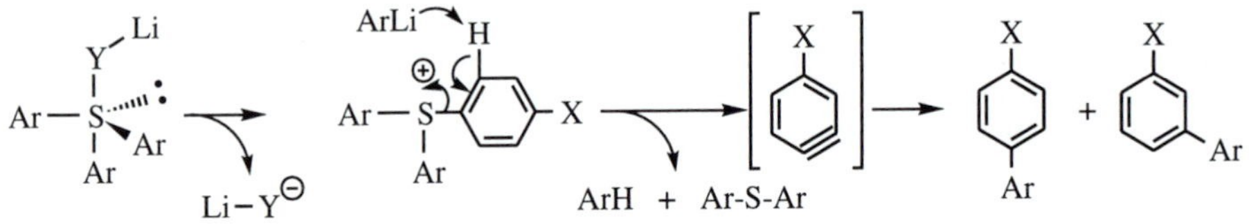

Step 2.B: Ligand coupling mechanism

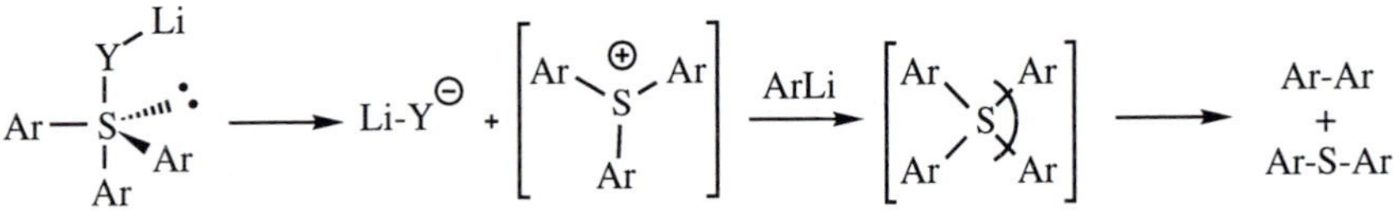

Step 2.C: Ligand exchange mechanism

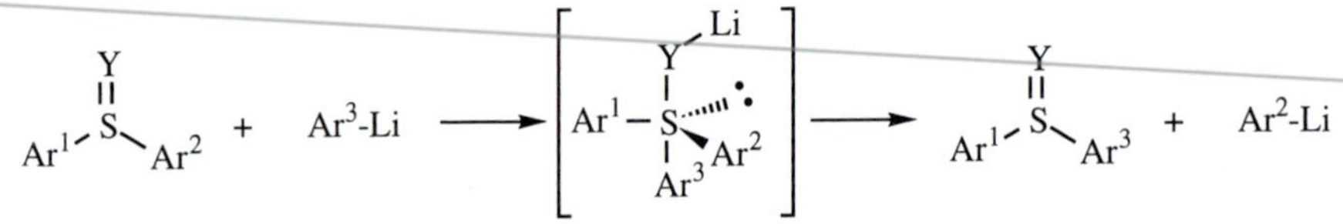

Scheme 3.11: Mechanisms involved in the reaction of diarylsulfoxides with aryllithium

The balance between the two pathways (aryne or ligand coupling) is dependent upon the nature of the group Y. In the case of the sulfoxides (Y = O), the aryne pathway is predominant. However, in the case of the *N*-tosyl sulfilimine system (Y = N-SO$_2$-C$_6$H$_4$-CH$_3$), the toluenesulfonamide anion has a greater leaving ability than the oxyanion. Therefore, the formation of the triarylsulfonium salt, leading to the ligand coupling product, is favoured [for example, in the case of sulfilimine (**112**)].[64]

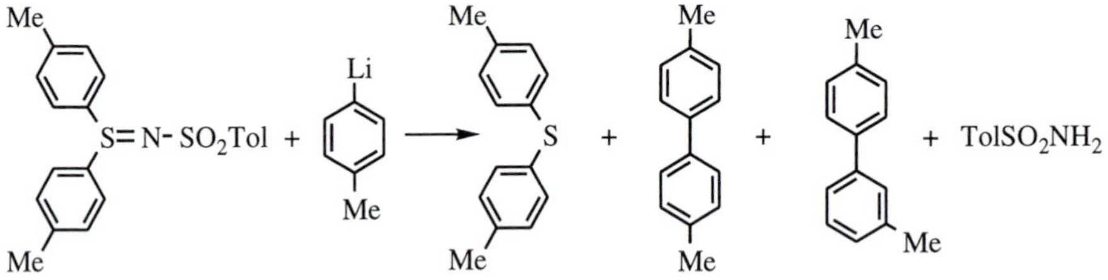

| (112) | (108) | (109), 72-82% | (110), 67% | (111), 2% | 58-65% |

The balance between the various pathways is also influenced by the nature of the aryl groups. The ligand coupling pathway becomes predominant when the aryl group is substituted with electron-withdrawing groups. Baker *et al.* showed that phenylsulfinyloxazoline derivatives react with arylmagnesium bromide to afford good to high yields of the biaryl product, accompanied with variable amounts of the ligand exchange product (**75**).[77,78,86] The phenylsulfoxide (**113**) failed to react with 2-methoxy-1-naphthylmagnesium bromide. The only recovered product was the ligand exchange product (**75**). Oae *et al.* have shown that 2-*tert*-butylsulfinylpyridine fails to undergo ligand exchange in the reactions with Grignard reagents, leading only to the ligand coupling product.[65] When this strategy was applied to the naphthyl case, a moderate yield of the derivative [(**115**), R[2] = 2-methoxy-1-naphthyl] was obtained in the reaction of 2-methoxy-1-naphthylmagnesium bromide with the *tert*-butylnaphthylsulfoxide derivative (**114**).[78,86]

Table 3.5: Ligand coupling reactions of naphthylsulfoxide derivatives (**113**) and (**114**)[77,78,86]

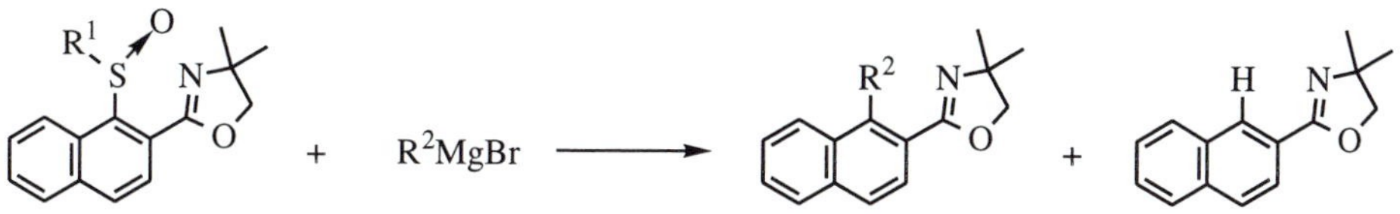

R[1]	R[2]	(**115**)	(**75**)
(**113**): Ph	Ph	72%	19%
(**113**): Ph	1-Naphthyl	67%	9%
(**113**): Ph	2-MeO-1-naphthyl	0%	only product
(**114**): *tert*-Bu	2-MeO-1-naphthyl	44%	

The ligand coupling reaction of homochiral sulfoxides [(**116**) - (**119**)] with 1-naphthylmagnesium bromide led to the atropoisomeric 1,1'-binaphthyl derivatives [(**120**) - (**123**)] in good yields (65-90%) and with good to high enantiomeric excess (60-95%).[78]

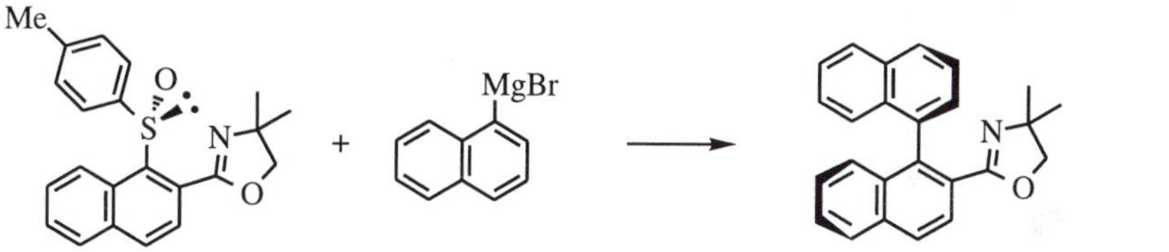

(**116**), (S)-sulfoxide

(**120**), (S)-binaphthyl 71% yield, 60% e.e.

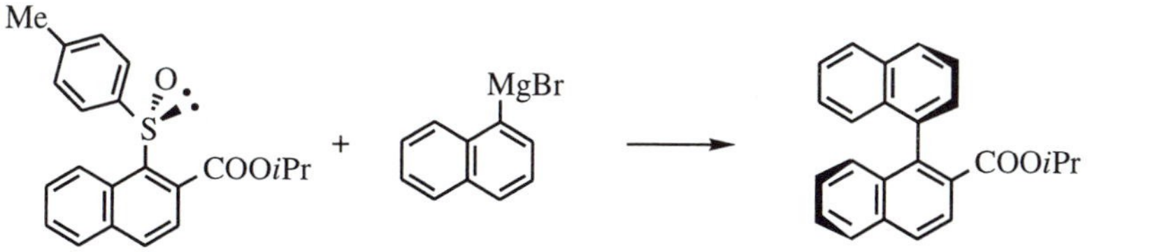

(**117**), (S)-sulfoxide

(**121**), (S)-binaphthyl 78% yield, 82% e.e.

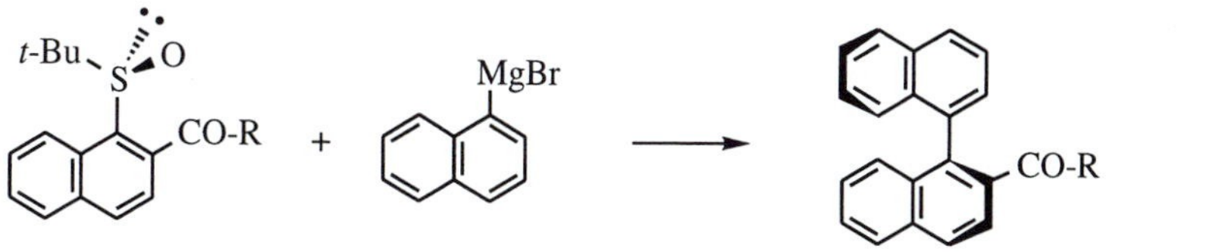

(**118**), (R)-sulfoxide, R = O-*i*Pr
(**119**), (R)-sulfoxide, R = NMe$_2$

(**122**), (R)-binaphthyl 90% yield, 95% e.e.
(**123**), (R)-binaphthyl 65% yield, 94.8% e.e.

3.2.4.2 Coupling reactions involving 1 or 2 heteroaryl groups

In a long series of papers, Oae and his group have studied the formation of biaryl compounds by ligand coupling occurring on a σ-sulfurane, formed by nucleophilic addition of an organometallic species on the sulfur atom of the sulfoxide substrate.[67] Depending on the nature of the two substituents of the sulfoxide substrate and of the alkyl or aryl group of the organometallic reagent, different pathways can play a role in the final outcome : 1 - direct ligand coupling, 2 - ligand coupling after pseudorotation, 3 - ligand exchange followed by nucleophile attack on a second molecule of sulfoxide and ligand coupling.

1 - Ligand coupling:

The incoming nucleophile (a 2-pyridyl group in Scheme 3.12) occupies directly the apical position, suitable for ligand coupling with the other group in equatorial position.

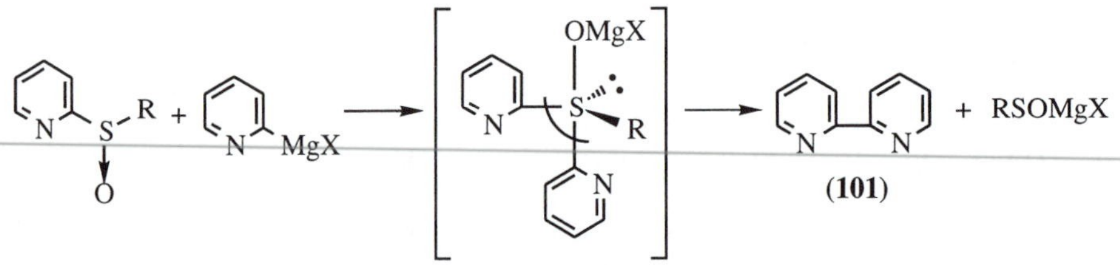

Scheme 3.12

2 - Ligand coupling after pseudorotation:

The incoming nucleophile R occupies, in the first instance, the apical position. Ligand coupling can take place only after pseudorotation (Ψ), bringing one equatorial 2-pyridyl group into the apical position, suitable for ligand coupling with the other group in equatorial position.

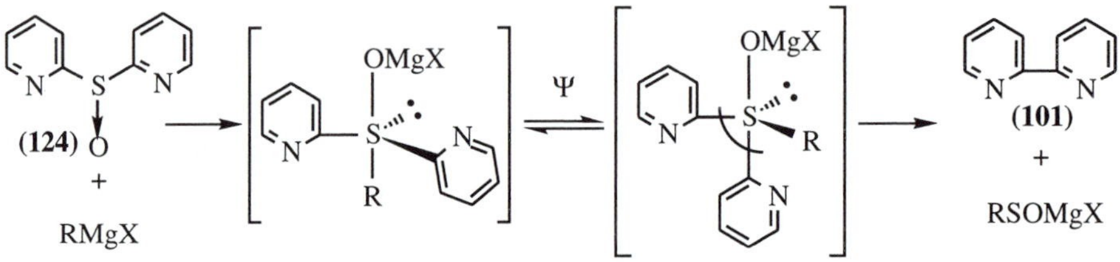

Scheme 3.13

3 - Ligand exchange followed by ligand coupling:

When ligand coupling cannot take place between two of the three substituents present on the first formed sulfurane, ligand exchange gives rise to the formation of a new organometallic species, such as 2-pyridyl Grignard derivative. (Scheme 3.14.a) This second species then reacts with the sulfoxide substrate affording a second sulfurane, which now contains two substituents prone to direct ligand coupling. (Scheme 3.14.b)

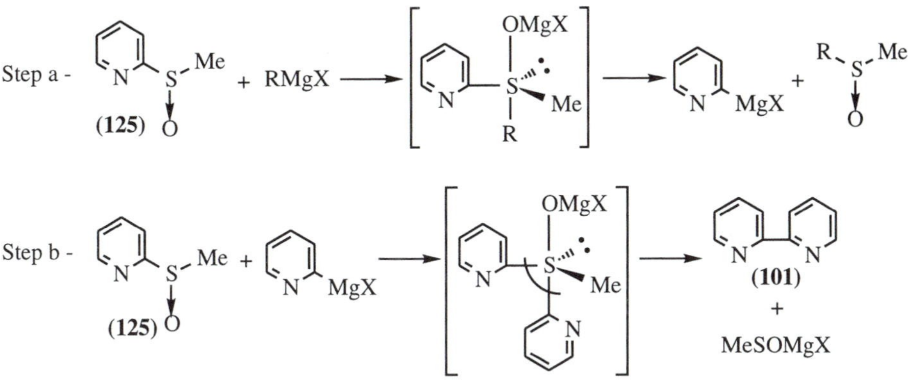

Scheme 3.14

These different mechanistic pathways are intervening in the various types of reactions of pyridyl sulfoxides [(124) - (127)] affording 2,2'-bipyridine (101). (Table 3.6) In the reaction of methyl 2-pyridylsulfoxide (125) with various Grignard reagents, the major or only product was 2,2'-bipyridine (101).[76,80,87] In the first step, ligand exchange takes place. This fact was demonstrated by trapping experiments with benzaldehyde, which led to α-(2-pyridyl)benzyl alcohol (15% yield). The ligand exchange product, 2-pyridylmagnesium halide, subsequently reacts with the methyl 2-pyridylsulfoxide (125) in excess to form a second sulfurane which then undergoes ligand coupling. Better yields are usually obtained when the reaction is performed in THF than in diethyl ether. When 2-pyridyllithium was used as the source of the incoming nucleophile, the reaction rate was very fast and the reaction occurred efficiently at - 18°C. In this case, the reaction proceeds by direct ligand coupling. A similar yield of the ligand coupling product was obtained upon treatment of bis(2-pyridyl)sulfoxide (123) with ethylmagnesium bromide. However, the reaction is then slower, as it requires pseudorotation before the ligand coupling step can take place.

Table 3.6: Preparation of 2,2'-bipyridine (101)[76,80,87]

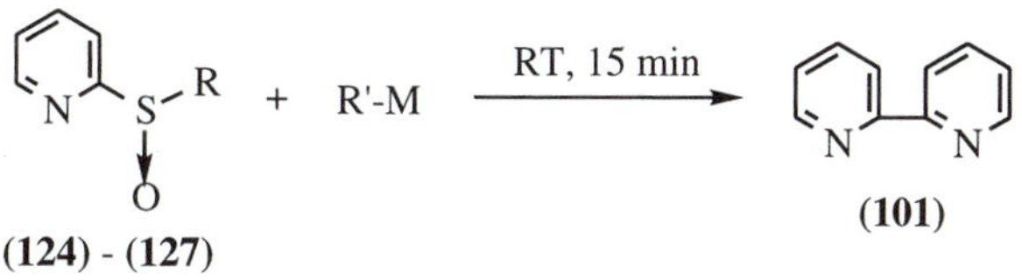

Substrate	R	R'-M	Solvent	(101) (%)	Ref
(125)	Me	MeMgBr	THF	73	76,87
(125)	Me	EtMgBr	THF	57	76,87
(125)	Me	EtMgBr	Et$_2$O	30	76,87
(125)	Me	PhMgBr	THF	79	76,87
(126)	Et	EtMgBr	THF	55	76,87
(127)	Ph	EtMgBr	THF	42	76,87
(127)	Ph	EtMgCl	THF	71	80
(125)	Me	2-PyLi	THF	59[a]	76,87
(124)	2-Py	EtMgBr	THF	63	76,87

a - reaction performed at - 18°C

When the pyridyl group is substituted on other positions of the ring, the reaction of organometallic reagents with alkyl 2-pyridylsulfoxides affords symmetrically substituted 2,2'-bipyridines.[76,87,88] (Table 3.7)

Table 3.7: Preparation of symmetrically di- or tetra-substituted 2,2'-bipyridine derivatives[76,87,88]

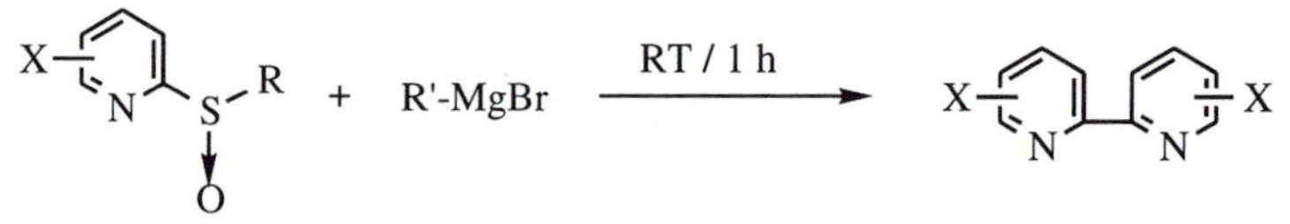

X	R	R'	Solvent	Yield (%)	Ref
5-Cl	Me	Et	Et$_2$O	40	76,87
6-Cl	Me	Et	Et$_2$O	55	76,87
3,5-Cl$_2$	Me	Et	Et$_2$O	52	76,87
6-Br	Me	Et	Et$_2$O	50	76,87
6-MeS	Me	Et	Et$_2$O	61	76,87
6-Cl	Et	Me	THF	66	88
6-Me	Et	Me	THF	53	88
6-Et	Et	Me	THF	60	88
6-THPOCH$_2$	Et	Me	THF	69	88

THP = tetrahydropyranyl

Table 3.8: Competition in the formation of 2-arylpyridine (**130**) and 2,2'-bipyridine (**101**)[76,89,90]

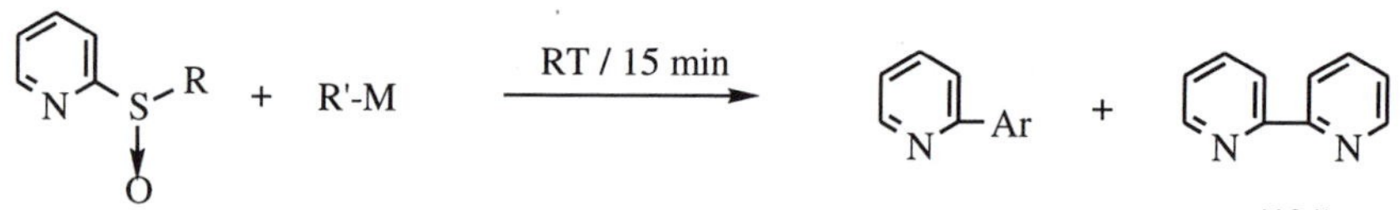

(**125**) - (**129**)   (**130**)   (**101**)

Substrate	R	R'M	Temp.	(**130**) (%)	(**101**) (%)	Ref
(**125**)	Me	PhMgBr	RT	9	68	76
(**126**)	Et	PhMgBr	RT	6	56	76
(**128**)	i-Pr	PhMgBr	RT	17	59	76
(**129**)	t-Bu	PhMgBr	RT	85	0	76
(**127**)	Ph	EtMgBr	RT	-	42	76
(**127**)	Ph	t-BuLi	- 95°C	30	26	89
(**127**)	Ph	PhLi	- 78°C	38	28	89
(**127**)	Ph	(**131**)	RT	87	-	90
(**127**)	Ph	(**132**)	RT	90	-	90

(**131**)   (**132**)

When a 2-pyridylsulfurane derived from the addition of an organometallic reagent on a sulfoxide substrate contains at least one phenyl group, two pathways can compete: a - direct ligand coupling giving (**130**) and b - the sequence ligand exchange-ligand coupling giving (**101**). (Table 3.8) When the phenyl group is introduced as the incoming nucleophile, the size of the second ligand of the sulfoxide substrate plays an important role on the relative ratios between (**101**) and (**130**). The ligand exchange route is preferred with non-hindered alkyl groups. But when a *tert*-butyl group is present on the sulfoxide, only the direct ligand coupling product [(**130**), Ar = Ph] is formed in high yield. However, when the *tert*-butyl group comes from the incoming nucleophile, then the two competing pathways take place: (**130**) and (**101**) are formed, although both in modest yields (30% and 26% respectively). In the case of a competition between two aryl groups, the sterically more hindered and more electron-rich 2-alkoxyphenyl and 2-phenoxyphenyl substituents were implied in the ligand coupling, giving the 2-(2'-substituted phenyl)pyridine in high yields.[76,89,90] (Table 3.8)

A range of 2-aryl 3-substituted pyridine atropoisomers were obtained in 75-90% yields (aryl = phenyl, 2-methoxyphenyl, 2-phenoxyphenyl and 1-naphthyl) by *ortho*-directed metallation with LDA (lithium diisopropylamide) - substitution (with RCOR' or Me₃SiCl), followed by reaction of the 3-substituted 2-pyridylsulfoxide with an organometallic reagent. The rotational barriers ranged from 15.6 to 19 kcal/mole. Poor to moderate yields were obtained when a (2-methoxynaphthyl)pyridine derivative was the end product (14-44%).[90,91]

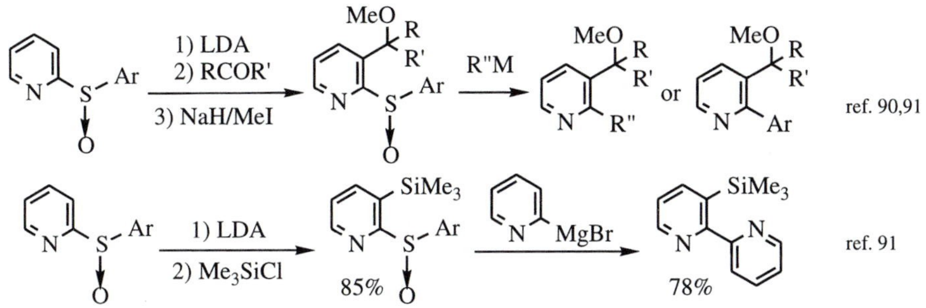

Unsymmetrically substituted bipyridines and oligopyridines can be easily obtained by proper choice of the pyridylsulfoxide and the pyridyllithium reagent.[88,92,93] Due to the high reaction rates, no homocoupling product, resulting from the sequence ligand exchange - ligand coupling, was observed. Generally good yields of the unsymmetrical bipyridyl derivatives were obtained. (Tables 3.9 - 3.11)

Table 3.9: Preparation of monosubstituted 2,2'-bipyridines[88]

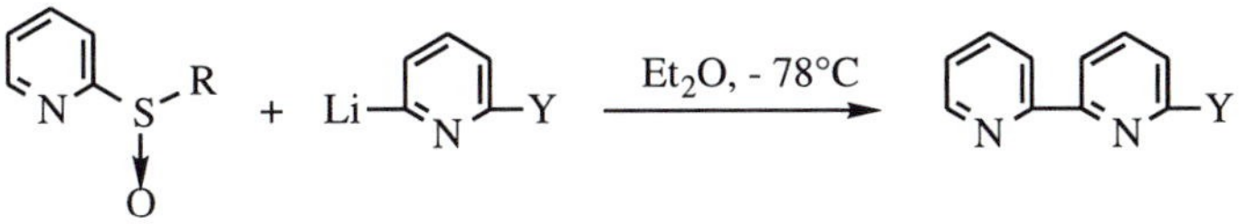

R	Y	Yield (%)
Me	Br	71
Me	Me	60
Me	CH₂OTBDMS	55

TBDMS = *tert*-butyldimethylsilyl

Table 3.10: Preparation of unsymmetrically disubstituted 2,2'-bipyridines[88,93]

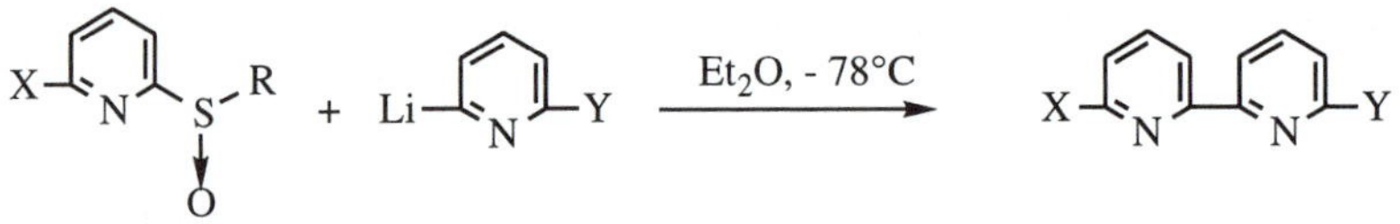

R	X	Y	Yield (%)	Ref
Et	Me	H, Br	57-60	88
Et	EtS	H	85	88
Et	THPOCH$_2$	Br	71	88
Et	TBDMSOCH$_2$	H	73	88
Et	MeOCH$_2$	CH$_2$OTHP	46	93

TBDMS = *tert*-butyldimethylsilyl; THP = tetrahydropyranyl

Table 3.11: Preparation of unsymmetrically substituted oligopyridines[92]

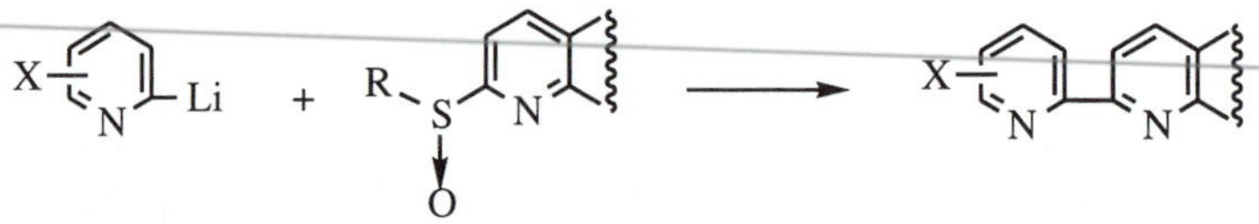

X	Heteroaryl	Yield (%)
6-Me	2-pyridyl	60
6-Me	2-quinolinyl	35
6-Me	6-(2,2'-bipyridyl)	54
5-Me	2-pyridyl	70
6-Br	2-pyridyl	80
6-Br	2-quinolinyl	80
6-Br	6-(2,2'-bipyridyl)	76

The sequence can be repeated to afford polypyridines, as in the case of the preparation of terpyridine (**133**) and tetrakispyridine (**134**) from 2,6-bis(ethylthio)pyridine.[88]

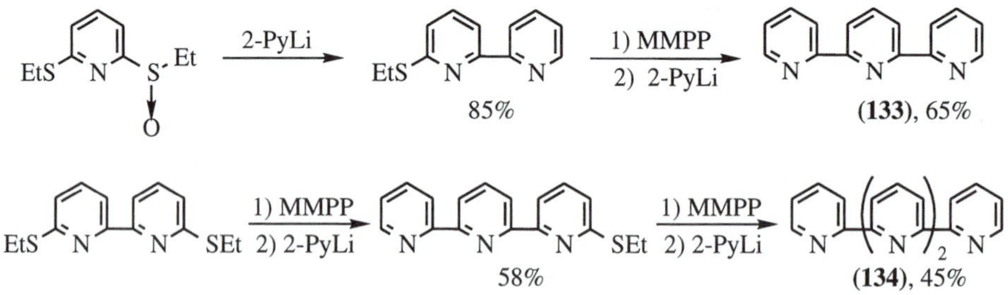

with 2-PyLi = 2-pyridyllithium and MMPP = magnesium monoperoxyphthalate

The lithium salts of the optically active (S)-pyridyl and (S)-bipyridylethanol derivatives [(**135**), n = 1 or 2] were treated with ethyl bipyridylsulfoxide (**136**) to afford good yields of the corresponding (S)-oligopyridyl ethanol compounds (**137**) and (**138**), the ligand coupling products.[94]

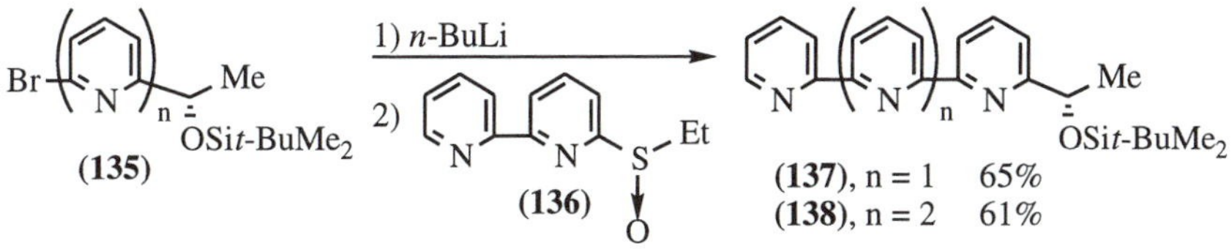

The isomeric 3-pyridyl and 4-pyridylsulfoxides behave differently. Due to the greater leaving abilities of 3-pyridylmagnesium halide compared to 2-pyridylmagnesium halide, the ligand exchange process is favoured over ligand coupling.[95] Indeed, reaction of 3-pyridyl and 4-pyridyl phenyl-sulfoxides (**139**) and (**141**) with phenylmagnesium bromide, followed by addition of an electrophile such as an aldehyde or a ketone, led to the addition products, (**140**) and (**142**) respectively, in poor to high yields.[96,97]

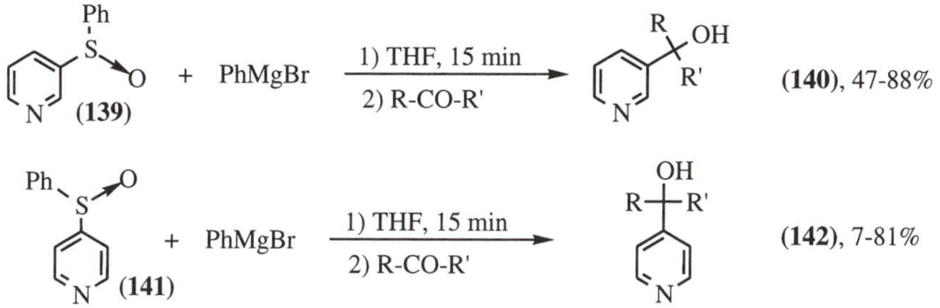

In spite of this easy ligand exchange process, ligand coupling took place in the reactions of 2-pyridyl, 3-pyridyl or 4-pyridylmagnesium bromides with 2-pyridyl, 3-pyridyl or 4-pyridylsulfoxide derivatives.[91,97] Modest to good yields were obtained in most cases. No reaction was observed in the combination of reagents involving the 3-pyridyl ligand on both partners of the reaction: 3-pyridyl magnesium bromide and 3-pyridylsulfoxide. The reaction of 3-pyridylsulfoxide (**139**) with 2- and 4-pyridyl Grignard led to mixtures of direct ligand coupling product and product of ligand coupling following an initial ligand exchange. In the case of the reaction with 2-pyridylmagnesium bromide, the ligand coupling product (**143**) and the 2,2'-bipyridine (**101**) were isolated in modest yields.[97]

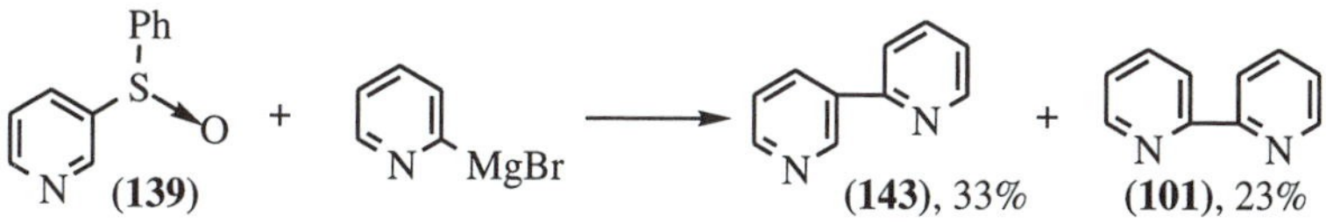

On the other hand, reaction of 3-pyridyl Grignard reagents with 2-pyridylsulfoxide (**127**) led directly to the ligand coupling product (**143**) in good yield.[97]

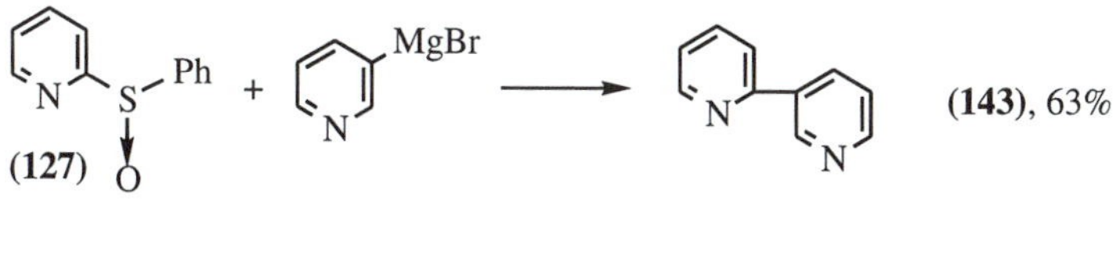

Reaction of 3-pyridyl Grignard reagents with 4-pyridylsulfoxide (141) led to mixtures of products resulting from direct ligand coupling (144) and from ligand exchange-ligand coupling sequence (145).[97]

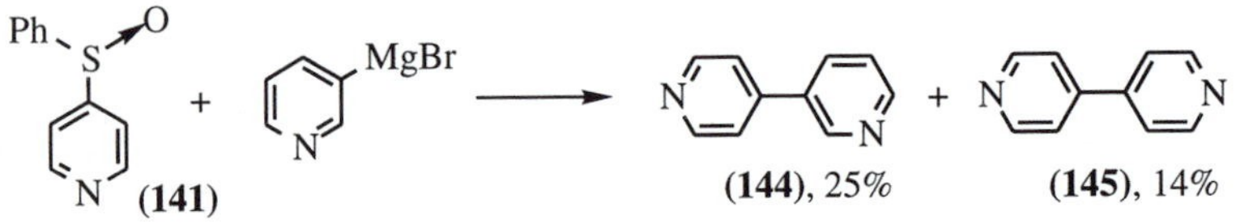

(141)   (144), 25%   (145), 14%

Better yields of the ligand coupling product were obtained when the reaction was performed on the more sterically hindered 4-(3-substituted pyridyl)sulfoxide derivatives (146) or (147).[91]

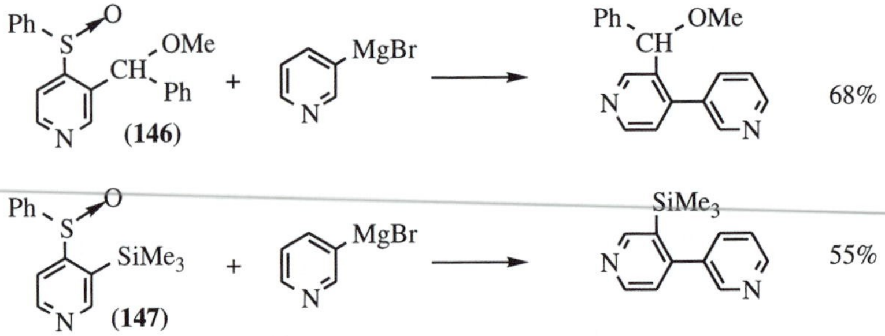

(146)   68%

(147)   55%

Other heteroaromatic sulfoxide systems can be involved in ligand coupling reactions. The more electron-withdrawing benzothiazolylsulfoxide (70) leads to benzothiazole, a ligand exchange product, and bibenzothiazole (71), resulting from the sequence ligand exchange-ligand coupling. It must be noted that this reaction was observed even when a ligand prone to facile ligand coupling, such as benzyl, is present on the intermediate sulfurane.[76]

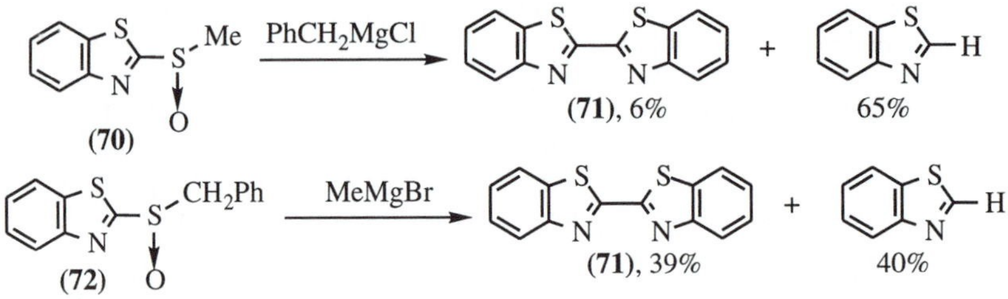

(70)   (71), 6%   65%

(72)   (71), 39%   40%

An efficient group for ligand coupling is the 2-pyrimidyl group. However, this ligand presents an odd behaviour. The 2-pyrimidyl group tends to couple preferentially with the phenyl group rather than with the benzyl group. Even when the benzyl group is introduced as the incoming nucleophile therefore occupying the apical position, pseudorotation takes place very significantly, and 2-phenyl-pyrimidine (67) is the major product.[72,75]

(65)   (66), 21%   (67), 40%

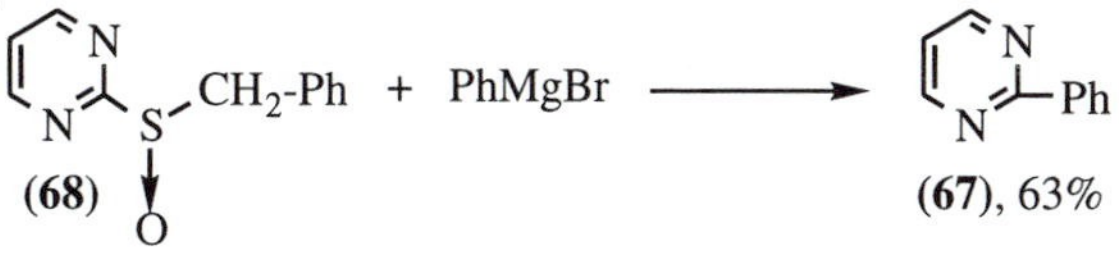

(68) + PhMgBr ⟶ (67), 63%

The 2-quinolinyl group behaves similarly to the 2-pyridyl group in ligand coupling reactions. In the case of a competition between phenyl and benzyl groups, only the 2-benzylquinoline (57) was obtained.[75]

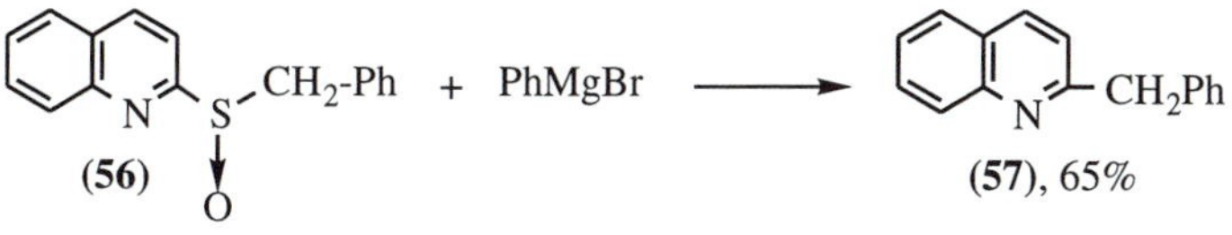

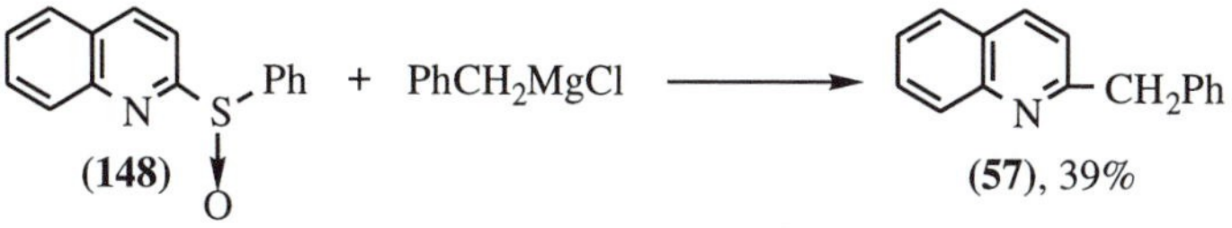

(56) + PhMgBr ⟶ (57), 65%

(148) + PhCH$_2$MgCl ⟶ (57), 39%

2-Phenylation of quinoline to give (150) was easily performed by reaction of ethyl 2-quinolinyl sulfoxide (149) with phenylmagnesium bromide. The ligand exchange-ligand coupling product, 2,2'-biquinoline (151), was nevertheless obtained as a minor by-product.[75]

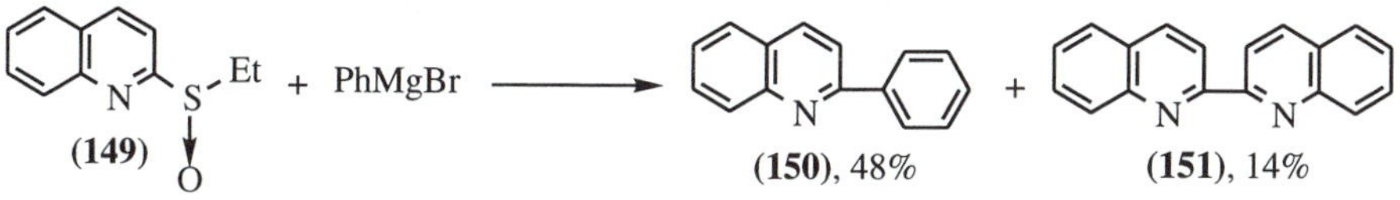

(149) + PhMgBr ⟶ (150), 48% + (151), 14%

The quinoline dimer (151) became the sole product when the alkyl 2-quinolinylsulfoxide was treated with an alkyl Grignard reagent. The reaction showed a remarkably high reaction rate. In the case of sulfoxide (86), it reached completion in less than 20 seconds at - 78°C, affording a near quantitative yield of the product (151).[81]

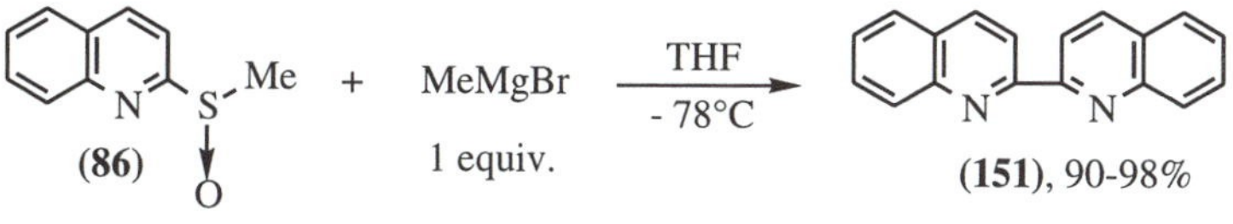

(86) + MeMgBr $\xrightarrow[- 78°C]{THF}$ (151), 90-98%
         1 equiv.

When an excess of alkyl Grignard reagent was used, the alkyl group added on the 2-position to afford good to high yields of the adduct (152). This adduct is thermally unstable and the alkyl group is lost at room temperature in a couple of days to regenerate (151).[81]

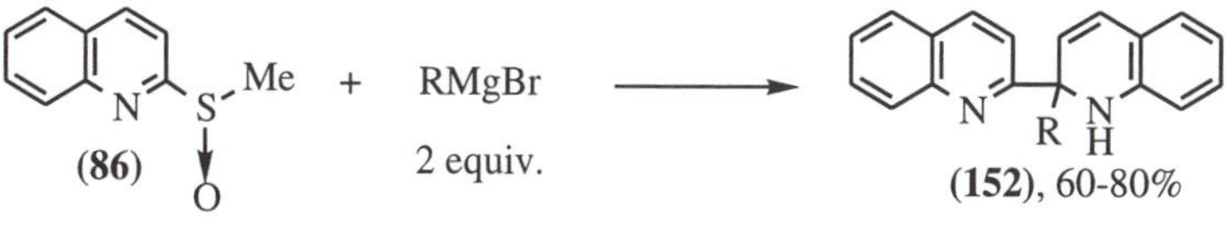

(86) + RMgBr ⟶ (152), 60-80%
         2 equiv.

The mixed heteroaromatic sulfoxide, 2-pyridyl 2-quinolinylsulfoxide (153), afforded only the direct ligand coupling product (154) upon reaction with an alkyl Grignard reagent.[81]

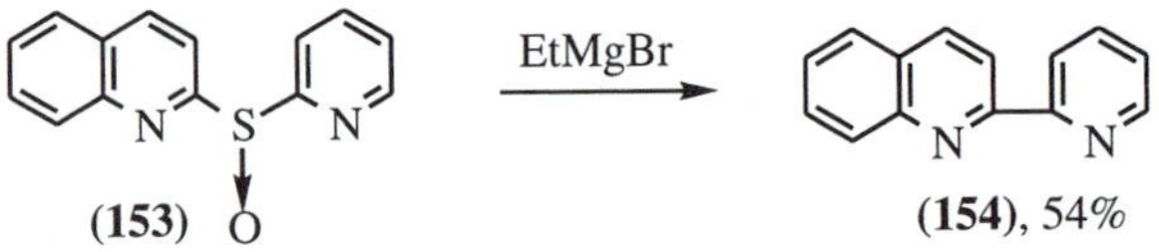

(153)   (154), 54%

In spite of the considerable difference in the rates of the reactions of organometallic reagents with 2-pyridyl and 2-quinolinylsulfoxides, a cross-coupling reaction was successfully performed with different combination of mixtures of sulfoxides [(155) + (156)].[98] This reaction is not selective and affords mixtures of cross-coupling products (157) and homocoupling products (158) and (159).

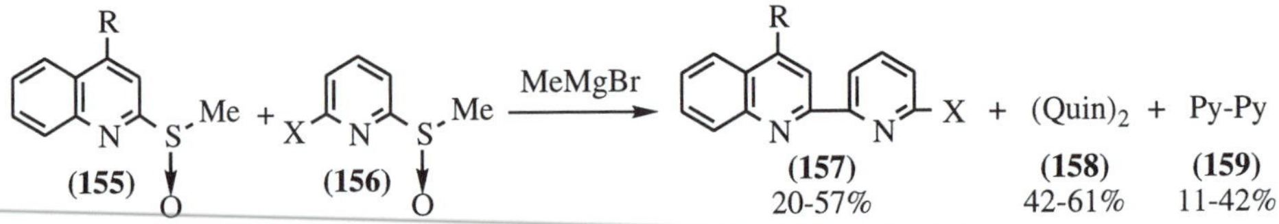

(155)   (156)   (157)   (158)   (159)
                                20-57%   42-61%   11-42%

with R = H, Me and X = H or Cl

Treatment of 2-pyridyl 2-thienylsulfoxide (160) with 2-thienyllithium (161) led to 2-(2-pyridyl)-thiophene (162) and a small amount of 2,5-bis(2-pyridyl)thiophene (163). The formation of the 2,5-bis(2-pyridyl)thiophene derivative (163) was explained by a transmetallation reaction between (162) and (161), leading to 5-pyridyl 2-thiophenyllithium, which subsequently reacts with the sulfoxide substrate (160).[99]

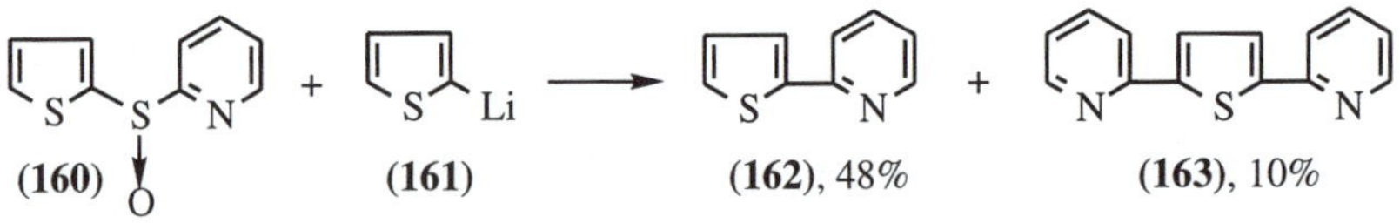

(160)   (161)   (162), 48%   (163), 10%

The reactions of the oxygen analogues [(164) with (165)] and of the selenium analogues [(167) with (168)] led only to the normal ligand coupling products [(166) and (169) respectively].[100,101]

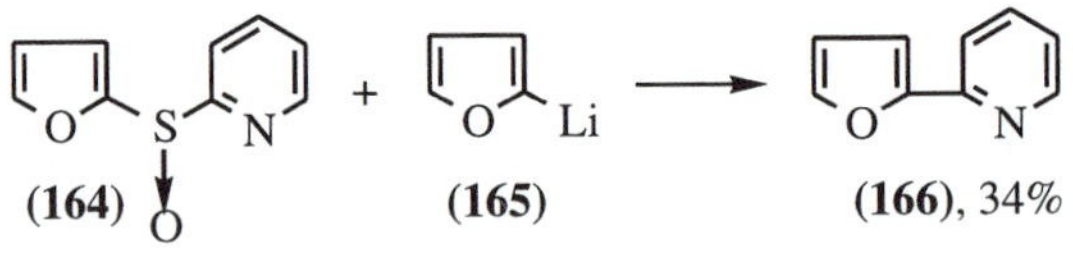

(164)   (165)   (166), 34%

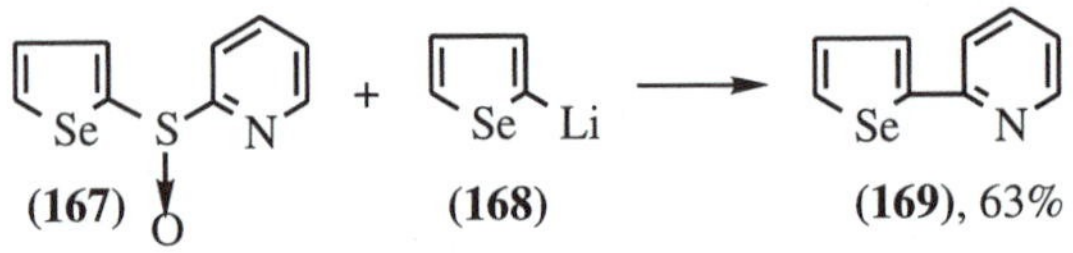

(167)   (168)   (169), 63%

In a study of the relative ligand coupling efficiencies between oxygen, sulfur and nitrogen containing heteroaromatic systems, the correlation between the ligand coupling abilities and the ^{13}C-NMR values of the *ipso* carbon atoms (δ in ppm are reported in Scheme 3.15) was not clearly demonstrated. The occurrence of the ligand coupling appeared to strongly depend upon the conformation of the initially formed sulfurane intermediate.[101]

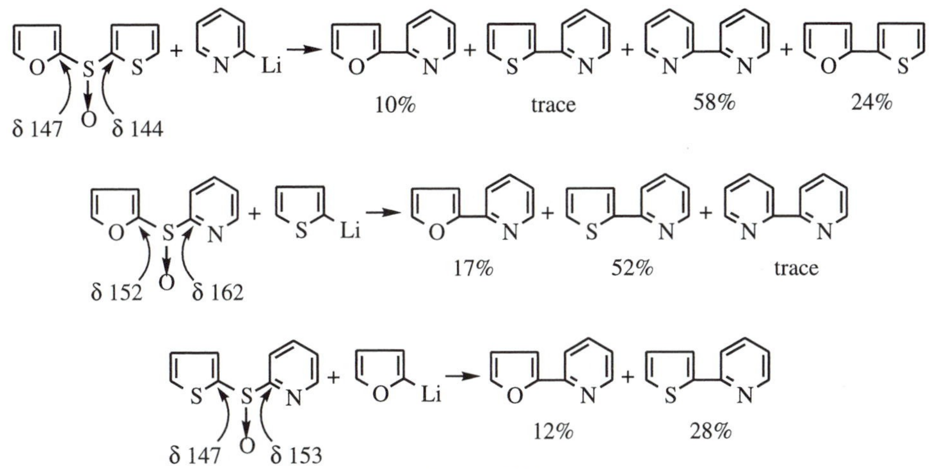

δ 147   δ 144   10%   trace   58%   24%

δ 152   δ 162   17%   52%   trace

δ 147   δ 153   12%   28%

Scheme 3.15

The reaction of di(2-furyl)sulfoxide with 2-furyllithium did not give any ligand coupling product over a wide range of temperatures.[101]

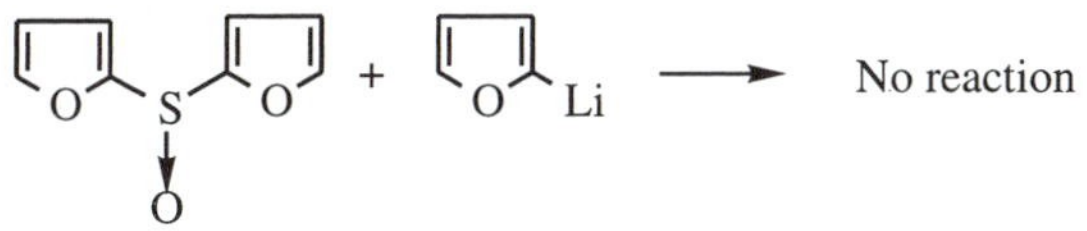

→   No reaction

By contrast, a moderate yield of 2,2'-bithiophene **(171)** was obtained in the case of the sulfur analogues [**(170)** with **(161)**].[99]

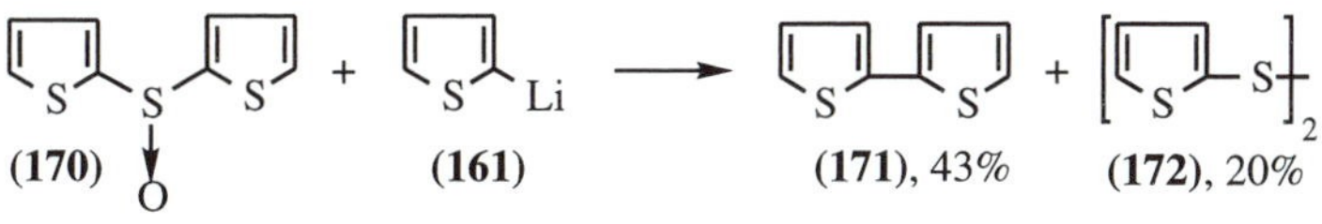

(170)   (161)   (171), 43%   (172), 20%

3.2.5 Aryl-alkyl coupling

When the intermediate sulfurane resulting from the addition of an organometallic reagent on an alkyl pyridylsulfoxide contains two alkyl groups, ligand exchange usually takes place. However, when a pyridylsulfoxide, containing a *tert*-butyl group originally bound to the sulfinyl sulfur atom **(129)**, is treated with ethylmagnesium bromide, no ligand exchange reaction happened. In fact, ligand coupling reaction was now observed, affording the 2-alkylpyridine derivatives **(173)** and **(174)** in moderate yields. The fact that *sec*-alkyl containing pyridylsulfoxides behaved similarly to afford the corresponding 2-alkylpyridine derivatives was also claimed. However, no example was reported.[72,80]

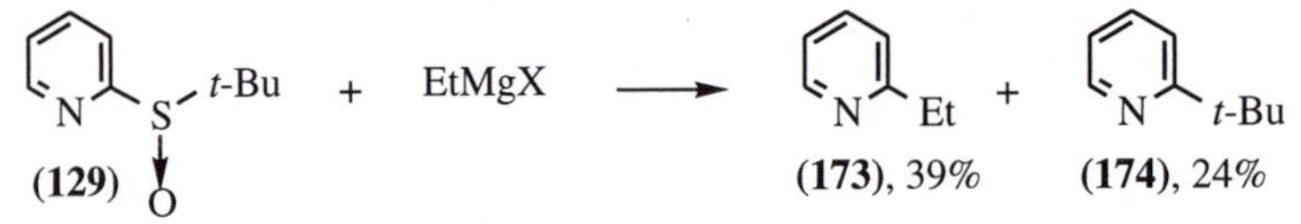

(129)   +   EtMgX   ⟶   (173), 39%   (174), 24%

A similar ligand coupling reaction involving an alkyl group was observed in the reaction of some 1-(2-oxazolinonaphthyl)sulfoxide derivatives with ethylmagnesium bromide.[78] When treated with ethylmagnesium bromide, the phenyl naphthylsulfoxide compound (175) afforded a modest yield of the ligand coupling product (77), the ligand exchange product (75) being the major product. More unexpectedly, the ethyl derivative (77) became largely predominant in the reaction involving the benzylsulfoxide (73) as well as the analogous 4-chlorobenzyl derivative (74).

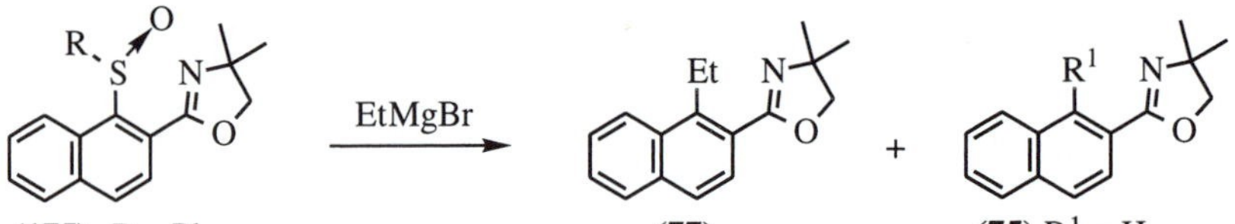

(175) R = Ph   (77)   (75) R^1 = H
(73) R = $PhCH_2$ (78) R^1 = $PhCH_2$
(74) R = $ClC_6H_4CH_2$ (79) R^1 = $ClC_6H_4CH_2$

(175)   ⟶   (77) (19%) + (75) (36%)
(73)   ⟶   (77) (69%) + (75) (10%) + (78) (13%)
(74)   ⟶   (77) (62%) + (75) (4%) + (79) (26%)

3.2.6 Vinyl-nucleophile coupling

Nucleophilic attack of methyllithium on allene sulfoxides offers a stereospecific route to the desulfurised allenes, the protonated ligand exchange product. Almost quantitative yields of (177) and (179) were obtained from (176) and (178) respectively.[102]

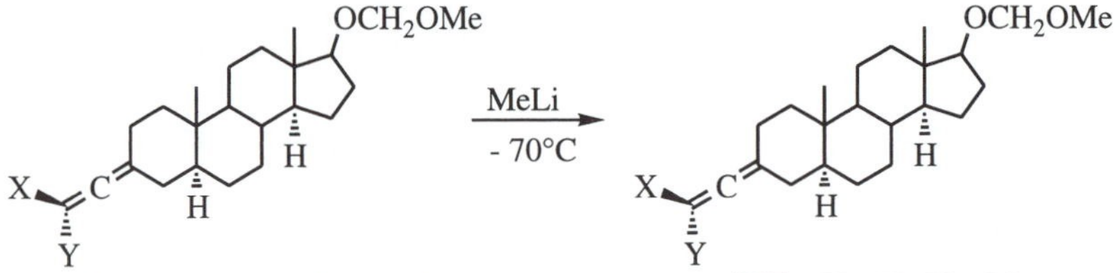

(176), X = S(O)-Ph; Y = Me   ⟶   (177), X = H; Y = Me
(178), X = Me; Y = S(O)-Ph   ⟶   (179), X = Me; Y = H

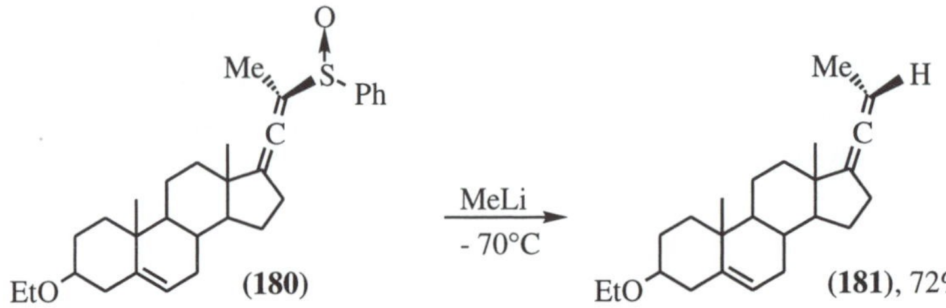

(180)   MeLi, -70°C   (181), 72%

When compound (180) was thermally converted to the isomeric diene (182) and then treated with methyllithium, a completely different outcome was observed. The major product was the phenyl diene (183), a ligand coupling product, isolated in 68% yield.

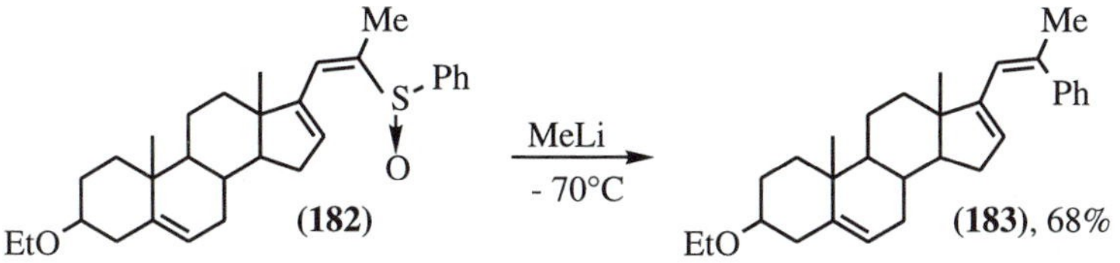

A related stereospecific desulfurisation of vinylsulfoxides was observed in the treatment of the vinylsulfoxides (184) and (185) by *tert*-butyllithium, in the presence of methanol. When the reduction was performed in the presence of deuteromethanol, the sulfoxide moiety was replaced by a deuterium atom.[103,104]

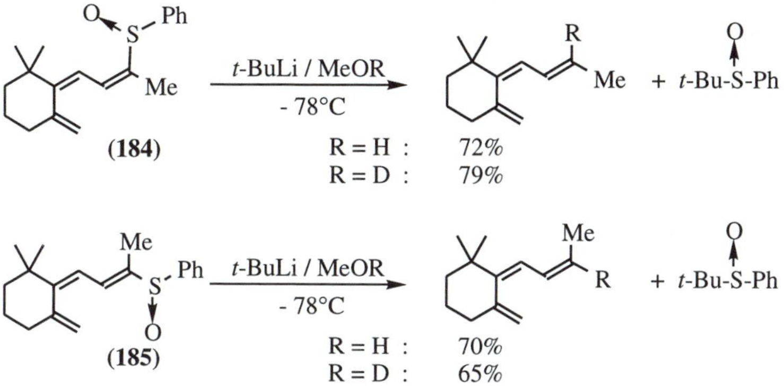

When applied to the allenylsulfoxide (186), this reaction led to the product of stereospecific reduction (187) with very high retention of configuration [(187) / (188) = 97 : 3]. A ligand exchange pathway resulting in the formation of allenyllithium was excluded by the fact that protonation of the lithium salt of the allene of (187) [R = Li, R' = Me] led only to the inversion product (188).[105]

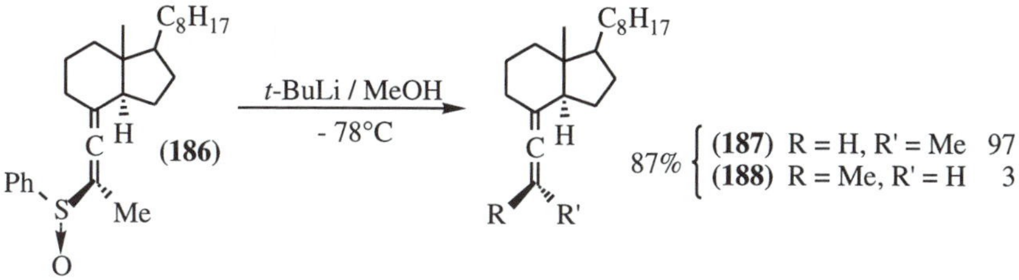

A ligand coupling pathway was suggested as a reasonable mechanism, although it was not possible to distinguish between path "a" (the ligand coupling way via α-elimination), and path "b" (following a β-elimination).[103,104] (Scheme 3.16)

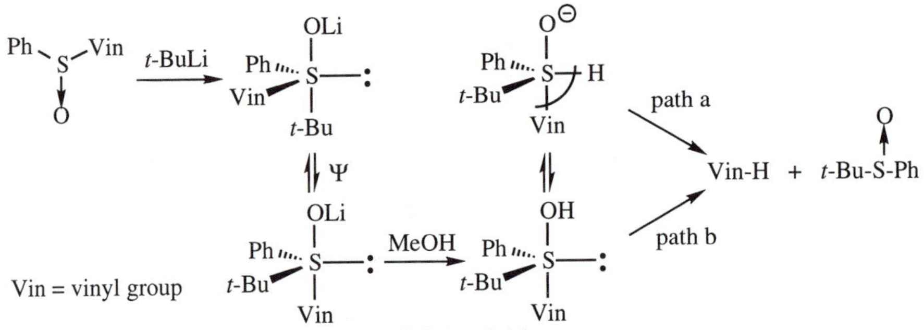

Scheme 3.16

Reactions of (E)- or (Z)-3-arylsulfinylpropenoates with Grignard reagents afforded substituted acrylates in good yields (up to 92%).[106] The usually very high stereospecificity is consistant with a ligand coupling process. (Scheme 3.17, path a) The complete absence of products resulting from pseudorotation could be explained by a very high rate for the ligand coupling compared to the rate of pseudorotation. However, the addition-elimination pathway was considered as more likely in this case of activated nucleophilic vinylic substitution. (Scheme 3.17, path b)[107]

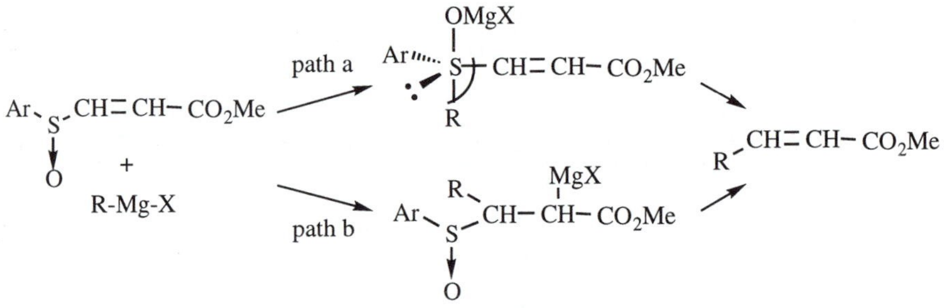

R = alkyl, aryl; Ar = Ph or 4-Tolyl

Scheme 3.17

3.2.7 Reactions with thionyl chloride

Symmetrical ligand coupling products can be obtained directly by reaction of thionyl chloride with organometallic reagents. Only organometallic derivatives of aryl and alcynyl groups led to coupling products. Alkyllithium or alkyl Grignard reagents as well as benzyl anions failed to give clean ligand coupling systems.[108] This reaction was first observed by Uchida *et al.* in 1972 who treated alkynyl and alkenyl Grignard reagents with thionyl chloride and obtained high yields of ligand coupling products.[109]

$$Ph-C\equiv C-MgBr \ + \ SOCl_2 \ \longrightarrow \ Ph-C\equiv C-C\equiv C-Ph \qquad 55-61\%$$

Better yields of coupling products were obtained when more electron-rich groups were present on the *para* position of the aryl group of the phenylethynyl reagent.[109]

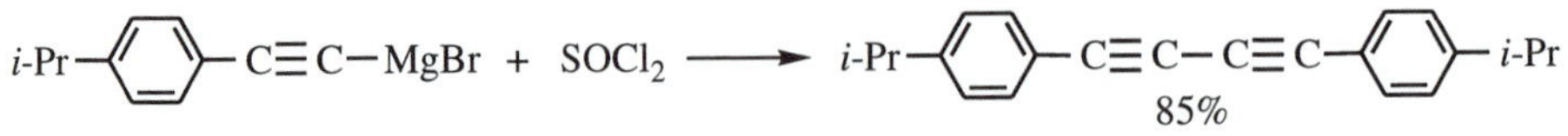

Trans-styryl magnesium bromide also afforded the coupling product in good yields.[109]

89%

However, when the organometallic reagent was prepared by reaction of *n*-butyllithium on β-bromo styrene, a more complex reaction system resulted. Only a very poor yield of 1,4-diphenylbutadiyne (**189**) was obtained, together with various amounts of phenylacetylene.[108,110]

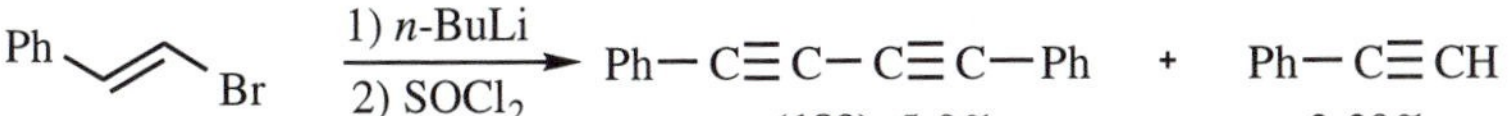

(**189**), 5-9% 3-39%

The formation of (**189**) by ligand coupling reaction became quantitative when lithium phenylacetylide was prepared and used as the organometallic reagent.

$$\text{Ph—C}\equiv\text{CH} \xrightarrow[\text{2) SOCl}_2]{\text{1) } n\text{-BuLi}} \text{Ph—C}\equiv\text{C—C}\equiv\text{C—Ph} \quad \textbf{(189)}\text{, 96\%}$$

A range of symmetrical biheteroaryl products have been directly prepared by reaction of the corresponding heteroaryllithium reagents with thionyl chloride. Depending on the nature of the heteroaryl group, the reaction can give the ligand coupling product (**190**), the symmetrical sulfoxide (**191**) or the symmetrical sulfide (**192**).[99,108,110,111] The ligand coupling products (**190**) were the only isolated products in the reactions involving either various pyridyllithium derivatives (47-77%),[111] or 2-quinolinyllithium (43%)[111] or 2-thienyllithium (48%).[99,108,110] With other heterocyclic lithium salts, more complex mixtures resulted from the reaction. As expected, 2-furyllithium afforded only the sulfoxide, which is not susceptible to ligand coupling.[101] (Table 3.12)

Table 3.12: Reactions of some heteroaryllithium reagents with thionyl chloride

$$\text{Ar-Li} + \text{SOCl}_2 \longrightarrow \underset{\textbf{(190)}}{\text{Ar-Ar}} + \underset{\textbf{(191)}}{\text{Ar-}\overset{\text{O}\uparrow}{\text{S}}\text{-Ar}} + \underset{\textbf{(192)}}{\text{Ar-S-Ar}}$$

Ar	(**190**) (%)	(**191**) (%)	(**192**) (%)	Ref
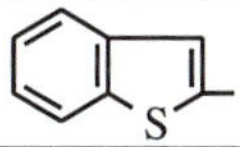	-	34	-	108,110
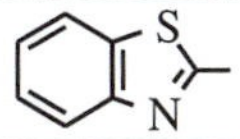	40 29	- 19	- 11	111 108,110
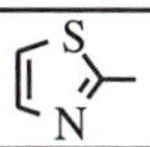	33	17	15	108,110
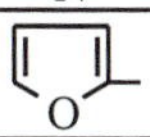	-	45	-	108,110
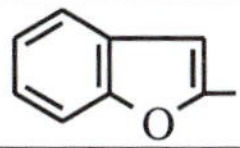	48	17	27	108,110

The reaction of *N,N'*-thionyldiimidazole with aryl Grignard reagents is a standard procedure for the synthesis of symmetrical sulfoxides.[112] When applied to thienyllithium, a high yield (84%) of the ligand coupling product (193) was serendipitously obtained instead of the expected di(2-thienyl) sulfoxide.[108,110]

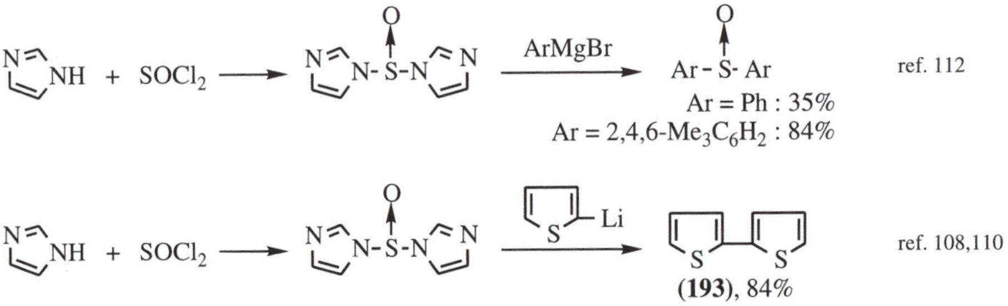

$$\text{imidazole-NH} + SOCl_2 \longrightarrow \text{N-S(=O)-N(imidazoles)} \xrightarrow{ArMgBr} Ar\text{-}\overset{O}{\underset{\uparrow}{S}}\text{-}Ar \qquad \text{ref. 112}$$

Ar = Ph : 35%
Ar = 2,4,6-Me$_3$C$_6$H$_2$: 84%

$$\text{imidazole-NH} + SOCl_2 \longrightarrow \text{N-S(=O)-N(imidazoles)} \xrightarrow{\text{2-thienyl-Li}} \text{(2,2'-bithienyl)} \qquad \text{ref. 108,110}$$

(193), 84%

The mechanism of the reactions of organometallic reagents with thionyl chloride involves a series of pseudorotations (Ψ), ligand exchange and ligand coupling reactions.[108] (Scheme 3.18)

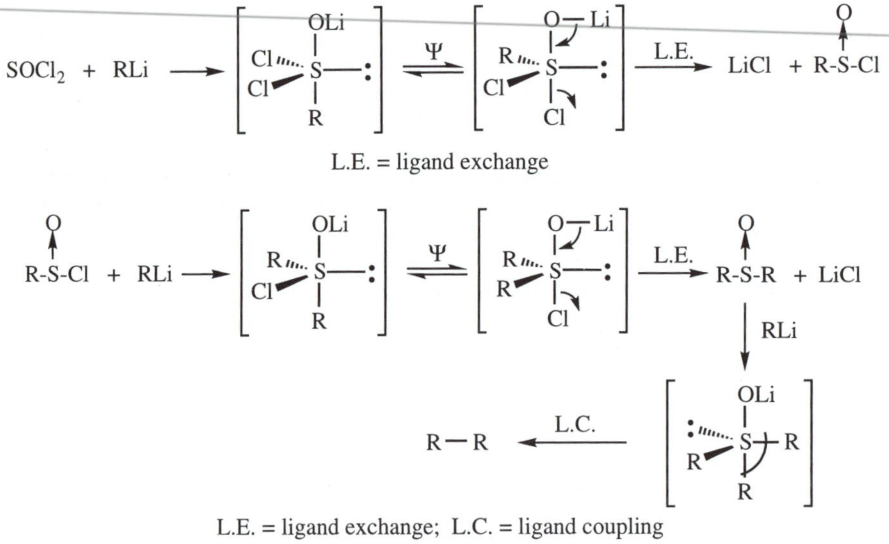

$$SOCl_2 + RLi \longrightarrow \left[\begin{array}{c} OLi \\ Cl_{\cdots}\!\!-\!S\!-\!: \\ Cl\!\!\nearrow \quad | \\ R \end{array} \right] \rightleftharpoons \left[\begin{array}{c} O\!-\!Li \\ R_{\cdots}\!\!-\!S\!-\!: \\ Cl\!\!\nearrow \quad | \\ Cl \end{array} \right] \xrightarrow{L.E.} LiCl + R\text{-}\overset{O}{\underset{\uparrow}{S}}\text{-}Cl$$

L.E. = ligand exchange

$$R\text{-}\overset{O}{\underset{\uparrow}{S}}\text{-}Cl + RLi \longrightarrow \left[\begin{array}{c} OLi \\ R_{\cdots}\!\!-\!S\!-\!: \\ Cl\!\!\nearrow \quad | \\ R \end{array} \right] \rightleftharpoons \left[\begin{array}{c} O\!-\!Li \\ R_{\cdots}\!\!-\!S\!-\!: \\ R\!\!\nearrow \quad | \\ Cl \end{array} \right] \xrightarrow{L.E.} R\text{-}\overset{O}{\underset{\uparrow}{S}}\text{-}R + LiCl$$

$$\downarrow RLi$$

$$R\!-\!R \xleftarrow{L.C.} \left[\begin{array}{c} OLi \\ :_{\cdots}\!\!-\!S\!\!+\!R \\ R\!\!\nearrow \quad | \\ R \end{array} \right]$$

L.E. = ligand exchange; L.C. = ligand coupling

Scheme 3.18

3.2.8 Hydrolysis of methoxymethylsulfoxides

The acid-catalysed cleavage of methoxymethylphenylsulfoxide (194) gave *S*-phenyl benzenethio-sulfinate (195).[113,114]

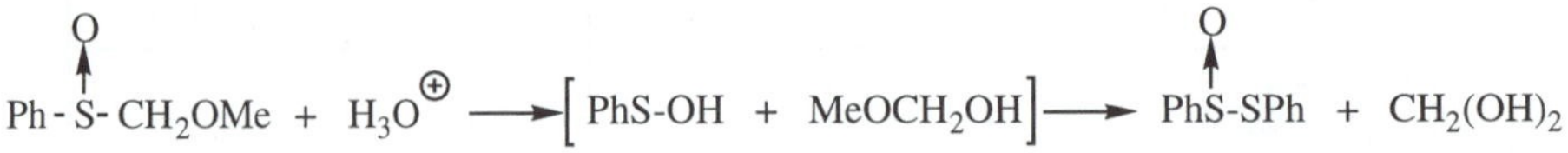

$$Ph\text{-}\overset{O}{\underset{\uparrow}{S}}\text{-}CH_2OMe + H_3O^{\oplus} \longrightarrow \left[PhS\text{-}OH + MeOCH_2OH \right] \longrightarrow Ph\overset{O}{\underset{\uparrow}{S}}\text{-}SPh + CH_2(OH)_2$$

(194)   (195)

The hydrolysis was accelerated by the presence of halide ions, and racemization of the substrate was not observed. Using [^{18}O]-sulfoxide as labelled substrate, 90% of the ^{18}O was retained in the *S*-phenyl benzenethiosulfinate, when the reaction was catalysed by hydrochloric acid or hydrobromic acid. When perchloric acid was used as a catalyst, only 50-60% of the ^{18}O was retained in the sulfinate. These results implied the possibility of different pathways depending on the nature of the catalyst. They were best accommodated by a general mechanism involving a ligand coupling reaction on a hypervalent sulfurane intermediate.[114] (Scheme 3.19)

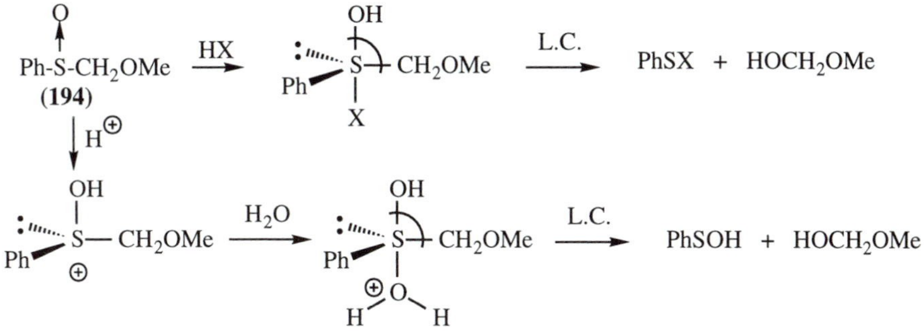

Scheme 3.19

3.3 SULFILIMINES

The properties and uses of sulfilimines, the aza analogues of sulfoxides, have not been as extensively studied as the chemistry of sulfoxides. However, the ligand coupling mechanism has also been shown to take place in some aspects of the chemistry of sulfilimines. Diaryl *N-p*-toluenesulfimide reacted with arylmagnesium halides to form diarylsulfide and biaryl in good yields.[115,116] Tracer experiments with [1-^{14}C]-diphenylsulfimide revealed that the reaction proceeds by the initial formation of a triphenylsulfonium salt, which subsequently reacts with phenylmagnesium bromide to afford a tetraarylsulfurane intermediate, undergoing ligand coupling. In one case, the intermediate triarylsulfonium salt was even isolated.[115]

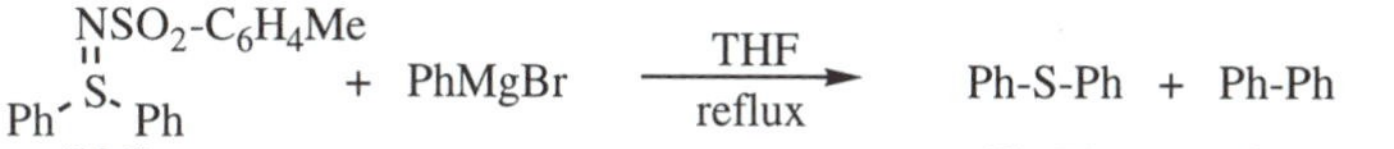

The toluenesulfonamide anion being expected to have a greater leaving ability than the oxyanion in the analogous sulfoxide reaction (see section 3.2.4.1), the rate of formation of the tetraarylsulfurane should be increased in comparison with the rate of decomposition by the arynic route. Accordingly, the yield of *m,p*-bitolyl (**111**) appeared considerably lower (1-2%) in the reaction of the *N*-tosylsulfilimine (**112**) compared to the yield (26%) obtained in the reaction with sulfoxide (**107**).[64]

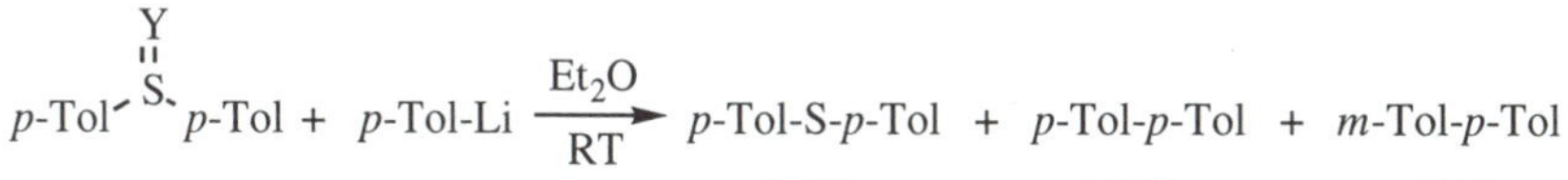

	(109)	(110)	(111)
(**107**) Y = O	66%	31%	26%
(**112**) Y = N-SO$_2$Tol	72-82%	66-67%	1-2%

The reduction of sulfilimines can be achieved by treatment of the *N*-sulfonylsulfimides with various reducing agents. These reactions occur through the formation of an intermediate sulfurane, which usually undergoes ligand coupling. This is the case of the reactions with lithium aluminum hydride (Scheme 3.20)[117] and with diphenyldisulfide (Scheme 3.21).[118] In the acid-catalysed reduction system using iodide ion as reducing agent, the tetracoordinate sulfurane is considered to partition by a variety of mechanisms to give an intermediate iodosulfonium ion, which rapidly undergoes a second reduction step to give the final products: the amine, the sulfide and iodine.[119,120]

. with lithium aluminum hydride:[117]

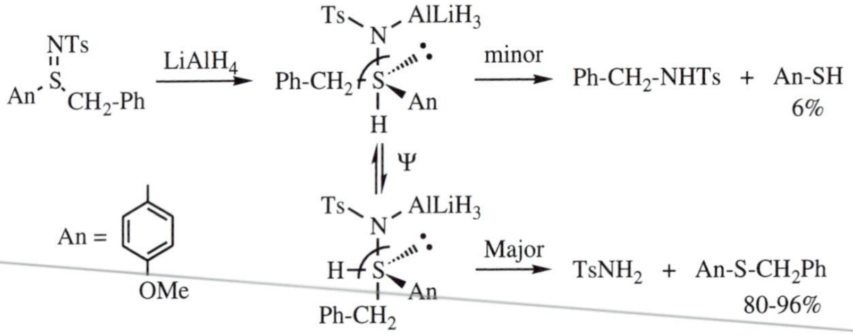

Scheme 3.20

. with diphenyldisulfide:[118]

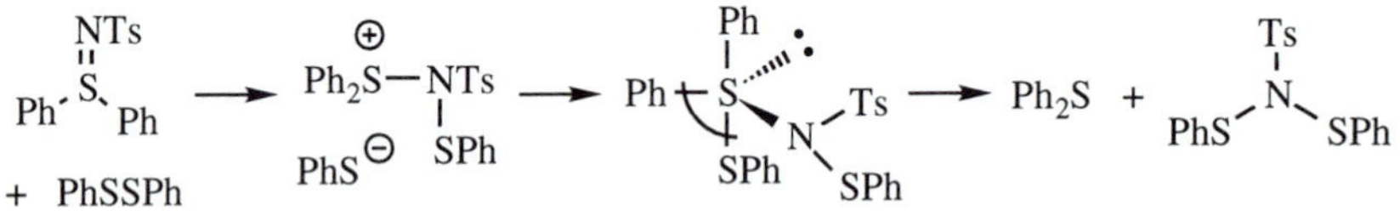

Scheme 3.21

The reaction with arenethiolate, once considered to be a ligand coupling,[121] was later shown to involve various radical species, as the reaction requires light to proceed.[122]

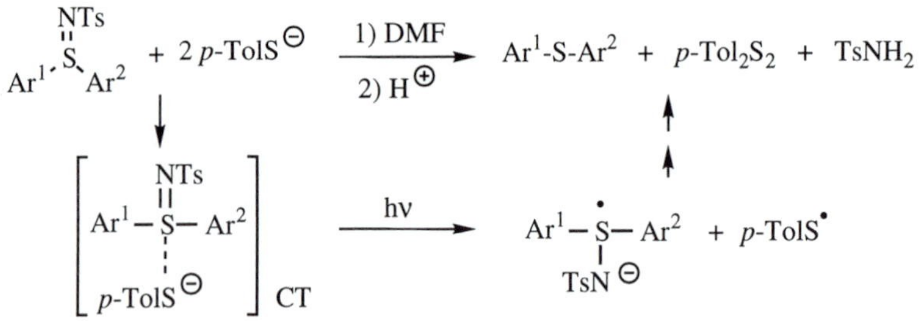

Scheme 3.22

N-sulfonylsulfimides react with various soft or hard nucleophiles to afford the reduced sulfide accompanied with different products, some of them being ligand coupling products. The nucleophiles which have been reported include cyanide ion, trivalent phosphorus compounds and harder nucleophiles such as alkoxide, hydroxide or azide. In the reaction with cyanide ion, *N*-sulfonyl sulfimides yielded *N*-tosylurea and the sulfide.[123,124]

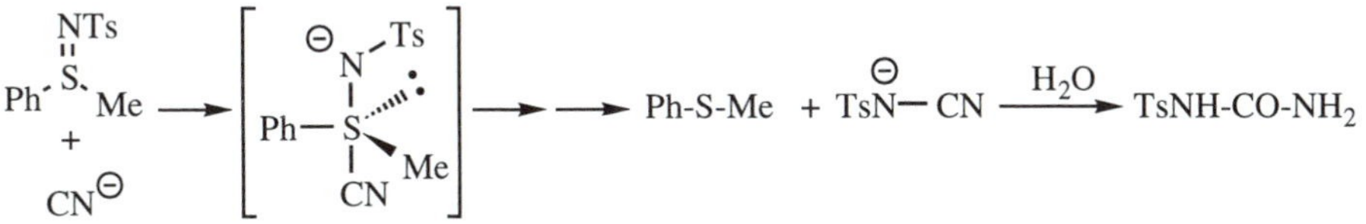

Scheme 3.23

The reaction of *N*-arylsulfonylsulfimides with a range of trivalent phosphorus compounds gave near quantitative yields of the phosphinimides and the sulfide.[125] Addition of protic solvents led to reduction products, which contain, besides the original sulfides, an alkyl-exchanged sulfide in which the alkyl group derived from the alcohol.[126,127] Other derivatives, (acid anhydride, amide, ester or thioester) were formed in the presence of carboxylic acid derivatives.[128,129] These reactions were favoured by the dipolar nature of the intermediately formed sulfurane.[127]

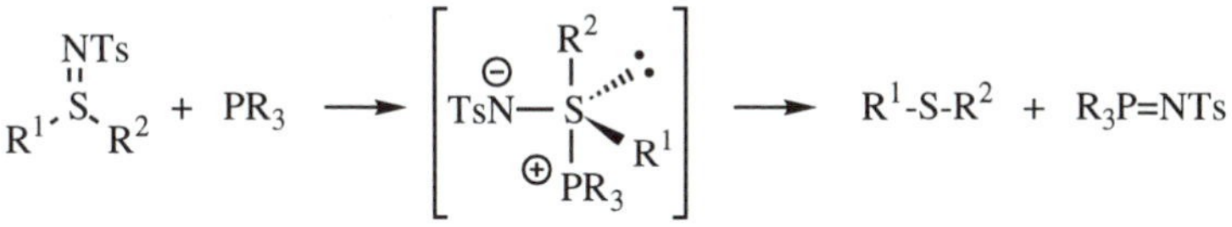

Scheme 3.24

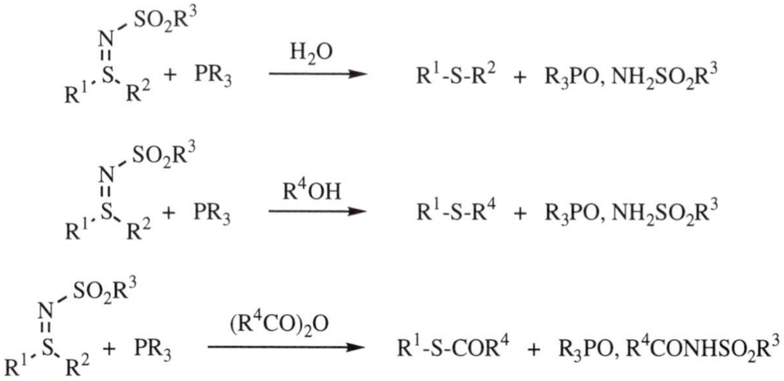

The reaction of *N*-tosylsulfilimines with harder nucleophiles such as alkoxide, hydroxide or azide gave mixtures of products arising from attack at sulfur and at the benzylic carbon. These latter products could also be formed by ligand coupling.[124] (Scheme 3.25)

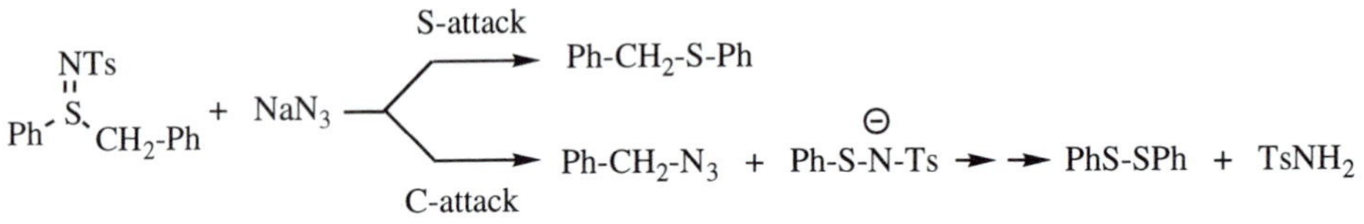

$$\underset{Ph}{\overset{\overset{\displaystyle NTs}{\overset{\|}{S}}}{\diagdown}}\underset{CH_2\text{-}Ph}{} + NaN_3$$

S-attack → Ph-CH$_2$-S-Ph

C-attack → Ph-CH$_2$-N$_3$ + Ph-S-N-Ts$^{\ominus}$ → → PhS-SPh + TsNH$_2$

Scheme 3.25

3.4 SULFONES

Substituted with appropriate leaving groups, activated aromatics and heteroaromatics react with bases and nucleophiles to lead to the products of aromatic nucleophilic substitution. Among the possible products, *ipso*-substitution is frequently observed. A variety of mechanistic pathways have been claimed. In the case of substrates with sulfonyl substituent as an activating group, ligand coupling has been suggested to explain a number of *ipso*-substitution reactions.[65] An early example is the reaction of phenyl (2-phenylalkynyl)sulfone with *n*-butyllithium, which afforded 1-phenyl 2-butyl acetylene (**197**) in quantitative yield.[130]

$$Ph-C\equiv C-SO_2Ph \quad \xrightarrow{\;n\text{-BuLi}\;} \quad Ph-C\equiv C-Bu \qquad (\textbf{197}),\,98\%$$

Treatment of 2- and 4-sulfonylpyridine derivatives with a range of Grignard compounds afforded different types of products.[131] The reaction of some 2-sulfonylpyridines with Grignard derivatives afforded directly the *ipso*-substituted products in generally high yields. When a chlorine atom is present on the α'-position of the pyridyl group, only the α-sulfonyl group is substituted. The reaction of 4-benzenesulfonylpyridine derivatives with Grignard compounds gave complex mixtures of products. Their reactions with aryl Grignard reagents led to the ligand coupling product in modest to moderate yields together with biaryl products derived from the Grignard ligand. The reaction with alkyl Grignard reagents, on the other hand, led to 4,4'-bipyridyl (**145**) and poor yields of the 4-alkyl pyridine. The 4,4'-bipyridyl could result from the sequence of reactions: ligand exchange, nucleophilic addition of the new 4-pyridyl Grignard and ligand coupling of the di(4-pyridyl)sulfur intermediate.[131]

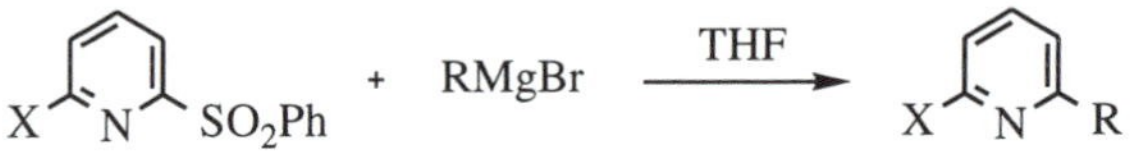

$$\underset{X}{\overset{}{\diagdown}}\underset{N}{\overset{}{\diagdown}} SO_2Ph \;+\; RMgBr \;\xrightarrow{\;THF\;}\; \underset{X}{\overset{}{\diagdown}}\underset{N}{\overset{}{\diagdown}} R$$

X = H, R = Ar, PhCH$_2$, *n*-hexyl, 5-hexenyl, *sec*-octyl   24-99%
X = Cl, R = ethyl, *n*-hexyl, 5-hexenyl   44-79%

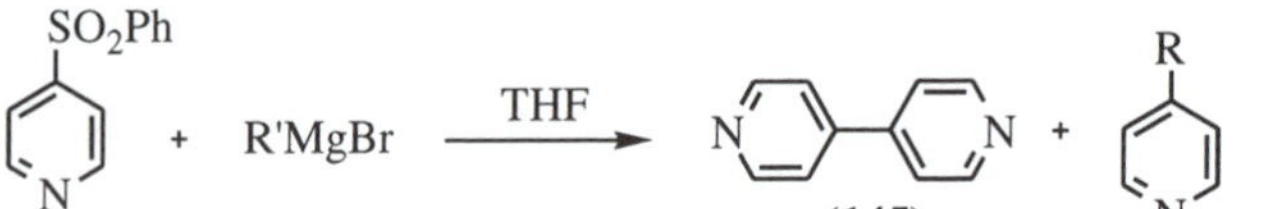

$$\underset{N}{\overset{\overset{\displaystyle SO_2Ph}{}}{\diagdown}} \;+\; R'MgBr \;\xrightarrow{\;THF\;}\; N\!\!-\!\!\diagdown\!\!-\!\!\diagdown\!\!-\!\!N \;+\; \underset{N}{\overset{\overset{\displaystyle R}{}}{\diagdown}}$$

(**145**)

	(145)	R-pyridine
R = Ph, *p*-Tol	-	25-51%
R = alkyl	23-54%	15-26%

Though the scope and mechanism of these reactions realizing the *ipso*-substitution of sulfonyl groups were not extensively investigated, the ligand coupling mechanism was suggested by Oae and Furukawa to best explain these results.[65] An alternative mechanism could be an electron transfer process (SET), a mechanism which is frequently observed in similar aromatic substitution systems.

However, this mechanism was excluded on the basis of the reaction of the 2-pyridylsulfone (**198**) with 5-hexenylmagnesium bromide which did not give any 2-(cyclopentylmethyl)pyridine (**199**), which would be expected for a radical process. The only isolated compound was the ligand coupling type product (**200**) in 99% yield.[131] In spite of these suggestions, the mechanism is more likely to be a simple addition on the C-2 atom of the pyridine ring, followed by elimination of phenylsulfinic acid.

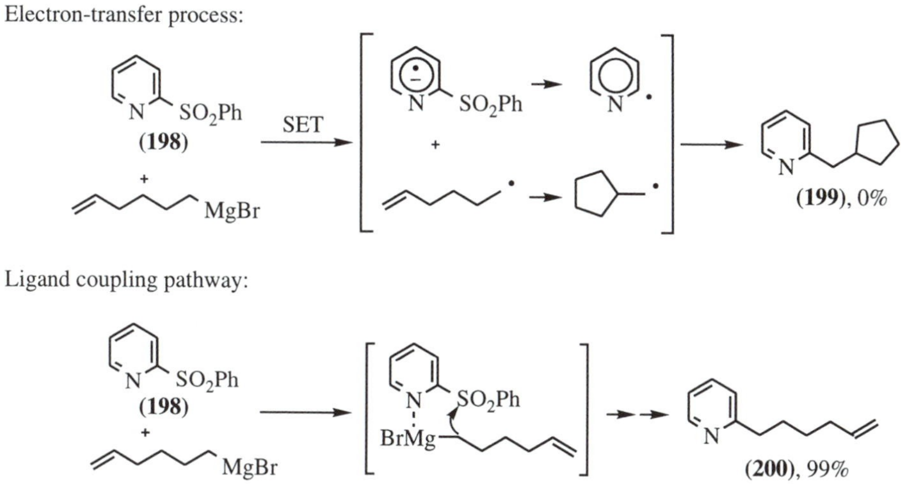

Electron-transfer process:

Ligand coupling pathway:

Scheme 3.26

3.5 POLYSULFANES

Compounds containing cumulated S-S bonds easily undergo interconversion reactions. Among the more common examples are: the decomposition of unstable sulfur allotropes in the solid state with formation of the stable S_8,[132] or the decomposition of S_8 in organic solvents with formation of S_6, S_7 and other homocyclic sulfur molecules,[133-135] as well as the decomposition of organic polysulfanes R_2S_n resulting in the formation of homologous molecules with larger or smaller values of n.[136,137] The early investigators favoured the formation of free radicals as primary intermediates.[136,137]

$$S_8 \quad \rightleftharpoons \quad S_6 + S_7 + \text{others}$$

However, Steudel suggested that hypervalent intermediates could be involved in the interconversion reactions.[138,139] Two types of hypervalent intermediate can be formed. The linear polysulfur chain can isomerise to a thiosulfoxide intermediate (**201**). (Scheme 3.27.A) In the second type, a sulfur atom of one molecule can insert into the S-S bond of a second molecule to form a trigonal-bipyramidal hypervalent intermediate (**202**). This intermediate (**202**) can either undergo directly ligand coupling, affording the macrocyclic S_{2n} molecule, (Scheme 3.27.B, path a) or pseudorotation can take place. Intermediate (**202**) is isomerised into intermediate (**203**) which then undergoes ligand coupling. This latter process overall constitues a "sulfur atom transfer" from one molecule to another. (Scheme 3.27.B, path b)

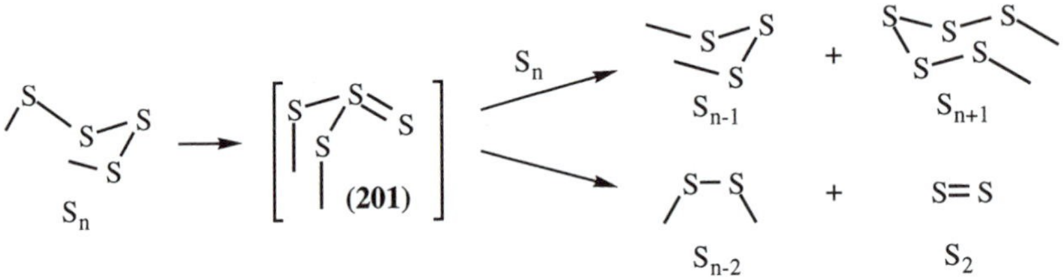

Scheme 3.27.A

or

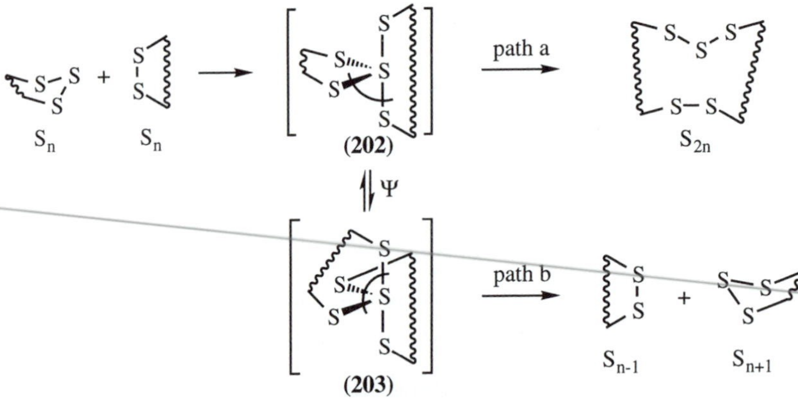

Scheme 3.27.B

This mechanism could also explain the interconversions observed in polysulfanes, such as the case of the trithiolane-pentathiepine equilibrium.[140,141]

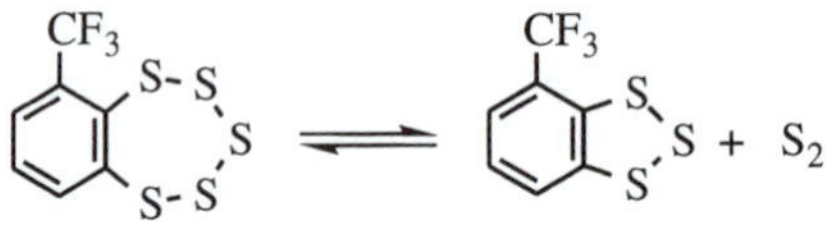

ref. 140

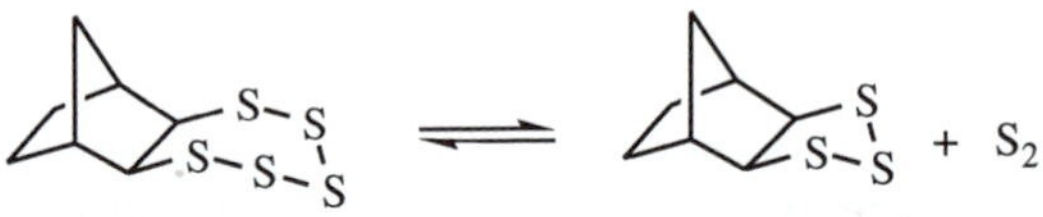

ref. 141

3.6 OXATHIETANES

Four-membered heterocyclic systems have been postulated to be involved as intermediates along the reaction pathway in a number of systems allowing olefin synthesis. This is the case of the Wittig reaction (phosphorus)[142] and a family of related reactions: the boron-Wittig (boron),[143] the Peterson (silicon)[144] and the Peterson-type reactions (germanium, tin, lead).[145,146] In all these systems, interaction of an organometallic species with a carbonyl substrate results in the formation of an olefin, the reaction proceeding through a slightly polar four-membered transition state (**204**). (Scheme 3.28)

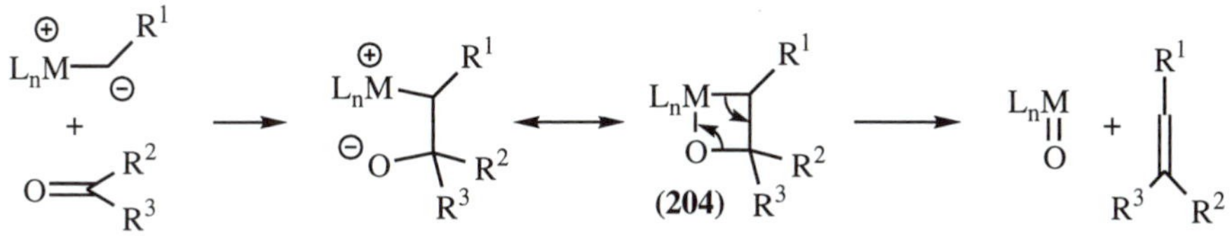

Scheme 3.28

In the case of the sulfur analogues, the reaction of sulfur ylides with carbonyl compounds, known as the Corey-Chaykovsky reaction, is a well-established procedure for the synthesis of oxiranes.[4,5] This reaction is usually considered to involve the formation of an *anti*-betaine (**205**) followed by a back side attack of an oxido anion on the β-carbon. (Scheme 3.29)

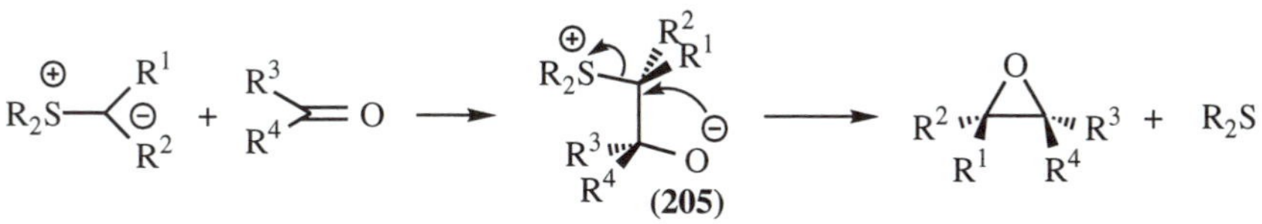

Scheme 3.29

By analogy with the olefination systems, and in particular the Wittig reaction using a phosphorus ylide, the intermediacy of an oxathietane along the reaction pathway might be considered as a possible alternative. Accordingly, a tetracoordinate $1,2\lambda^4$-oxathietane (**206**) was synthesised and its structure determined by X-ray crystallography. Thermal decomposition led to a mixture of products resulting from fragmentation back to the starting ketone, isomerisation via a 1,3-proton shift, as the major decomposition pathways. However, the formation of an oxirane (**207**) by a minor pathway suggested that such a $1,2\lambda^4$-oxathietane may be an intermediate in the Corey-Chaykovsky reaction.[147,148]

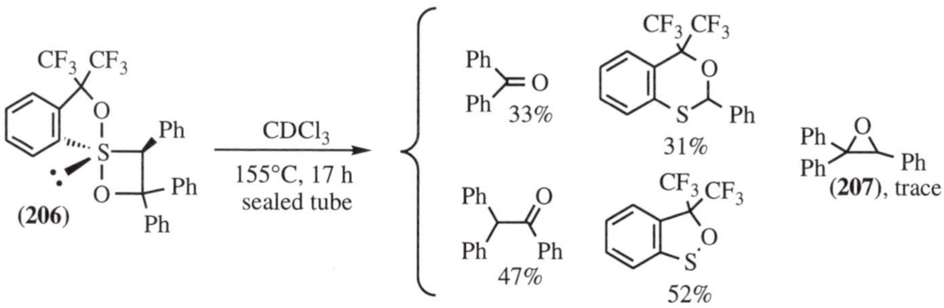

This suggestion was supported by the thermolysis of a $1,2\lambda^6$-oxathietane (**208**), which afforded a good yield of the corresponding epoxide (**209**).[149]

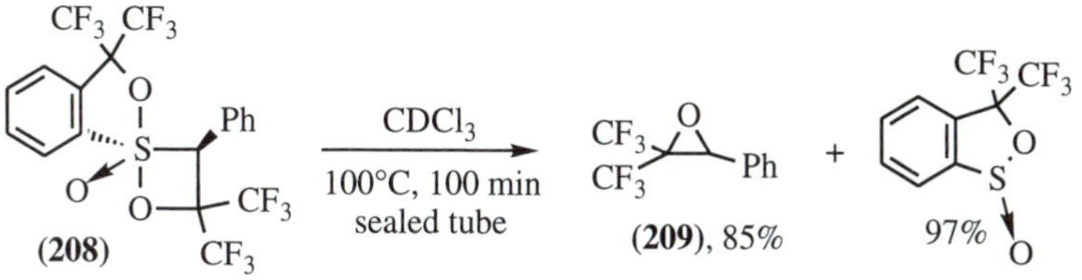

This suggestion was further supported by the synthesis and study of the behaviour of the two C-4 stereoisomers of a $1,2\lambda^6$-oxathietane (210a) and (210b) bearing two different substituents on the 4-position (CF_3 and Ph). Their thermal decomposition led to the selective formation of the corresponding epoxides (211a) and (211b) with retention of configuration. However, addition of lithium bromide to this thermolysis system led to a different stereochemical outcome with the formation of a mixture of the epoxide stereoisomers (211a) + (211b). Therefore the thermolysis of the oxathietanes, a salt-free Corey-Chaykovsky type reaction, is a carbon-oxygen ligand coupling reaction with retention of configuration. By contrast, the Corey-Chaykovsky reaction involves the *anti*-betaine intermediate (205) which evolves by a back side attack and does not involve a cyclic four membered heterocyclic intermediate.[150]

(210a) and (211a) : $R^1 = CF_3$ and $R^2 = $ Ph

(210b) and (211b) : $R^1 = $ Ph and $R^2 = CF_3$

3.7 REFERENCES

1. For a recent general survey of structure and mechanism of organosulfur: Oae, S.; Doi, J.T. *Organic Sulfur Chemistry - Structure and Mechanism*; CRC Press: Boca Raton, **1991**.
2. For a recent general survey of organosulfur in synthesis: Metzner, P.; Thuillier, A. *Sulfur Reagents in Organic Synthesis*; Academic Press: London, **1995**.
3. Trost, B.M. *Fortsch. Chem. Forsch.* **1973**, *41*, 1-29.
4. Corey, E.J.; Chaykovsky, M. *J. Am. Chem. Soc.* **1962**, *84*, 3782-3783.
5. Aubé, J. Epoxidation and Related Processes, in *Comprehensive Organic Synthesis*; Trost, B.M.; Fleming, I., Eds.; Pergamon Press: Oxford, **1991**; Vol. 1, Ch. 3.2, p. 820.
6. Corey, E.J.; Oppolzer, W. *J. Am. Chem. Soc.* **1964**, *86*, 1899-1900.
7. Trost, B.M.; Atkins, R.C. *Tetrahedron Lett.* **1968**, 1225-1229.
8. Trost, B.M.; Atkins, R.C.; Hoffman, L. *J. Am. Chem. Soc.* **1973**, *95*, 1285-1295.
9. Wittig, G.; Fritz, H. *Liebigs Ann. Chem.* **1952**, *577*, 39-46.
10. Sheppard, W.A. *J. Am. Chem. Soc.* **1962**, *84*, 3058-3063.
11. Franzen, V.; Joschek, H.I.; Mertz, C. *Liebigs Ann. Chem.* **1962**, *654*, 82-91.
12. Khim, Y.H.; Oae, S. *Bull. Chem. Soc. Jpn.* **1969**, *42*, 1968-1971.
13. Trost, B.M.; LaRochelle, R.W.; Atkins, R.C. *J. Am. Chem. Soc.* **1969**, *91*, 2175-2177.
14. LaRochelle, R.W.; Trost, B.M. *J. Am. Chem. Soc.* **1971**, *93*, 6077-6086.
15. Scardiglia, F.; Roberts, J.D. *Tetrahedron* **1968**, *3*, 197-208.
16. de Graff, G.B.R.; den Hertog, H.J.; Melger, W.Ch. *Tetrahedron Lett.* **1965**, 963-968.
17. Friedman, L.; Chlebowski, J.F. *J. Am. Chem. Soc.* **1969**, *91*, 4864-4871.
18. Trost, B.M.; Arndt, H.C. *J. Am. Chem. Soc.* **1973**, *95*, 5288-5298.
19. Harrington, D.; Weston, J.; Jacobus, J.; Mislow, K. *J. Chem. Soc., Chem. Comm.* **1972**, 1079-1080.
20. Hori, M.; Kataoka, T.; Shimizu, H.; Miyagaki, M. *Chem. Lett.* **1972**, 515-520.
21. Hori, M.; Kataoka, T.; Shimizu, H.; Miyagaki, M. *Chem. Pharm. Bull.* **1974**, *22*, 1711-1720.
22. Hori, M.; Kataoka, T.; Shimizu, H.; Miyagaki, M. *Chem. Pharm. Bull.* **1974**, *22*, 2004-2013.

23. Hori, M.; Kataoka, T.; Shimizu, H.; Miyagaki, M.; Murase, M. *Chem. Pharm. Bull.* **1974**, *22*, 2014-2019.
24. Hori, M.; Kataoka, T.; Shimizu, H.; Miyagaki, M. *Chem. Pharm. Bull.* **1974**, *22*, 2020-2029.
25. Hori, M.; Kataoka, T.; Shimizu, H.; Miyagaki, M.; Takagi, T. *Chem. Pharm. Bull.* **1974**, *22*, 2030-2041.
26. Sheppard, W.A. *J. Am. Chem. Soc.* **1971**, *93*, 5597-5598.
27. Sheppard, W.A.; Foster, S.S. *J. Fluorine Chem.* **1972/1973**, *2*, 53-62.
28. Ogawa, S.; Matsunaga, Y.; Sato, S.; Erata, T.; Furukawa, N. *Tetrahedron Lett.* **1992**, *33*, 93-96.
29. Ogawa, S.; Sato, S.; Furukawa, N. *Tetrahedron Lett.* **1992**, *33*, 7925-7928.
30. Sato, S.; Furukawa, N. *Tetrahedron Lett.* **1995**, *36*, 2803-2806.
31. Ogawa, S.; Matsunaga, Y.; Sato, S.; Iida, I.; Furukawa, N. *J. Chem. Soc., Chem. Commun.* **1992**, 1141-1142.
32. Oae, S.; Ishihara, H.; Yoshihara, M. *Khim. Geterotsikl. Soedin.* **1995**, 1053-1058; *Chem. Heterocycl. Compd. (Engl. Transl.)* **1995**, *31*, 917-921.
33. Oae, S.; Ishihara, H.; Yoshihara, M. *Zh. Org. Khim.* **1996**, *32*, 282-286.
34. Franzen, V.; Mertz, C. *Liebigs Ann. Chem.* **1961**, *643*, 24-29.
35. Wiegand, G.H.; Mc Ewen, W.E. *Tetrahedron Lett.* **1965**, 2639-2642.
36. Wiegand, G.H.; Mc Ewen, W.E. *J. Org. Chem.* **1968**, *33*, 2671-2675.
37. Srivastava, S.K.; Chauhan, P.M.S.; Bhaduri, A.P. *Chem. Commun.* **1996**, 2679-2680.
38. Knapczyk, J.W.; Wiegand, G.H.; Mc Ewen, W.E. *Tetrahedron Lett.* **1965**, 2971-2977.
39. Oae, S.; Khim, Y.H. *Bull. Chem. Soc. Jpn.* **1969**, *42*, 3528-3535.
40. Knapczyk, J.W.; Mc Ewen, W.E. *J. Am. Chem. Soc.* **1969**, *91*, 145-150.
41. Lai, C.C.; Mc Ewen, W.E. *Tetrahedron Lett.* **1971**, 3271-3274.
42. Furukawa, N.; Matsunaga, Y.; Sato, S. *Synlett* **1993**, 655-656.
43. Kitamura, T.; Miyaji, M.-a.; Soda, S.-i.; Taniguchi, H. *J. Chem. Soc., Chem. Commun.* **1995**, 1375-1376.
44. Kitamura, T.; Soda, S.-i.; Morizane, K.; Fujiwara, Y.; Taniguchi, H. *J. Chem. Soc., Perkin Trans. 2* **1996**, 473-474.
45. Trost, B.M.; LaRochelle, R.W. *Tetrahedron Lett.* **1968**, 3327-3330.
46. LaRochelle, R.W.; Trost, B.M.; Krepski, L. *J. Org. Chem.* **1971**, *36*, 1126-1136.
47. Hyne, J.B.; Jensen, J.H. *Can. J. Chem.* **1962**, *40*, 1394-1398.
48. Sneen, R.A.; Felt, G.R.; Dickason, W.C. *J. Am. Chem. Soc.* **1973**, *95*, 638-639.
49. Islam, M.N.; Leffek, K.T. *J. Chem. Soc., Perkin Trans. 2* **1977**, 958-962.
50. Buckley, N.; Oppenheimer, N.J. *J. Org. Chem.* **1994**, *59*, 5715-5723.
51. Buckley, N.; Maltby, D.; Burlingame, A.L.; Oppenheimer, N.J. *J. Org. Chem.* **1996**, *61*, 2753-2762.
52. Buckley, N.; Oppenheimer, N.J. *J. Org. Chem.* **1996**, *61*, 7360-7372.
53. Young, P.R.; Ruekberg, B.P. *J. Mol. Struct. (Theochem)* **1989**, *186*, 85-99.
54. Umemoto, T. *Chem. Rev.* **1996**, *96*, 1757-1777.
55. Yagupol'skii, L.M.; Kondratenko, N.V.; Timofeeva, G.N. *Zh. Org. Khim.* **1984**, *20*, 115-118.
56. Umemoto, T.; Ishihara, S. *Tetrahedron Lett.* **1990**, *31*, 3579-3582.
57. Ono, T.; Umemoto, T. *J. Fluorine Chem.* **1995**, *74*, 77-82.
58. Ono, T.; Umemoto, T. *J. Fluorine Chem.* **1991**, *54*, 204.
59. Umemoto, T.; Ishihara, S. *J. Am. Chem. Soc.* **1993**, *115*, 2156-2164.
60. Ono, T.; Umemoto, T. *J. Fluorine Chem.* **1996**, *80*, 163-166.
61. Umemoto, T.; Adachi, K. *J. Org. Chem.* **1994**, *59*, 5692-5699.
62. For a recent review of the application of sulfoxides to enantioselective synthesis, see: Carreño, M.C. *Chem. Rev.* **1995**, *95*, 1717-1760.
63. Andersen, K.K.; Yeager, S.A.; Peynircioglu, N.B. *Tetrahedron Lett.* **1970**, 2485-2488.

64. Ackerman, B.K.; Andersen, K.K.; Karup-Nielsen, I.; Peynircioglu, N.B.; Yeager, S.A. *J. Org. Chem.* **1974**, *39*, 964-968.
65. Oae, S.; Furukawa, N. in *Adv. Heterocyclic Chem.*; Katritzky, A.R., Ed.; Academic Press: San Diego, **1990**; Vol. 48, 1-63.
66. Oae, S.; Uchida, Y. *Acc. Chem. Res.* **1991**, *24*, 202-208.
67. Oae, S. *Reviews on Heteroatom Chemistry* **1991**, *4*, 195-225.
68. Oae, S.; Kawai, T.; Furukawa, N. *Tetrahedron Lett.* **1984**, *25*, 69-72.
69. Oae, S.; Kawai, T.; Furukawa, N.; Iwasaki, F. *J. Chem. Soc., Perkin Trans. 2* **1987**, 405-411.
70. Wakabayashi, S.; Ishida, M.; Takeda, T.; Oae, S. *Tetrahedron Lett.* **1988**, *29*, 4441-4444.
71. Oae, S.; Takeda, T.; Kawai, T.; Furukawa, N. *Phosphorus Sulfur* **1987**, *34*, 133-137.
72. Oae, S.; Takeda, T.; Uenishi, J.; Wakabayashi, S. *Phosphorus, Sulfur, and Silicon* **1996**, *115*, 179-182.
73. Uenishi, J.; Yamamoto, A.; Takeda, T.; Wakabayashi, S.; Oae, S. *Heteroatom Chem.* **1992**, *3*, 73-79.
74. Oae, S.; Takeda, T.; Wakabayashi, S.; Iwasaki, F.; Yamazaki, N.; Katsube, Y. *J. Chem. Soc., Perkin Trans. 2* **1990**, 273-276.
75. Oae, S.; Takeda, T.; Wakabayashi, S. *Heterocycles* **1989**, *28*, 99-102.
76. Oae, S.; Kawai, T.; Furukawa, N. *Phosphorus Sulfur* **1987**, *34*, 123-132.
77. Baker, R.W.; Pocock, G.R.; Sargent, M.V. *J. Chem. Soc., Chem. Commun.* **1993**, 1489-1491.
78. Baker, R.W.; Hockless, D.C.R.; Pocock, G.R.; Sargent, M.V.; Skelton, B.W.; Sobolev, A.N.; Twiss, E.; White, A.H. *J. Chem. Soc., Perkin Trans. 1* **1995**, 2615-2629.
79. Baker, R.W.; Hambley, T.W.; Turner, P. *J. Chem. Soc., Chem. Commun.* **1995**, 2509-2510.
80. Kawai, T.; Kodera, Y.; Furukawa, N.; Oae, S.; Ishida, M.; Takeda, T.; Wakabayashi, S. *Phosphorus Sulfur* **1987**, *34*, 139-148.
81. Wakabayashi, S.; Kubo, Y.; Takeda, T.; Uenishi, J.; Oae, S. *Bull. Chem. Soc. Jpn.* **1989**, *62*, 2338-2341.
82. Oae, S.; Takeda, T.; Wakabayashi, S. *Tetrahedron Lett.* **1988**, *29*, 4445-4448.
83. Wakabayashi, S.; Kiyohara, Y.; Kameda, S.; Uenishi, J.; Oae, S. *Heteroatom Chem.* **1990**, *1*, 225-232.
84. Andersen, K.K.; Yeager, S.A. *J. Org. Chem.* **1963**, *28*, 865-867.
85. Wildi, B.S.; Taylor, S.W.; Potratz, H.A. *J. Am. Chem. Soc.* **1951**, *73*, 1965-1967.
86. Baker, R.W.; Pocock, G.R.; Sargent, M.V.; Twiss, E. *Tetrahedron: Asymmetry* **1993**, *4*, 2423-2426.
87. Kawai, T.; Furukawa, N.; Oae, S. *Tetrahedron Lett.* **1984**, *25*, 2549-2552.
88. Uenishi, J.; Tanaka, T.; Wakabayashi, S.; Oae, S.; Tsukube, H. *Tetrahedron Lett.* **1990**, *31*, 4625-4628.
89. Furukawa, N.; Ogawa, S.; Matsumura, K.; Fujihara, H. *J. Org. Chem.* **1991**, *56*, 6341-6348.
90. Shibutani, T.; Fujihara, H.; Furukawa, N. *Tetrahedron Lett.* **1991**, *32*, 2943-2946.
91. Furukawa, N.; Shibutani, T.; Fujihara, H. *Tetrahedron Lett.* **1989**, *30*, 7091-7094.
92. Uenishi, J.; Tanaka, K.; Nishiwaki, K.; Wakabayashi, S.; Oae, S.; Tsukube, H. *J. Org. Chem.* **1993**, *58*, 4382-4388.
93. Tsukube, H.; Uenishi, J.; Higaki, H.; Kikkawa, K.; Tanaka, K.; Wakabayashi, S.; Oae, S. *J. Org. Chem.* **1993**, *58*, 4389-4397.
94. Uenishi, J.; Nishiwaki, K.; Hata, S.; Nakamura, K. *Tetrahedron Lett.* **1994**, *35*, 7973-7976.
95. Shibutani, T.; Fujihara, H.; Furukawa, N.; Oae, S. *Heteroatom Chem.* **1991**, *2*, 521-531.
96. Furukawa, N.; Shibutani, T.; Matsumura, K.; Fujihara, H.; Oae, S. *Tetrahedron Lett.* **1986**, *27*, 3899-3902.
97. Furukawa, N.; Shibutani, T.; Fujihara, H. *Tetrahedron Lett.* **1987**, *28*, 5845-5848.
98. Wakabayashi, S.; Tanaka, T.; Kubo, Y.; Uenishi, J.; Oae, S. *Bull. Chem. Soc. Jpn.* **1989**, *62*, 3848-3850.

99. Oae, S.; Inubushi, Y.; Yoshihara, M. *Heteroatom Chem.* **1993**, *4*, 185-188.
100. Oae, S.; Inubushi, Y.; Yoshihara, M. *Heteroatom Chem.* **1994**, *5*, 223-228.
101. Inubushi, Y.; Yoshihara, M.; Oae, S. *Heteroatom Chem.* **1996**, *7*, 299-306.
102. Neef, G.; Eder, U.; Seeger, A. *Tetrahedron Lett.* **1980**, *21*, 903-906.
103. Theobald, P.G.; Okamura, W.H. *Tetrahedron Lett.* **1987**, *28*, 6565-6568.
104. Theobald, P.G.; Okamura, W.H. *J. Org. Chem.* **1990**, *55*, 741-750.
105. van Kruchten, E.M.G.A.; Haces, A.; Okamura, W.H. *Tetrahedron Lett.* **1983**, *24*, 3939-3942.
106. Cardellicchio, C.; Cicciomessere, A.R.; Naso, F.; Tortorella, P. *Gazzetta Chim. Ital.* **1996**, *126*, 555-558.
107. Lucchini, V.; Modena, G.; Pasquato, L. *J. Am. Chem. Soc.* **1995**, *117*, 2297-2300. Modena, G. *Acc. Chem. Res.* **1971**, *4*, 73-80. Rappoport, Z. *Acc. Chem. Res.* **1992**, *25*, 474-479.
108. Oae, S.; Inubushi, Y.; Yoshihara, M. *Phosphorus, Sulfur, and Silicon* **1995**, *103*, 101-110.
109. Uchida, A.; Nakazawa, T.; Kondo, I.; Iwata, N.; Matsuda, S. *J. Org. Chem.* **1972**, *37*, 3749-3750.
110. Oae, S.; Inubushi, Y.; Yoshihara, M.; Uchida, Y. *Phosphorus, Sulfur, and Silicon* **1994**, *95-96*, 361-365.
111. Uchida, Y.; Echikawa, N.; Oae, S. *Heteroatom Chem.* **1994**, *5*, 409-413.
112. Bast, S.; Andersen, K.K. *J. Org. Chem.* **1968**, *33*, 846-847.
113. Okuyama, T.; Toyoda, M.; Fueno, T. *Bull. Chem. Soc. Jpn.* **1990**, *63*, 1316-1321.
114. Okuyama, T.; Fueno, T. *Bull. Chem. Soc. Jpn.* **1990**, *63*, 3111-3116.
115. Manya, P.; Sekera, A.; Rumpf, P. *Bull. Soc. Chim. Fr.* **1971**, 286-294.
116. Oae, S.; Yoshimura, T.; Furukawa, N. *Bull. Chem. Soc. Jpn.* **1972**, *45*, 2019-2022.
117. Kim, K.S.; Jung, I.B.; Kim, Y.H.; Oae, S. *Tetrahedron Lett.* **1989**, *30*, 1087-1090.
118. Oae, S.; Tsuchida, Y.; Tsujihara, K.; Furukawa, N. *Bull. Chem. Soc. Jpn.* **1972**, *45*, 2856-2858.
119. Young, P.R.; Huang, H.C. *J. Am. Chem. Soc.* **1987**, *109*, 1805-1809.
120. Young, P.R.; Huang, H.C. *J. Am. Chem. Soc.* **1987**, *109*, 1810-1813.
121. Oae, S.; Aida, T.; Nakajima, K.; Furukawa, N. *Tetrahedron* **1974**, *30*, 947-955.
122. Fujimori, K.; Togo, H.; Pelchers, Y.; Nagata, T.; Furukawa, N.; Oae, S. *Tetrahedron Lett.* **1985**, *26*, 775-778.
123. Oae, S.; Aida, T.; Tsujihara, K.; Furukawa, N. *Tetrahedron Lett.* **1971**, 1145-1148.
124. Oae, S.; Aida, T.; Furukawa, N. *Int. J. Sulfur Chem.* **1973**, *8*, 401-409.
125. Aida, T.; Furukawa, N.; Oae, S. *Chem. Lett.* **1973**, 805-808.
126. Aida, T.; Furukawa, N.; Oae, S. *Chem. Lett.* **1974**, 121-124.
127. Aida, T.; Furukawa, N.; Oae, S. *J. Chem. Soc., Perkin Trans. 2* **1976**, 1438-1444.
128. Aida, T.; Furukawa, N.; Oae, S. *Chem. Lett.* **1975**, 29-32.
129. Oae, S.; Aida, T.; Furukawa, N. *Chem. Pharm. Bull.* **1975**, *23*, 3011-3016.
130. Smorada, R.L.; Truce, W.E. *J. Org. Chem.* **1979**, *44*, 3444-3445.
131. Furukawa, N.; Tsuruoka, M.; Fujihara, H. *Heterocycles* **1986**, *24*, 3337-3340.
132. Steudel, R.; Passlack-Stephan, S.; Holdt, G. *Z. Anorg. Allg. Chem.* **1984**, *517*, 7-42.
133. Steudel, R.; Strauss, R.; Koch, L. *Angew. Chem., Int. Ed. Engl.* **1985**, *24*, 59-60.
134. Steudel, R.; Strauss, R. *Z. Naturforsch., Teil B* **1982**, *37*, 1219-1220.
135. Tebbe, F.N.; Wasserman, E.; Peet, W.G.; Vatvars, A.; Hayman, A.C. *J. Am. Chem. Soc.* **1982**, *104*, 4971-4972.
136. Kende, I.; Pickering, T.L.; Tobolsky, A.V. *J. Am. Chem. Soc.* **1965**, *87*, 5582-5586.
137. Pickering, T.L.; Saunders, K.J.; Tobolsky, A.V. *J. Am. Chem. Soc.* **1967**, *89*, 2364-2367.
138. Steudel, R. *Top. Curr. Chem.* **1982**, *102*, 149-176.
139. Laitinen, R.S.; Pakkanen, T.A.; Steudel, R. *J. Am. Chem. Soc.* **1987**, *109*, 710-714.
140. Chenard, B.L.; Harlow, R.L.; Johnson, A.L.; Vladerchick, S.A. *J. Am. Chem. Soc.* **1985**, *107*, 3871-3879.
141. Bartlett, P.D.; Ghosh, T. *J. Org. Chem.* **1987**, *52*, 4937-4943.

142. Maryanoff, B.E.; Reitz, A.B. *Chem. Rev.* **1989**, *89*, 863-927.
143. Pelter, A. *Pure Appl. Chem.* **1994**, *66*, 223-233.
144. Ager, D.J. *Org. React. (New York)* **1990**, *38*, 1-223.
145. Kauffmann, T. *Angew. Chem., Int. Ed. Engl.* **1982**, *21*, 410-429.
146. Pereyre, M.; Quintard, J.-P.; Rahm, A. In *Tin in Organic Synthesis*; Butterworths: London, **1987**; pp. 176-177.
147. Kawashima, T.; Ohno, F.; Okazaki, R. *Angew. Chem., Int. Ed. Engl.* **1994**, *33*, 2094-2095.
148. Kawashima, T.; Okazaki, R. *Synlett* **1996**, 600-608.
149. Ohno, F.; Kawashima, T.; Okazaki, R. *J. Am. Chem. Soc.* **1996**, *118*, 697-698.
150. Kawashima, T.; Ohno, F.; Okazaki, R.; Ikeda, H.; Inagaki, S. *J. Am. Chem. Soc.* **1996**, *118*, 12455-12456.

Chapter 4

Ligand Coupling Involving Organophosphorus Compounds

In the rich chemistry of organophosphorus compounds, a number of mechanistic pathways have been shown to occur. Among them, the ligand coupling mechanism is only marginally claimed to explain some reaction products. However, a number of theoretical studies have been performed on model compounds to study the ligand coupling pathways. (Chapter 2)

4.1 DECOMPOSITION OF PENTAVALENT PHOSPHORANE DERIVATIVES

Stable phosphoranes with five σ ligands are generally prepared by action of an organometallic reagent on an halogenophosphorane. The decomposition of the phosphoranes constitutes in itself an isolated ligand coupling reaction, which otherwise is part of a multistage process. Early studies of the pyrolysis of pentaphenylphosphorane, PPh$_5$, found evidence of the cleavage of C-P bonds, although no triphenylphosphane, PPh$_3$, was isolated.[1] In a following study, it was observed that PPh$_3$ along with Ph-Ph was formed.[2] The presence of a free radical mechanistic component was ascertained by the observation of the initiation of styrene polymerisation. The high yield of benzene which was measured shows that the ligand coupling is just a minor pathway, the main one being the free radical decomposition reaction.

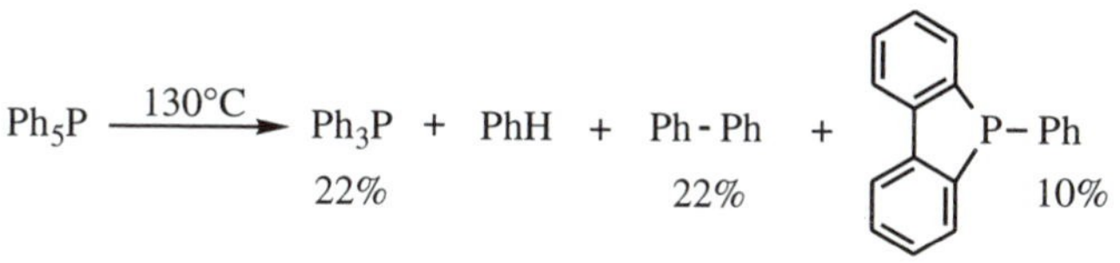

$$Ph_5P \xrightarrow{130°C} Ph_3P + PhH + Ph\text{-}Ph + \text{[spirocyclic P-Ph]}$$

22%   22%   10%

Subsequently, a variety of polyaromatic phosphoranes were found to undergo C-P bond scission to form tertiary phosphanes along with new carbon-carbon bond formation.[3-5] Thermolysis of spirocyclic pentaaryl or alkyltetraarylphosphoranes occurs under milder conditions to afford two main types of ligand coupling products. With a small R group (such as methyl or phenyl), the ligand coupling takes place between two units of the spirocyclic system.

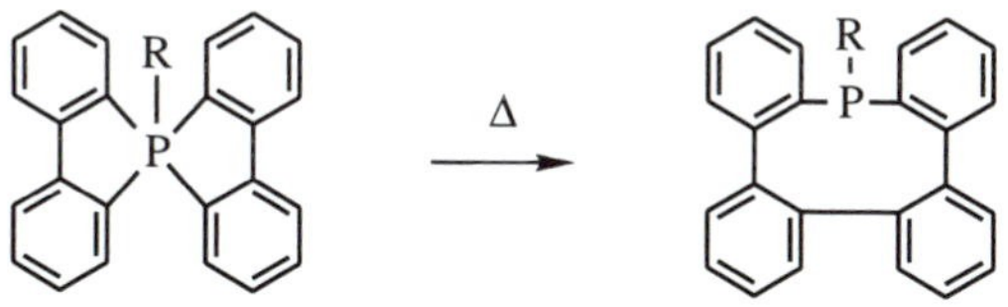

With a large R group (such as 8-quinolinyl or 9-anthryl), release of the steric strain favours a ligand coupling process between one unit of the spirocyclic system and the fifth ligand to afford a 2,2'-biphenyldiyl system.

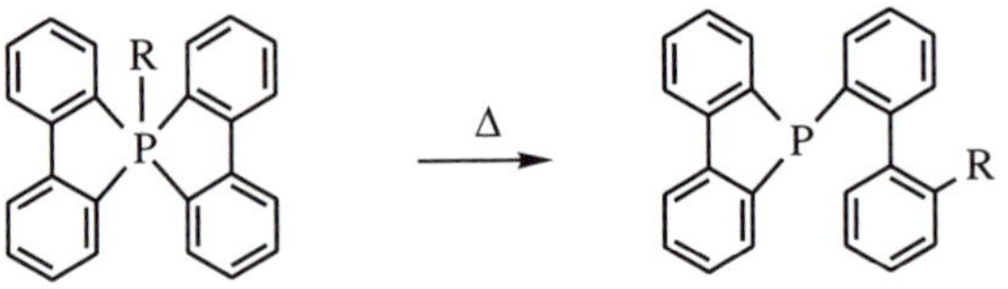

Pentaphenylphosphorane reacts with phenol or alcohols to yield the corresponding alkoxy or aryloxytetraphenylphosphoranes. Thermal degradation of these ethers is very dependent upon the nature of the R group. In the case of the phenoxy derivative, two degradative pathways take place. The major one involves a proton transfer from one phenyl to the phenoxy ligand, affording triphenyl-phosphane, phenol and resinous products derived from benzyne. The minor pathway follows a ligand coupling process, leading to diphenylether.[6] With the alkoxy derivatives, the predominant pathway is the ligand coupling with formation of alkylphenylethers.[7]

Table 4.1: Decomposition of alkoxy- and aryloxy-tetraphenylphosphoranes at 190 - 200°C[6,7]

$$Ph_4POR \xrightarrow{190\text{-}200°C} Ph_3P + PhOR + ROH + PhH$$

R	PPh$_3$ (%)	PhOR (%)	ROH (%)	PhH (%)	Others (%)
Me	97	72	20	4	resinous
i-Pr	86	50	32	8	resinous
PhCH$_2$	95	91	-	11	PhCHO, 7
Ph	82	15	84	-	resinous

In the case of alkoxy methyltriphenylphosphorane Ph$_3$PMe(OR), with R = Me, Et, or i-Pr, a similar thermal degradation was also observed. For example, the methoxy compound (**1**) decomposed into anisole, which was isolated in 87% yield.[8]

The cyclic phosphonite (**2**) reacts with aromatic cyclic disulfides in acetonitrile to produce a sulfur-containing phosphorane. The system is reversible and the direction depends on the nature of the solvent. In acetonitrile, the phosphorane (**3**) precipitates. But in other solvents such as benzonitrile, chloroform, methylene dichloride, benzene or THF, the reverse reaction takes place and is complete overnight at room temperature. The practical interest of this reaction is quite limited as it occurs only with one phosphonite (**2**).[9]

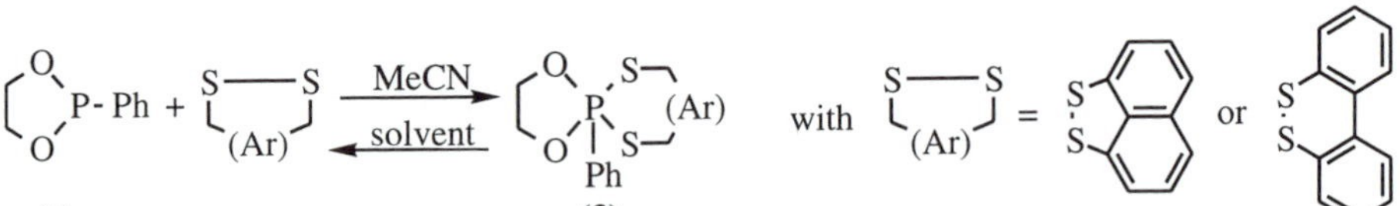

The codeposition reaction between phosphane, PH_3, and fluorine, F_2, in excess argon gave three products identified from matrix infrared spectra at 16K: the difluorophosphorane (**4**), H_3PF_2, which decomposed to the monofluorophosphane (**5**), H_2PF, by loss of HF and to the difluorophosphane, HPF_2, by loss of H_2.[10] Difluorophosphorane (**4**) was prepared as a crystalline material by selective low temperature fluorination of phosphane with xenon difluoride in CF_2HCl at - 78°C.[11] In the presence of anhydrous $AlCl_3$, the unstable monofluorophosphane (**5**) was detected in the gas phase by IR-spectroscopy.

$$H_3PF_2 \ + \ 1/x \ AlCl_3 \ \longrightarrow \ H_2PF \ + \ HCl \ + \ 1/x \ AlCl_{3-x}F_x$$
$$\text{(4)} \qquad\qquad\qquad\qquad\qquad \text{(5)}$$

In the presence of anhydrous KF, difluorophosphorane decomposes to give only phosphane PH_3. Under basic conditions, in the presence of NH_3 or NMe_3, the dehydrofluorination product (**5**) is observed as a transient species and reacts further by elimination of HF to yield a polymeric form of $(PH)_n$. In the presence of electrophilic halides, such as BF_3 or $TiCl_4$, H_3PF_2 afforded the monochlorophosphane H_2PCl.[12]

4.2 REACTIONS INVOLVING PHOSPHORUS (V) REAGENTS

The unusual reactivity of pyridylphosphane derivatives[13] was first described by Davies and Mann, who observed in 1944 that when the pyridylphosphine sulfide (**6**) was heated with excess methyl iodide, only tetramethylphosphonium iodide was isolated in almost theoretical yield.[14]

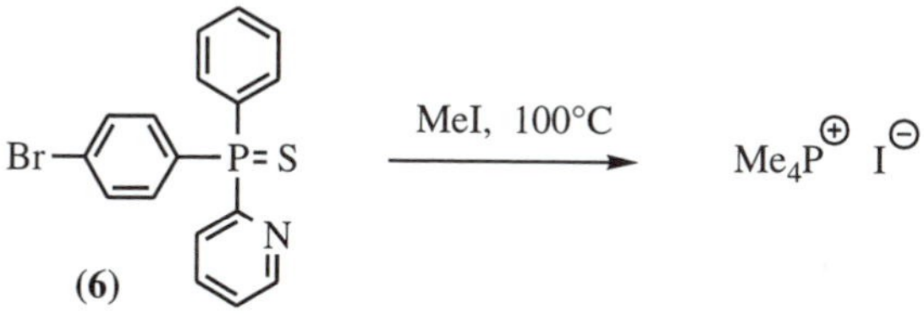

A later study showed that tri- and pentavalent di- or tri-pyridylphosphane derivatives (**7**) - (**10**) react with methyl iodide to yield 2,2'-dipyridine dimethiodide.[15]

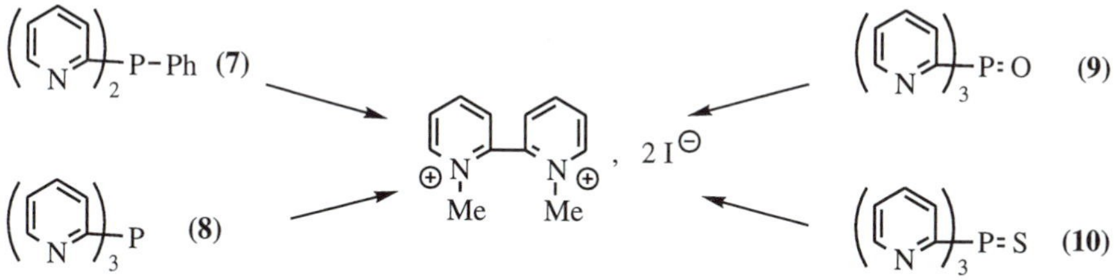

reaction conditions: MeI, MeOH, 100°C

4.2.1 Heteroaryl-heteroaryl coupling

4.2.1.1 Phosphine oxides

The first synthetically useful application of ligand coupling of (2-pyridyl)phosphine oxides was discovered by Newkome *et al.* in their studies on the synthesis of phosphorus containing macrocycles.[16] Treatment of bis[2-(6-bromopyridyl)]phenylphosphane with sodium hexaethyl glycolate afforded the expected phosphorus containing macrocycle. But when the bis[2-(6-substituted pyridyl)]phenylphosphine oxides (11) or (13) were treated with sodium alcoholate, ring contraction occurred with extrusion of a phosphorus moiety, to afford only the corresponding dipyridyl derivatives (12) or (14).

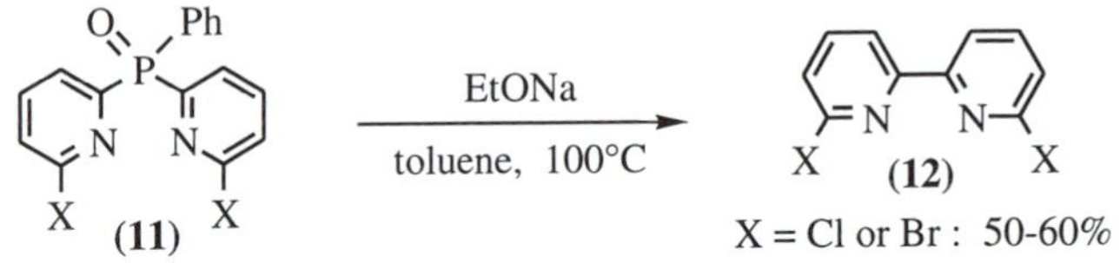

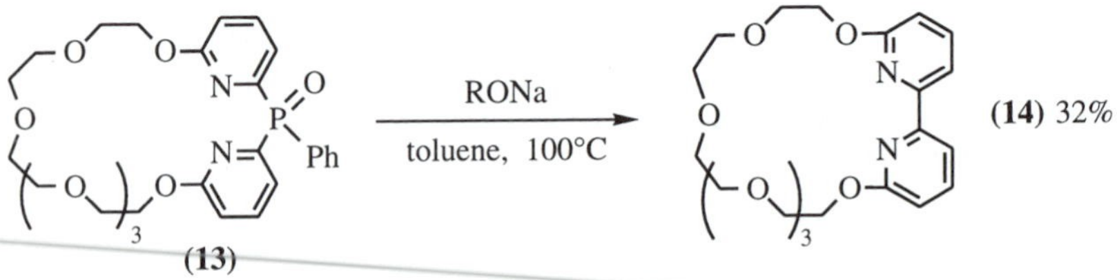

X = Cl or Br : 50-60%

(14) 32%

Oae and co-workers have extensively examined this reaction. Nucleophilic attack of an heteroaryllithium or an organometallic reagent on the phosphorus atom produces an intermediate pentacoordinate derivative. Coupling between an equatorial and an axial substituent leads to the products.[17,18] (Scheme 4.1)

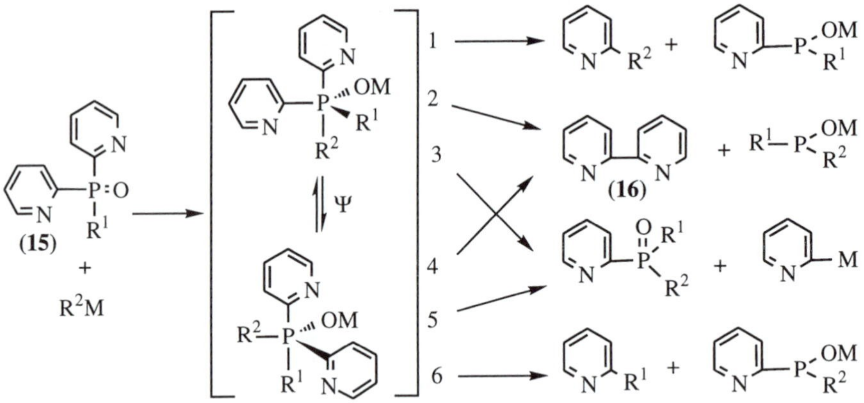

R^1 = 2-pyridyl or phenyl; R^2 = Me, Ph, Ph-CH$_2$ or 4-MeC$_6$H$_4$-CH$_2$; M = Li or MgX
Pathways 1, 2, 4 and 6 are ligand coupling reactions.
Pathways 3 and 5 are ligand exchange reactions.

Scheme 4.1

Phosphine oxides bearing three 2-pyridyl groups, (15), (R^1 = CH$_3$) react with organometallic reagents to afford mostly 2,2'-bipyridine (16) (23-65% yields), together with minor amounts of 2-substituted pyridines and pyridine. With phosphine oxides bearing two 2-pyridyl groups, the reaction with methyl or phenyl Grignard or lithium reagents gives predominantly 2,2'-bipyridine (16). But with benzylic Grignard reagents, the 2-benzylpyridine becomes the predominant ligand coupling product. For example, reaction of 4-methylphenyl bis(2-pyridyl)phosphine oxide (15), (R^1 = PhCH$_2$) with benzylmagnesium chloride afforded 65% of 2-benzylpyridine.[17,18]

The alkaline hydrolysis of phosphonium salts gives phosphine oxides and hydrocarbons resulting from the most stable carbanions in each case.[19,20] Benzyl tris(2-pyridyl)phosphonium bromide (17) undergoes rapid alkaline hydrolysis at room temperature to give good yields of benzylbis(2-pyridyl) phosphine oxide and pyridine.[18]

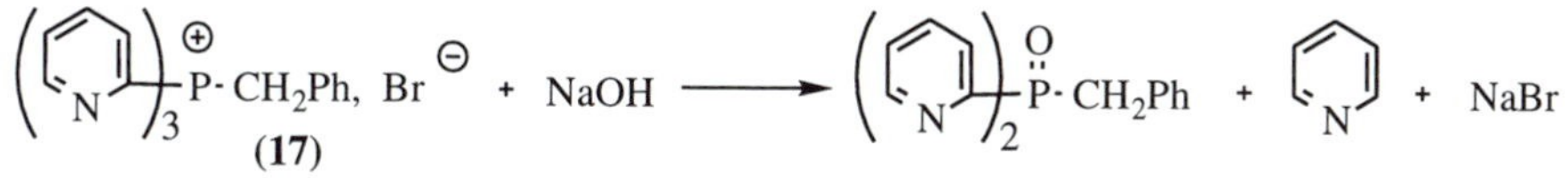

Upon treatment with acidic or neutral solvents, such as water or alcohols, phosphonium salts (18) or phosphine oxides (19) bearing at least two 2-pyridyl or substituted pyridyl groups afford a mixture of the corresponding 2,2'-bipyridine and pyridine in overall moderate to excellent yields.[18,21] (Schemes 4.2 and 4.3)

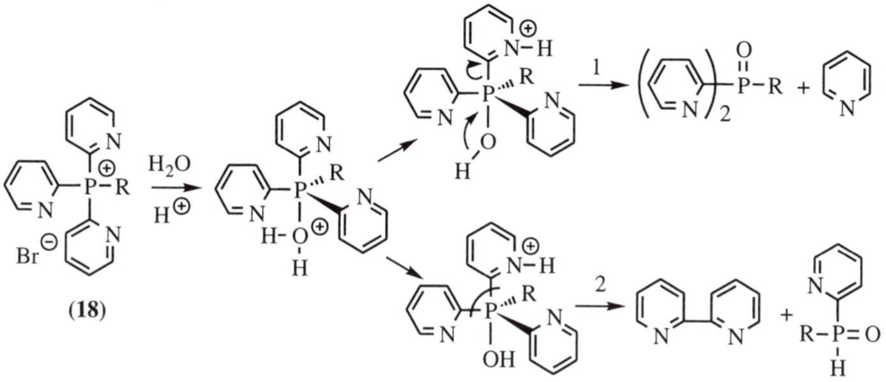

Scheme 4.2: Acid catalysed hydrolysis of phosphonium salts

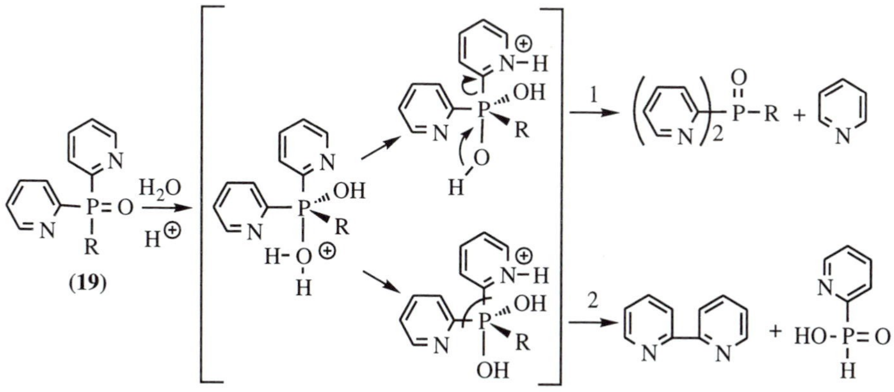

Scheme 4.3: Acid catalysed hydrolysis of phosphine oxides

In both cases, formation of a pentacoordinate intermediate takes place, followed by ligand coupling or ligand exchange. In the hydrolysis of phosphonium salts as well as in the hydrolysis of the phosphine oxides, two concomitant pathways, ligand exchange (pathway 1) and ligand coupling (pathway 2), are competing. They result in the formation of the substituted 2-pyridylphosphine oxides or phosphinic acid derivatives and 2,2'-bipyridine together with pyridine. The absence of 2-benzyl-pyridine in the hydrolysis of the mixed benzyltris(2-pyridyl)phosphonium derivative (17) or the corresponding benzylbis(2-pyridyl)phosphine oxide derivative [(19), R = PhCH$_2$] indicates that, in aqueous media, pseudorotation of the benzyl group to the axial position does not take place.[18,21]

Reaction of tris(2-pyridyl)phosphane (8) with chlorine in methylene dichloride or acetonitrile gave chlorotris(2-pyridyl)phosphonium chloride or its covalent isomer tris(2-pyridyl)phosphonium dichloride (20). Treatment of the dichloride derivative with dilute HCl afforded the coupling product, 2,2'-bipyridine (16), whereas tris(2-pyridyl)phosphine oxide (9) was isolated from the reaction with sodium hydroxide.

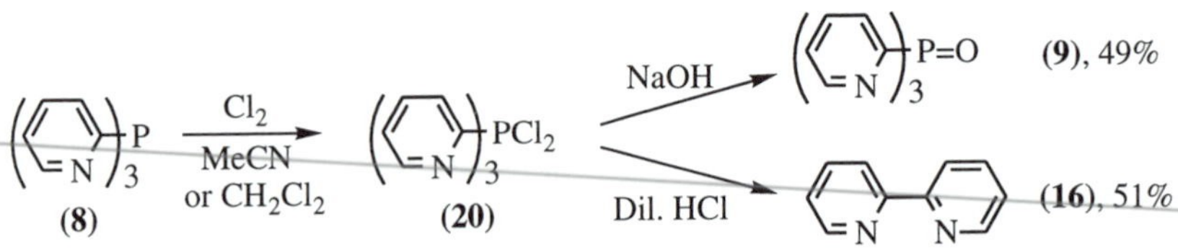

Treatment of tris(2-pyridyl)phosphane or phosphine oxide with chlorine in acetonitrile in the presence of a small amount of water or an alcohol gave the 5-chloro-2,2'-bipyridine (21) as the major ligand coupling product.

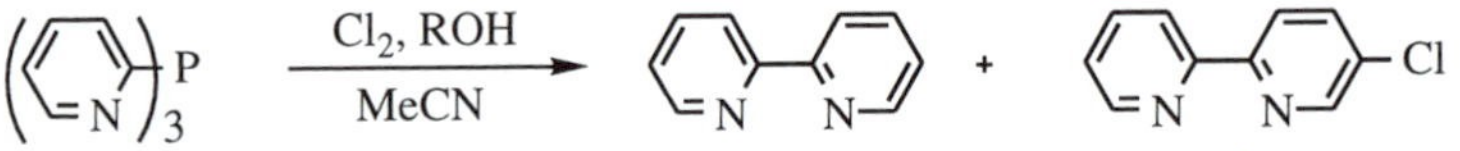

molar ratio ROH / (8) = 10

R =	(16)	(21)
H	14%	73%
Me	7%	61%
t-Bu	2%	84%

Good yields of ligand coupling products were also obtained in the reaction of tris(2-pyridyl) phosphane or phosphine oxide with chlorine in methanol.

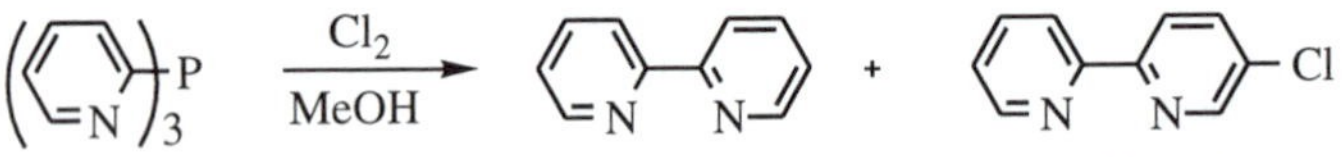

In these reactions, the first formed chlorotris(2-pyridyl)phosphonium chloride salt (20) is converted in the presence of an hydroxylic solvent to the methoxyphosphonium salt and then to the pentacoordinated intermediate. The phosphorus groups behave as strong electron-donating group to the axial pyridine. This ligand can undergo electrophilic substitution at the 5-position to yield the chlorinated intermediates, which undergo a ligand coupling process to lead to product (21). The corresponding bromo derivative was formed in the analogous reaction using bromine.[22] (Scheme 4.4)

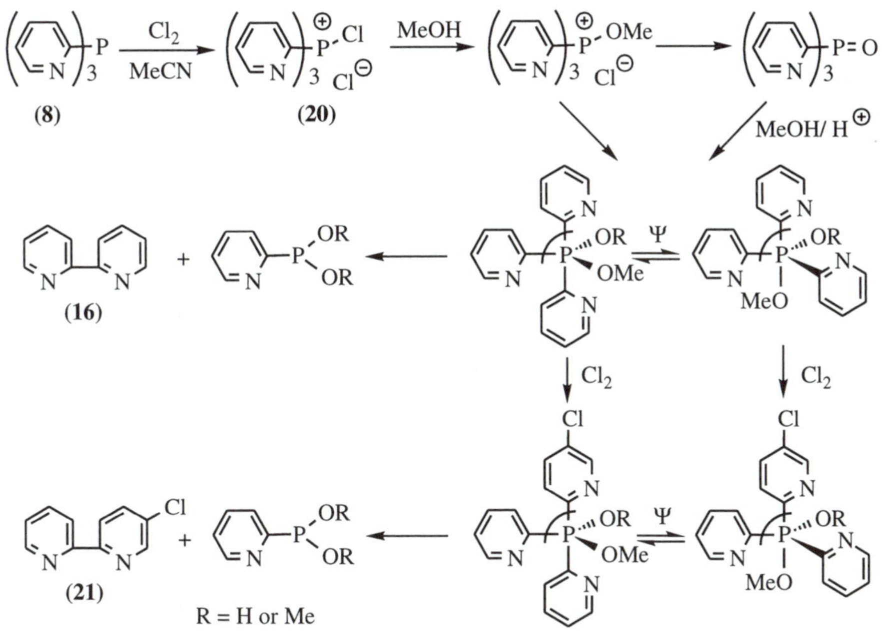

Scheme 4.4

Other electrophilic substitutions on the 5-position, for example deuteration or diazo coupling, were observed to take place very easily. The presence of an electron-donating substituent on the 6-position of the pyridine ring did not affect the reaction. In fact, it improved the yield of the ligand coupling product in the case of the diazo coupling reaction.[23]

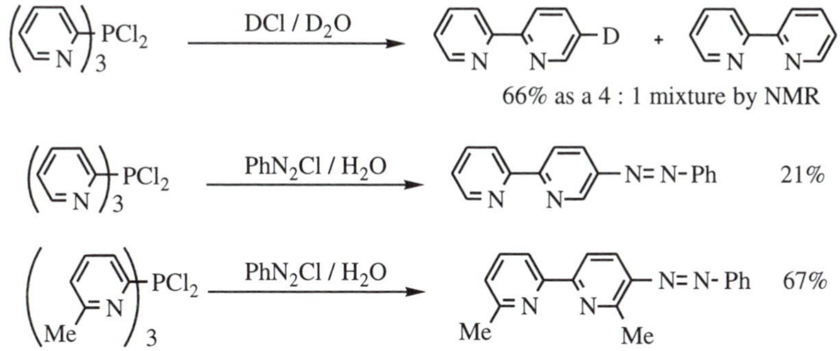

66% as a 4 : 1 mixture by NMR

21%

67%

4.2.1.2 Phosphinimines

When the *N*-trimethylsilyltris(2-pyridyl)phosphinimine (**22**) was treated with one equivalent of methyllithium at - 78°C, a lithium complex (**23**) binding a dipyridyl moiety with a trivalent amino-methyl (2-pyridyl)phosphane was isolated. This compound results from a ligand coupling process taking place on the pentacoordinate phosphorus intermediate formed by addition of methyllithium on the imine bond.

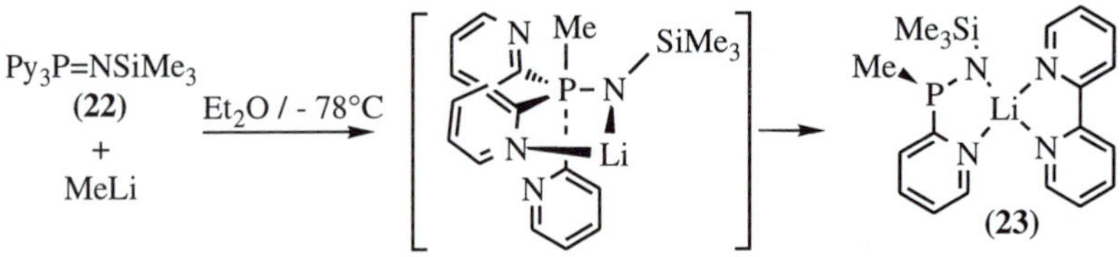

It must be noted that when the analogous triphenylphosphinimine **(24)** was treated under similar conditions, no such ligand coupling was observed. After three days at 25°C, only the dimeric organolithium complex could be isolated (53%), and its structure was shown by X-ray crystallography to be **(25)**. In this case, deprotonation at the *ortho* position of one phenyl took place, instead of the nucleophilic attack of MeLi on the phosphorus atom. Two iminophosphorane units are chelated to a single central lithium ion by the two *ortho*-carbon atoms and the two nitrogen donor atoms. A second lithium ion is also coordinated to the two *ortho*-carbon atoms and to the oxygen atom of a molecule of diethyl ether.[24]

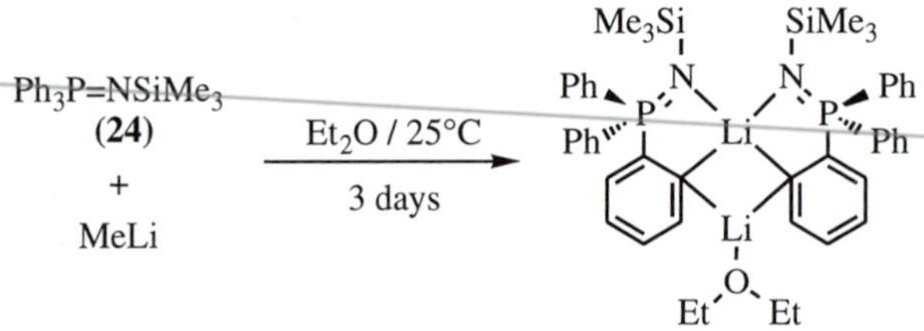

4.2.1.3 Phosphorus oxychloride

Reaction of the lithium salts of various substituted pyridines with phosphorus oxychloride afforded cleanly the 2,2'-bipyridine derivatives. In the case of 2,6-dibromopyridine, the reaction led to a good yield of 6,6'-bipyridine (70%).[25]

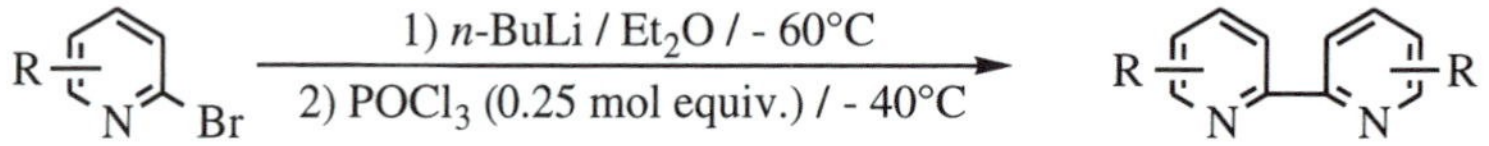

R =	H	51%
	4-Me	58%
	6-Me	62%
	6-Ph	58%
	6-Br	70%

The reaction was also extended to the use of 2-benzothiazolyllithium and 2-quinolyllithium.[25]

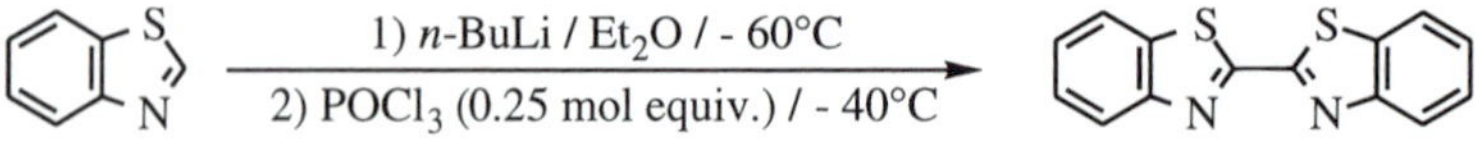

55%

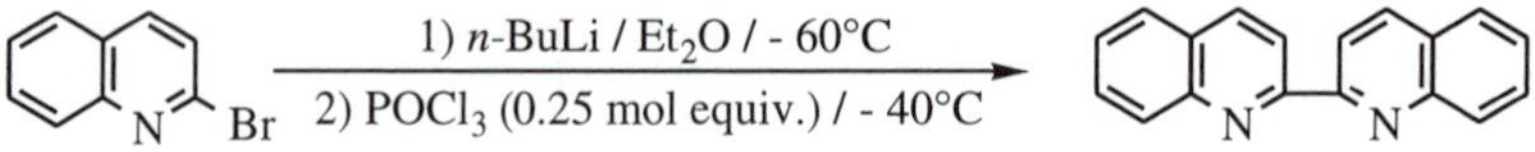

44%

4.2.2 Phenyl-phenyl

Phenyl-phenyl coupling does not generally take place with pentavalent phosphorus compounds, whether with phosphoranes or with phosphine oxides. The reaction of organolithium derivatives with triarylphosphine oxides usually leads to mixtures of products resulting of ligand exchange and disproportionation.[26] However, when the phenyl ring is substituted with electron-withdrawing groups, such a process can be observed. For example, alkaline hydrolysis of tetraphenylphosphonium salts, bearing at least one electron-withdrawing group in the *para* position leads to moderate yields of ligand coupling products. In the reaction mixture, a range of other products were formed by a ligand exchange pathway. Among them, triphenylphosphine oxide and monosubstituted benzenes deriving from the substituted phenyl ligand were present. In the case of the *ortho* derivatives, only the phosphine oxides were obtained by ligand exchange.[27,28]

para-substituted:

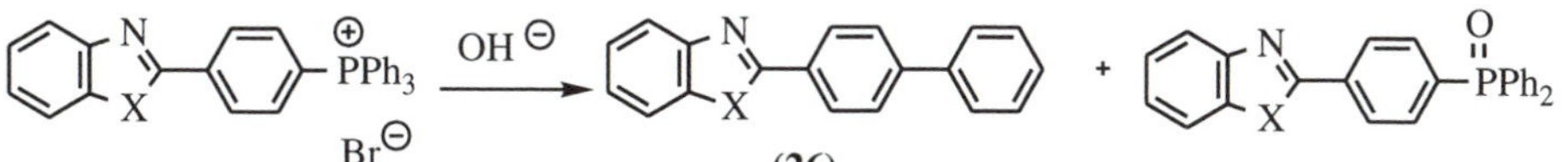

(26)

Yield of (26) : X = NH, 0%; X = NMe, 5%; X = O, 30%; X = S, 30%

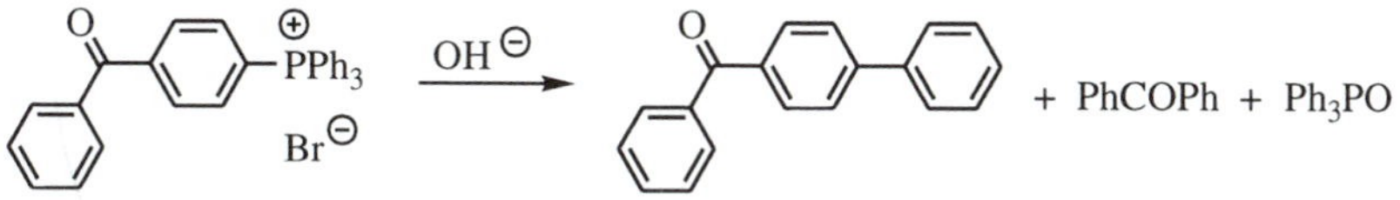

ortho-substituted:

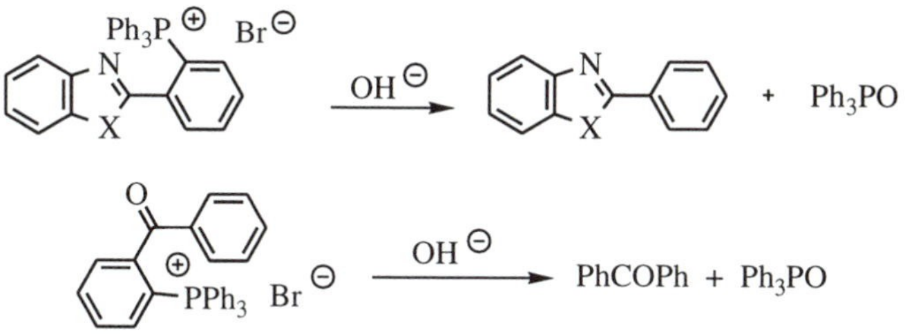

The scope of this ligand coupling reaction is rather limited, as analogues substituted with *para*-trifluoromethyl or phenyl groups did not afford any ligand coupling products, and with the *para*-cyano analogue only traces of 4-cyanobiphenyl were detected.

4.2.3 Aryl-vinyl coupling

While reaction of tetraphenylphosphonium salts with aryllithium or arylmagnesium salts affords the stable pentaarylphosphoranes,[1] reaction of vinyllithium with tetraphenylphosphonium salts did not lead to tetraphenylvinylphosphorane, but instead only the ligand coupling product, styrene, was isolated. Moreover, reaction with both *cis*- and *trans*-2-propenyllithium afforded only the *cis*- and *trans*-propenylbenzenes with 100% conservation of the stereochemistry of the starting propenyllithium.[29]

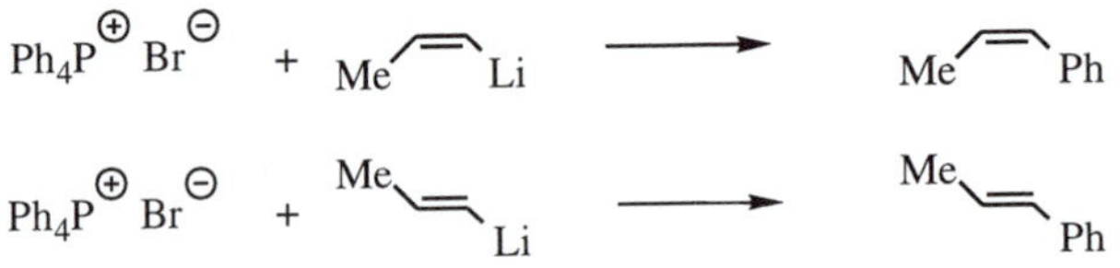

$Ph_4P^{\oplus} \; Br^{\ominus}$ + Me—Li   ⟶   Me—Ph

$Ph_4P^{\oplus} \; Br^{\ominus}$ + Me—Li   ⟶   Me—Ph

4.3 REACTIONS INVOLVING PHOSPHORUS (III) REAGENTS

Simple heterocyclic-substituted phosphanes are most generally prepared by reaction of the appropriate organometallic reagent with phosphorus trihalide.[30] However side reactions were sometimes observed, which resulted in poor overall yields. Thus the tris(2-pyridyl)phosphane was prepared in 13% yield from the reaction of 2-pyridylmagnesium bromide with PCl_3.[18] Using butyl lithium-bromine exchange as a means to generate pyridyllithium, a mixture of 2,2'-bipyridine and butyldi(2-pyridyl)phosphane was obtained.[31] In a series of papers,[32-34] Holms *et al.* have described the synthesis of a number of substituted pyridylphosphanes. The occurrence of ligand coupling products was also frequently observed.[32,35] Oae *et al.* developed this unwanted side reaction into a useful synthesis of 2,2'-dipyridyl derivatives. Treatment of triarylphosphanes bearing 2-pyridyl groups with organolithium reagents in THF afforded ligand coupling and ligand exchange products.[36] When tris(2-pyridyl)phosphane was treated with 2 equivalents of 2-pyridyllithium, a quantitative yield of 2,2'-bipyridine was obtained.[37]

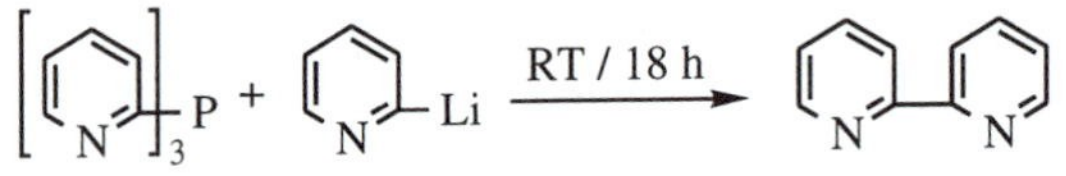

RT / 18 h

Moderate to good yields of 2,2'-bipyridine can be obtained by direct treatment of 2-pyridyllithium with PCl_3, together with small amounts of the corresponding triarylphosphanes.[25] The dibromoterpyridine (27) could be prepared in 53% yield by a one-pot reaction between 2,6-dibromopyridine (4 molar equiv.) with butyllithium (4 molar equiv.) followed by PCl_3.[38]

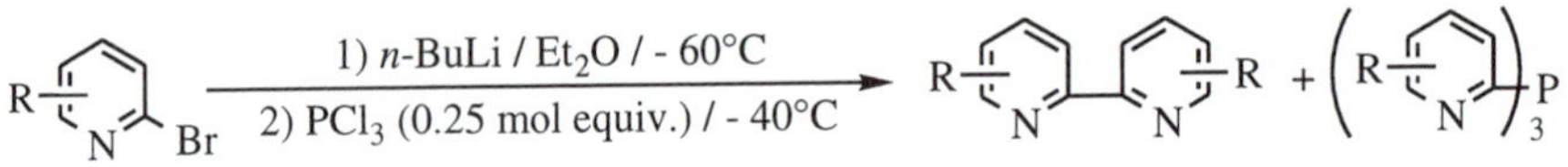

1) *n*-BuLi / Et_2O / - 60°C
2) PCl_3 (0.25 mol equiv.) / - 40°C

R =		
H	50%	18%
4-Me	45%	29%
6-Me	16%	72%
6-Ph	58%	35%

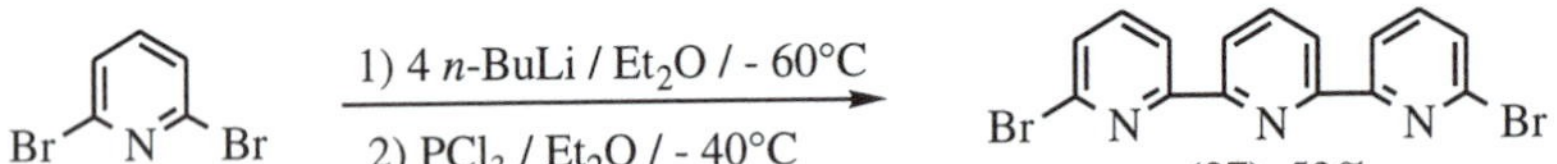

1) 4 *n*-BuLi / Et_2O / - 60°C
2) PCl_3 / Et_2O / - 40°C

(27), 53%

Reaction of triarylphosphanes with lithium metal affords the lithium diarylphosphide. Similarly, tris(2-pyridyl)phosphane reacts with lithium metal in THF with cleavage of a P-pyridyl bond and ligand coupling to 2,2'-bipyridine and lithium bis(2-pyridyl)phosphide. After hydrolysis, the bis(2-pyridyl)phosphane was isolated in 66% yield, but no yield was reported for the bipyridine.[39]

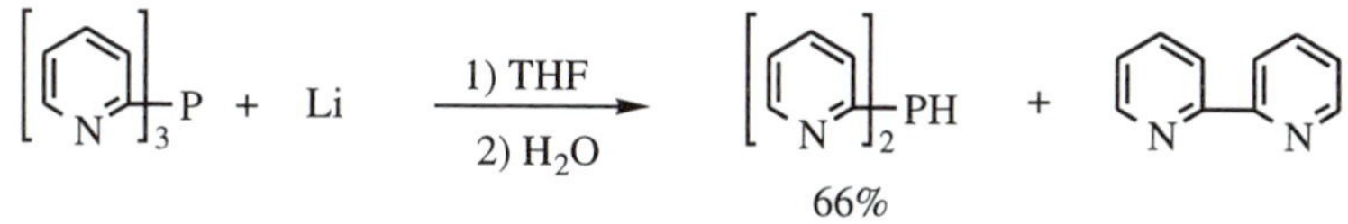

66%

Moore and Whitesides reported that the reaction of 2-thiazolyllithium with PCl₃ afforded the trithiazolylphosphane in 64%.[40] However, when they tried to extend this reaction to the preparation of tris(2-benzothiazolyl)phosphane (28) from 2-benzothiazolyllithium, they failed. This phosphane (28) was eventually obtained by reaction of PCl₃ with 2-trimethylsilylbenzothiazole.[40] Later, when Uchida *et al.* reexamined the reaction of 2-benzothiazolyllithium with PCl₃, they discovered that 2,2'-bibenzo-thiazolyl (29) was formed as a major product (61% based on PCl₃). This compound was also formed in 66% yield when tris(2-benzothiazolyl)phosphane was treated with one equivalent of PhLi. 2-Phenylbenzothiazole was also present in a small amount (9%).[37]

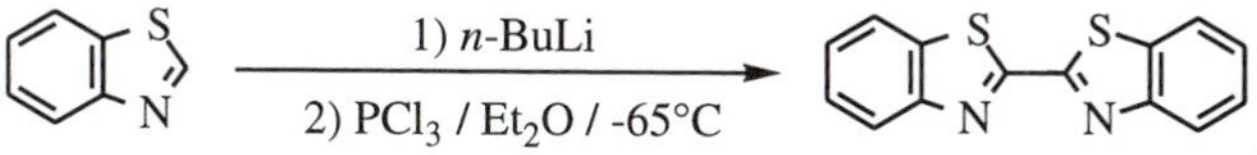

(29), 61%

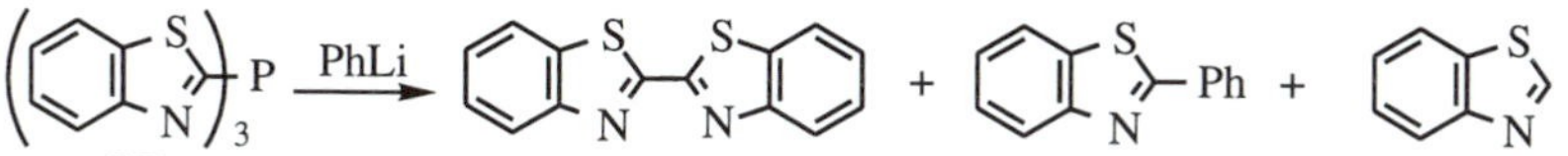

(28)   66%   9%   56%

Ligand exchange is a process known to occur in the reaction of tertiary phosphanes with organolithium reagents.[41,42] However the mechanism which was postulated to explain the ligand coupling products involves nucleophilic attack of the lithium reagent on the phosphorus atom to give a pentacoordinate intermediate. The equatorial 2-pyridyl or heteroaryl group then couples with the axial group to yield 2,2'-bipyridine or bi-heteroaryl derivatives. Other nucleophiles may also be effective, as treatment of the tris-2-(6-cyanopyridyl)phosphane (30) with methylmagnesium iodide did not lead to the expected 6-acetylpyridine derivative, but instead the 6,6'-diacetyl-2,2'-dipyridyl (31) was predominantly formed.[35] The acid-catalysed hydrolysis of the acetal-containing phosphane (32) requires also the *in situ* oxidation to a pentavalent species, eventually leading to the ligand coupling product (33).[32]

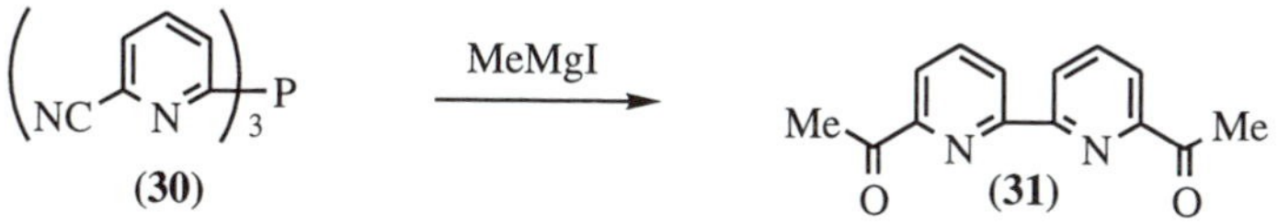

(30)   MeMgI   (31)

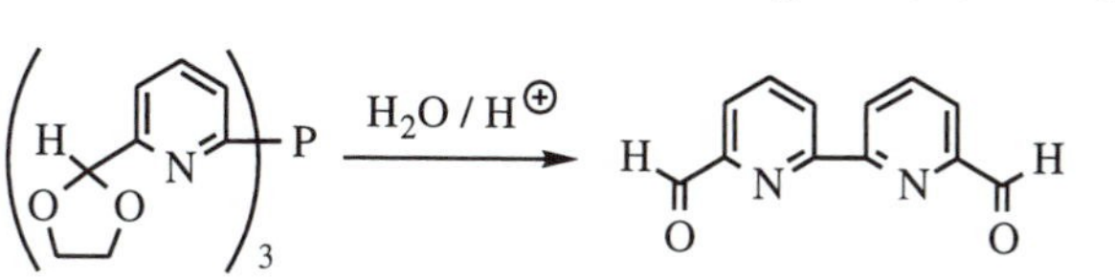

(32)   H₂O / H⊕   (33)

4.4 REFERENCES

1. Wittig, G.; Rieber, M. *Liebigs Ann. Chem.* **1949**, *562,* 187-192.
2. Wittig, G.; Geissler, G. *Liebigs Ann. Chem.* **1953**, *580,* 44-57.

3. Wittig, G.; Maercker, A. *Chem. Ber.* **1964**, *97*, 747-768.
4. Hellwinkel, D. *Chem. Ber.* **1965**, *98*, 576-587.
5. Hellwinkel, D.; Lindner, W. *Chem. Ber.* **1976**, *109*, 1497-1505.
6. Razuvaev, G.A.; Osanova, N.A. *J. Organomet. Chem.* **1972**, *38*, 77-82.
7. Razuvaev, G.A.; Osanova, N.A.; Brilkina, T.G.; Zinovjeva, T.I.; Sharutin, V.V. *J. Organomet. Chem.* **1975**, *99*, 93-106.
8. Schmidbaur, H.; Stühler, H.; Buchner, W. *Chem. Ber.* **1973**, *106*, 1238-1250.
9. Kimura, Y.; Kokura, T.; Saegusa, T. *J. Org. Chem.* **1983**, *48*, 3815-3816.
10. Andrews, L.; Withnall, R. *Inorg. Chem.* **1989**, *28*, 494-499.
11. Beckers, H. *Z. Anorg. Allg. Chem.* **1993**, *619*, 1869-1879.
12. Beckers, H. *Z. Anorg. Allg. Chem.* **1993**, *619*, 1880-1886.
13. Newkome, G.R. *Chem. Rev.* **1993**, *93*, 2067-2089.
14. Davies, W.C.; Mann, F.G. *J. Chem. Soc.* **1944**, 276-283.
15. Mann, F.G.; Watson, J. *J. Org. Chem.* **1948**, *13*, 502-531.
16. Newkome, G.R.; Hager, D.C. *J. Am. Chem. Soc.* **1978**, *100*, 5567-5568.
17. Uchida, Y.; Onoue, K.; Tada, N.; Nagao, F.; Oae, S. *Tetrahedron Lett.* **1989**, *30*, 567-570.
18. Uchida, Y.; Onoue, K.; Tada, N.; Nagao, F.; Kozawa, H.; Oae, S. *Heteroatom Chem.* **1990**, *1*, 295-306.
19. Fenton, G.W.; Ingold, C.K. *J. Chem. Soc.* **1929**, *1*, 2342-2357.
20. Berlin, K.D.; Butler, G.B. *Chem. Rev.* **1960**, *60*, 243-260.
21. Uchida, Y.; Kozawa, H.; Oae, S. *Tetrahedron Lett.* **1989**, *30*, 6365-6368.
22. Uchida, Y.; Kajita, R.; Kawasaki, Y.; Oae, S. *Tetrahedron Lett.* **1995**, *36*, 4077-4080.
23. Uchida, Y.; Kajita, R.; Kawasaki, Y.; Oae, S. *Phosphorus, Sulfur, and Silicon* **1994**, *93-94*, 405-409.
24. Steiner, A.; Stalke, D. *Angew. Chem., Int. Ed. Engl.* **1995**, *34*, 1752-1755.
25. Uchida, Y.; Echikawa, N.; Oae, S. *Heteroatom Chem.* **1994**, *5*, 409-413.
26. Furukawa, N.; Ogawa, S.; Matsumura, K.; Fujihara, H. *J. Org. Chem.* **1991**, *56*, 6341-6348.
27. Allen, D.W.; Benke, P. *Phosphorus, Sulfur, and Silicon* **1994**, *86*, 259-262.
28. Allen, D.W.; Benke, P. *J. Chem. Soc., Perkin Trans. 1* **1995**, 2789-2794.
29. Seyferth, D.; Fogel, J.; Heeren, J.K. *J. Am. Chem. Soc.* **1966**, *88*, 2207-2212.
30. Redmore, D. *Chem. Rev.* **1971**, *71*, 315-337.
31. Jakobson, H.J. *J. Mol. Spectrosc.* **1970**, *34*, 245-256.
32. Parks, J.E.; Wagner, B.E.; Holm, R.H. *J. Organomet. Chem.* **1973**, *56*, 53-66.
33. Larsen, E.; LaMar, G.N.; Wagner, B.E.; Parks, J.E.; Holm, R.H. *Inorg. Chem.* **1972**, *11*, 2652-2668.
34. Parks, J.E.; Wagner, B.E.; Holm, R.H. *Inorg. Chem.* **1971**, *10*, 2472-2478.
35. Parks, J.E. *PhD Dissertation*, Univ. of Wisconsin, **1973**.
36. Uchida, Y.; Kawai, M.; Masauji, H.; Oae, S. *Heteroatom Chem.* **1993**, *4*, 421-426.
37. Uchida, Y.; Takaya, Y.; Oae, S. *Heterocycles* **1990**, *30*, 347-351.
38. Uchida, Y.; Okabe, M.; Kobayashi, H.; Oae, S. *Synthesis* **1995**, 939-940.
39. Steiner, A.; Stalke, D. *Organometallics* **1995**, *14*, 2422-2429.
40. Moore, S.S.; Whitesides, G.M. *J. Org. Chem.* **1982**, *47*, 1489-1493.
41. Kyba, E.P. *J. Am. Chem. Soc.* **1975**, *97*, 2554-2555.
42. Kyba, E.P. *J. Am. Chem. Soc.* **1976**, *98*, 4805-4809.

Chapter 5

Ligand Coupling Involving Organoiodine Compounds

Known for a long time, the hypervalent iodine compounds have found a number of applications in organic synthesis mostly as a result of the activity of a number of chemists in the past two decades. As such, the chemistry of these compounds has been extensively reviewed in the past recent years, as well for their intrinsic behaviour as for their interest in organic synthesis.[1-25] As a result, this chapter has become one of the most important for ligand coupling reactions. However, as an extension of the early observations concerning the decomposition of diaryliodonium salts, the importance of this mechanism was completely ignored until recently. As Sandin[26] had found that diaryliodonium salts decomposed by formation of free radicals, most of nucleophilic aromatic substitution reactions were considered by Beringer and his coworkers[27-30] to imply the intermediacy of free radical species formed by decomposition of a 9-I-2 intermediate.[31,32] It is only in the mid 1980's that Barton *et al.* underlined the importance of the competition between two processes in the nucleophilic arylation of enolates.[33] By using an efficient free radical trap, 1,1-diphenylethylene, they showed that the *C*-arylation of a β-ketoester is a ligand coupling reaction whereas the observation of free radical derived products is in fact a detrimental competitive side reaction which is inhibited by 1,1-diphenylethylene. The importance of this observation starts only now to be recognised, even though the free radical pathway is still sometimes considered as a preferred alternative to the ligand coupling nucleophilic substitution.[34] The ligand coupling mechanism was suggested by Moriarty to be involved in a wide range of the reactions of organoiodine compounds.[12]

5.1 DIARYLIODONIUM SALTS

5.1.1 Diaryliodonium halides : decomposition

The thermolysis of iodonium salts possessing a nucleophilic counteranion can be realised either in solution or in the molten state under rather drastic conditions. It affords the arylated derivative of the anion together with an aryl iodide.

$$Ar_2I^{\oplus} X^{\ominus} \xrightarrow{\Delta} ArX + ArI \qquad \text{with } X = Cl, Br \text{ or } I$$

The synthetic interest of this reaction is somewhat limited. Among the synthetically useful reactions, 2- and 3-haloindoles can be obtained from the 3-indolyliodonium salts by reaction with chloride or bromide ions in DMSO.[35] Thermolysis of dibenzoiodolium salts (**1**) leads to the diiodo compounds (**2**) in good yields.[36-38]

(2), 98%

The mechanistic aspects of the thermolysis reaction have been a long standing matter of debate. The reaction has been interpreted for decades in two ways: either as a homolytic mechanism or as a heterolytic mechanism. The nucleophilic-type substitution mechanism was suggested in 1936 by Lucas et al.[39] while, at the same time, Sandin et al.[26] suggested the occurrence of the free radical mechanism. These two major alternatives led to a number of mechanistic studies.[35,38,40-48] In particular, the thermolysis of unsymmetrically substituted iodonium salts bearing *ortho*-methyl groups (3) and (4) showed an unexpected preferential attack of the halide on the more substituted phenyl ring.[45,47] This *ortho*-effect does not agree with either the free radical mechanism or the classical nucleophilic aromatic substitution. It is now generally admitted that the occurrence of a ligand coupling mechanism is the major pathway in the thermolysis of diaryliodonium salts bearing a nucleophilic counteranion. This mechanism was implicitly suggested by Lancer and Wiegand.[47] It appeared to Koser as the most likely in his review in 1983.[5] In 1985, Grushin et al. suggested its occurrence to explain the reactions of phenyl (B-carboranyl)iodonium salts[49] and it is only in 1992 that these authors generalised the intervention of ligand coupling mechanisms in iodonium chemistry in competition with free radical processes.[16,50]

Table 5.1: Relative ratios of aryl halides in the thermal decomposition of *ortho*-methyl substituted diaryliodonium salts[45,47]

(3)			
	X = Cl	84	16
	Br	87	13
	131I	65	35

(4)			
	X = Cl	95	5
	Br	96	4
	131I	84	16

5.1.2 Reactions of diaryliodonium salts with inorganic anions

Among the reactions of diaryliodonium salts with inorganic compounds, their hydrolysis has been studied in detail.[40,41,43,51-53] The hydrolysis reaction proceeds quite slowly in boiling water. However, in the presence of aqueous alkali, the hydrolysis takes place in a few days for diphenyliodonium bromide to afford a mixture of iodobenzene, bromobenzene and phenol. With weakly nucleophilic anions, the reaction affords iodobenzene, phenol and a small amount of diphenylether. The reaction was considered to involve an unstable hydroxydiaryliodane, which is likely to be in equilibrium with diaryliodonium hydroxide. Decomposition of the hydroxyiodane leads to the products.[53]

$$\text{Ar}_2\text{I}^{\oplus}\ \text{X}^{\ominus}\ +\ \text{OH}^{\ominus}\ \longrightarrow\ \text{Ar}_2\text{I}^{\oplus}\ \text{OH}^{\ominus}\ \rightleftharpoons\ \text{Ar}_2\text{IOH}\ \longrightarrow\ \text{ArOH}\ +\ \text{ArI}$$

Several inorganic anions have been arylated by diaryliodonium salts, possessing a non-nucleophilic counteranion. A number of examples are reported in table 5.2.

Table 5.2: Arylation of inorganic anions by diaryliodonium salts

Nucleophile	Product	Yield (%)	Ref
NaNO$_2$	ArNO$_2$	35-88	40,54-59
NaN$_3$	ArN$_3$	95-98	58,59
Na$_2$SO$_3$	ArSO$_3$H	60-95	30,40,57
NaSCN	ArSCN or ArNCS	17-98	58-60
NaCN	ArCN	23-58	40,58,61
NaF	ArF	39-96	59,62,63
Na$_2$Te	Ar$_2$Te	76-92	64
KSP(O)(OR)$_2$	ArSP(O)(OR)$_2$	86-95	65
NaSeP(O)(OEt)$_2$	Ar$_2$Se$_2$	62-82	66
NaP(O)(OR)$_2$	ArP(O)(OR)$_2$	79-93	67

5.1.3 Reactions of diaryliodonium salts with carbon nucleophiles

5.1.3.1 Reactions with organometallic reagents

Iodonium salts react with aryl Grignard reagents and with aryllithium reagents to afford biaryl products in varying yields. The reaction proceeds through triaryliodane intermediates, which undergo more or less spontaneously ligand coupling. Triphenyliodane (**7**), prepared by reaction of phenyllithium at - 80°C with either diphenyliodonium iodide (**5**) or phenyliodine dichloride (**6**), decomposes easily at - 10°C into biphenyl (**8**).[68] (Scheme 5.1) The cyclic aryliodinanes, 5-aryl-5*H*-dibenzoiodoles, are more stable. They are synthesised at 0°C and persist for hours to several days at room temperature.[69-72] Due to their rigidity, the cyclic iodinanes require more severe conditions for their decomposition, which can follow different pathways. At room temperature, 5-phenyl-5*H*-dibenzoiodole (**9**) gave a complex mixture of products, involving various free radical intermediates. In hexane under reflux, 2-iodo-*ortho*-terphenyl (**10**) was formed as the major product (80%).[70] The ligand coupling pathway was considered more likely than the free radical pathway, although it does not agree with the Grushin ligand coupling model for iodonium compounds.[16] (Scheme 5.2) The diorganoiodonium salt itself can be an intermediate of the overall sequence of reactions. This is the case of the reaction of aryliodonium dichlorides with arylmetallic reagents, which afforded directly biaryl products.[73,74]

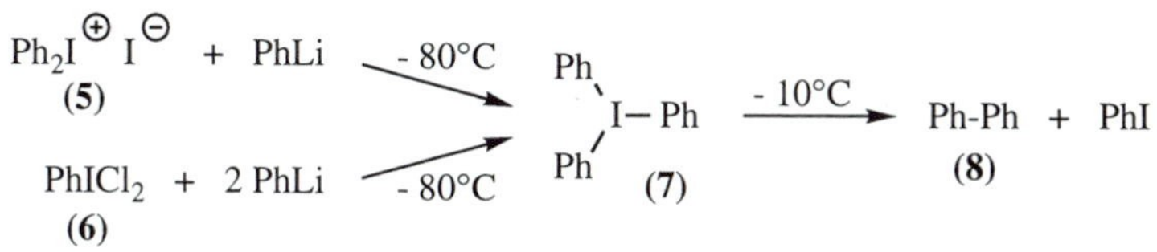

$$\text{Ph}_2\text{I}^{\oplus}\ \text{I}^{\ominus}\ +\ \text{PhLi}\ \xrightarrow{-80°C}$$

(**5**)

$$\text{PhICl}_2\ +\ 2\ \text{PhLi}\ \xrightarrow{-80°C}\ \overset{\text{Ph}}{\underset{\text{Ph}}{}}\text{I}-\text{Ph}\ \xrightarrow{-10°C}\ \text{Ph-Ph}\ +\ \text{PhI}$$

(**6**)   (**7**)   (**8**)

Scheme 5.1

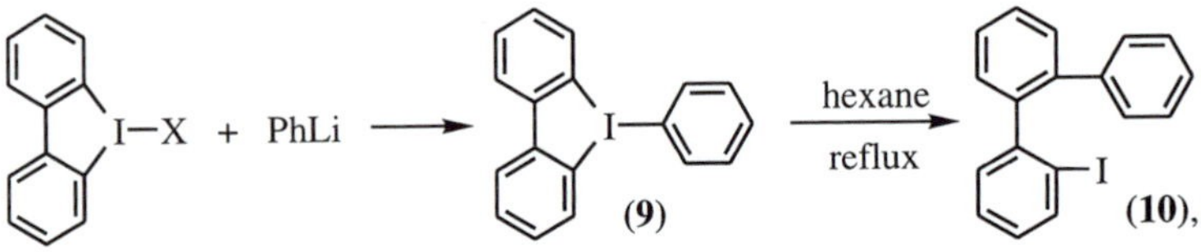

Scheme 5.2

The reaction of diaryliodonium salts with alkylmetal reagents affords in general modest yields of the arylalkanes, together with by-products resulting from a free radical pathway.[40,73,75] However, when iodobenzene dichloride was treated with butyllithium, good yields of octan were obtained. The careful mechanistic study eventually showed that, in this case, the reaction did not proceed by ligand coupling, but by a classical S_N2 mechanism.[76]

The behaviour of 3-indolylphenyliodonium trifluoroacetate (11) appeared unusual. It reacted with a number of alkyl and aryllithium reagents to lead to the 3-substituted indoles (12).[77]

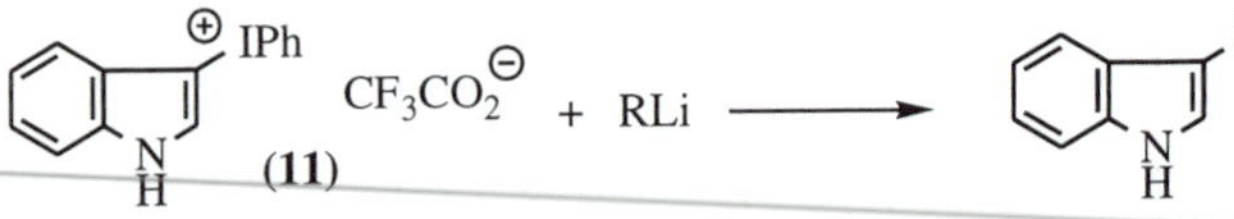

The reaction of iron pentacarbonyl with alkyl or phenyllithium gives acyltetracarbonylferrates, which react *in situ* with diaryliodonium salts to afford moderate to good yields of arylketones.[78]

$$Li[RCOFe(CO)_4] \ + \ Ar_2I^{\oplus} \ X^{\ominus} \longrightarrow \ R\text{-}CO\text{-}Ar \ + \ ArI$$
$$\qquad\qquad\qquad\qquad\qquad\qquad\qquad 48\text{-}85\% \qquad 85\text{-}92\%$$

5.1.3.2 Reactions with ketonic and ester compounds

Diaryliodonium salts react more or less easily with carbonyl compounds to afford the *C*-arylated derivatives. Depending upon the nature of the substrate, different experimental conditions have been used. These reactions are generally performed in alcoholic solvents (*t*-BuOH, *t*-AmOH ....) or in DMF, at temperatures ranging from low (- 78°C) to reflux of the solvent. Arylation of simple ketones has been obtained either by reaction of the ketone enolates with an appropriate diaryliodonium salt,[79-81] or by reaction of the ketone enol silyl ether with diaryliodonium fluoride.[82] Phenylation of the potassium enolate of acetone (13) with diphenyliodonium bromide (14) afforded a modest yield of the monophenylation product, but the stimulation with solvated electrons led to overreaction due to the subsequent reaction of the iodobenzene, a good $S_{RN}1$ arylating agent under these conditions.[80]

$$CH_3COCH_2^{\ominus} \ K^{\oplus} \ + \ Ph_2I^{\oplus} \ Br^{\ominus} \xrightarrow[\text{- 78°C}]{NH_3} \ CH_3COCH_2\text{-}Ph \quad 33\%$$
$$\quad\ (13) \qquad\qquad\quad (14)$$

Variable results were obtained when the enolates of more or less sterically hindered acetophenone derivatives were treated with diphenyliodonium salts in alcoholic solvents. Phenylation was observed together with side reactions such as polymerisation of free radical species.[79]

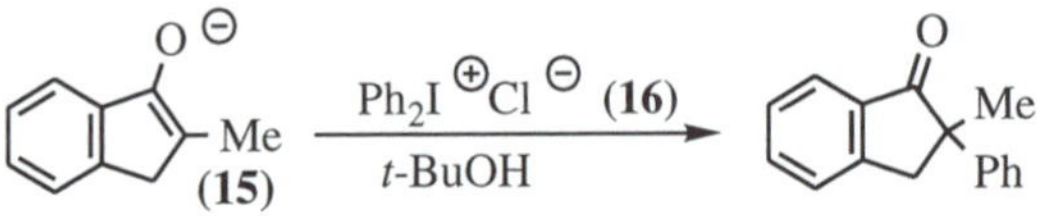

The phenylation of hydrocodone (**18**) was cleanly performed in a good yield by reaction of the lithium enolate with diphenyliodonium iodide (**5**).[81]

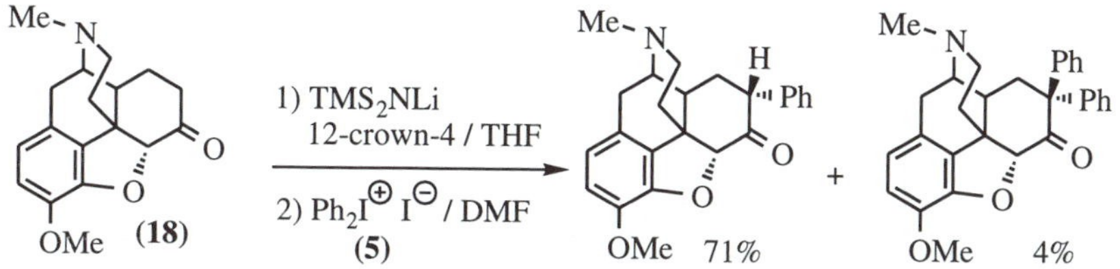

The occurrence of two competing pathways was also observed in the reaction of some enol silyl ethers with diphenyliodonium fluoride (**19**), which afforded the mono- or the di-phenylated products in moderate to good yields. This *C*-phenylation reaction is likely to occur by a ligand coupling process. However, with some hindered silyl enol ethers, for example (**20**), diketone dimers (**21**) were also isolated, indicative of a free-radical component in the overall mechanistic picture.[82]

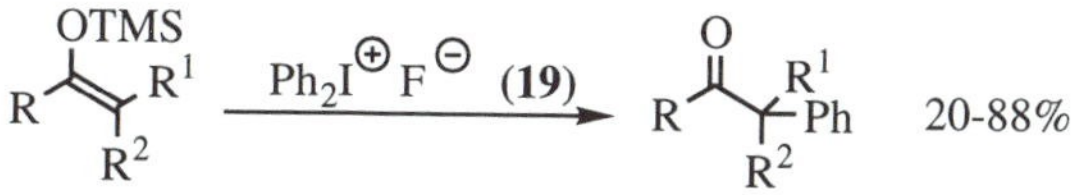

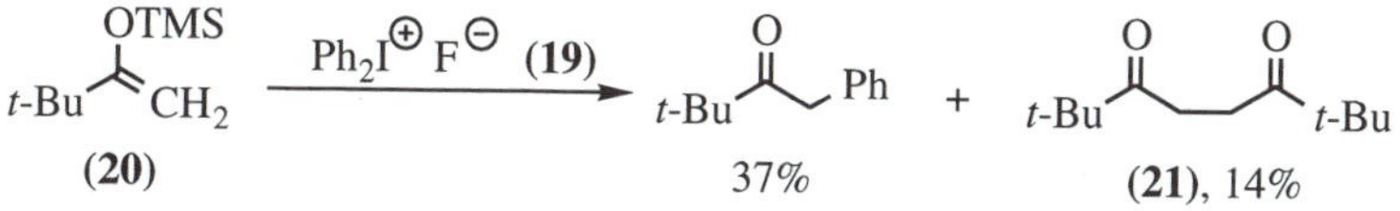

The major part of the reactions of α-arylation which have been reported were performed on substrates containing active methylene groups, such as β-diketones, β-ketoesters, β-ketonitriles and malonic acid derivatives. Less activated compounds, such as β-ketosulfides have also been efficiently arylated on the α-carbon. (Table 5.3) The yield of the arylated product can be increased by addition of 1,1-diphenylethylene, acting as a free radical trap. For example, in the reaction of phenylation of ethyl cyclohexanonecarboxylate, addition of 1,1-diphenylethylene reduced the radical chain decomposition in such a way that an 80% yield was obtained instead of 55% in absence of 1,1-diphenylethylene.[33]

Table 5.3: Examples of α-arylation of activated carbonyl derivatives.

Substrate	Product	Yield (%)	Ref
MeCOCH$_2$CO$_2$Et	MeCOCHPhCO$_2$Et	-	83
RCH(CO$_2$Et) R = H, alkyl, Ph, NHAc	RCH(CO$_2$Et)	23-68	29
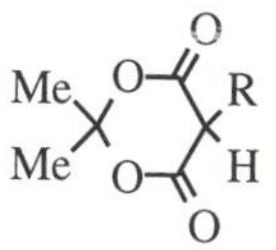 R = H, alkyl, Ar	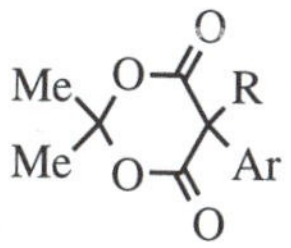	59-95	84

Substrate	Product	Yield (%)	Ref
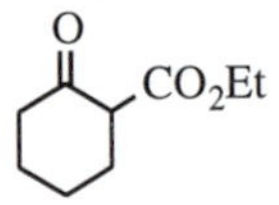 (2-oxocyclohexane CO₂Et)	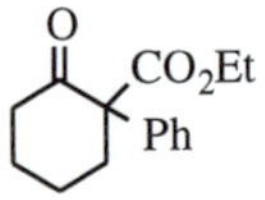 (2-oxocyclohexane CO₂Et, Ph)	60[a] 55[b] 80[c]	85 33 33
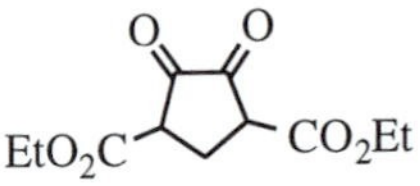 (cyclopentanedione, EtO₂C, CO₂Et)	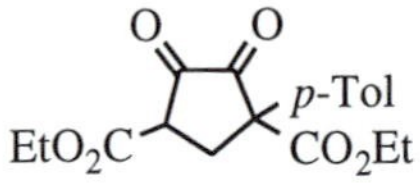 (cyclopentanedione, EtO₂C, *p*-Tol, CO₂Et)	-	86
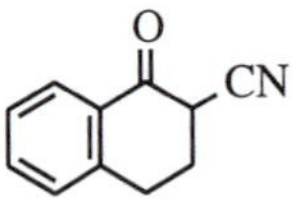 (tetralone CN)	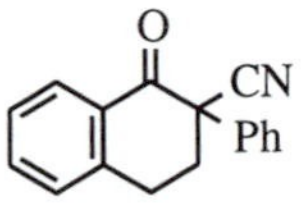 (tetralone CN, Ph)	50	79
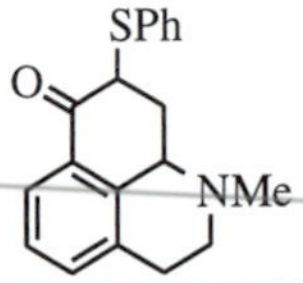 (SPh, O, NMe)	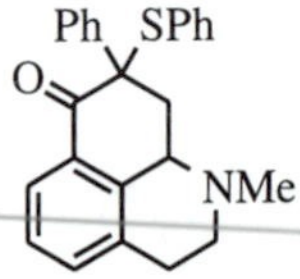 (Ph, SPh, O, NMe)	70	87

a - reaction with diphenyliodonium chloride and sodium *tert*-butoxide in *tert*-butanol under reflux

b - reaction with diphenyliodonium acetate and *N-tert*-butyl-*N',N',N",N"*-tetramethylguanidine in *tert*-butanol under reflux

c - reaction with diphenyliodonium acetate, *N-tert*-butyl-*N',N',N",N"*-tetramethylguanidine and 1,1-diphenylethylene in *tert*-butanol under reflux

In the case of β-diketones, a mixture of *C*- and *O*-arylated products were obtained in the reaction of 1,3-diketones with diaryliodonium salts, the *O*-aryl product being obtained generally in low yields. This ambient reactivity of 1,3-diketone derivatives was observed in different cases: sterically hindered reactants,[28,88] polybenzoylmethane derivatives[27] and dimedone.[27,89]

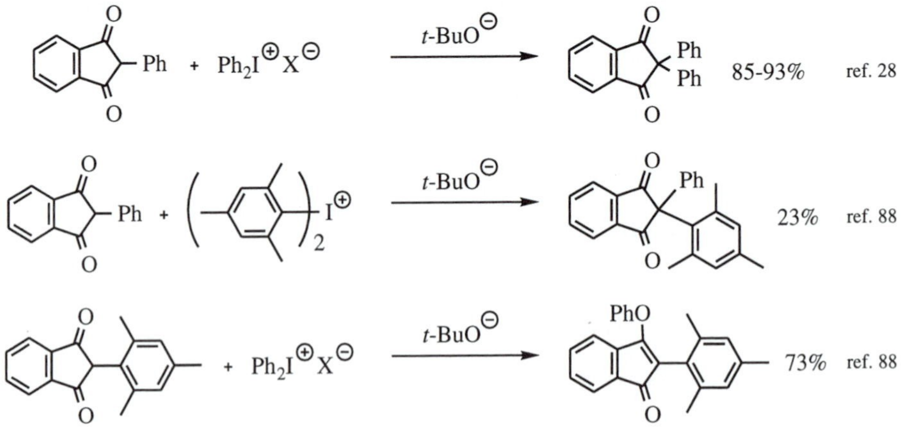

$$\text{(indandione-Ph)} + Ph_2I^{\oplus}X^{\ominus} \xrightarrow{t\text{-BuO}^{\ominus}} \text{(indandione-Ph,Ph)} \quad 85\text{-}93\% \quad \text{ref. 28}$$

$$\text{(indandione-Ph)} + \left(\text{(mesityl)}\right)_2 I^{\oplus} \xrightarrow{t\text{-BuO}^{\ominus}} \text{(indandione-Ph,mesityl)} \quad 23\% \quad \text{ref. 88}$$

$$\text{(indandione-mesityl)} + Ph_2I^{\oplus}X^{\ominus} \xrightarrow{t\text{-BuO}^{\ominus}} \text{(PhO-indandione-mesityl)} \quad 73\% \quad \text{ref. 88}$$

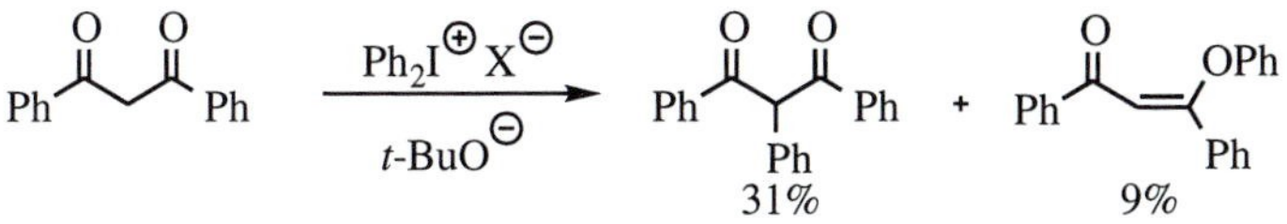

ref. 27

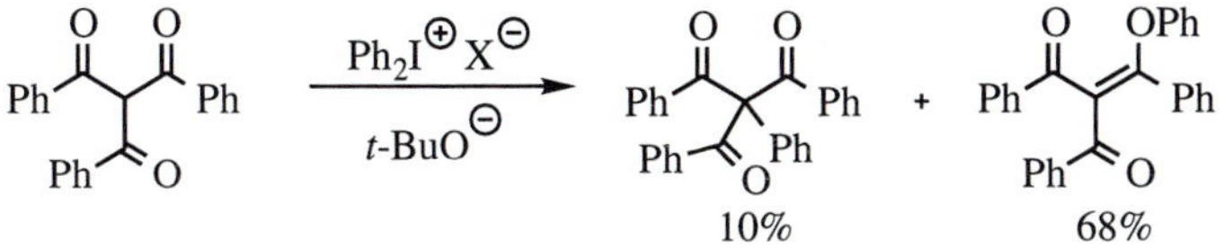

31% 9%

ref. 27

10% 68%

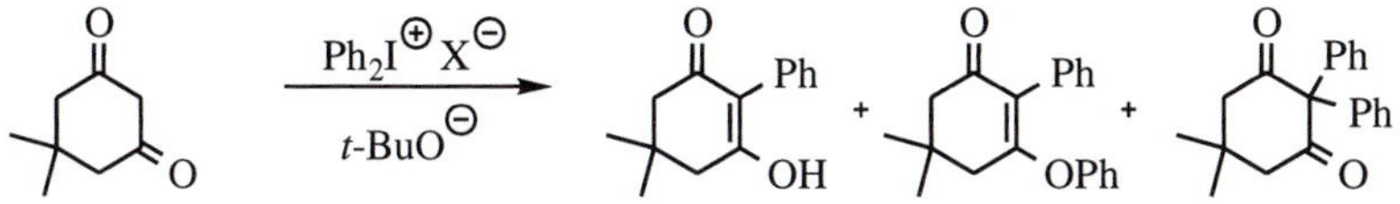

ref. 27,89

Arylation at the γ-position has been reported in the case of a number of β-diketones, when the dianions of these substrates were treated with diaryliodonium salts.[90,91]

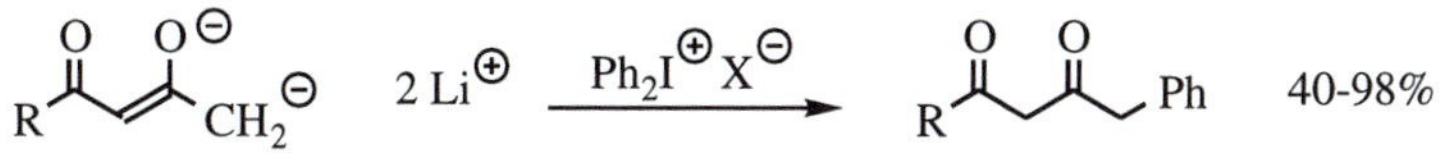

40-98%

An intramolecular variation of the arylation reaction of a β-diketone led to cyclised products (24). The reaction involved the treatment of an α-(aryl)alkyl β-dicarbonyl derivative (22) with phenyl-iodine(III) bistrifluoroacetate (23), followed by base catalysed ligand coupling reaction.[92]

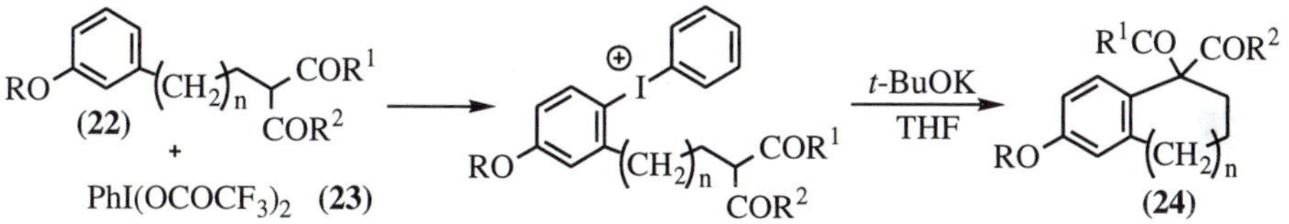

(22)

+

PhI(OCOCF$_3$)$_2$ (23)

t-BuOK
THF

(24)

42-88% yields for the two steps

5.1.3.3 Reactions with nitro compounds

Diphenyliodonium tosylate (25) reacted with the sodium salt of nitroalkanes to afford relatively good yields of exclusively the *C*-phenyl products. The reaction took place at room temperature in DMF, but required a higher temperature for the less reactive ethyl α-nitrocaproate (55°C).[93] Lower yields were obtained when the reaction was performed with diphenyliodonium bromide in methanol.[32]

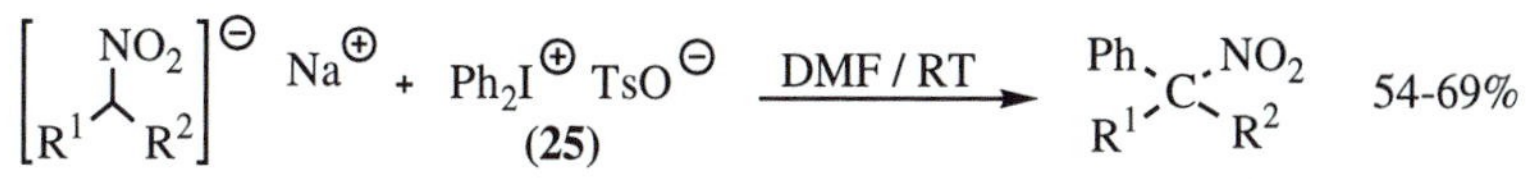

DMF / RT

(25)

54-69%

The potassium salts of 1,1-dinitroethane and 1,1-dinitropropane are also phenylated by diphenyl-iodonium tosylate (25) in *tert*-butanol under reflux (yields 67-68%). Poor yields were obtained when the reaction was performed in DMF (5%).[94]

5.1.3.4 Reactions with enamines

The reaction of cyclohexanone enamines (**26**) with diaryliodonium salts gave only poor yields of the α-aryl ketones (**27**) after hydrolysis.[95]

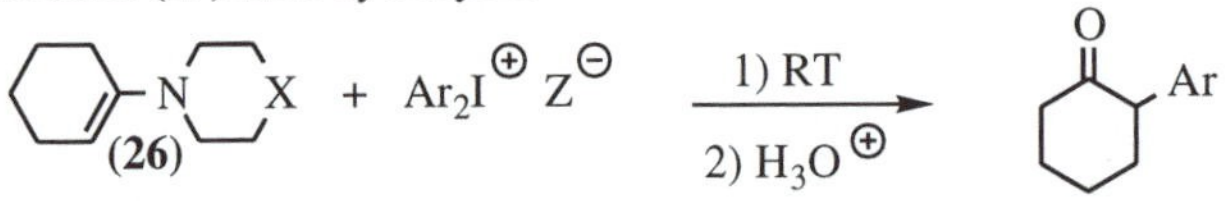

5.1.3.5 Miscellaneous reactions

The reaction of the 3-phenyliodonium salt of quinolin-2,4-diol (**28**) with various nucleophilic reagents led to the ligand coupling products in good yields.[96] (Scheme 5.3)

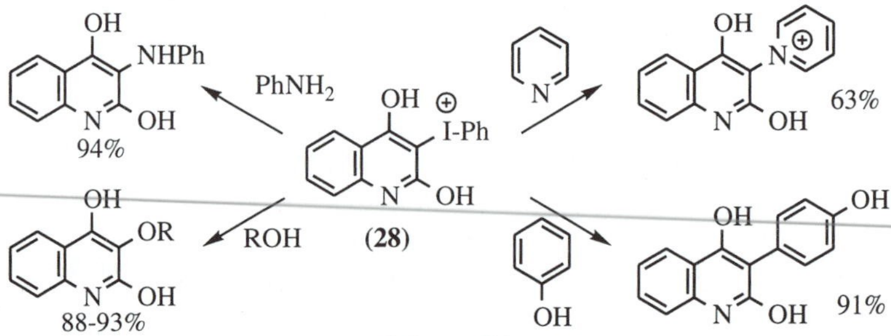

Scheme 5.3

The reaction of diaryliodonium salts with carbon monoxide in the presence of methanol led to the formation of methyl benzoate derivatives.[97]

$$Ar_2I^{\oplus} X^{\ominus} + CO + MeOH \longrightarrow ArCO_2Me \quad 49\text{-}60\%$$

5.1.4 Reactions of diaryliodonium salts with oxygen nucleophiles

5.1.4.1 Reactions with alcohols

The *O*-arylation of methanol by reaction of diaryliodonium bromides with sodium methoxide was first reported by Beringer *et al.* to afford good yields of anisole derivatives (43-71%).[40] Later studies showed that the reactions of diaryliodonium salts with metal alkoxides led to a range of products.[57,98-100] The alkyl aryl ether is accompanied by side products such as arenes, aldol resins or ketones, which are sometimes preponderant. Two major pathways can explain the occurrence of these products: a radical chain reaction and a ligand coupling reaction. The generally modest yields of alkyl aryl ethers can be improved by addition of a crown ether.[101] The unwanted radical pathway can be stopped by addition of DPE (1,1-diphenylethylene), acting as a radical scavenger. This results in a significant increase of the yield of the ligand coupling product, the alkyl aryl ether.[99]

$$p\text{-MeC}_6H_4IPh^{\oplus} BF_4^{\ominus} + EtONa \xrightarrow{\text{EtOH, 71°C}} PhOEt + p\text{-MeC}_6H_4OEt$$
$$9.5\% \qquad 2.6\%$$

$$p\text{-MeC}_6H_4IPh^{\oplus} BF_4^{\ominus} + EtONa \xrightarrow[\text{Ph}_2C=CH_2 \text{ (1 equiv.)}]{\text{EtOH, 71°C}} PhOEt + p\text{-MeC}_6H_4OEt$$
$$55\% \qquad 22\%$$

In the case of unsymmetrical diaryliodonium salts (the anisyl, tolyl or nitrophenyl derivatives of phenyliodonium salts[99,100] or some thienyl aryliodonium salts[57]), the product composition is always in favour of the ligand coupling product involving the less electron-rich arene group. A good yield of 3-methoxyindole (60%) was obtained in the reaction of phenyl 3-indolyliodonium chloride with methanol in the presence of boron trifluoride-etherate.[77]

$$p\text{-MeOC}_6\text{H}_4\overset{\oplus}{\text{I}}\text{Ph}\ \overset{\ominus}{\text{BF}_4} + \text{EtONa} \longrightarrow \underset{15\%}{\text{PhOEt}} + \underset{1\%}{p\text{-MeOC}_6\text{H}_4\text{OEt}}$$

$$p\text{-MeC}_6\text{H}_4\overset{\oplus}{\text{I}}\text{C}_6\text{H}_4 m\text{-NO}_2\ \overset{\ominus}{\text{BF}_4} + \text{EtONa} \longrightarrow \underset{4\%}{p\text{-MeC}_6\text{H}_4\text{OEt}} + \underset{79\%}{m\text{-NO}_2\text{C}_6\text{H}_4\text{OEt}}$$

5.1.4.2 Reactions with carboxylic acid salts

Diaryliodonium salts react with sodium benzoate to give the aryl benzoates in moderate to good yields (40-85%).[40] Treatment of biphenyleneiodonium sulfate (29) with sodium methoxide failed to give 2-iodo-2'-methoxybiphenyl. However, this product was obtained by an indirect sequence involving the formation of 2-iodo-2'-acetoxybiphenyl (30) by reaction of the biphenyleneiodonium sulfate with sodium acetate in boiling glacial acetic acid. Nearly quantitative yields were obtained when the reaction was catalysed by copper salts.[102]

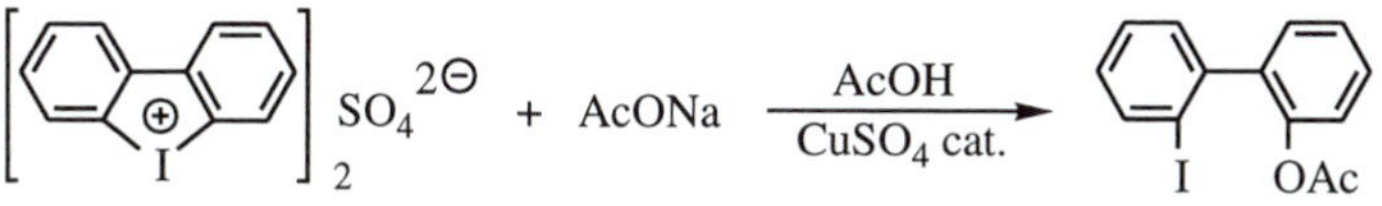

(29)   (30), 99%

5.1.4.3 Reactions with phenols

The reaction of phenols with diaryliodonium salts leads to the diaryl ethers in modest to good yields, depending on the substrate and the reaction conditions.[31,40,51,103] A number of studies have dealt with this reaction as it was involved in the synthetic sequence towards diaryl ethers of biological importance, such as bisbenzylisoquinoline alkaloids[104] or thyronine and tyrosine derivatives.[105] The reaction requires relatively harsh conditions, as it is performed under reflux of the solvent. In the case of unsymmetrical diaryliodonium salts, the less electron-rich aryl group of the iodonium salt is preferentially transferred.

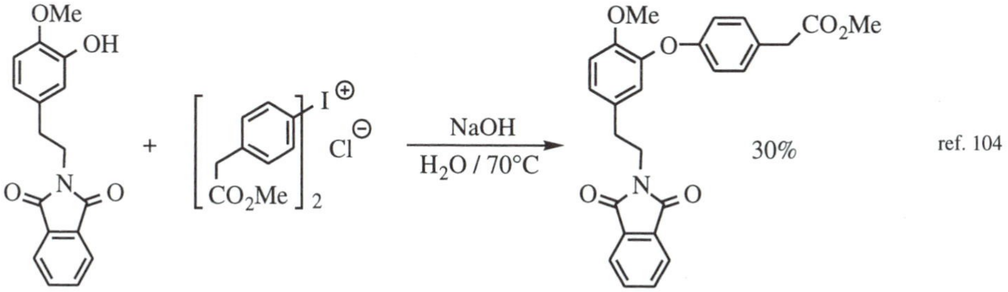

30%   ref. 104

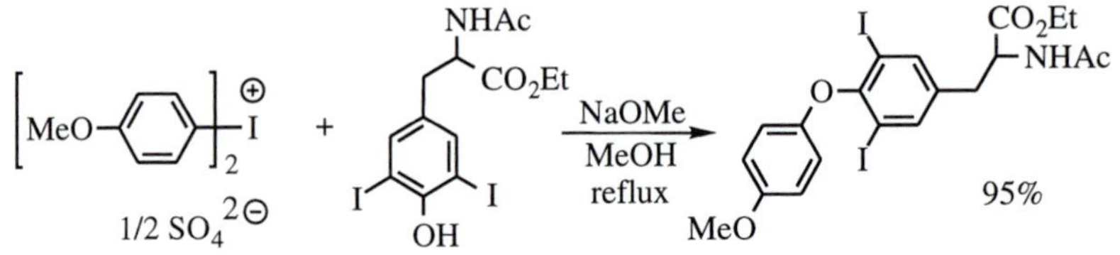

ref. 105

A range of unsymmetrical polychlorinated diarylethers have been obtained by this reaction in modest to good yields.[106,107]

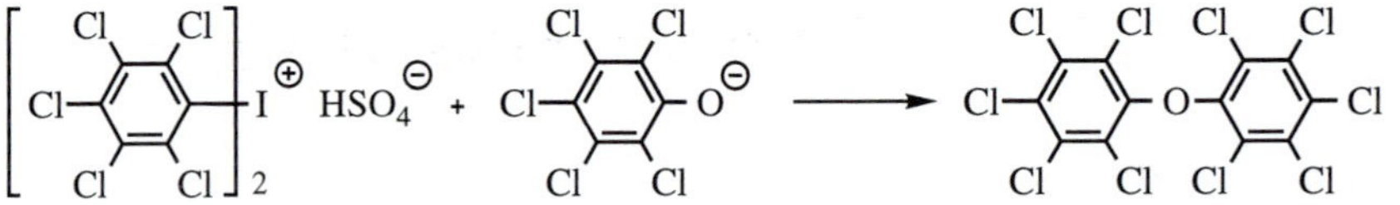

30%

5.1.4.4 Reactions with nitrogen-containing compounds arylated at oxygen

Hydroxamic acids react with diaryliodonium salts to afford the *O*-phenyl derivatives.[108,109] A number of heterocyclic compounds containing *N*-hydroxyamino groups were selectively arylated on the oxygen atom.[110-116] (Table 5.4) In the case of oximate anions, ambident behaviour was observed, with predominant *O*-arylation.[117] With heterocyclic oximes, the *O*-aryl ethers were mainly formed and served as precursors to prepare unstable aryl fulminates.[118,119] (Table 5.5) In the reaction of quinolone derivatives with diaryliodonium salts, the products of *O*- or *C*-arylation were obtained.[120,121] (Table 5.6)

Table 5.4: *O*-Arylation of *N*-hydroxyamino compounds

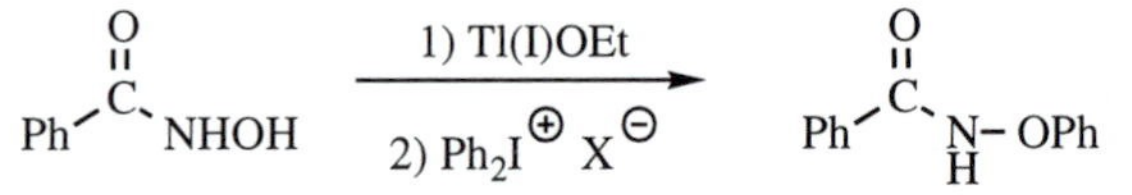	66%	ref. 109
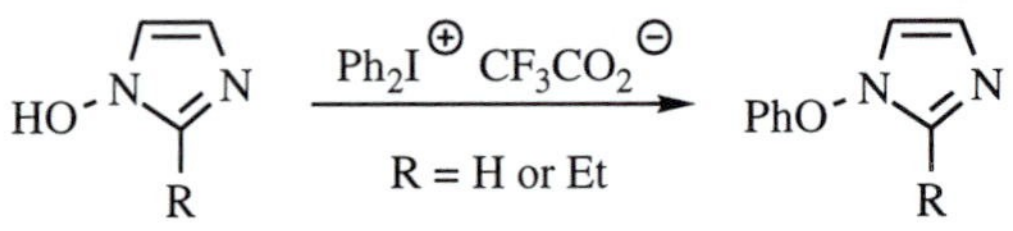	70-72%	ref. 110
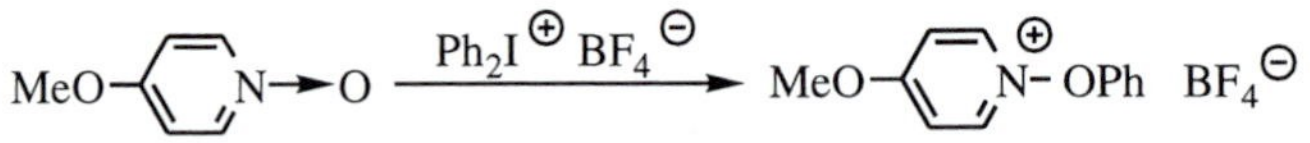	95%	ref. 111
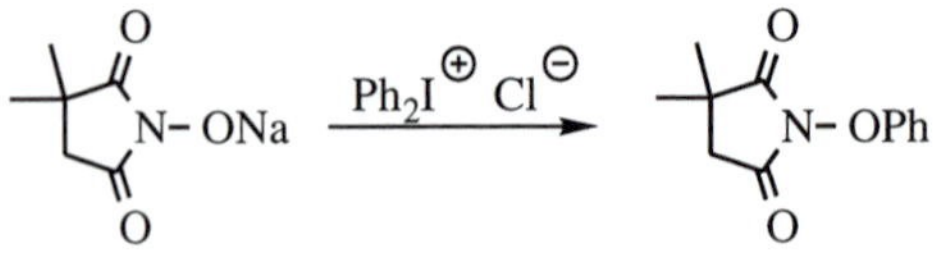		ref. 115
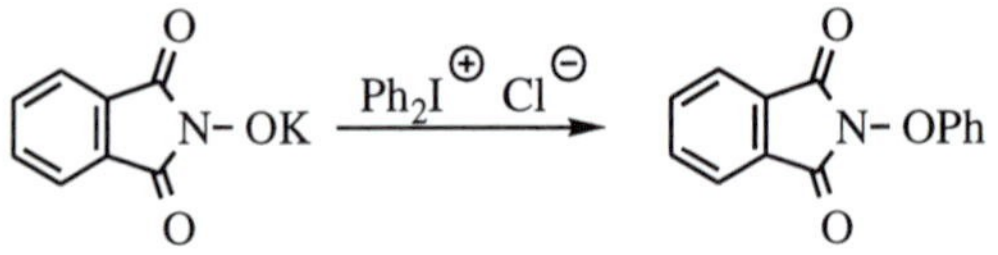	66%	ref. 114

Table 5.5: *O*-Arylation of oximes

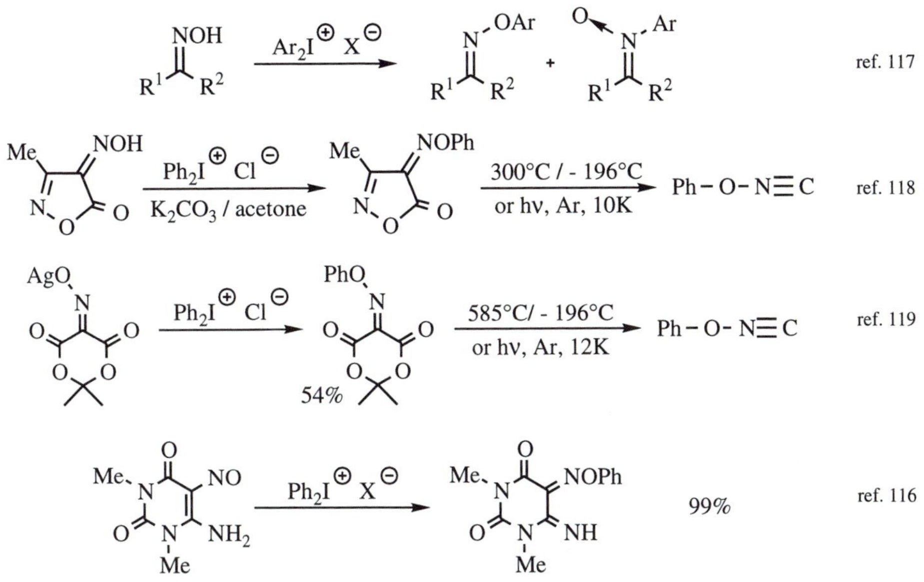

Table 5.6: *O*-Arylation of quinolin-2-one derivatives

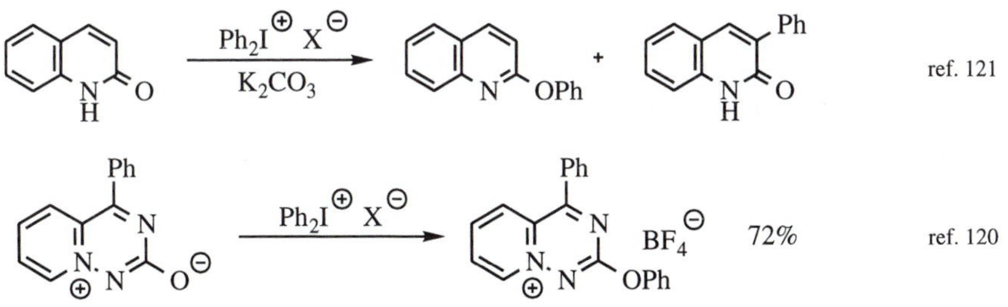

5.1.5 Reactions of diaryliodonium salts with nitrogen nucleophiles

The reaction of amines with diaryliodonium salts does not give, in general, synthetically useful yields of the aniline derivatives. Competing pathways take place to afford the ligand coupling product and the formation of free radicals which evolve into different by-products. The outcome is dependent upon the nature of the amine and of the reagent and its counterion.[40,122] The reaction of *N,N*-diphenylamine was reported to afford triphenylamine in 20% yield.[123,124]

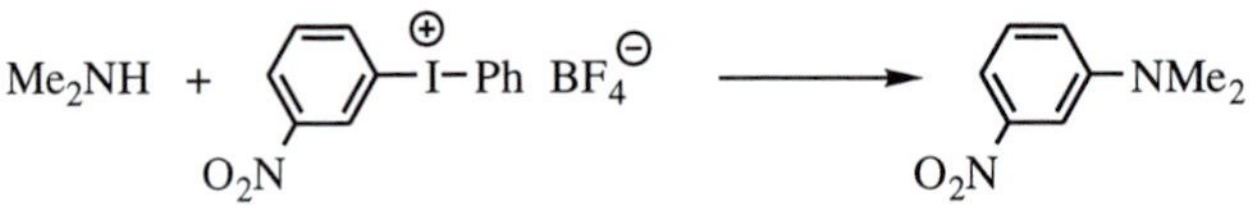

$$\text{Me}_2\text{NH} + \underset{\text{O}_2\text{N}}{\text{[Ar]}}-\overset{\oplus}{\text{I}}-\text{Ph } \text{BF}_4^{\ominus} \longrightarrow \underset{\text{O}_2\text{N}}{\text{[Ar]}}-\text{NMe}_2 \quad 43\% \qquad \text{ref. 40}$$

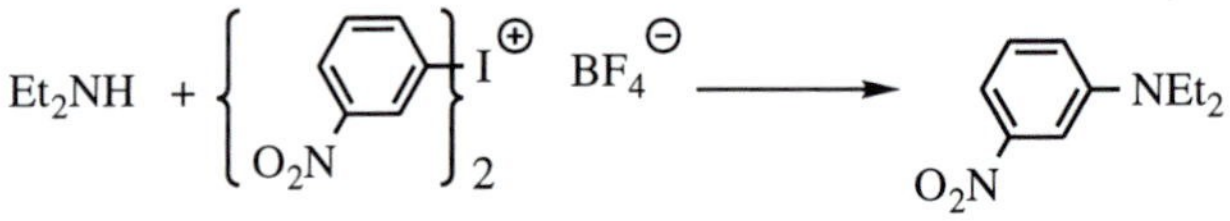

$$\text{Et}_2\text{NH} + \left\{ \underset{\text{O}_2\text{N}}{\text{[Ar]}}\overset{\oplus}{\text{I}} \right\}_2 \text{BF}_4^{\ominus} \longrightarrow \underset{\text{O}_2\text{N}}{\text{[Ar]}}-\text{NEt}_2 \quad 4\% \qquad \text{ref. 40}$$

$$\text{Ph-NH}_2 + \text{Ph}_2\overset{\oplus}{\text{I}} \text{Br}^{\ominus} \xrightarrow{\text{H}_2\text{O}} \text{Ph-Br} \qquad \text{only product} \qquad \text{ref. 122}$$

$$\text{Ph-NH}_2 + \text{Ph}_2\overset{\oplus}{\text{I}} \text{BF}_4^{\ominus} \xrightarrow{i\text{-PrOH}} \text{Ph-NH-Ph} \qquad 45\text{-}50\% \qquad \text{ref. 122,124}$$

The outcome of the reaction of pyridine with diaryliodonium salts was dependent upon the nature of the counterion. With a non-nucleophilic counterion as the tetrafluoroborate, the *N*-phenylpyridinium salt (31) was obtained in high yield.[123] With bromide, no reaction occurred and with chloride, ring arylation was observed as a result of the formation of phenyl free radicals.[125]

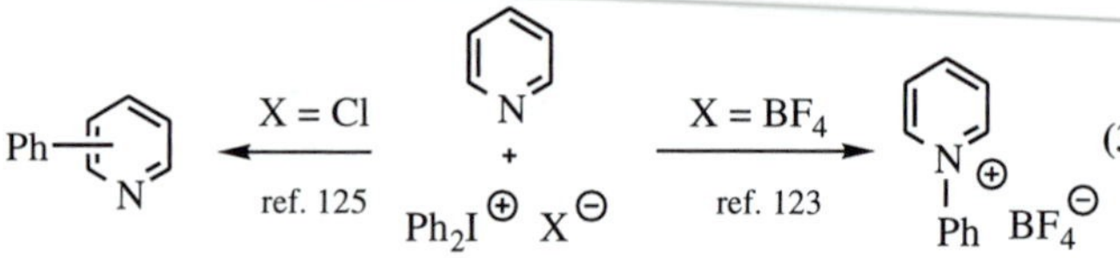

A range of amidic nitrogen functions have been *N*-arylated by diaryliodonium salts. The reaction conditions vary with the nature of the substrate, and moderate to good yields of the mono- or di-arylated products (32) - (34) can be obtained.[40,126]

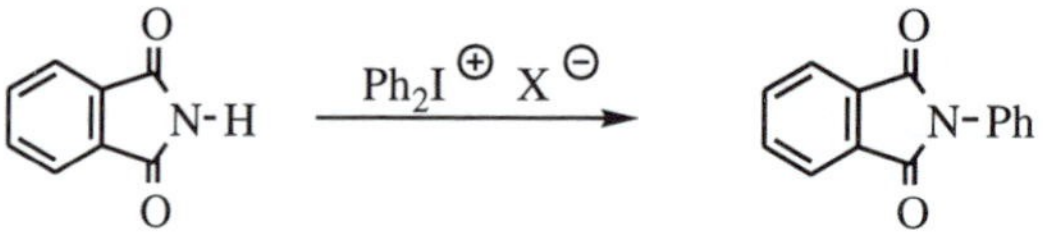

(32), 33%   ref. 126

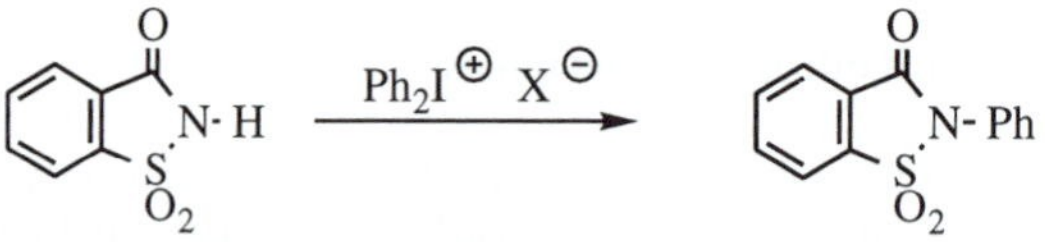

(33), 75%   ref. 40

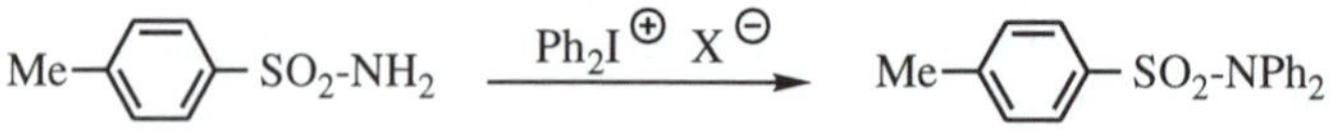

$$\text{Me}-\text{[Ar]}-\text{SO}_2\text{-NH}_2 \xrightarrow{\text{Ph}_2\overset{\oplus}{\text{I}} \text{X}^{\ominus}} \text{Me}-\text{[Ar]}-\text{SO}_2\text{-NPh}_2 \quad (34),\ 45\% \qquad \text{ref. 40}$$

Although the reaction of hydroxamic acids with diaryliodonium salts affords the *O*-phenyl derivatives,[108,109] (see Section 5.1.1.3.3) the reaction of the closely related *N*-hydroxycarbamates led exclusively to the *N*-aryl derivatives.[127]

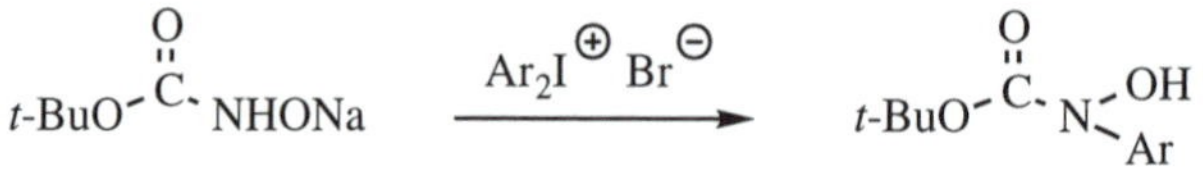

$$t\text{-BuO}-\overset{\overset{\text{O}}{\|}}{\text{C}}-\text{NHONa} \xrightarrow{\text{Ar}_2\overset{\oplus}{\text{I}} \text{Br}^{\ominus}} t\text{-BuO}-\overset{\overset{\text{O}}{\|}}{\text{C}}-\text{N}\overset{\text{OH}}{\underset{\text{Ar}}{}} \quad 23\text{-}90\% \qquad \text{ref. 127}$$

The overall process may be further complicated, as the treatment of *N*-alkyl-*N*-hydroxycarbamate (35) led to an intermediate *N*-aryl derivative which underwent a series of rearrangements and acyl migrations to afford eventually the 2-hydroxyanilide derivative (36).[128]

A variety of heterocyclic systems afforded the *N*-aryl derivatives.[120,129-132] Aryne intermediates were sometimes involved. Indeed, the reaction of the sodium anion of 5-phenyltetrazole (37) with di-*p*-tolyliodonium bromide gave the two 1,5- and 3,5-diarylated derivatives as mixtures of *meta*- and *para*-tolyl regioisomers in a 1:1 ratio.[129]

5.1.6 Reactions of diaryliodonium salts with organochalcogen nucleophiles

Reactions of diaryliodonium salts with various organochalcogen anions led to the arylchalcogen derivatives. (Table 5.7) The anions of sulfur compounds afford generally the *S*-aryl derivatives. Sometimes the intermediate iodanes can be isolated. Examples are the reaction of dithiocarbamate and xanthate salts with diaryliodonium.[133,134] Upon heating, the dithiocarbamate derived iodane afforded the aryldithiocarbamates.[133] Thermal decomposition of the xanthate derived iodanes gave mixtures of arylxanthates and diaryldisulfides.[134]

Thiolates react with diaryliodonium salts to afford the corresponding aryl sulfide derivatives in fair to good yields.[135,136] The intermediate arylthioiodanes, Ar_2ISR, have been sometimes isolated and their thermal decomposition led to the unsymmetrically substituted sulfides.[134,136] A small range of heteroaromatic sulfides afforded the *S*-aryl compounds.[134,137,138] Under more severe conditions, the diarylsulfides react further with diaryliodonium salts to afford the corresponding triarylsulfonium salts either under thermal activation[123,139,140] or under photochemical activation.[141] Upon copper catalysis, good to quantitative yields are then obtained at room temperature.[142-145] Good yields of 1-aryl-benzo[b]thiophenium triflates were similarly obtained by the copper catalysed arylation of benzo[b] thiophene with diphenyliodonium triflate.[146] Moreover, the reaction was extended to the double arylation of the more readily available arylthiols.[147] Similarly, the reaction of areneselenolates with diaryliodonium salts led to diarylselenides[148] and triarylselenonium salts were easily obtained by copper catalysed arylation of diarylselenides.[142]

The arylation of the phosphorus atom of organophosphorus compounds has been described in the case of triarylphosphine. Best yields are obtained when the reaction is performed under photochemical activation, although thermal activation afforded also the tetraarylphosphonium salts.[149,150]

Table 5.7: Arylation of organochalcogen anions by diaryliodonium salts

Nucleophile	Product	Yield (%)	Ref
RSNa	RSAr	-	135,145
PhSR	Ph$_2$SR $^\oplus$	81-98*	151
Ar^{1}Ar^{2}S	Ar^{1}Ar^{2}Ar^{3}S $^\oplus$	65-99*	142
ArSH	Ph$_2$SAr $^\oplus$	48-87*	147
Ar^{1}SO$_2$Na	Ar^{1}SO$_2$Ar^2	22-95	40,59,60
RC(O)SK	RC(O)SAr	42-76	152
ArC(S)SNa	ArC(S)SAr	65-88	153
R^{1}R^{2}NC(S)SNa	R^{1}R^{2}NC(S)SAr	56-90	154,155
Ar^{1}SeNa	Ar^{1}SeAr^2	58-75	148
Ph$_2$Se	Ph$_2$SeAr $^\oplus$	49-90*	142
Ar^{1}TeNa	Ar^{1}TeAr^2	46-81	156

* copper-catalysed reaction

5.1.7 Copper-catalysed reactions of diaryliodonium salts

The arylation of nucleophiles by reaction with diaryliodonium salts can be greatly facilitated by copper catalysis. This effect was observed by Beringer *et al.* in the thermal decomposition of diaryliodonium halides[41] as well as by Caserio *et al.* in the hydrolysis of diaryliodonium salts.[52] The thermal decomposition of diphenyliodonium chloride shows a reduced activation energy upon copper catalysis : Ea = 19 kcal/mole in diethylene glycol in the presence of CuCl instead of 31 kcal/mole in the absence of catalyst.[41] From the synthetic point of view, the copper-catalysed arylation with diaryliodonium salts has been particularly useful in the case of a number of reactions involving heteroatomic nucleophiles, in particular for *S-*, *Se-*, *O-* and *N-*arylation reactions.

The *S-*arylation of sulfides is greatly improved by copper catalysis. The uncatalysed reaction requires heating under reflux to afford modest to moderate yields of the *S-*aryl sulfonium derivatives. By contrast, upon copper catalysis, the reaction proceeds at room temperature to afford good to high yields of the *S-*aryl sulfonium derivatives.[142-146,151] Moreover, the reaction was extended to the double arylation of the more readily available arylthiols.[147] Similarly, the reaction of areneselenolates with diaryliodonium salts led to diarylselenides[148] and triarylselenonium salts were easily obtained by copper catalysed arylation of diarylselenides.[142]

ref. 145

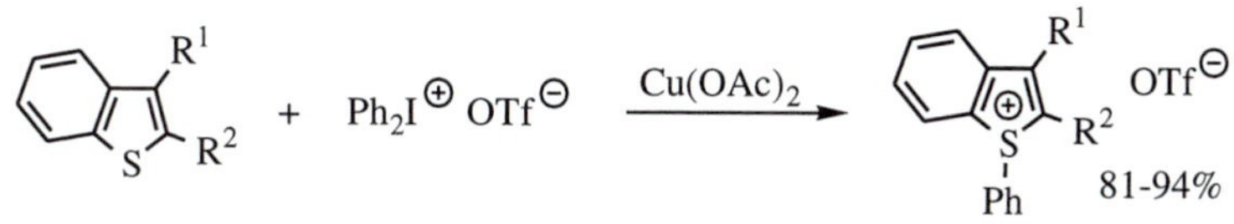

ref. 146

$$2\,Ar_2I^{\oplus}X^{\ominus} + Ar'SH + R_3N \xrightarrow[\text{or Cu(II)}]{\text{Cu(I)}} Ar_2\overset{\oplus}{S}\text{-}Ar'\ X^{\ominus} + 2\,ArI + R_3NH^{\oplus}X^{\ominus}$$

48-87% ref. 147

$$Ar_2I^{\oplus}\ AsF_6^{\ominus} + Ph_2Se \xrightarrow{Cu(OBz)_2} Ph_2\overset{\oplus}{Se}\text{-}Ar\ AsF_6^{\ominus} \quad 49\text{-}90\%$$

ref. 142

The catalytic effect of copper salts on the *O*-arylation of alcohols has been observed, but its synthetic importance remains limited.[61,157] The formation of 2-iodo-2'-acetoxybiphenyl (**30**) by reaction of the biphenyleneiodonium sulfate (**29**) with sodium acetate in boiling glacial acetic acid was quantitative when the reaction was catalysed by copper salts.[102] (see 5.1.4.2) In the case of phenols, the *O*-arylation by diaryliodonium salts is preferentially performed in a weakly basic medium (generally triethylamine) in the presence of copper bronze in dichloromethane or methanol at room temperature or at mild temperatures (around 50°C). Under these conditions, the *O*-aryl ethers are obtained easily in good yields.[158-160]

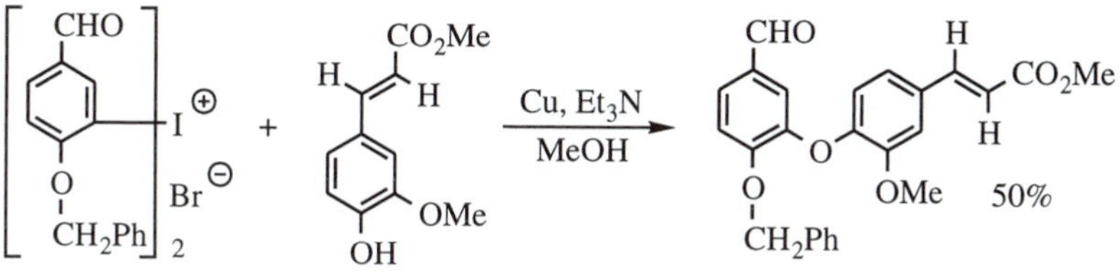

ref. 159

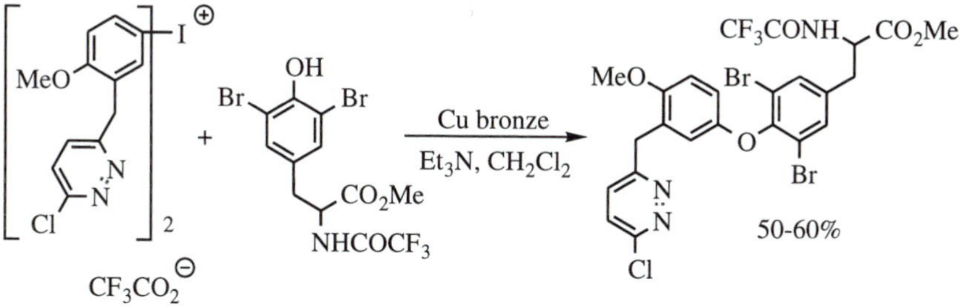

ref. 160

The *N*-arylation of amines also greatly benefits from copper catalysis, mostly in the case of anilines.[61,161,162] The reaction involves the treatment of the aniline by the diaryliodonium salt in 2-propanol or in DMF, at temperatures ranging from 80 to 100°C. *N*-Arylanthranilic acid derivatives were obtained in high yields by the copper(II) catalysed reaction of the appropriate aniline with various diphenyliodonium-2-carboxylate derivatives.[161,162] The less electron-rich ligand is transferred to the anilines nearly exclusively. Although diphenyliodonium-2-carboxylate (**38**) is a good source of benzyne at higher temperatures under neutral conditions,[163-165] arynic intermediates are not involved in this copper catalysed reaction.

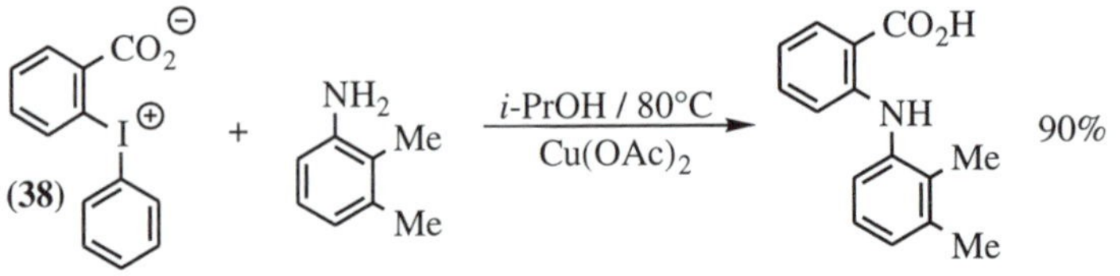

The copper catalysis can also influence the regiospecificity of a reaction involving ambient nucleophiles. For example, the uncatalysed reaction of thiocyanate affords arylthiocyanates, ArSCN. Upon copper catalysis, the same system leads to arylisocyanates, ArNCS, as the major products.[60] The copper-catalysed reaction of iodonium salts with organostannanes using copper(I) iodide as a catalyst in DMF at room temperature led to cross-coupling products in high yields. When the reaction was performed in the presence of carbon monoxide under mild conditions, it led to the synthesis of organoarylketones in good yields. Aqueous conditions were compatible to effect the same mild reactions when the boron reagents, in the presence of a base, were used instead of the stannane reagents.[166]

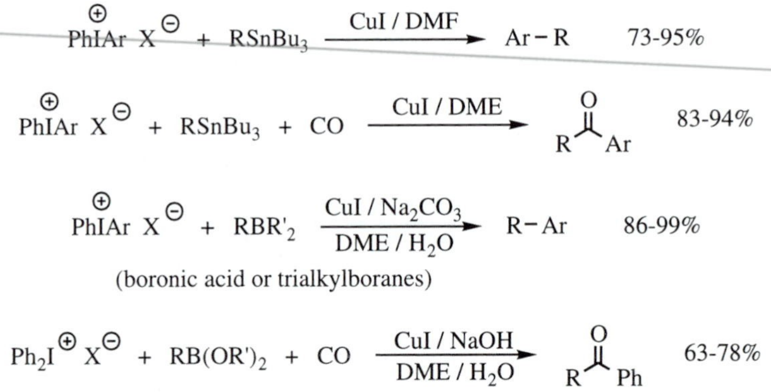

5.2 ARYLVINYLIODONIUM SALTS

Phenyl vinyliodonium salts are good precursors of vinyl cation equivalents due to the high nucleofugality of the phenyliodonio group. This has been evaluated to be 10^6 times as high as that of the triflate group.[167] Therefore, the chemistry of alkenyliodonium salts involves mostly reactions with nucleophiles leading to alkenylated products. The phenyliodonio group is generally replaced by the nucleophile with retention of configuration. These reactions are best explained by the ligand coupling mechanism. (Scheme 5.4)

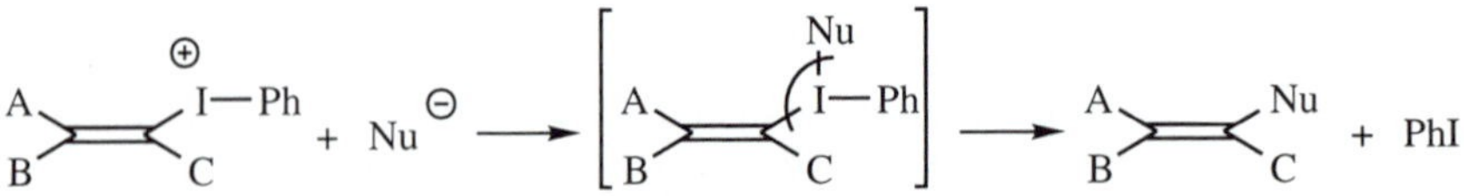

Scheme 5.4

However, other important mechanisms have been observed to play a role in the chemistry of alkenyliodonium compounds: Michael-type addition, alkylidenecarbene generation, and β-elimination affording alkynes.

5.2.1 Reactions with halide nucleophiles

Thermal decomposition of vinyliodonium salts (39) leads to the halide derivatives (40).[168] In the case of substituted vinyliodonium salts such as propenyliodonium salts (41), thermolysis gave the product of inversion of configuration (42).[169] However, reaction with copper chloride at room temperature afforded by contrast the chloroalkene (43) with retention of configuration.

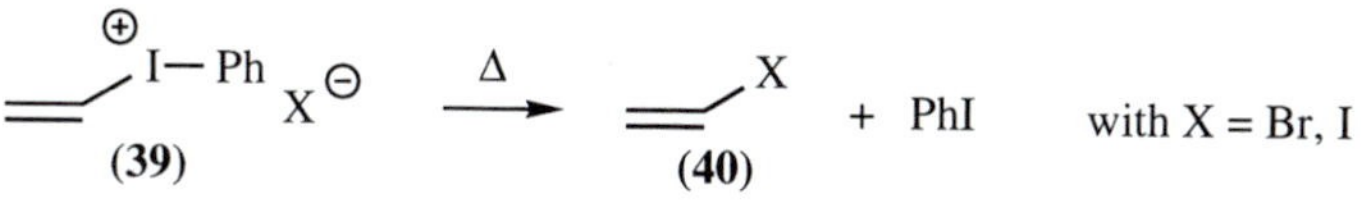

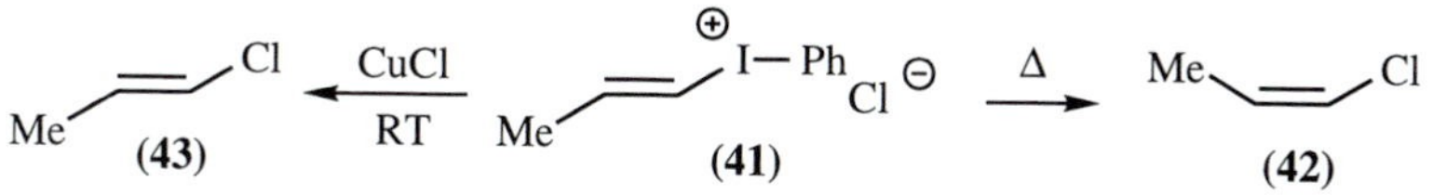

The outcome of the reaction of alkenyliodonium salts with halide salts depends on the substitution pattern of the iodonium substrate. Thus, treatment of (E)-alkenyliodonium salts with tetra(n-butyl)ammonium halides (X = Cl, Br or I) afforded exclusively the corresponding (Z)-alkenyl halides, the product of inversion of configuration.[170] The mechanism of these reactions with configuration inversion still remains a matter of debate, an S_N2-type transition being postulated. When the same reaction was performed with a combination of cuprous halides and potassium halides at room temperature in the dark, complete retention of stereochemistry was observed.

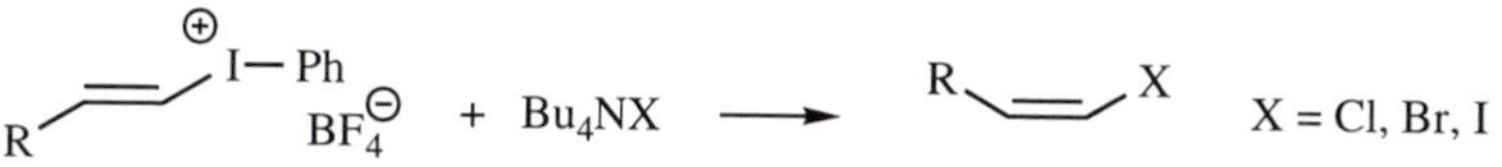

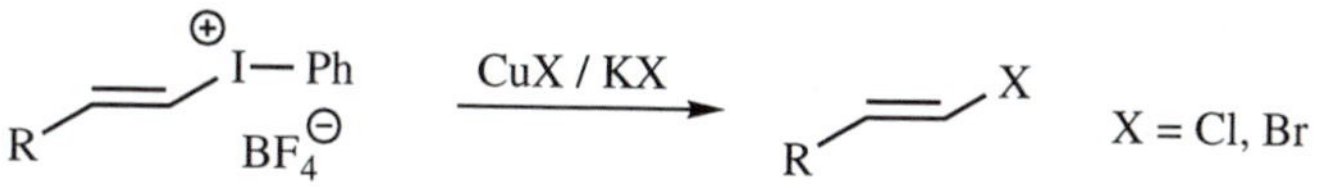

When the β-carbon of the vinyl group bears two substituents, retention of configuration is now observed as a result of a ligand coupling mechanism. However, an alternative mechanism involving addition-elimination (Ad-E) cannot be formally excluded.[171,172]

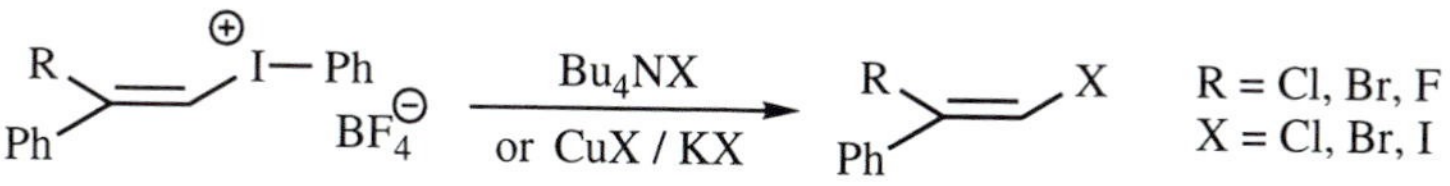

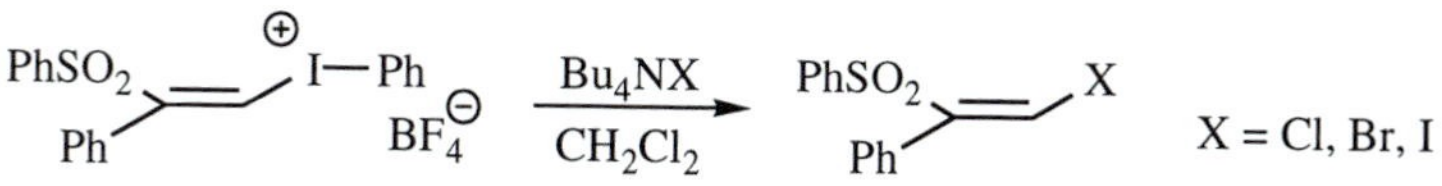

The reaction of the cyclic cyclohexenyliodonium salt (44) with potassium halides in the presence of copper halides gave good to high yields of the vinyl halides (45).[173,174]

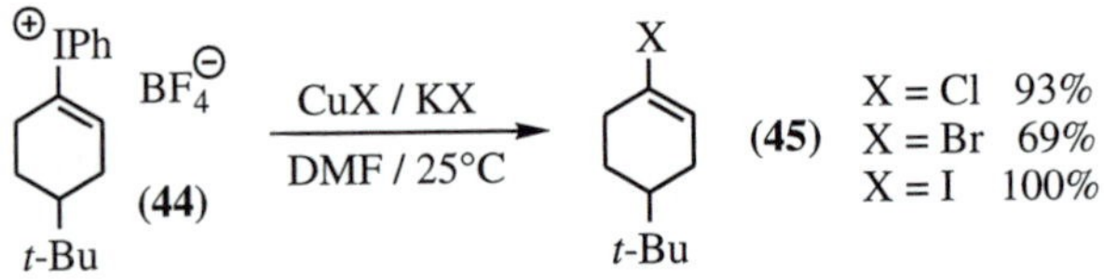

$$X = Cl \quad 93\%$$
$$(45) \quad X = Br \quad 69\%$$
$$X = I \quad 100\%$$

A hindered vinyl iodide, such as (47), was prepared in high yield by treatment of the iodonium salt (46) with the copper(I) iodide - potassium iodide system.[175]

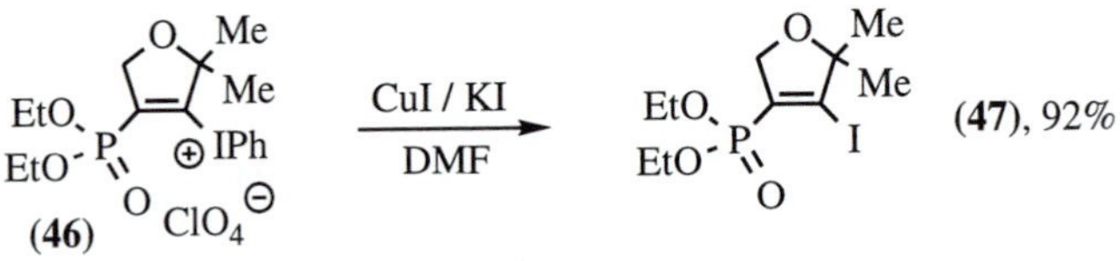

Similarly, alkenyldiiodonium salts (48) reacted with sodium halide and copper(I) halide salts to afford the vinyl dihalide derivatives (49) in modest to good yields.[176]

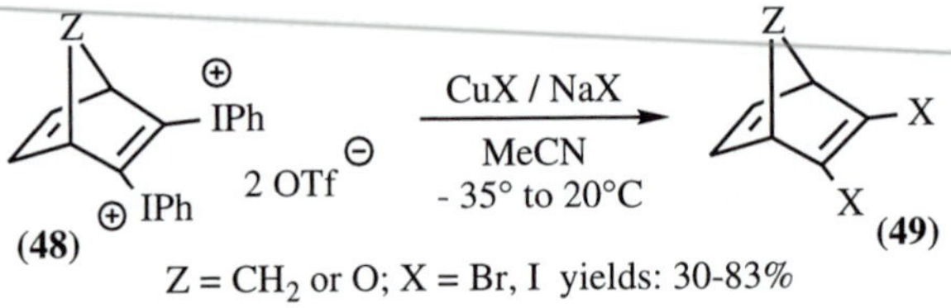

$$Z = CH_2 \text{ or } O; \; X = Br, I \quad \text{yields: } 30\text{-}83\%$$

5.2.2 Reactions with carbon nucleophiles

Two types of carbon nucleophiles have been used with vinyliodonium salts: carbanions and alkylcopper reagents. The reagents of the latter type have shown the most interesting synthetic possibilities, although they have not been extensively used.

The reaction of β-diketone enolates with vinyliodonium salts led to the α-vinyl-β-diketone derivatives in good yields, but the number of known examples is rather limited.[174,177]

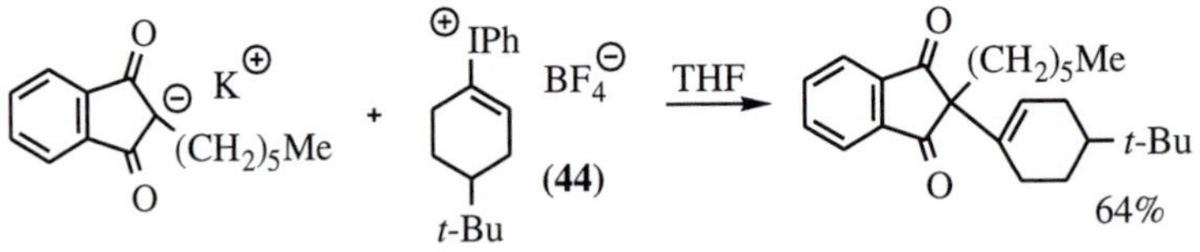

ref. 174

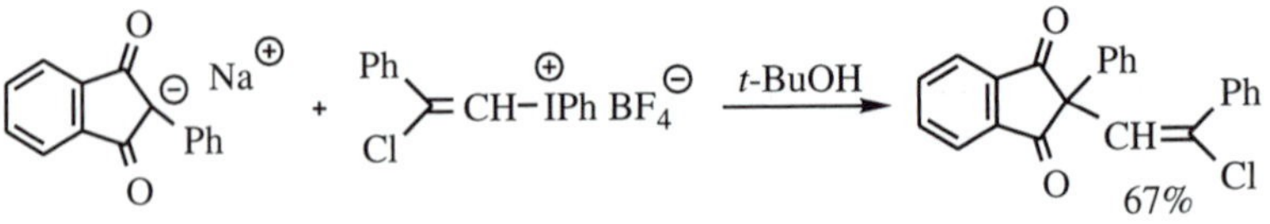

ref. 177

The reaction of the iodonium salt (44) with *n*-butyllithium did not afford significant amounts of the ligand coupling product [only 0.2% of the 1-*n*-butylcyclohexene derivative (50)], but led to a complex mixture of hydrolysis products.[174]

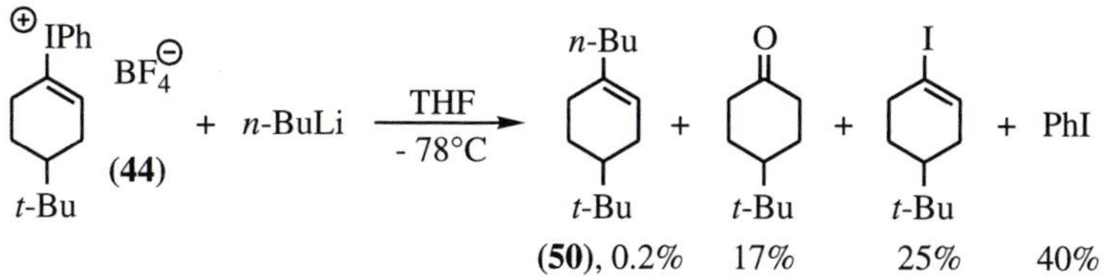

Treatment of the linear β-halovinyliodine dichloride **(51)** with aryllithium is an indirect method for the synthesis of diaryliodonium salts, ligand exchange being the main process.[178-180]

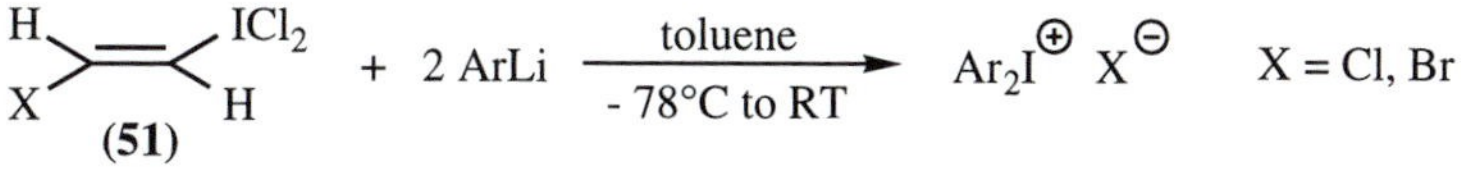

The reaction is believed to proceed stepwise to form first an arylvinyliodonium salt **(52)**, which reacts a second time with aryllithium to afford a diarylvinyliodane **(53)**. This iodane undergoes β-elimination to afford eventually the diaryliodonium compound and acetylene.[178]

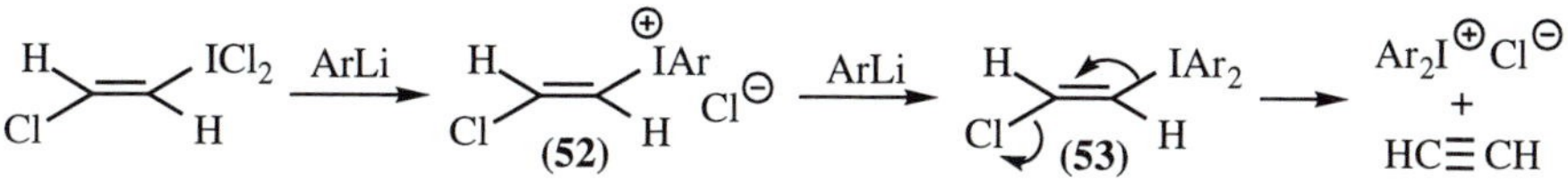

Similarly, [β-(trifluoromethylsulfonyloxy)vinyl]phenyliodonium triflate **(54)** reacted with aryllithium to give arylphenyliodonium triflates by ligand exchange.[181]

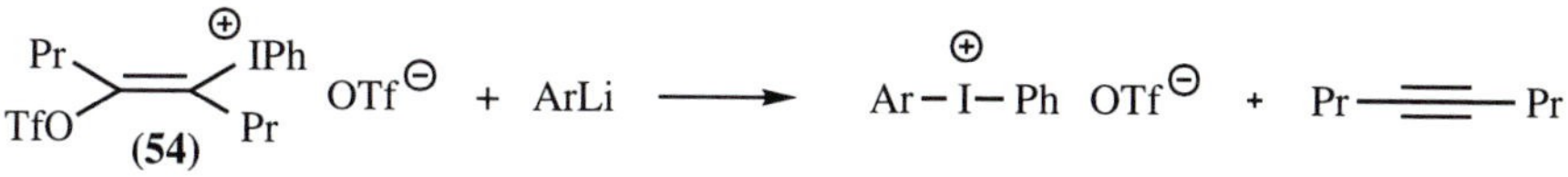

Different types of organocopper reagents have been used as efficient nucleophiles for vinyliodonium salts. With the cyclohexenyliodonium salt **(44)**, the conversions to 1-cyano, 1-alkyl and 1-arylcyclohexenes are examples of these copper(I) assisted reactions.[174]

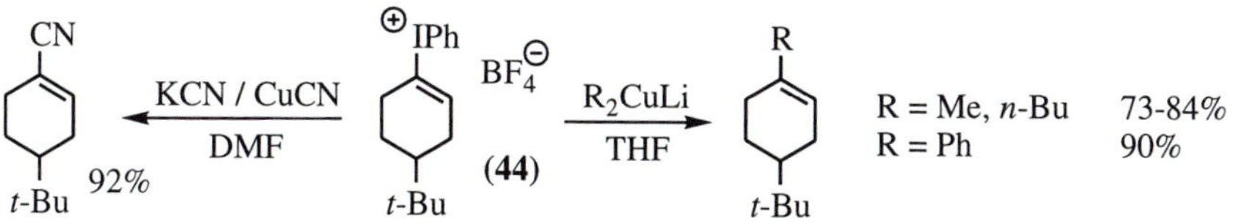

This type of copper-catalysed reaction was also extended to the alkynylation of alkenyldiiodonium salts **(48)**, through reaction with lithium alkynyl cuprates.[182]

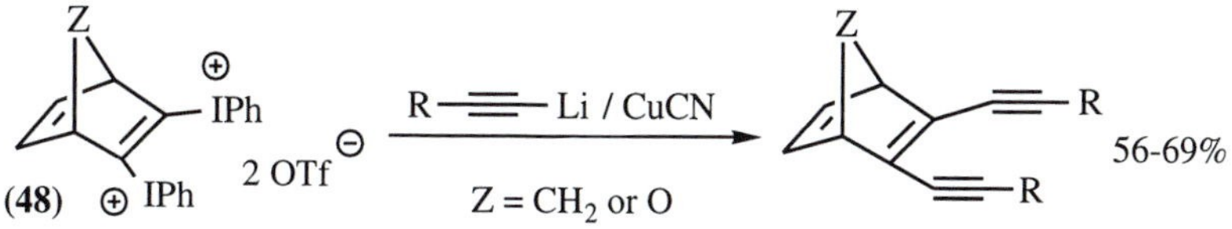

Pyrrole and indole derived reagents also reacted with vinyliodonium subtrates in copper catalysed reactions.[183,184]

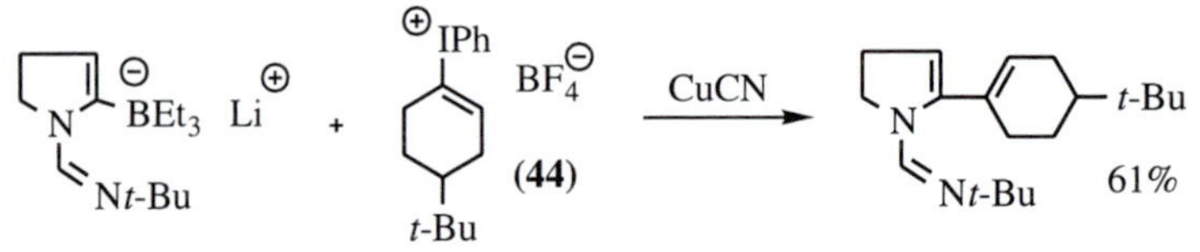

ref. 183

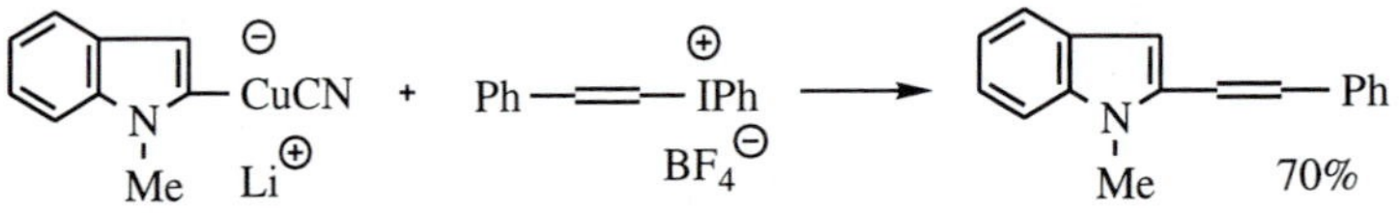

(44)

61%

ref. 184

70%

The copper catalysed vinylation of styreneboronic esters (55) can be realised with styryliodonium salts (56) under aqueous conditions, in the presence of a base, and affords high yields of 1,4-diphenyl butadiene (57).[166]

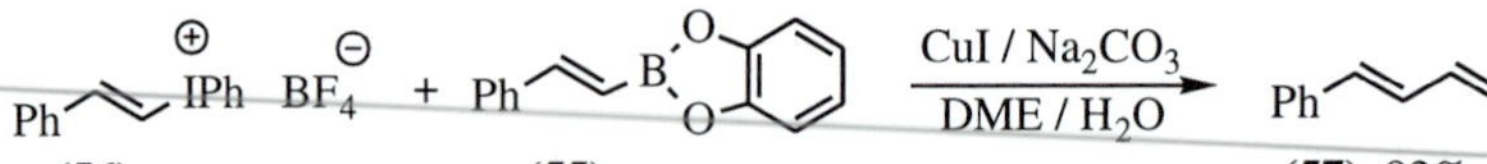

(56) + (55) → (57), 92%

CuI / Na_2CO_3
DME / H_2O

The copper catalysed reactions, either halogenations or phenylations, afford the product of retention of configuration. A ligand coupling mechanism has been suggested to explain these results.[174] Oxidative coupling of the cuprate and the vinyliodonium ions gives λ^3-iodanes (58). In a first step, these iodanes undergo ligand coupling with elimination of iodobenzene to form a second intermediate (59), a copper(III) species. This copper(III) intermediate subsequently undergoes a second ligand coupling, to afford eventually the overall reaction product (60). (Scheme 5.5)

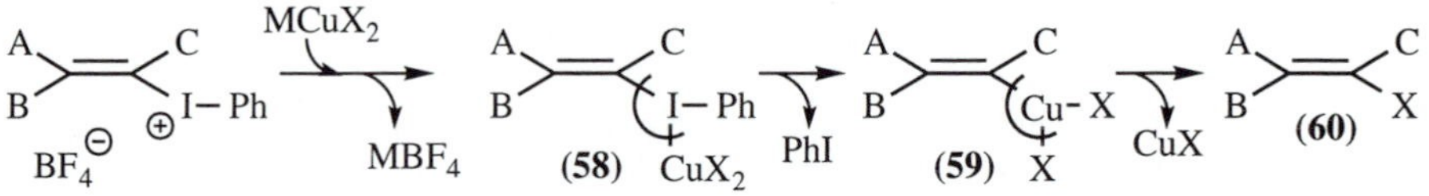

Scheme 5.5

5.2.3 Reactions with heteroatomic nucleophiles

A limited number of reactions of vinyliodonium salts with oxygen nucleophiles has been reported. These reactions usually lead to the ligand coupling products.[168,175,176,185]

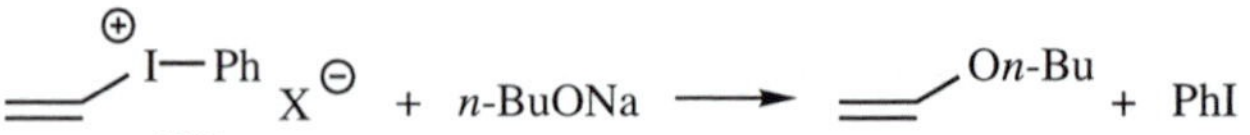

(39) + n-BuONa → On-Bu + PhI   ref. 168

+ AgOTf (cat.) → OTf   15-20%   ref. 185

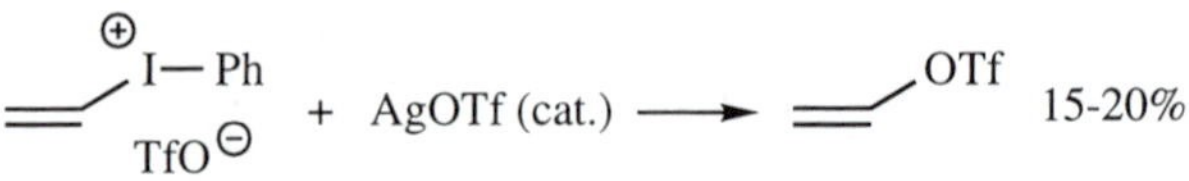

(48) → PhCOONa → OCOPh / OCOPh 25% ref. 176

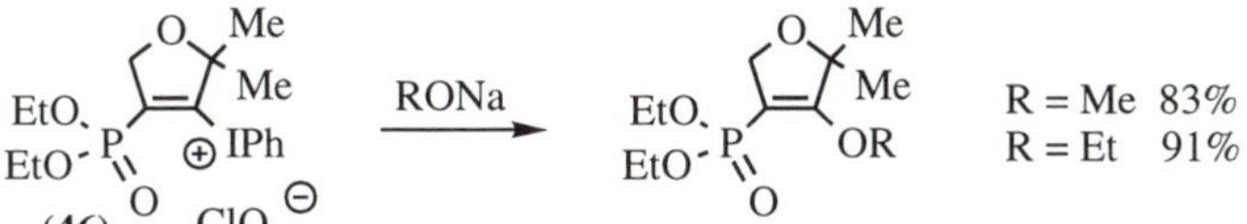

(46) → RONa → R = Me 83% / R = Et 91% ref. 175

In the case of the reaction of ethenylphenyliodonium iodide [(**39**), X = I] with silver(I) oxide, the formation of a vinylhydroxyiodane (**61**) was suggested to take place, this intermediate collapsing by ligand coupling to afford acetaldehyde (**62**).[22,168] In the case of the reaction of the cyclic iodolium salt (**63**), a different mechanism takes place as ring opening happens nonstereospecifically to give a mixture of the geometric isomers of stilbene derivatives (**64**).[186]

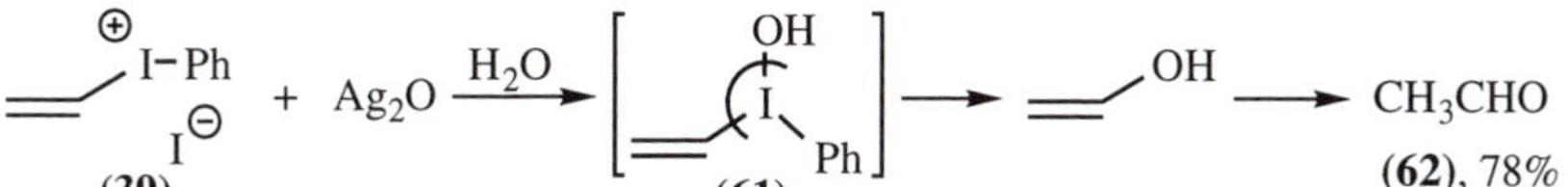

(39) + Ag$_2$O $\xrightarrow{H_2O}$ [(61)] → → CH$_3$CHO (62), 78% ref. 168

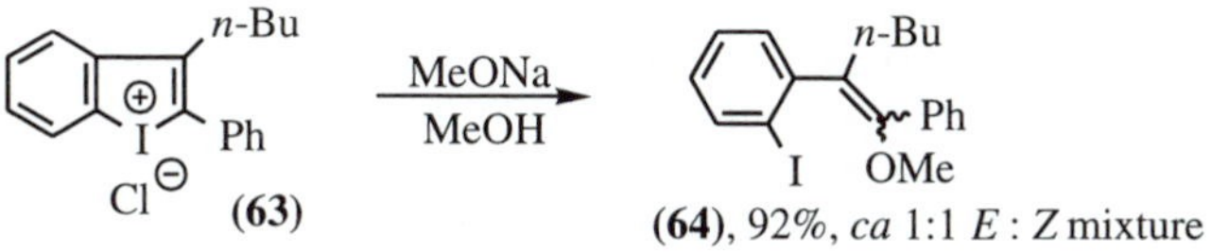

(63) $\xrightarrow[\text{MeOH}]{\text{MeONa}}$ (64), 92%, *ca* 1:1 *E* : *Z* mixture ref. 186

The lactonic vinyliodonium salt (**65**) undergoes ring opening and ligand coupling upon treatment with water or methanol. The reaction was explained to proceed first by ring opening of the lactone by solvolysis. The intermediately formed α-phenyliodonioketone (**66**) can undergo either a direct S$_N$ displacement or this can result from the formation of an hydroxy- or alkoxy-iodane collapsing by ligand coupling.[187]

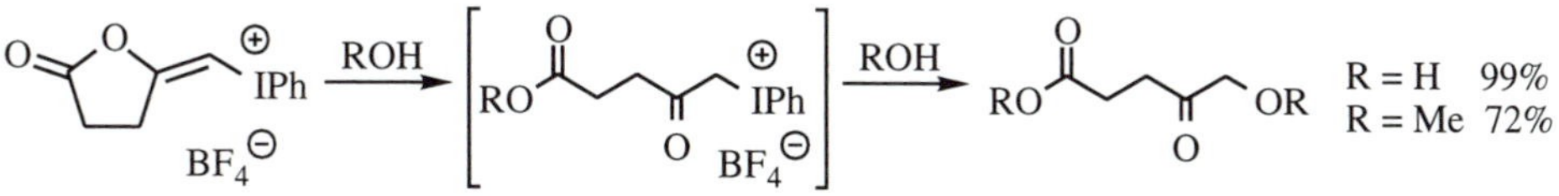

(65) $\xrightarrow{\text{ROH}}$ [(66)] $\xrightarrow{\text{ROH}}$ R = H 99% / R = Me 72%

Three types of reactions involving nitrogen nucleophiles have been reported. One is the reaction of vinyliodonium salts (**44**) with sodium nitrite which affords nitroalkene derivatives.[174,188] For the cyclohexenyliodonium salt, copper(II) sulfate catalysis was required for the reaction to proceed efficiently. The second one is the reaction of the vinyllactone (**65**) with diethylamine, which, similarly to the reactions with oxygen nucleophiles, reacted by ring opening followed by displacement of the phenyliodonio group to yield the bis(diethylamino) derivative.[187] The third type is the reaction of sodium azide with the cyclic dihydrofuryliodonium perchlorate (**46**), which afforded quantitative yield of the vinyl azide.[175]

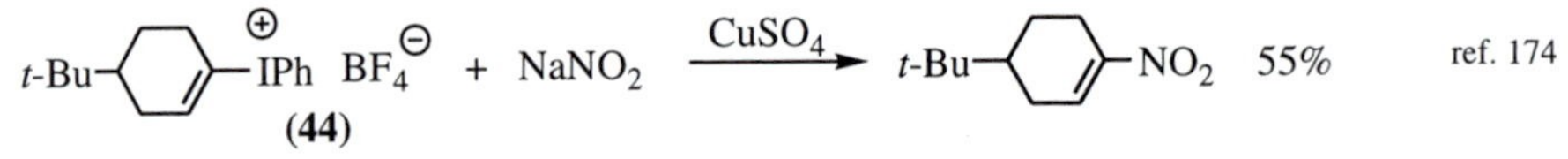

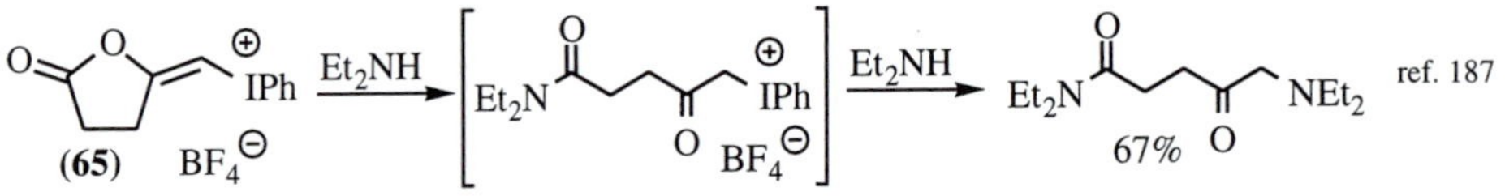

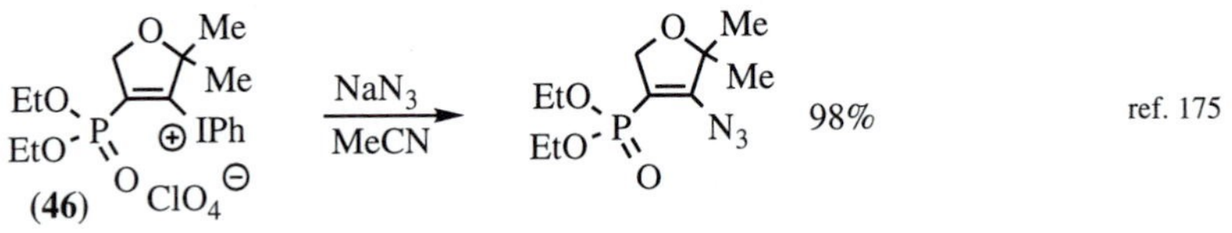

The cyclohexenyliodonium salt (44) reacted with sodium thiophenoxide to give the corresponding vinylphenylsulfide.[174]

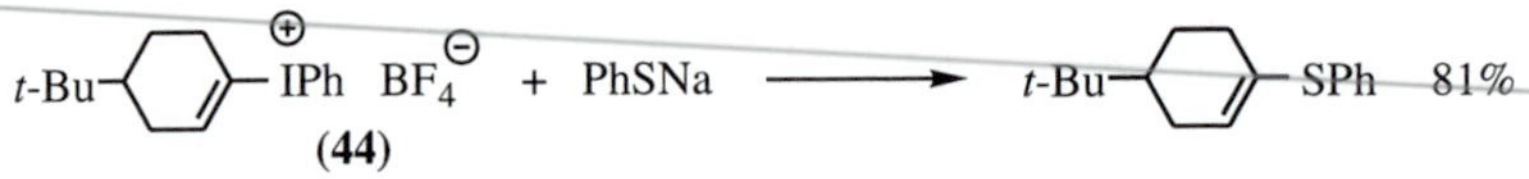

A second type of sulfur nucleophile has also been tested with cyclic as well as with linear vinyliodonium salts. The cyclohexenyliodonium salt (44) reacted with sodium phenylsulfinate to give the corresponding vinylphenylsulfone in a poor yield (29%). In the presence of 18-crown-6, a significant change was obtained, as the yield of sulfone was raised to 80%.[189] Acyclic (Z)-β-phenylsulfonyl alkenyliodonium tetrafluoroborates (67) undergo nucleophilic vinylic substitution upon treatment with sodium phenylsulfinate to afford good yields of the (Z)-1,2-bis(phenylsulfonyl)alkenes (68) with complete retention of stereochemistry.[189]

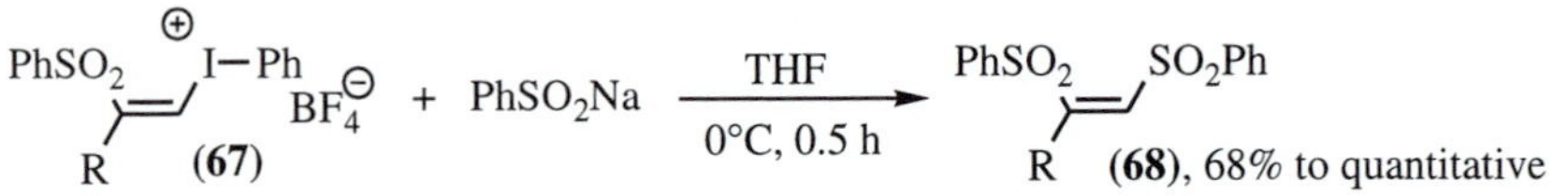

In the case of the (Z)-β-haloalkenyliodonium tetrafluoroborates (69), the reaction with sodium phenylsulfinate led to the formation of (Z)-1,2-bis(phenylsulfonyl)alkenes (70). The reaction proceeds in a first step by a Michael-type addition of the nucleophile to the alkenyliodonium at the C-β atom. Elimination of chloride ion generates the (Z)-β-phenylsulfonylalkenyliodonium salt (67). In a second step, the sodium phenylsulfinate reacts with the iodonium center to afford the 1,2-bis(phenyl-sulfonyl) alkenes in fairly good yields. Trace amounts of (Z)-chlorosulfones (71) were also observed. When this reaction was performed with the (Z)-2-bromo-1-decenyliodonium salt (72), a large amount of the corresponding (Z)-bromosulfone (74) was obtained. All these reactions were exclusively stereoselective with retention of configuration.[190]

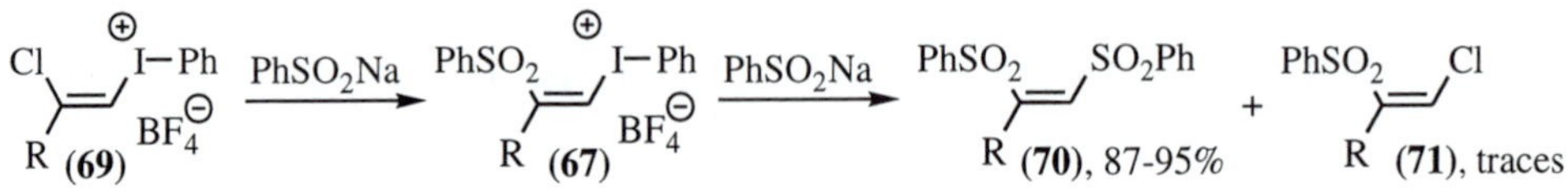

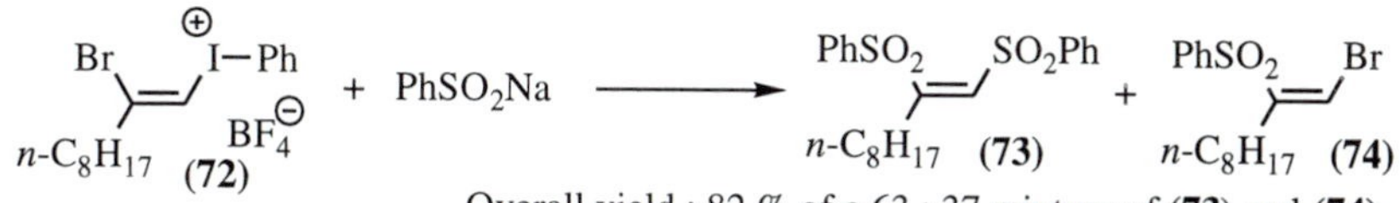

Overall yield : 82 % of a 63 : 37 mixture of (73) and (74)

The reaction of the cyclic hindered iodonium perchlorate (46) with triphenylphosphine led unexpectedly to the vinyl iodide (75) as the major product, along with the expected vinylphosphonium salt (76). This result was explained by the formation of an iodophosphonium intermediate (77) which follows two different pathways for the ligand coupling process.[175] (Scheme 5.6)

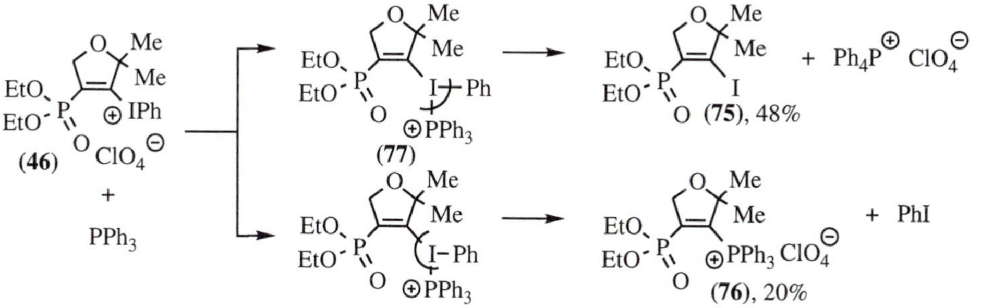

Scheme 5.6

A similar reaction was also obtained in the case of the vinyldiiodonium triflate (78) which afforded exclusively the diiodide (79) upon treatment with triphenylphosphine.[176]

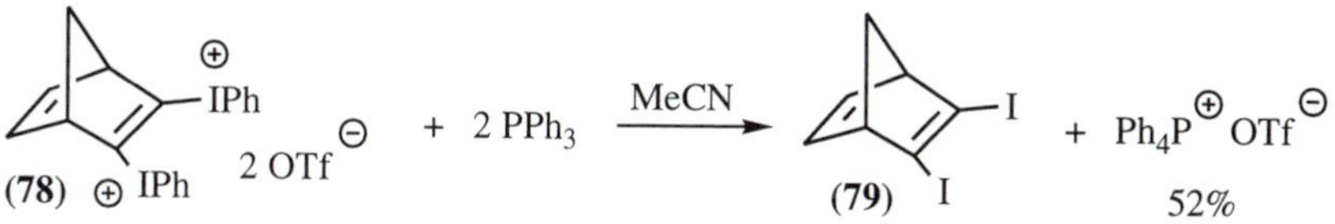

5.3 ALKYNYLARYLIODONIUM SALTS

After an early report in the 1960's on some reactions of alkynylphenyliodonium salts,[177] the chemistry of these species laid dormant until the mid 1980's. Since then, this field has experienced an intense development. In 1965, Beringer and Galton reported the alkynylation of 2-substituted indane-1,3-dione in good yield, by reaction of the sodium salt of indanedione with phenyl(phenylalkynyl) iodonium chloride (80).[177]

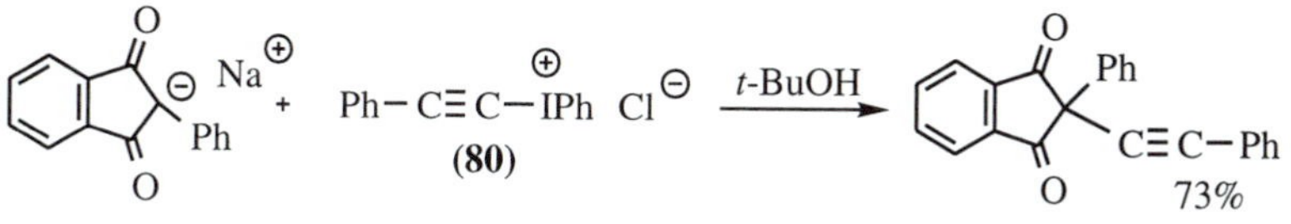

The particular reactivity of this type of compounds was underlying already in this first report by Beringer. If the alkynyl cation equivalent behaviour was well demonstrated, the Michael addition behaviour was as well already observed during the attempted anion exchange with sodium tetrafluoroborate.[177]

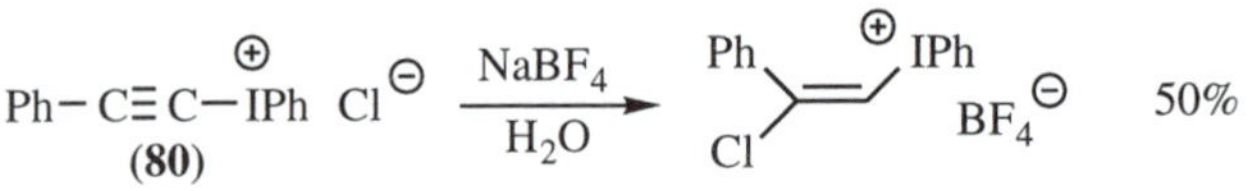

$$Ph-C\equiv C-\overset{\oplus}{I}Ph \ Cl^{\ominus} \xrightarrow[H_2O]{NaBF_4} \underset{Cl}{\overset{Ph}{\diagdown}}=\underset{}{\overset{\overset{\oplus}{I}Ph}{\diagup}} BF_4^{\ominus} \quad 50\%$$
(80)

The seminal paper which led to the revival of alkynyliodonium chemistry was published by Ochiai *et al.* in 1986. The reaction of β-diketone enolates with alkynyliodonium tetrafluoroborates containing an alkyl chain led to the formation of cyclopentene derivatives such as (81).[191]

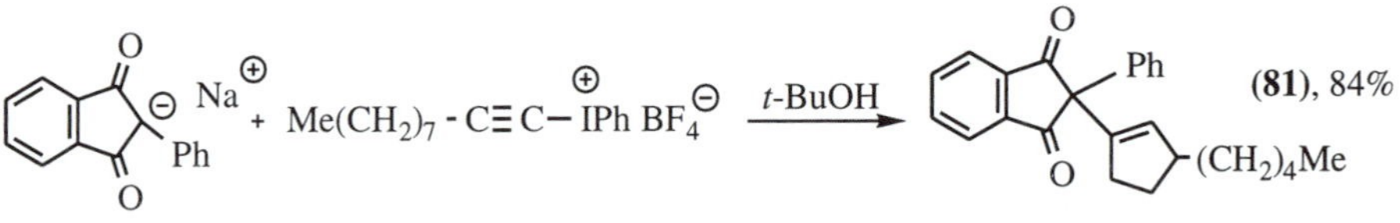

(81), 84%

When the alkynylation reaction described by Beringer[177] was reproduced using (β-^{13}C-phenylethynyl) phenyliodonium tetrafluoroborate (82) as a reagent, it appeared that the substituted indanedione was largely enriched with ^{13}C at the α-carbon of the ethynyl group. Only 6% of the alkynylation product could be ascribed to a classical Ad-E pathway or rather the alternative ligand coupling pathway.[191]

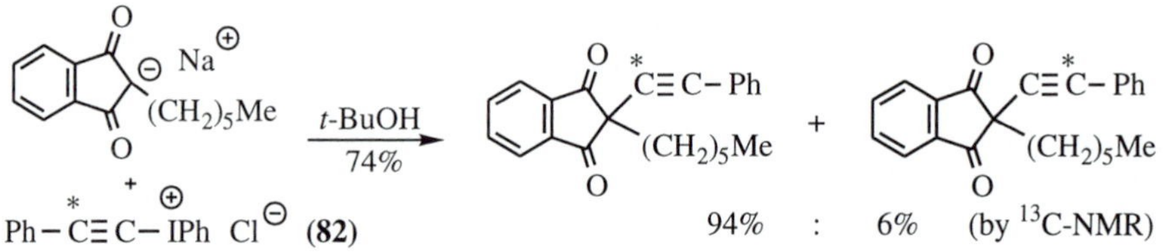

94% : 6% (by ^{13}C-NMR)

These two observations were explained by the intermediacy of carbene species. Nucleophilic β-addition generates a vinyliodonium ylide - iodoallene intermediate (83), which loses iodobenzene giving the alkylidenecarbene (84). If either the substituent R or the nucleophile Nu has a high migratory aptitude, a 1,2-shift leads to the rearranged alkyne products. However, when the substituent of the alkylidenecarbene has a γ-hydrogen and if both the substituent and nucleophile have a poor migratory aptitude, intramolecular 1,5-C-H insertion takes place to form the cyclopentene derivatives. If a viable external trap is present, intermolecular insertion then occurs preferentially.[14,17,21,22,24]

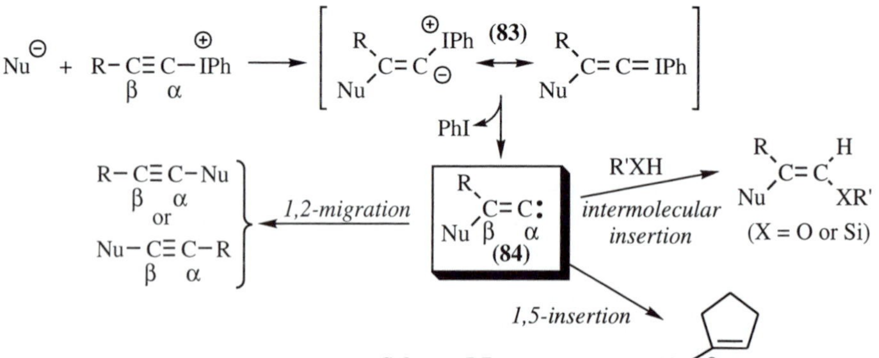

Scheme 5.7

Following this clear demonstration, a large series of nucleophilic reactions of alkynyl aryliodonium salts were subsequently discovered.[14,17,21,22,24] They have for the most part been explained by the sequence: Michael addition - carbene formation. The nucleophiles which have been used are derived from sulfur, phosphorus, oxygen, nitrogen, halides and enolates. However, a small number of reactions appeared to be compatible with a ligand coupling mechanism. The decomposition of alkynyliodonium tosylates and mesylates (83) afforded a mixture of iodoalkynes, phenylsulfonates (84) and a small amount of alkynylsulfonates (85). This reaction takes place either under thermal activation or under cuprous triflate or silver(I) sulfonates catalysis in acetonitrile at room temperature.[192-194]

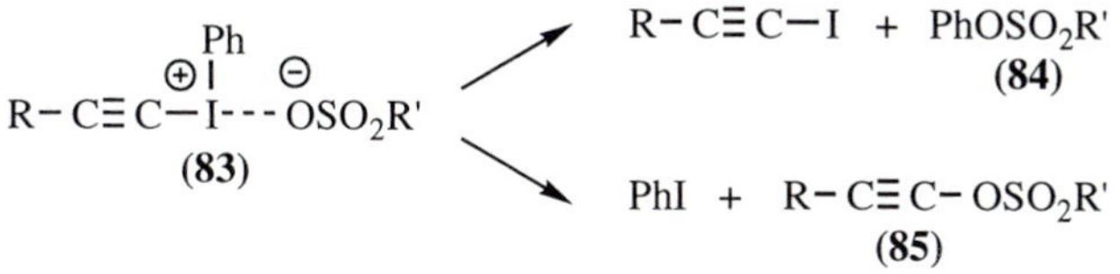

$$R-C{\equiv}C\overset{\overset{\displaystyle Ph}{\overset{\displaystyle \oplus|}{|}}}{I}{\cdots}\overset{\ominus}{O}SO_2R' \quad \begin{array}{c} \nearrow \quad R-C{\equiv}C-I \;+\; PhOSO_2R' \\ (84) \\ \\ \searrow \quad PhI \;+\; R-C{\equiv}C-OSO_2R' \\ (85) \end{array}$$

(83)

A second type is the reaction of alkynylphenyliodonium tosylates with benzoate loaded ion exchange resin. Alkynyl benzoates (86) and phenyl iodide were the products which could result from ligand coupling on the intermediate alkynylphenyliodonium benzoates. However, isolation of these intermediates has not been possible.[195,196] (Scheme 5.8) The same reaction with diethyl phosphate led to the corresponding phosphate esters (87).[195,197]

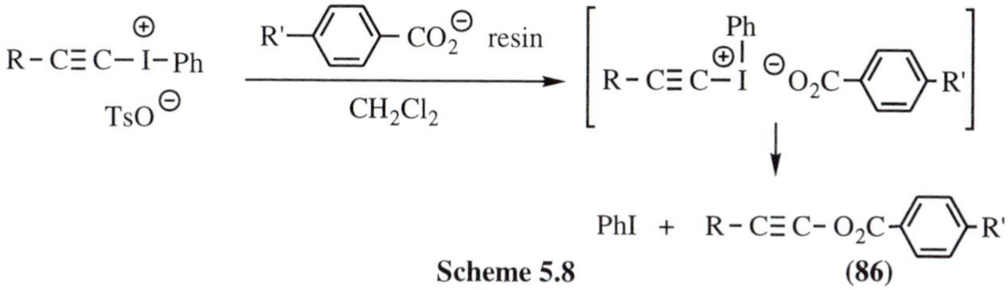

$$\begin{array}{c} R-C{\equiv}C-\overset{\oplus}{I}-Ph \\ \overset{\ominus}{T}sO \end{array} \quad \xrightarrow[\text{CH}_2\text{Cl}_2]{\quad R'{-}\!\!\bigcirc\!\!-CO_2^{\ominus}\ \text{resin}\quad} \quad \left[\ R-C{\equiv}C-\overset{\overset{\displaystyle Ph}{\overset{\displaystyle \oplus|}{|}}}{I}\ \overset{\ominus}{O}_2C{-}\!\!\bigcirc\!\!-R'\ \right]$$

$$\downarrow$$

$$PhI \;+\; R-C{\equiv}C-O_2C{-}\!\!\bigcirc\!\!-R'$$

Scheme 5.8 (86)

$$\begin{array}{c} R-C{\equiv}C-\overset{\oplus}{I}-Ph\ \overset{\ominus}{T}sO \end{array} \;+\; (EtO)_2\overset{\overset{\displaystyle O}{\|}}{P}-O^{\ominus} \quad\longrightarrow\quad R-C{\equiv}C-O-\overset{\overset{\displaystyle O}{\|}}{P}(OEt)_2 \quad (87)$$

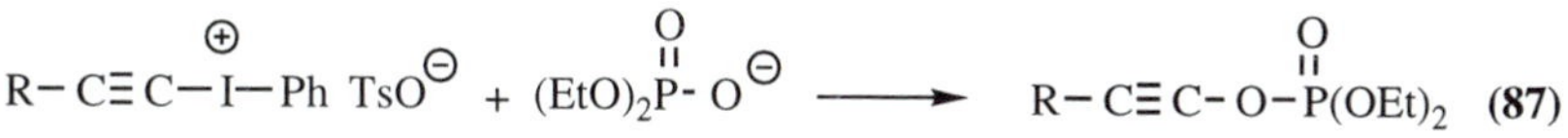

The onium transfer reaction between alkynylphenyliodonium tetrafluoroborates and triphenylarsine afforded high yields of 1-alkynyltriphenylarsonium tetrafluoroborates.[198] However, this reaction appeared to be mechanistically at variance with the generally admitted patterns of reactivity of alkynyliodonium salts. Reaction of phenyl(phenylethynyl-2-^{13}C)iodonium tetrafluoroborate (88) (99% enriched) led to the arsonium salts (89) with more than 95% of ^{13}C enrichment on the β-carbon atom. Although the Michael carbene pathway was not totally excluded, the ligand coupling pathway seemed therefore more important.

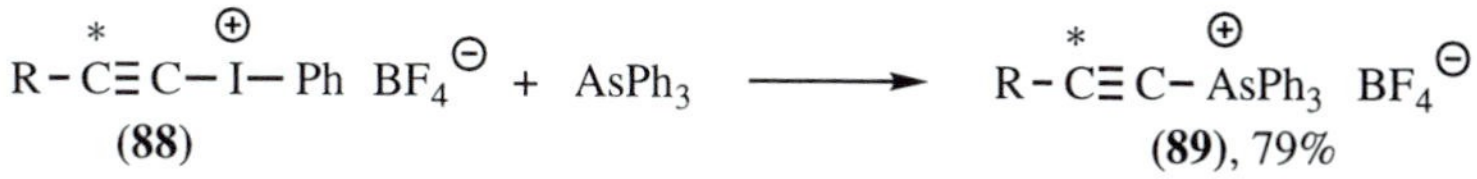

$$\begin{array}{c} \overset{*}{R}-C{\equiv}C-\overset{\oplus}{I}-Ph\ \overset{\ominus}{B}F_4 \;+\; AsPh_3 \quad\longrightarrow\quad \overset{*}{R}-C{\equiv}C-\overset{\oplus}{A}sPh_3\ \overset{\ominus}{B}F_4 \\ (88) \qquad\qquad\qquad\qquad\qquad\qquad\qquad (89),\ 79\% \end{array}$$

In the case of carbon nucleophiles, the behaviour depends on the nature of the substrate. β-Dicarbonyl enolates react with alkynyliodonium salts to afford a transient carbene, which evolves either by migration towards alkynyl-substituted product, or by insertion towards cyclopentene annellation.[14,17,21,22,24,177,191] It has been reported that "*alkynylation of simple enolates does not occur*",[14] but the concerned experiments have not been reported. In the case of the reaction of heteroaryllithium reagents with alkynylphenyliodonium salts, ligand exchange was the only observed evolution.[199]

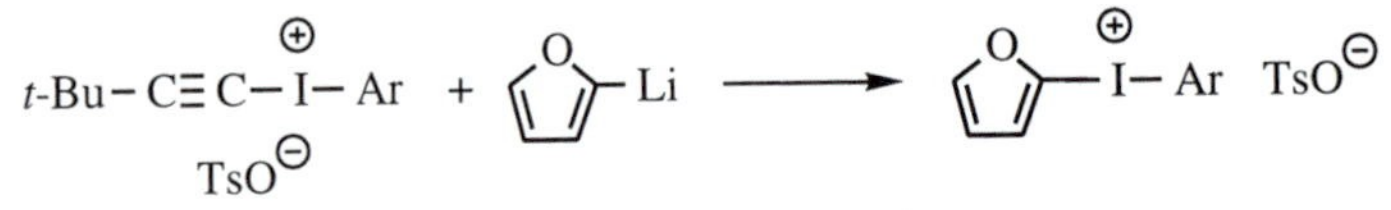

$$t\text{-Bu}-C\equiv C-\overset{\oplus}{I}-Ar \quad + \quad \text{(furan)}-Li \quad \longrightarrow \quad \text{(furan)}-\overset{\oplus}{I}-Ar \quad TsO^{\ominus}$$
$$TsO^{\ominus}$$

Under particular experimental conditions, ligand coupling may be observed between two alkyne ligands simultaneously present on the intermediate iodane. Indeed, addition of lithium acetylide to phenyliodine(III) dicarboxylates (90) led to modest to moderate yields of alkynyl carboxylates (91). By contrast, inverse addition [that is addition of PhI(OCOR)₂ to the acetylide] led to the conjugated diynes (92) as the major product.[196] (Scheme 5.9)

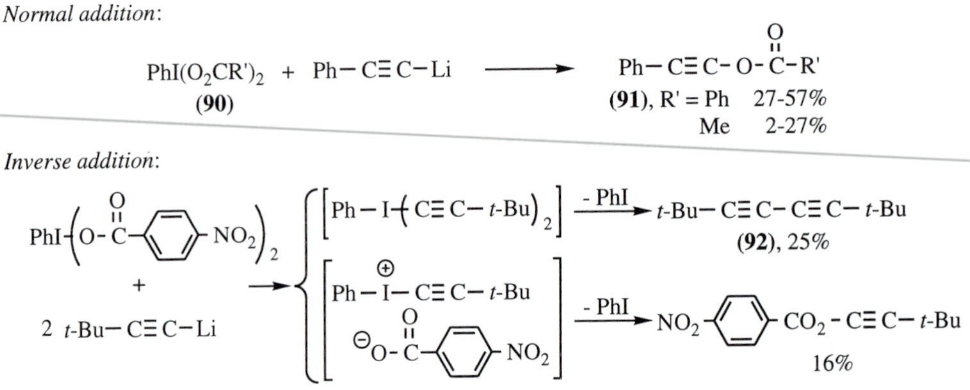

Normal addition:

$$PhI(O_2CR')_2 \quad + \quad Ph-C\equiv C-Li \quad \longrightarrow \quad Ph-C\equiv C-O-\overset{O}{\overset{\|}{C}}-R'$$
$$(90) \qquad\qquad\qquad\qquad\qquad (91), R' = Ph \quad 27\text{-}57\%$$
$$Me \quad 2\text{-}27\%$$

Inverse addition:

$$PhI\left(O-\overset{O}{\overset{\|}{C}}-\underset{}{\bigcirc}-NO_2\right)_2$$
$$+$$
$$2 \ t\text{-Bu}-C\equiv C-Li$$

$$\left[Ph-I\left(C\equiv C-t\text{-Bu}\right)_2\right] \xrightarrow{-PhI} t\text{-Bu}-C\equiv C-C\equiv C-t\text{-Bu}$$
$$(92), 25\%$$

$$\left[Ph-\overset{\oplus}{I}-C\equiv C-t\text{-Bu}\right] \xrightarrow{-PhI} NO_2-\underset{}{\bigcirc}-CO_2-C\equiv C-t\text{-Bu}$$
$$\overset{\ominus}{O}-\overset{O}{\overset{\|}{C}}-\underset{}{\bigcirc}-NO_2$$
$$16\%$$

Scheme 5.9

A related system involving also tricoordinate iodine intermediates is the reaction of trimethylsilyl phenylacetylene with aryl or alkyliodine difluorides which gave coupling products. Although no yields were reported, the relative ratios indicated that pentafluorophenyliodine difluoride led to the diyne as the main product [(93) : (94) = 3.25] whereas perfluoropropyliodine difluoride led to the preferential formation of 1-perfluoropropyl-2-phenylacetylene, (94) R = C₃F₇.[200]

$$RIF_2 \quad + \quad 2 \ TMS-C\equiv C-Ph$$
$$\downarrow KF \ / \ 18\text{-crown-}6 \ / \ CH_2Cl_2$$
$$Ph-C\equiv C-C\equiv C-Ph \quad + \quad R-C\equiv C-Ph \quad + \quad RI \quad + \quad I-C\equiv C-Ph$$
$$(93) \qquad\qquad\qquad (94)$$

Vinyl copper reagents undergo alkynylation by reaction with alkynylaryliodonium tosylates to give conjugated vinylalkynes (95) in a highly stereoselective reaction (> 98%) proceeding with retention of configuration. The system was suggested to involve an oxidative addition of the alkynyl species affording a first Cu(III) - iodine intermediate (96), which undergoes a ligand coupling. The newly formed alkynylvinyl copper species (97) then undergoes a second ligand coupling to afford eventually the vinylalkyne.[201] (Scheme 5.10)

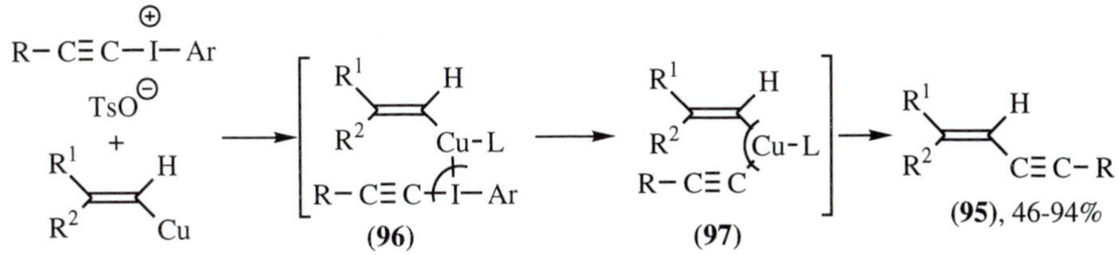

R−C≡C−I−Ar $\overset{\oplus}{}$
TsO$\overset{\ominus}{}$
+
$\overset{R^1}{\underset{R^2}{}}$C=C$\overset{H}{\underset{Cu}{}}$ ⟶
$\left[\begin{array}{c} R^1\\ R^2 \end{array} C=C \begin{array}{c} H\\ Cu\text{-}L \end{array} \\ R-C≡C\overset{+}{\underset{}{}}I-Ar \\ (96) \right.$ ⟶ $\left. \begin{array}{c} R^1\\ R^2 \end{array} C=C \begin{array}{c} H\\ Cu\text{-}L \end{array} \\ R-C≡C \\ (97) \right]$ ⟶ $\overset{R^1}{\underset{R^2}{}}$C=C$\overset{H}{\underset{C≡C-R}{}}$

(95), 46-94%

Scheme 5.10

A similar behaviour was observed in the reaction of lithium dialkynylcuprates with alkynylaryl iodonium tosylates leading to conjugated diynes. Unsymmetrical diynes (**98**) were also prepared by this method. However, the selectivity was moderate, due to the competitive formation of the symmetrical diynes (**99**).[202] This reaction has been recently applied to the synthesis of various liquid-crystalline diaryacetylenes.[203,204]

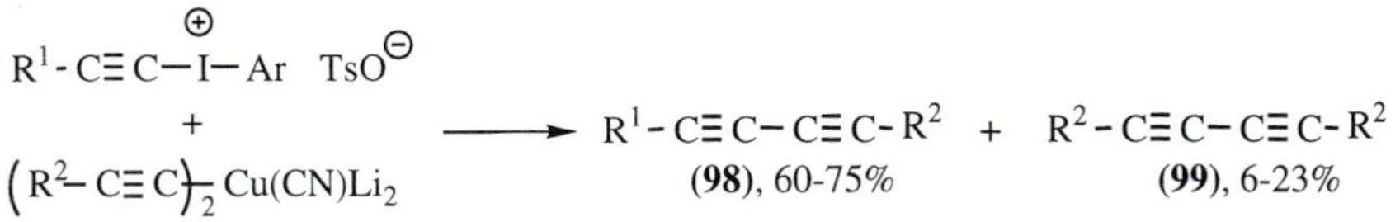

$$R^1\text{-}C≡C\overset{\oplus}{-}I\text{-}Ar \quad TsO\overset{\ominus}{}$$
$$+$$
$$\left(R^2\text{-}C≡C\right)_2 Cu(CN)Li_2$$
⟶ $R^1\text{-}C≡C\text{-}C≡C\text{-}R^2$ + $R^2\text{-}C≡C\text{-}C≡C\text{-}R^2$
(**98**), 60-75%   (**99**), 6-23%

Lithium diphenylcuprates and lithium dialkylcuprates reacted similarly with alkynylphenyl iodonium tosylates to afford phenyl- and alkyl-substituted alkynes respectively.[202]

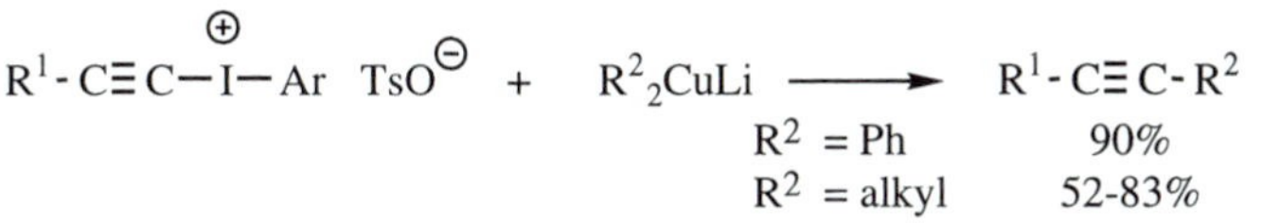

$$R^1\text{-}C≡C\overset{\oplus}{-}I\text{-}Ar \quad TsO\overset{\ominus}{} \quad + \quad R^2_2CuLi \quad ⟶ \quad R^1\text{-}C≡C\text{-}R^2$$
$$R^2 = Ph \qquad 90\%$$
$$R^2 = alkyl \qquad 52\text{-}83\%$$

5.4 ARYLPERFLUOROALKYLIODONIUM SALTS

(Polyfluoroalkyl)aryliodonium salts react with nucleophiles to afford the polyfluoroalkylated nucleophiles under mild conditions. The products are almost exclusively derived from the fluoroalkyl moiety. Two types of fluoroalkyl chain R_f have been involved in these reactions: the perfluoroalkyl and the α,α-dihydroperfluoroalkyl compounds.[3, 11,205,206]

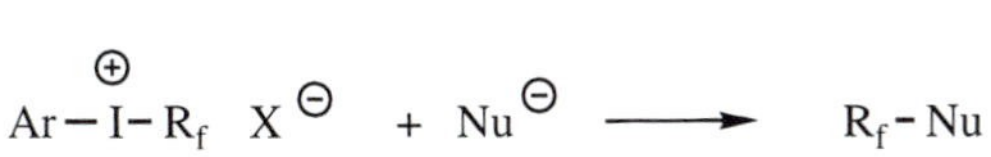

$$Ar\overset{\oplus}{-}I\text{-}R_f \quad X\overset{\ominus}{} \quad + \quad Nu\overset{\ominus}{} \quad ⟶ \quad R_f\text{-}Nu$$

5.4.1 Perfluoroalkyl aryliodonium salts

Various types of carbanionic reagents undergo easily perfluoroalkylation through reaction with (perfluoroalkyl)aryliodonium salts by a ligand coupling mechanism. The reactivity of (perfluoroalkyl)-aryliodonium salts showed a dependency upon the nature of the anionic counterpart. In their seminal work, Yagupolskii *et al.* reported that a variety of nucleophiles can be easily perfluoroalkylated by reaction with (perfluoroalkyl)-*p*-tolyliodonium chlorides, (**100**) and (**101**), in a polar solvent under mild conditions (Table 5.8).[207] Good to excellent yields were obtained with carbanions. In the case of the aniline substrates, only products of alkylation on the *para* position of the ring were formed in relatively modest yields. It is likely that, in this case, the reaction does not proceed by ligand coupling but by a normal electrophilic aromatic substitution. This is similar to the reaction of a range of aromatic substrates which have been perfluoroalkylated by (perfluoroalkyl)aryliodonium triflates (FITS reagents).[208] The analogous (perfluoroalkyl)phenyliodonium tetrafluoroborate salts appeared to be more effective perfluoroalkylating reagents than the chlorides.[209]

Table 5.8: Perfluoroalkylation with perfluoroalkyl-*p*-tolyliodonium chlorides (**100**) and (**101**)[207]

$$Me-\text{C}_6\text{H}_4-I\overset{R_f}{\underset{Cl}{\cdots}} + Nu^{\ominus} \xrightarrow{DMF} R_f-Nu$$

(**100**): R = C_3F_7 ; (**101**): R = C_6F_{13}

Iodonium reagent	Nucleophile	Product	Yield (%)
(**100**)	ArSNa	ArS-C_3F_7	56-96
(**100**)	PhSeNa	PhSe-C_3F_7	87
(**100**)	KSCN	C_3F_7-SCN	59
(**100**)	KSeCN	C_3F_7-SeCN	45
(**101**)	NaNO$_2$	C_6F_{13}-NO$_2$	60
(**100**)	PhNRMe	*p*-C_3F_7-C_6H_4NRMe	33-35

A third type of derivatives of iodonium salts, the perfluoroalkyl phenyliodonium trifluoromethane sulfonates (FITS), showed an enhanced reactivity compared to the chloride and the tetrafluoroborate salts. In a series of papers, Umemoto *et al.* have extensively studied the synthesis of these compounds and their reactivity towards a wide range of more or less powerful nucleophiles.[3,206] Different mechanistic pathways have been demonstrated to occur in these reactions. Ligand coupling appears to be involved in a number of nucleophilic displacements implying various types of carbanions, thiols and other nucleophilic substrates.

5.4.1.1 Perfluoroalkylation of carbanions

FITS reagents reacted with various carbanions to afford under mild conditions the corresponding fluoroalkylated derivatives. The yields are quite dependent upon the nature of the carbanion and of the counterion. (Table 5.9)

Table 5.9: Perfluoroalkylation of carbanions with FITS reagents[210,211]

$$\text{C}_6\text{H}_5-I\overset{R_f}{\underset{OTf}{\cdots}} + Nu^{\ominus} \longrightarrow R_f-Nu + PhI$$

(**102**): R = C_8F_{17} ; (**103**): R = C_3F_7

Iodonium reagent	Nucleophile	Product	Yield (%)
(**102**)	*n*-C_8H_{17}-MgCl	*n*-C_8H_{17}-C_8F_{17}	58
(**102**)	*n*-C_8H_{17}-Cu	*n*-C_8H_{17}-C_8F_{17}	26
(**102**)	*n*-C_8H_{17}-Li	*n*-C_8H_{17}-C_8F_{17}	9
(**103**)	PhCH$_2$-MgCl	PhCH$_2$-C_3F_7	82
(**102**)	CH$_2$=CHMgBr	CH$_2$=CH-C_8F_{17} + Ph-C_8F_{17}	8 17
(**102**)	Ph−C≡C−Li	Ph−C≡C−C_8F_{17}	66

Alkyl- and alkynyl-magnesium halides afforded perfluoroalkylated products in good yields, but the alkenyl Grignard reagents afforded only poor yields of ligand coupling products. In the reaction with vinylmagnesium halide, iodobenzene formed as a by-product in the first step underwent trans-metallation with the vinylmagnesium halide. The resulting phenyl Grignard reacted with the remaining iodonium salt to afford a small amount of perfluoroalkylbenzene, as a by-product of the reaction. Alkyllithium gave poor yields, but good yields were obtained with alkynyllithium reagents.[210,211]

The reaction of FITS compounds with the sodium salt of 2-methylcyclohexan-1,3-dione led to the *O*-perfluoroalkylated and the *C*-perfluoroalkylated products. The ratio *O*-derivatives : *C*-derivatives was strongly dependent upon the reaction temperature, the *C*-perfluoroalkylated product predominating at lower temperature. Similar results were obtained with ethyl 2-methylacetoacetate. In the case of the sodium salt of diethyl 2-methylmalonate and of 2-nitropropane, only the *C*-perfluoroalkylated products were isolated.[211]

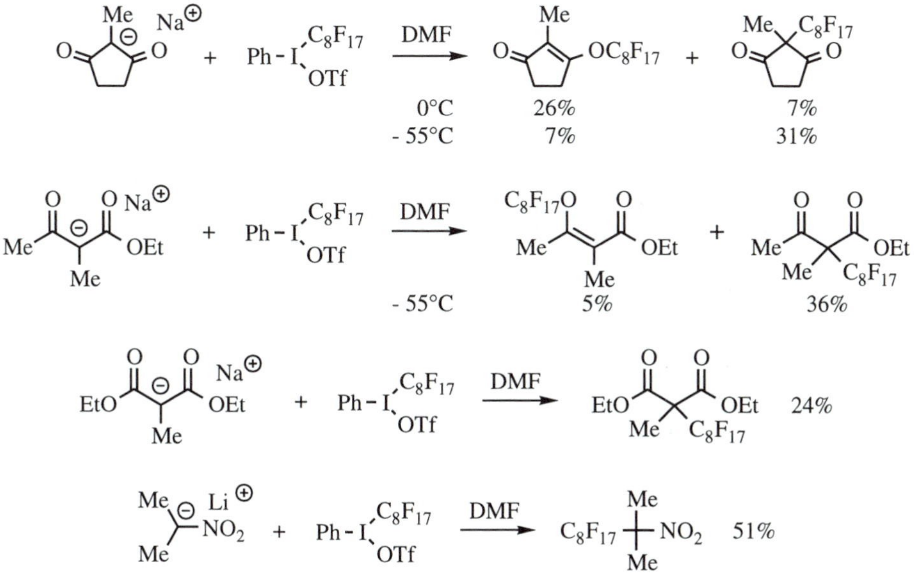

Treatment of the anion of acetophenone with (perfluoroalkyl)phenyliodonium triflate (**102**) led to a poor overall yield of alkylated products. Moreover, the normal alkylation product was contaminated with the product of dehydrofluorination.

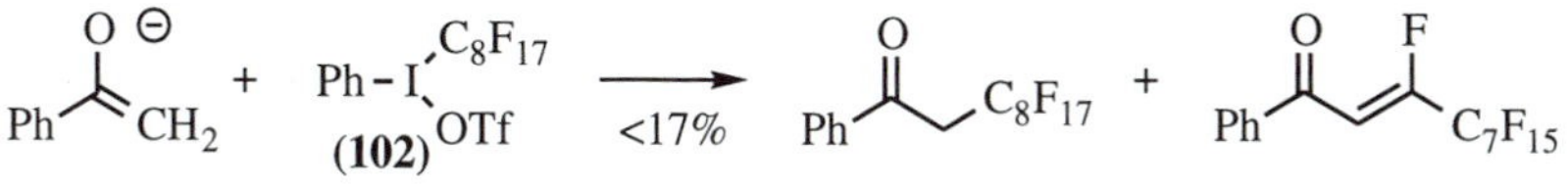

From a synthetic point of view, the synthesis of α-(perfluoroalkyl)ketones is preferably performed by the reaction of (perfluoroalkyl)phenyliodonium triflates with silyl enol ethers, which affords high yields of the perfluoroalkylated products (**104**) under mild conditions. The reaction of β,γ-unsaturated silyl enol ethers with (perfluoroalkyl)phenyliodonium triflates led to the γ-perfluoroalkyl-α,β-unsaturated carbonyl derivatives (**105**) in good to high yields.[212]

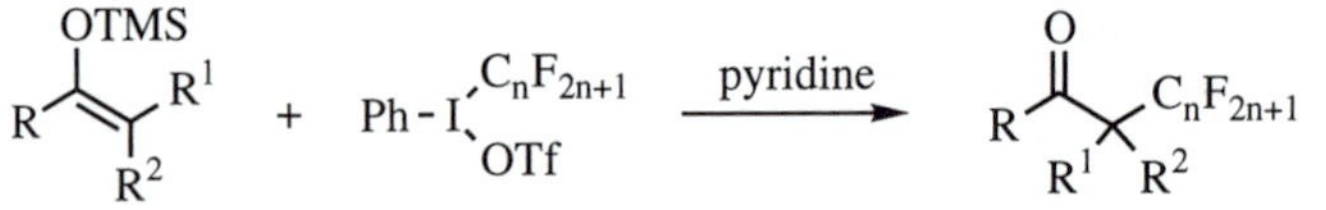

$$\text{(104), 59-88\%}$$

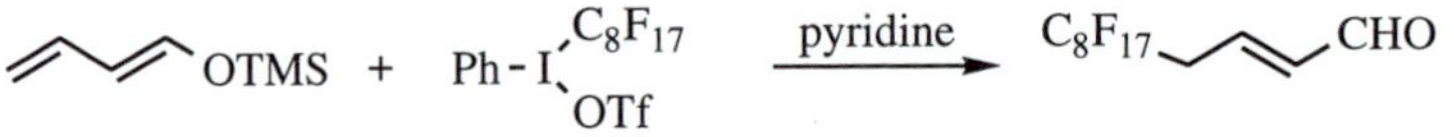

$$\text{(105), 85\%}$$

In these reactions, the ligand coupling mechanism is not involved. A different mechanism operates: the formation of an intermediate cationic species which is likely to result from the initial formation of a π-complex (106).[212] Such a mechanism has been invoked to explain the perfluoroalkylation of a number of ethylenic and acetylenic compounds.[213-215]

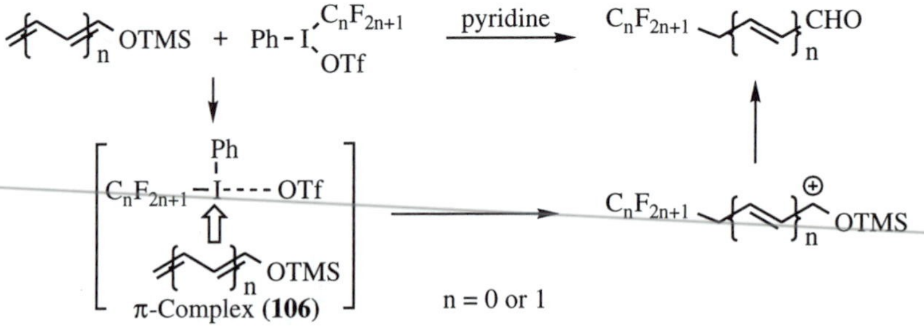

$$\text{π-Complex (106)} \qquad n = 0 \text{ or } 1$$

5.4.1.2 Perfluoroalkylation of thiols

FITS reagents reacted very easily with thiols in the presence of pyridine to give the perfluoroalkyl sulfides in good to high yields. *S*-perfluoroalkylation was selective in the case of polyfunctional substrates. In the presence of groups such as hydroxy, carboxy, alkoxycarbonyl or dialkylamino, the sulfur atom was always selectively substituted.[216] (Table 5.10)

Table 5.10: Perfluoroalkylation of thiols with FITS reagents[216]

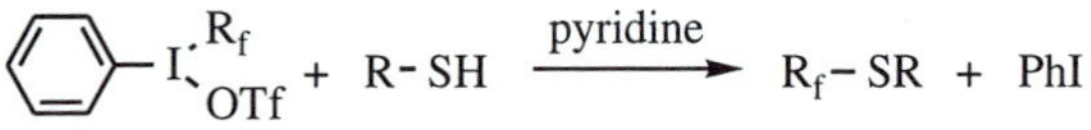

$$\text{(102): R = C}_8\text{F}_{17} \text{ ; (103): R = C}_3\text{F}_7 \text{ ; (107): R = C}_2\text{F}_5$$

Iodonium reagent	Nucleophile	Product	Yield (%)
(103)	$HS\text{-}CH_2CO_2Bu$	$C_3F_7\text{-}S\text{-}CH_2CO_2Bu$	88
(102)	$HS\text{-}CH_2CO_2H$	$C_8F_{17}\text{-}S\text{-}CH_2CO_2H$	93
(102)	$HS\text{-}CH_2CH_2OH$	$C_8F_{17}\text{-}S\text{-}CH_2CH_2OH$	83
(102)	$HS\text{-}CH_2CH_2SH$	$C_8F_{17}\text{-}S\text{-}CH_2CH_2\text{-}SH$ + $C_8F_{17}\text{-}SCH_2CH_2S\text{-}C_8F_{17}$	46 (as 1.3:1 mixture)
(102)	$HS\text{-}CH_2CH_2NMe_2$	$C_8F_{17}\text{-}S\text{-}CH_2CH_2NMe_2$	52
(107)	Ph-SH	$Ph\text{-}S\text{-}C_2F_5$	59

5.4.1.3 Perfluoroalkylation of phenols

The reaction of phenol with FITS reagents led to a mixture of *ortho*-, *meta*- and *para*-perfluoroalkylphenols, formed by electrophilic aromatic substitution. In the case of the relatively hindered electron-rich phenols, a different outcome was observed: *O*-perfluoroalkylation was observed together with the *C*-perfluoroalkylation products. At high temperature, the amount of the *O*-substituted product became predominant.[217]

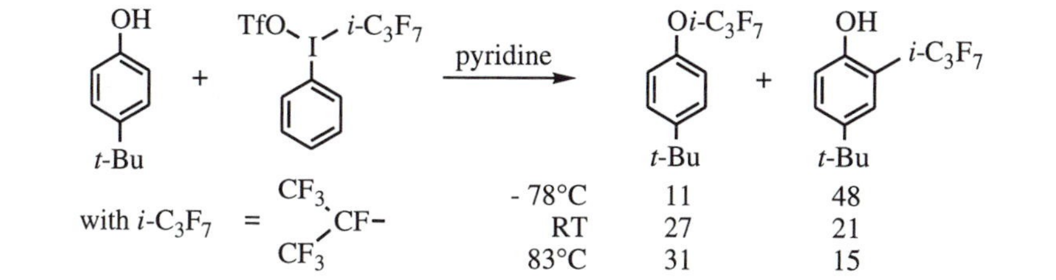

OH
TfO
I
i-C3F7
t-Bu
pyridine
Oi-C3F7
t-Bu
OH
i-C3F7
t-Bu
with i-C3F7 =
CF3
CF-
CF3
- 78°C 11 48
RT 27 21
83°C 31 15

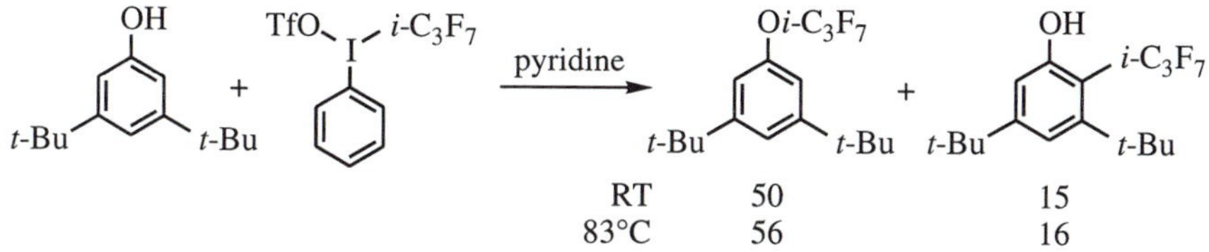

OH
t-Bu
t-Bu
TfO
I
i-C3F7
pyridine
Oi-C3F7
t-Bu
t-Bu
OH
i-C3F7
t-Bu
t-Bu
RT 50 15
83°C 56 16

The formation of the *C*-perfluoroalkylated product may be explained by an ionic reaction, through electrophilic aromatic substitution. The *O*-perfluoroalkylation was attributed to the greater thermodynamic stability of the *O*-perfluoroalkyl products compared to the *C*-perfluoroalkyl products. However, it may also be explained by a shift in the balance between two mechanisms: the electrophilic substitution leading to the *C*-perfluoroalkyl products, and the ligand coupling process occurring on an intermediate aryloxyiodane which would lead to the *O*-perfluoroalkyl products. In the case of the more highly hindered phenol, 2,4,6-tri-*tert*-butylphenol **(108)**, a completely different mixture of products was obtained. Due to the observation of a transient blue colouration in the early stages of the reaction, it was suggested that this reaction proceeds through the formation of an intermediate radical cation which is generated by electron transfer between the phenol and the FITS reagent.[217]

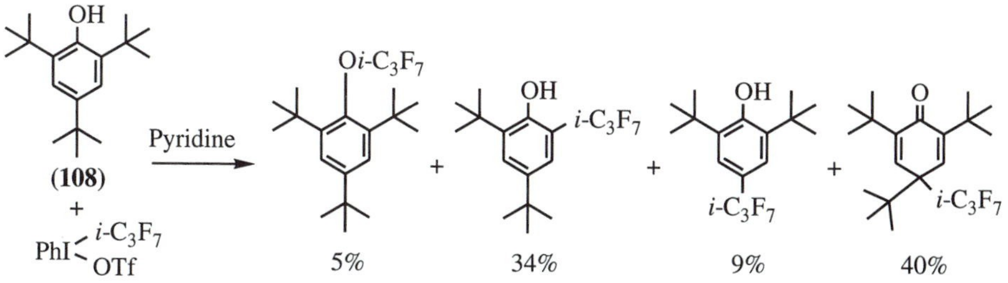

OH
(108)
+
PhI
i-C3F7
OTf
Pyridine
Oi-C3F7
OH
i-C3F7
OH
i-C3F7
O
i-C3F7
5%
34%
9%
40%

5.4.1.4 Miscellaneous perfluoroalkylation reactions

The reaction of (trimethyl)organylstannanes with (perfluoroalkylaryl)iodonium salts gave **(109),** the products of substitution of the stannyl group by the perfluoroalkyl group. As the reaction of (*Z*)-trimethylstyryltin with heptafluoropropyl-*p*-tolyliodonium salts afforded exclusively (*E*)-heptafluoropropylstyrene in moderate yields (30-40%), the reaction was suggested to involve free radicals.[218]

$$\text{Me}_3\text{Sn-R} \ + \ \overset{\displaystyle C_3F_7\diagdown \overset{\displaystyle X}{\underset{\displaystyle Ar}{\text{I}}}}{} \ \longrightarrow \ C_3F_7\text{-R} \ + \ \text{ArI} \ + \ \text{Me}_3\text{SnX}$$
$$\textbf{(109)},\ 17\text{-}77\%$$

Some other perfluoroalkylation of nucleophilic substrates which could occur by a ligand coupling mechanism have also been reported : the reaction of the sodium salt of phtalhydrazide **(110)**[206] and of the sodium salt of 2,3-dihydroxyquinoxaline **(111)**.[208] Benzaldehyde oxime was reported to give only the *O*-perfluoroalkylated product in a poor 13% yield, and alkoxide anions remained unaffected.[206]

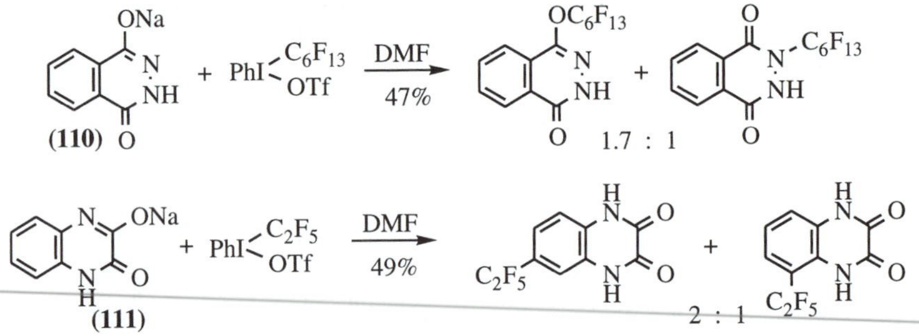

5.4.2 α,α-Dihydroperfluoroalkyl aryliodonium salts

Like perfluoroalkylation, electrophilic α,α-dihydroperfluoroalkylation can be efficiently performed by the use of the appropriate (α,α-dihydroperfluoroalkyl)phenyliodonium salts. A wide range of nucleophiles were alkylated under mild conditions. Most of the reactions can occur by a ligand coupling taking place on an intermediate trisubstituted iodane(III) compound. This is the case of the reactions with carbanions, amines, alkoxides, phenoxides, thiolates and carboxylic acid ammonium salts. Dialkylation of aniline was also obtained by using two molar equivalents of the iodonium reagent.[219,220] (Table 5.11)

Table 5.11: α,α-Dihydroperfluoroalkylation of various types of nucleophiles[219]

$$\text{PhI}\diagdown\overset{\displaystyle \text{CH}_2\text{-(CF}_2)_n\text{-CF}_3}{\underset{\displaystyle \text{OTf}}{}} \ + \ \text{Nu}^\ominus \ \longrightarrow \ \text{Nu-CH}_2\text{-(CF}_2)_n\text{-CF}_3 \ + \ \text{TfO}^\ominus$$

Substrate	Iodonium, n=	Product	Yield (%)
PhCH₂CH₂MgBr	6	Ph(CH₂)₃-(CF₂)₆CF₃	36
Me−C(CO₂Et)(CO₂Et) Na⁺	0	Me−C(CO₂Et)(CF₃CH₂)(CO₂Et)	25
PhOLi	0	PhO-CH₂-CF₃	79
Ph(CH₂)₂OLi	0	Ph(CH₂)₂O-CH₂-CF₃	61
PhCO₂H (a)	0	PhCO₂-CH₂-CF₃	99
PhNH₂ (b)	0	PhNH-CH₂-CF₃	98
PhNH₂ (c)	0	PhN(CH₂-CF₃)₂	70

(a) in the presence of 2,4,6-collidine
(b) one equivalent of the iodonium salt in the presence of 2,4,6-collidine
(c) two equivalents of the iodonium salt in the presence of 2,6-di-*tert*-butyl-4-methylpyridine

α,α-Dihydroperfluoroethyl onium salts were easily and selectively obtained in the case of tertiary amines, pyridine, quinoline, diphenylsulfide and triphenylphosphine.[221] (Table 5.12)

Table 5.12: Synthesis of onium derivatives by α,α-dihydroperfluoroethylation[221]

Substrate	Product	Yield (%)
PhNMe$_2$	Me–N(+)(Ph)(Me)–CH$_2$CF$_3$ TfO(−)	89
(pyridine)	(pyridinium)–N(+)–CH$_2$CF$_3$ TfO(−)	83
Me–(2,6-lutidine)–N	Me–(ring)(Me)(Me)–N(+)–CH$_2$CF$_3$ TfO(−)	63
Ph$_2$S	Ph(Ph)–S(+)–CH$_2$CF$_3$ TfO(−)	90
Ph$_3$P	Ph$_3$P(+)·CH$_2$CF$_3$ TfO(−)	95

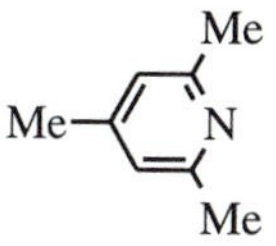

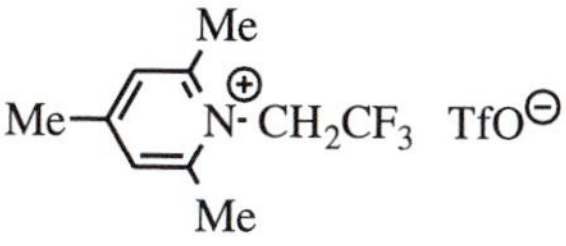

In the case of polyfunctional substrates, *N*-trifluoroethylation was selectively realised in the case of amino-alcohols **(112)** or **(113)**.[222] *N*-trifluoroethylation can also be performed in the case of diamines, as for example the bistrifluoroethylation of 1,10-diaza-18-crown-6 **(114)**.[223]

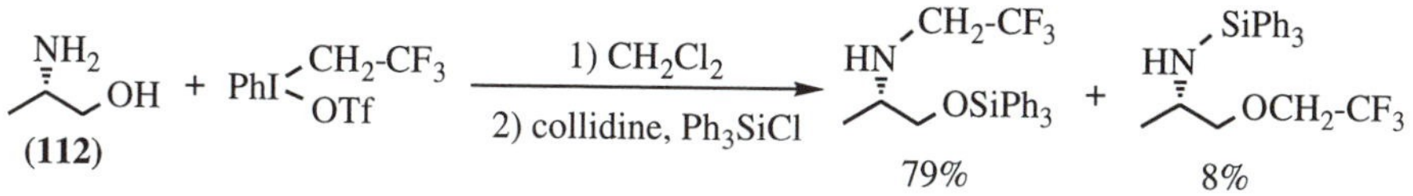

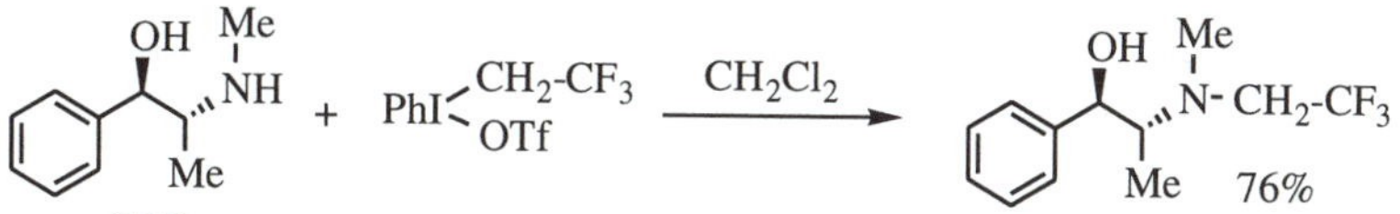

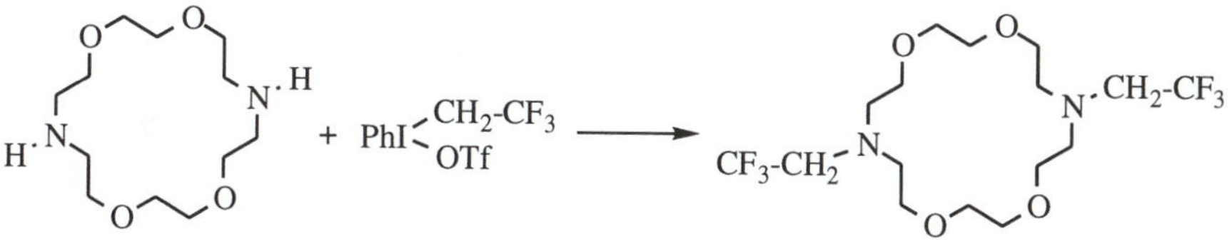

The (α,α,ω-trihydroperfluoroalkyl)phenyliodonium tetrafluoroborates (115) and (118) reacted with nucleophilic substrates to afford the polyfluoroalkylated derivatives. This is the case of *p*-chloro-thiophenol, aniline and pyridine. In the reaction with *N,N*-dimethylaniline, the *N*-polyfluoroalkyl product (116) was predominant, but nevertheless a small amount of ring-alkylation product (117) was also obtained.[224]

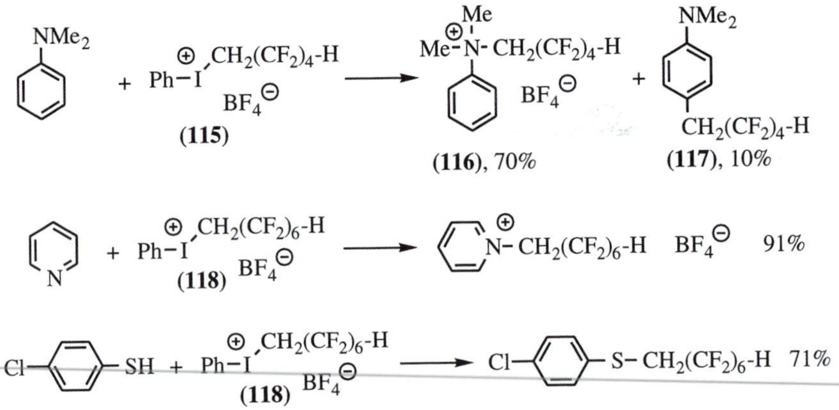

The reaction of enol trimethylsilyl ethers of carbonyl compounds with (α,α-dihydroperfluoroalkyl) phenyliodonium triflates required promotion by potassium fluoride to proceed at room temperature to give the β-perfluoroalkyl carbonyl compounds in good yields.[225] In the case of the silyl enol ether of an α,β-unsaturated ketone (119), the δ-perfluoroalkyl-α,β-unsaturated carbonyl compound (120) was the only product formed. The reaction is likely to follow a path similar to the one used in the reaction of silyl enol ethers with (perfluoroalkyl)phenyliodonium salts. In a first step, a π-complex is formed which evolves into the cationic product of α- or γ-addition, followed by desilylation to the carbonyl reaction product.

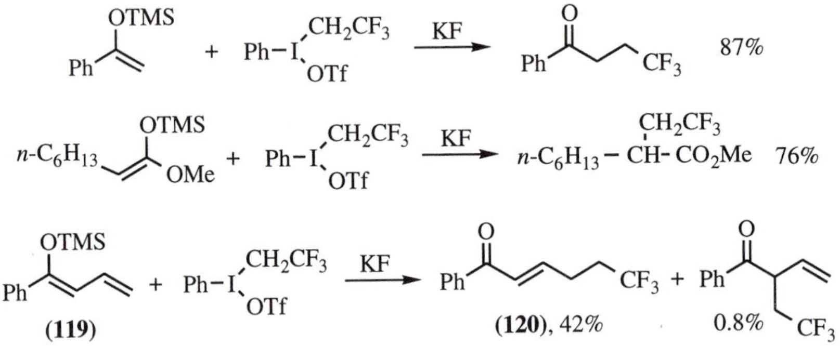

5.5 ARYLIODINE(III) DERIVATIVES

Aryliodine(III) derivatives have found a wide range of application in organic synthesis, particularly for oxidative functionalization of alkenes, amines, carbonyl derivatives, phenols.[6,7,12,13,15,25] Among the various mechanisms which have been invoked to explain these transformations, the ligand coupling mechanism has been implied in a number of these reactions: carbon-carbon bond formation, carbonyl α-functionalisation, hydrazone α-functionalisation.

5.5.1 Carbon-carbon coupling

Symmetrically substituted 1,4-diketones can be obtained by reaction of carbonyl triethylstannyl enol ethers with diacetoxyiodobenzene.[226]

The reaction was extended to a wider range of substrates by using the silyl instead of the stannyl ether. Indeed, treatment of ketone trimethylsilyl enol ethers with iodosylbenzene in the presence of boron trifluoride-diethyl ether gave the 1,4-butanedione derivatives in generally good yields. In the first stage of the reaction, the trimethylsilyl enol ether reacts with the iodosylbenzene-BF_3 complex to form an unstable α-phenyliodonioketone derivative. This intermediate reacts again with the nucleophilic silyl enol ether to form a tricoordinate iodane, which undergoes ligand coupling to afford the 1,4-butanedione compound.[227-229] (Scheme 5.11)

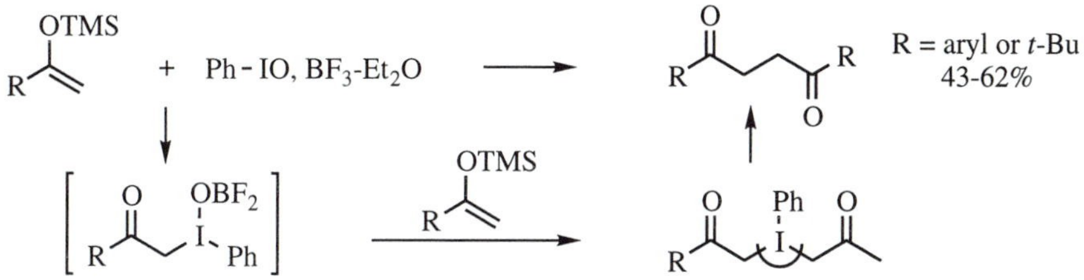

Scheme 5.11

This ligand coupling method was quite efficient for acetophenone and *tert*-butylketone derivatives. However, coupling of the TMS ether of cyclohexanone (121) failed. The oxidative coupling of the TMS ether of cyclohexanone (121) to give 2,2'-bicyclohexanone was successful only when this silyl enol ether was treated with the iodosobenzene-tetrafluoroborate complex.[230,231]

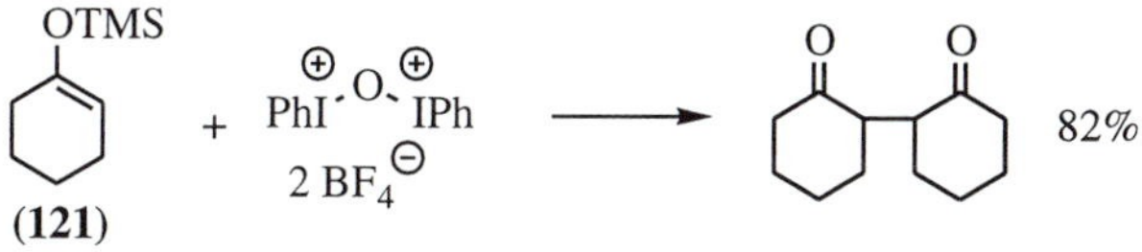

The reaction of the trimethylsilyl enol ether of various acetophenone derivatives (122) with the electrophilic PhIO-HBF$_4$ at - 78°C generated a highly reactive iodonium salt, an α-ketomethyl phenyliodonium salt (123). Treatment of this salt with a second different trimethylsilyl enol ether (124) led to unsymmetrical 1,4-butanediones (125), as major products.[232,233] (Scheme 5.12)

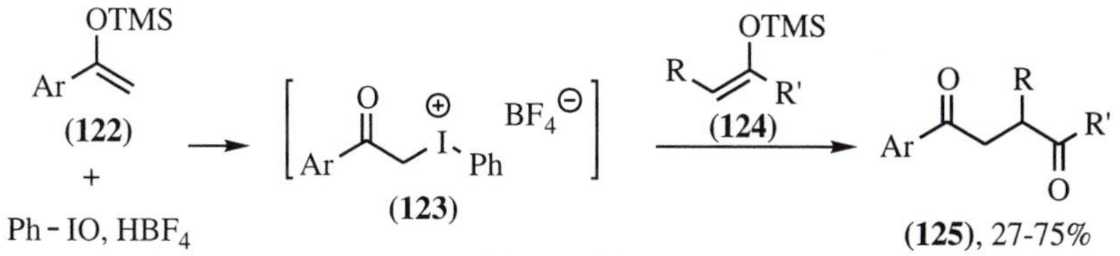

Scheme 5.12

Reaction of the (α-ketomethyl)phenyliodonium salts (126) with alkenes or allyltrimethylsilane led to allylic alkylation products in good yields. In the case of the reaction of the tetrasubstituted alkene, 2,3-dimethylbut-2-ene (127), the dihydrofuran derivative (128) was obtained.[232,233] (Scheme 5.13)

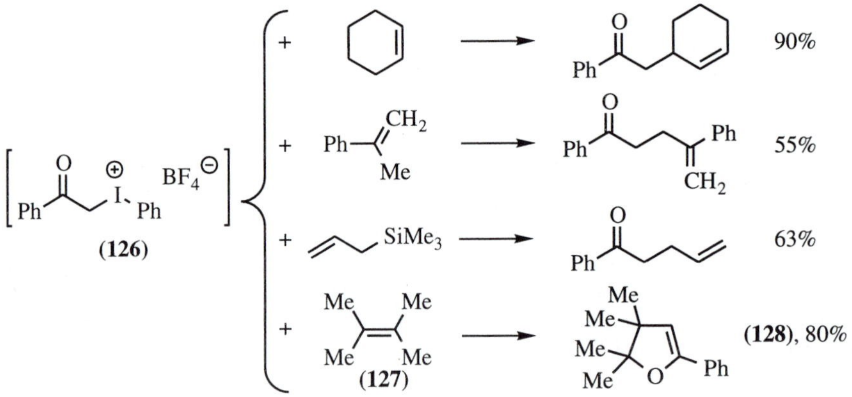

Scheme 5.13

The reaction was suggested to proceed by a carbocationic mechanism followed by ligand coupling. The electrophilic attack of the iodine atom on the double bond leads first to an intermediate cation (129) which loses either a proton or a trimethylsilyl cation. The tricoordinate iodane intermediates of types (130), (131) or (132) eventually undergo ligand coupling to afford the final reaction products.[233] (Scheme 5.14)

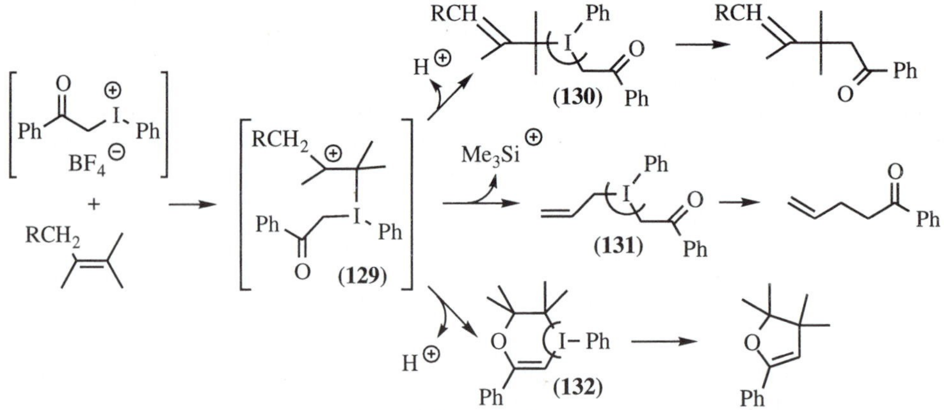

Scheme 5.14

Closely related reactions are the self-coupling of β-dicarbonyl compounds upon treatment either with iodosylbenzene in the presence of boron trifluoride-diethyl ether in neutral solvents,[234] or with iodobenzene diacetate in acetic acid.[235]

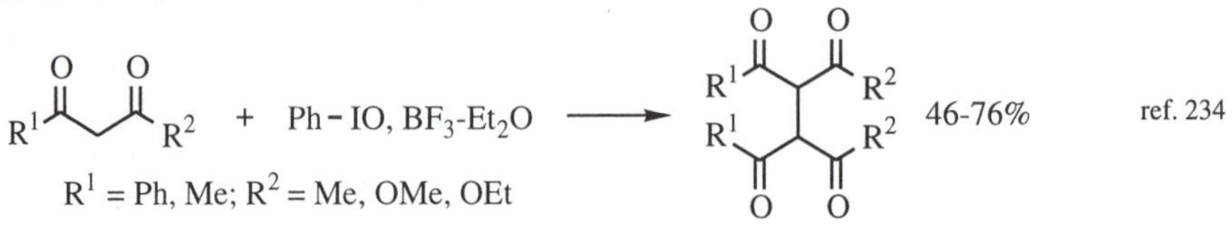

$R^1 = Ph, Me; R^2 = Me, OMe, OEt$

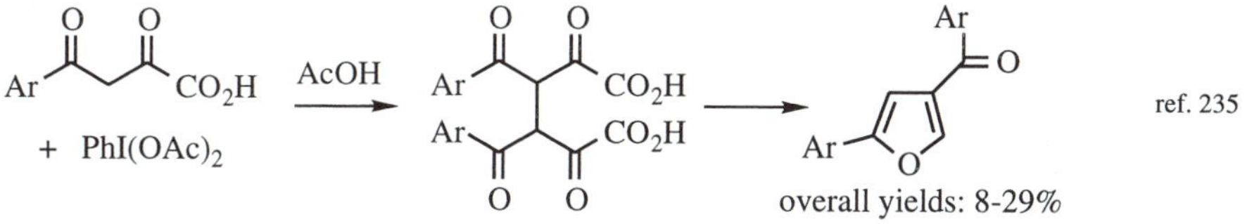

$$Ar \overset{O}{\underset{}{\|}} \overset{O}{\underset{}{\|}} CO_2H \; + \; PhI(OAc)_2 \xrightarrow{\; AcOH \;} Ar \overset{O}{\underset{}{\|}} \overset{O}{\underset{}{\|}} \overset{CO_2H}{\underset{Ar}{\overset{}{}}} \overset{CO_2H}{\underset{O}{\overset{}{}}} \longrightarrow \; \text{furan} \quad \text{ref. 235}$$

overall yields: 8-29%

These reactions were explained by the formation of an iodonium salt intermediate (133), which reacts with a second molecule of the substrate to yield a tricoordinate species (134). By a ligand coupling process, this iodane compound leads to the self-coupling product.[12,234] (Scheme 5.15)

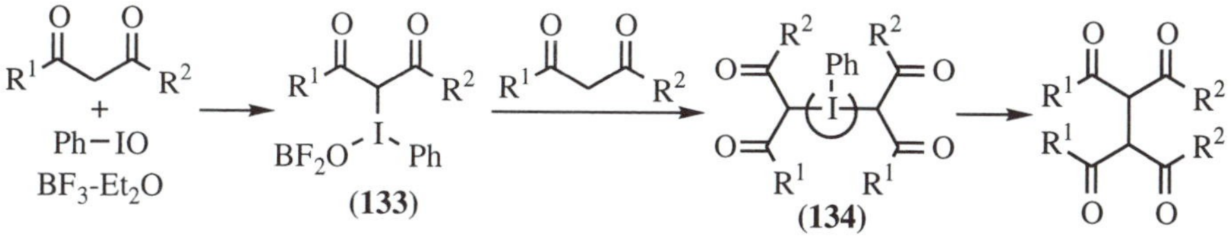

$$R^1 \overset{O}{\underset{}{\|}} \overset{O}{\underset{}{\|}} R^2 \; + \; Ph-IO \quad BF_3\text{-}Et_2O \longrightarrow R^1 \overset{O}{\underset{}{\|}} \overset{O}{\underset{}{\|}} R^2 \quad BF_2O^{-}\overset{I}{}Ph \;\; (133)$$

(134)

Scheme 5.15

The reaction of an allylmetal derivative of group IV (M = Si, Sn or Ge) with aromatic substrates in the presence of the complex iodosylbenzene-boron trifluoride-diethyl ether results in carbon-carbon bond formation realizing the allylation of the aromatic substrate in generally good yields.[236,237]

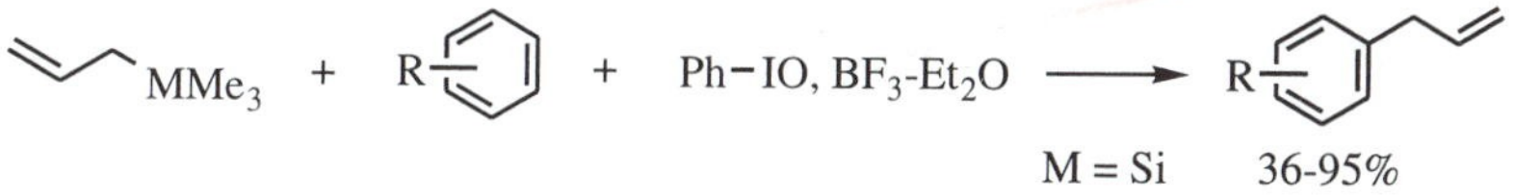

$$\text{allyl-}MMe_3 \; + \; R\text{-}\text{Ar} \; + \; Ph-IO, \; BF_3\text{-}Et_2O \longrightarrow R\text{-}\text{Ar-allyl}$$

M = Si   36-95%

The reaction was explained by a nucleophilic attack of the allylmetal on the electron-deficient iodine atom. The resulting allyliodine(III) intermediate (135) then reacts with the aromatic substrate to afford a tricoordinate allyliodine(III) compound (136), which undergoes ligand coupling to give the allylated aromatic derivative. (Scheme 5.16)

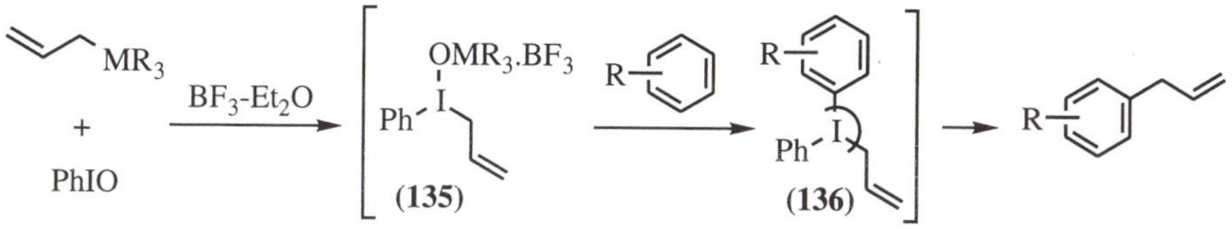

$$\text{allyl-}MR_3 \; + \; PhIO \xrightarrow{\; BF_3\text{-}Et_2O \;} \left[\begin{array}{c} OMR_3.BF_3 \\ Ph\text{-}I\text{-allyl} \\ (135) \end{array} \xrightarrow{R\text{-}\text{Ar}} \begin{array}{c} \text{(136)} \\ Ph\text{-}I \end{array} \right] \longrightarrow R\text{-}\text{Ar-allyl}$$

Scheme 5.16

When functionalised, allylsilanes can react intramolecularly under the reaction conditions. For example, treatment of the hydroxyallylsilanes (137) with the complex PhIO - BF₃ - diethyl ether led to the formation of 5- and 6-membered β-methylene cyclic ethers (139) derived from intramolecular cyclisation of the intermediate (138).[238]

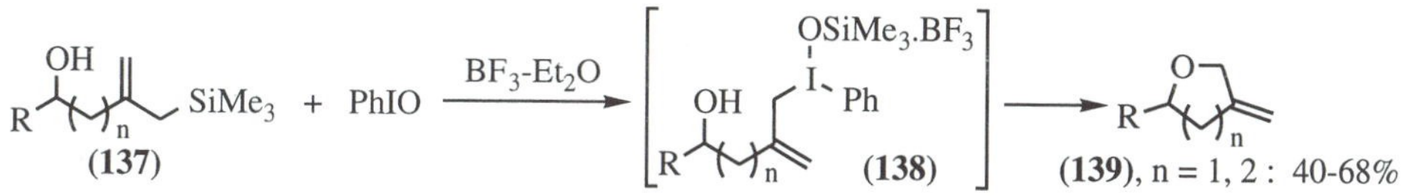

$$R\overset{OH}{\underset{n}{\|}}SiMe_3 \; + \; PhIO \xrightarrow{\; BF_3\text{-}Et_2O \;} \left[\begin{array}{c} OSiMe_3.BF_3 \\ R\overset{OH}{\underset{n}{\|}}\;I\text{-}Ph \\ (138) \end{array} \right] \longrightarrow \begin{array}{c} (139), \; n = 1, 2 : \; 40\text{-}68\% \end{array}$$

(137)

5.5.2 Carbon-heteroatom coupling

Reactions of alkenes with hypervalent iodine compounds lead mostly to vicinally functionalised alkanes. This is the case with PhI(OAc)$_2$, PhIO, PhI(OH)OTs, PhI(OTf)O(TfO)IPh and other related reagents.[230,231,233,239-247] For example, treatment of alkenes with PhI(OH)OTs, (HTIB), affords *vic* bis(tosyloxy)alkanes with a *syn* stereospecificity.[239,241] It is generally admitted that this reaction proceeds by the electrophilic attack of the hypervalent iodine species on the ethylenic double bond to afford a carbonium ion intermediate (140). This intermediate undergoes two consecutive SN2 substitution reactions to eventually give the final products.[13] (Scheme 5.17)

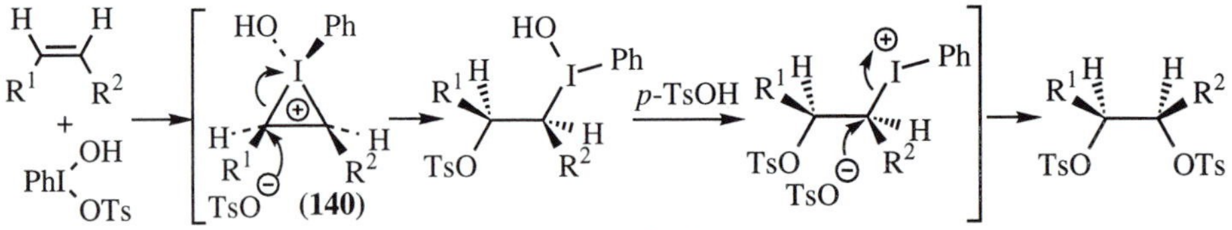

Scheme 5.17

An alternative mechanism which has been envisioned as compatible with the overall stereochemical outcome invokes the formation of an iodine(V) intermediate (141) undergoing two consecutive ligand coupling steps to afford the vicinal *syn* product.[13] (Scheme 5.18)

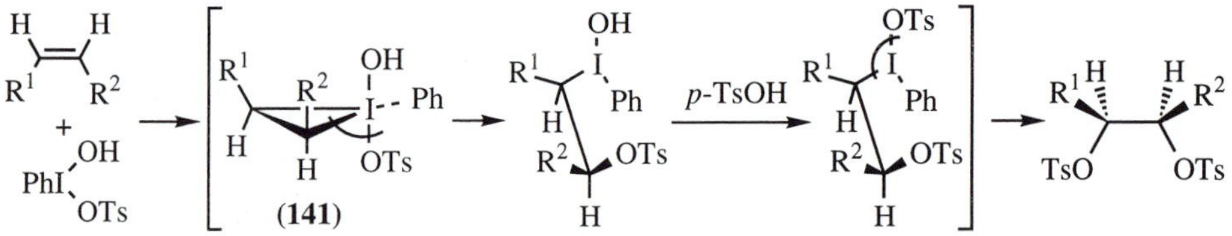

Scheme 5.18

Depending on the nature of the substrate, the polar mechanism may become the only likely possibility. This is the case when non stereospecific functionalisation is observed, as in the reaction of HTIB with stilbenes (142). Another case is the formation of rearranged products, such as in the reaction of 1,1-diphenylethylene (143) with HTIB.[240,241] (Scheme 5.19)

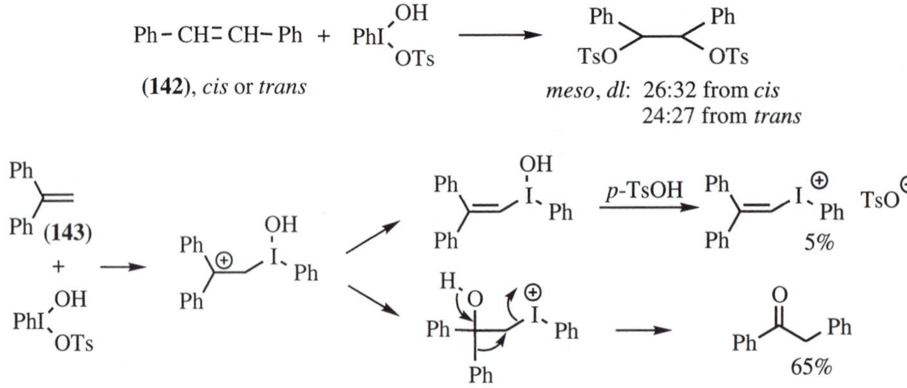

Ph–CH=CH–Ph + PhI(OH)OTs ⟶

(142), *cis* or *trans*

meso, dl: 26:32 from *cis*
24:27 from *trans*

Scheme 5.19

The α-functionalisation of ketones through reaction with iodine(III) derivatives has led to the introduction of a number of functional groups on the α-position: acetoxy[248] and particularly organosulfonyloxy.[239,249-251] HTIB is a very efficient reagent for the introduction of a tosyloxy group on the α-position of ketones in non-hydroxylic solvents.[239] Its analog, hydroxy(mesyloxy) iodobenzene, introduces an α-mesyloxy group under the same type of reaction conditions.[249,251] The ultrasound promoted reaction of HTIB with various ketones provided a direct and mild method of α-tosyloxylation of ketones in generally good yields. The α-tosyloxy derivatives of cyclopentanone and cycloheptanone, otherwise difficult to obtain, were prepared in relatively good yields by this method.[252]

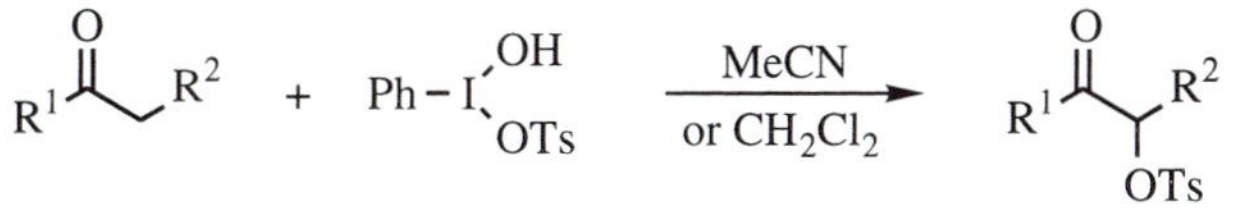

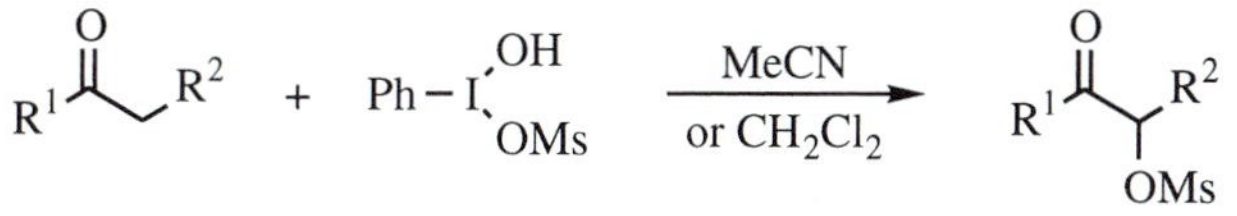

The reaction involves first the electrophilic addition of a hypervalent iodine reagent to the enol tautomer of the ketone to give an α-phenyliodonioketone (**144**). The second step is generally assumed to be a classical S_N2 displacement yielding the final product.[13] (Scheme 5.20)

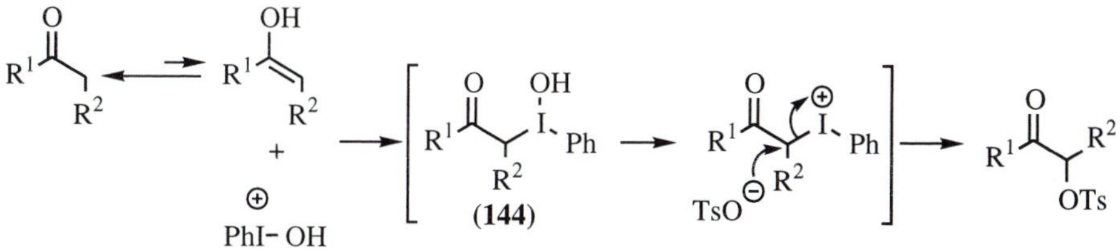

Scheme 5.20

However, by analogy with the reactions of alkenes, an alternative mechanism via the formation of a pentacoordinate cyclic periodinane (**145**) is also plausible. Oxidative addition of HTIB or its analogs to the enol affords the periodinane. Ligand coupling yields a new elusive intermediate, an α-hydroxy-iodinane (**146**), which undergoes elimination of iodobenzene.[13] (Scheme 5.21)

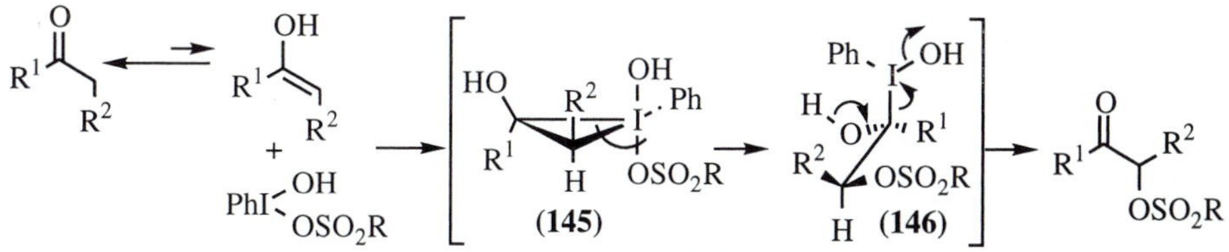

Scheme 5.21

The direct functionalisation of ketones is not limited to simple sulfonates. The α-(+)-10-camphorsulfonyloxy group has been also introduced into various types of ketones or carbonyl compounds bearing an active methylene group by using the HTIB analog, [α-(+)-10-camphorsulfonyl-oxy]hydroxyiodobenzene (**147**).[250]

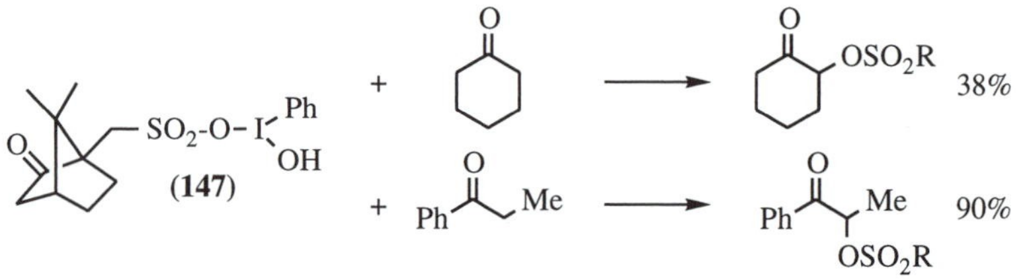

A major drawback of the direct α-functionalisation of ketones is the poor regioselectivity of the reaction. For example, 2-butanone reacts with HTIB in acetonitrile under reflux to afford a 1.57:1 mixture of 3-tosyloxy-2-butanone (**148**) and 1-tosyloxy-2-butanone (**149**).[239]

This drawback was circumvented by treating the preformed silyl enol ethers with phenyliodine(III) reagents. In this way, a variety of groups were introduced regiospecifically: hydroxy,[228,253-256] alkoxy,[228,253,257] trifluoromethanesulfonyloxy[258] and other sulfonyloxy groups.[259] In the case of the 6-methylcyclohexyl trimethylsilyl-1,2-enol ether (**150**), only one regioisomer was formed in high yield.

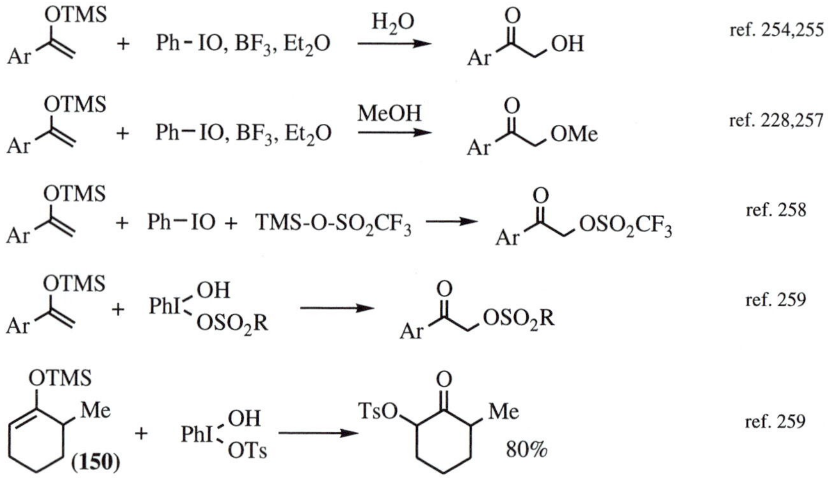

A second advantage of the silyl enol ether method is that better yields were obtained in the reaction employing the silyl enol ether than in the reaction by direct functionalisation of the ketone.[239,259]

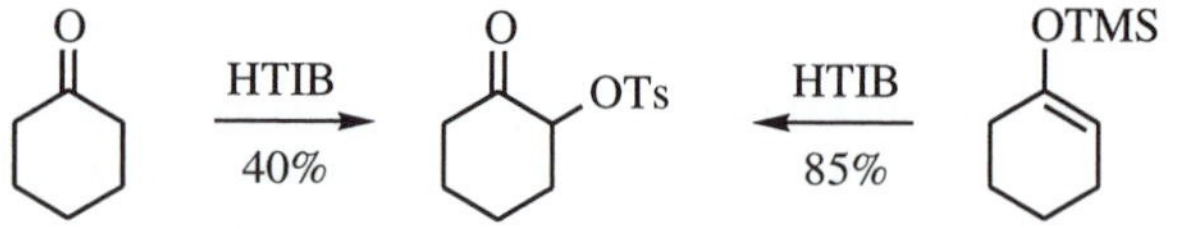

Acid-sensitive or easily oxidised substrates can be functionalised in good conditions by the reaction of their silyl enol ethers with the appropriate phenyliodine(III) reagent.[259]

α-Alkoxylation of ketones has also been performed in good to excellent yields by treatment of the α-metalloketones (151), themselves derived from the silyl enol ethers, with iodosobenzene in the presence of an alcohol. A metal-iodine transmetallation is likely to occur, giving the same type of acylmethylene iodine(III) intermediates which undergo ligand coupling to give the end products.[260]

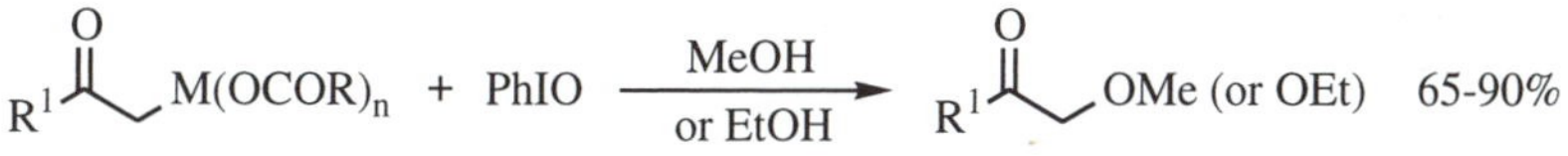

$$R^1 \overset{O}{\underset{\|}{C}}\!\!-\!M(OCOR)_n \ + \ PhIO \ \xrightarrow[\text{or EtOH}]{\text{MeOH}} \ R^1 \overset{O}{\underset{\|}{C}}\!\!-\!OMe \ (\text{or OEt}) \quad 65\text{-}90\%$$

(151), M = Pb, Tl or Hg

Simple esters as well as ε-caprolactone afforded their α-sulfonyloxy derivatives in good yields (60-85% yields) when their silyl enol ethers (152) were treated with PhI(OH)OTs or PhI(OH)OMs.[259]

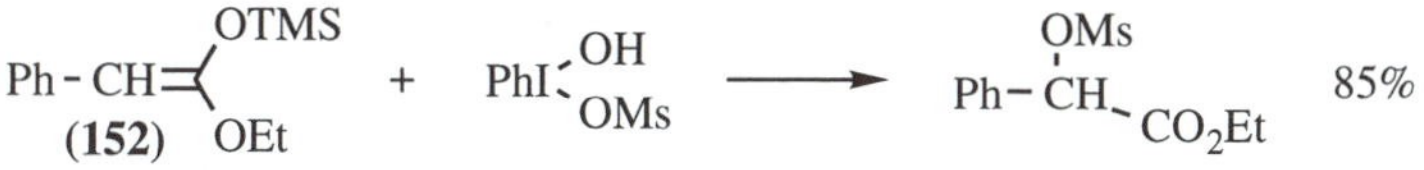

Treatment of silyl enol ethers (153) with *N*-tosyliminoiodobenzene (154), a nitrogen analog of iodosylbenzene, in acetonitrile led to α-tosylamination of the carbonyl compound giving the α-*N*-tosyl aminoketones (155). This reaction was explained by the intermediacy of a tricoordinate amino-iodane(III) (156), which subsequently undergoes ligand coupling.[261] A related system was described by Evans *et al.*, who reported that the treatment of alkenes or silyl enol ethers with *N*-tosylimino-iodobenzene (154), in the presence of cupric triflate as catalyst, affords the corresponding *N*-tosyl-aziridines (157).[262-265] In the case of the enol ethers, hydrolysis led to the α-*N*-tosylaminoketones (155). However, in this case, the mechanism is totally different as the reaction results from the addition of the nitrene onto the double bonds giving intermediately an aziridine.

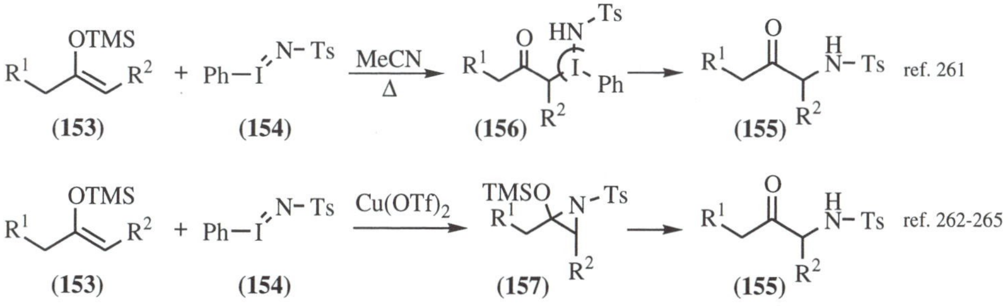

The reaction of 1,3-dicarbonyl compounds with iodosylbenzene, hydroxy(tosyloxy)iodobenzene or hydroxy(mesyloxy)iodobenzene, gave similar results to those obtained with monocarbonyl compounds. However, only the 2-functionalised derivatives were obtained.[234,249,266] Excellent results were observed in the case of substrates containing a perfluoroalkyl chain. Treatment of the fluoro dicarbonyl compounds (158) with HTIB afforded the α-tosyloxy compounds (159) which were isolated as their dihydrate.

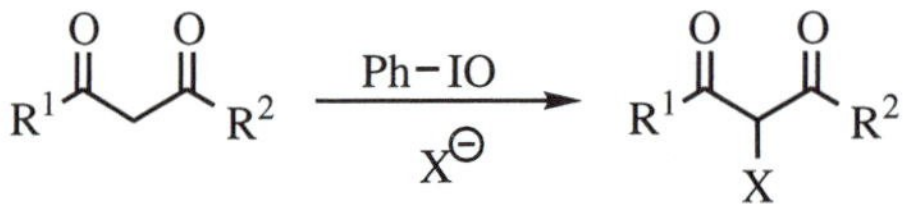

R^1 = Me or Ph; R^2 = Me: X = OSO$_2$Me, 73-83%; X = OMe, 67%; X = N$_3$, 70-76%
R^1 = Me or Ph; R^2 = OMe: X = OSO$_2$Me, 76%; X = OR, 59-63%; X = N$_3$, 48-52%

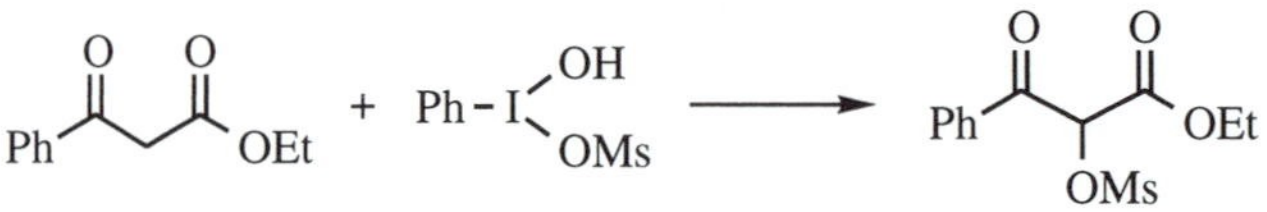

ref. 249

ref. 249

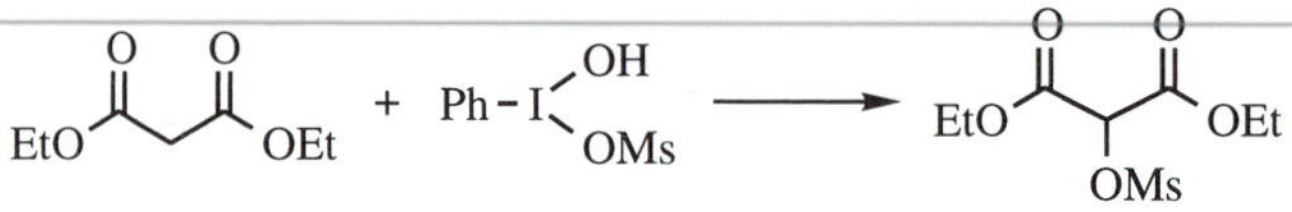

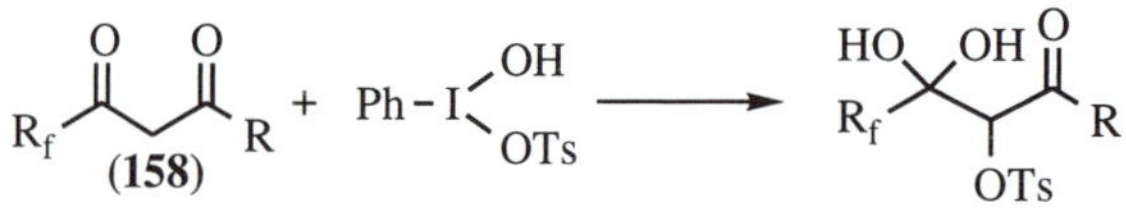

(159), 79-97%   ref. 266

A new reaction of diazo compounds was discovered when α-aryliodonio diazo compounds were prepared and treated with nucleophiles. α-Aryliodonio diazo compounds (160) are readily accessible by treatment of bis(pyridino)aryliodine(III) salts with alkyl diazoacetates, or by the Lewis acid mediated (Me$_3$SiOTf) reaction of diacetoxyiodobenzene with alkyl diazoacetates. Nucleophilic substitution at the α-carbon was carried out by reaction with a series of neutral nucleophiles under mild neutral conditions to lead to the α-oniosubstituted diazo compounds (161).[267]

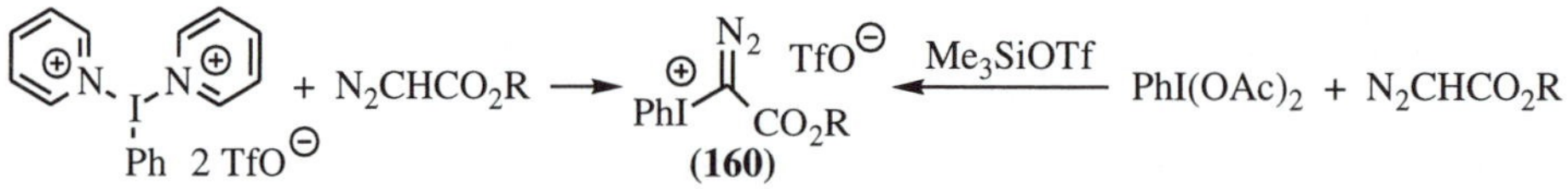

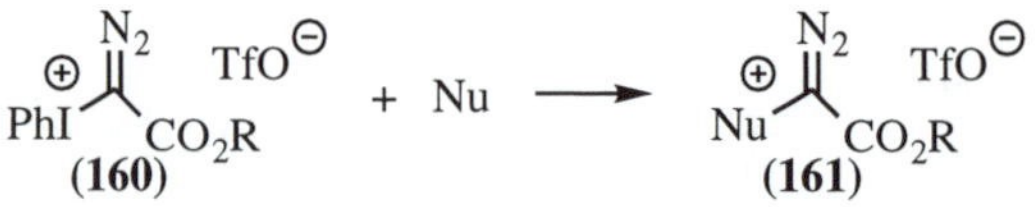

Nu$^\oplus$	$\langle\!\!\bigcirc\!\!\rangle$N$^-$$^\oplus$	Me$_2$S$^\oplus$	Ph$_3$As$^\oplus$	Ph$_3$Sb$^\oplus$	Et$_3$N$^\oplus$
Yield	74%	90%	95%	86%	53%

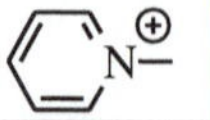

In the case of triphenylphosphine as nucleophile, the reaction proceeded further to afford the bis-substituted product (**162**).

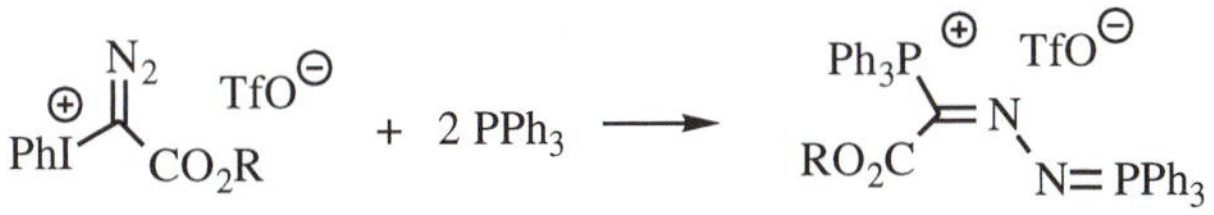

$$ \underset{PhI}{\overset{\oplus}{}} \overset{N_2}{\underset{CO_2R}{\|}} \, TfO^{\ominus} \; + \; 2\,PPh_3 \; \longrightarrow \; \underset{RO_2C}{\overset{Ph_3P^\oplus \; TfO^{\ominus}}{}} {=} \underset{N= PPh_3}{\overset{N}{}} \qquad (162),\,72\% $$

A somewhat anecdotic reaction involving a multistep mechanism is the unexpected result observed in the treatment of propellane (**163**), tricyclo[1.1.1.0^{1,3}]pentane, with dichloroiodobenzene at low temperature (0 to - 20°C) in the presence of chloroform under irradiation. The main isolated product appeared to be 1-trichloromethyl-3-iodobicyclo[1.1.1]pentane (**164**).[268]

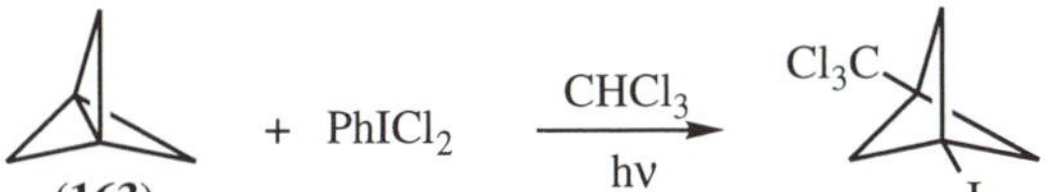

$$ (163) \;+\; PhICl_2 \;\xrightarrow[hv]{CHCl_3}\; (164),\,23\% $$

To explain the presence of a trichloromethyl and an iodine atom at the bridgehead atoms, two photochemically generated radicals were postulated to react with propellane leading to an unstable organoiodine(III) derivative (**165**). This iodane intermediate underwent ligand coupling to afford eventually the iodopropellane derivative.[268] (Scheme 5.22)

$$ PhICl_2 \;\xrightarrow{hv}\; Ph\overset{\bullet}{I}Cl \;+\; Cl^\bullet $$

$$ CHCl_3 \;\xrightarrow{hv}\; H^\bullet \;+\; Cl_3C^\bullet $$

$$ Cl^\bullet \;+\; CHCl_3 \;\longrightarrow\; HCl \;+\; Cl_3C^\bullet $$

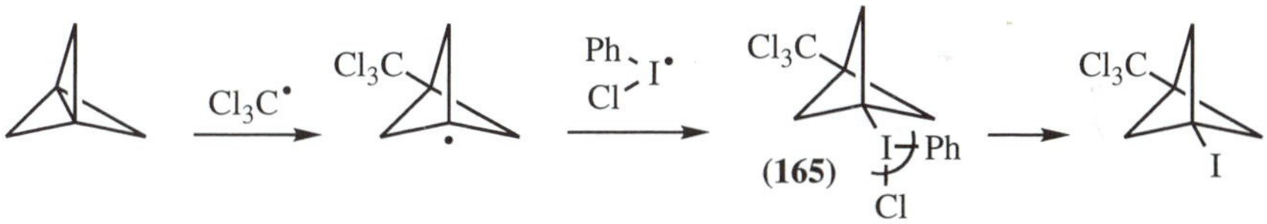

Scheme 5.22

The reaction of alkyl iodides with aryliodine(III) dicarboxylates, particularly the bis(trifluoro-acetate), afforded as the main product the ester (**166**) derived from substitution of iodine of the alkyl iodide by an acyloxy group.[269]

$$ PhI(OCOCF_3)_2 \;+\; R'I \;\longrightarrow\; PhI \;+\; RCO_2I \;+\; RCO_2R' $$
$$ (166),\,32\text{-}100\% $$

The reaction starts with the nucleophilic attack from the halogen of the alkyl iodide on the iodine of the aryliodine compound. The unstable intermediate (**167**) can be either covalent or ionic. If ionic, the weakly nucleophilic counteranion attacks the positive halogen atom to produce the ester and another intermediate (**168**) which undergoes ligand coupling to give iodobenzene and the acylhypoiodite, CF_3CO_2I. Alternatively, if the first formed intermediate is covalent, it can undergo a first ligand coupling process to afford iodobenzene and an alkyliodine(III) dicarboxylate (**166**) which then decomposes, again by ligand coupling, to afford the ester and CF_3CO_2I similarly to the products formed in the ionic sequence.[269] (Scheme 5.23)

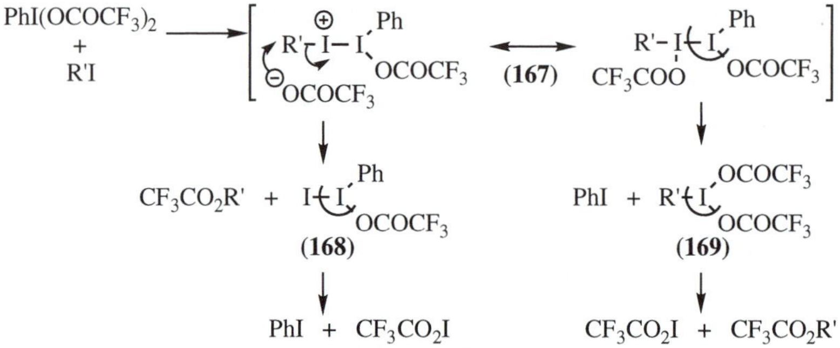

Scheme 5.23

5.5.3 Oxidation of nitrogen compounds

5.5.3.1 Oximes

The reactivity of bis(trifluoroacetoxy)iodobenzene (**23**) towards oximes is of interest as several products are formed depending on the substrate.[270] A ligand coupling mechanism may be suggested for the reaction of aliphatic ketoximes with bis(trifluoroacetoxy)iodobenzene, which gives fair yields of the stable, intensely blue 1-nitroso-1-trifluoroacetoxyalkanes.[271]

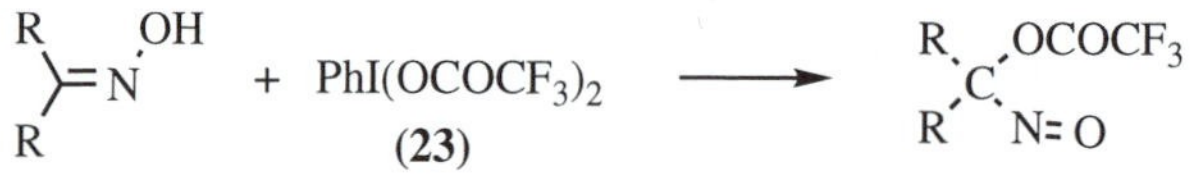

5.5.3.2 Hydrazone derivatives

Nitrogen-containing derivatives of carbonyl compounds react with diacetoxyiodobenzene or with bis(trifluoroacetoxy)iodobenzene to afford different types of products. The most frequently isolated product is the regenerated carbonyl compound. However, in some instances, the unstable azo intermediate can be isolated.[272,273] The reaction proceeds in two steps. Interaction of the hydrazone with the diacyloxyiodobenzene leads to an intermediate hydrazino(phenyl)iodine(III) derivative which then undergoes an intramolecular rearrangement, of the type B *ipso*-allylic ligand coupling, to produce the unstable phenylazo compound. Usually this compound is hydrolysed *in situ* going directly to the carbonyl compound. However, in the case of the reaction of the hydrazones of α-ketoesters with bis(trifluoroacetoxy)iodobenzene, the intermediate α-trifluoroacetoxy-α-azo esters were isolated as relatively stable compounds. (Scheme 5.24)

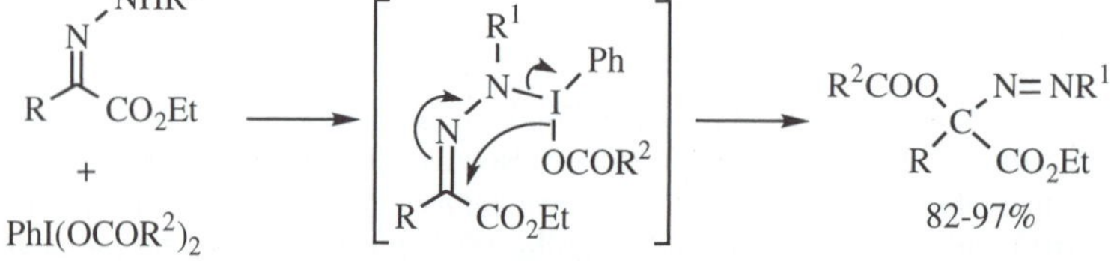

Scheme 5.24

This type of reactivity was also observed in the reaction of acylhydrazones with iodobenzene diacetate. In the case of *ortho*-hydroxyarylketone acylhydrazones, this reaction with iodobenzene diacetate in methylene dichloride at room temperature gave, by a sequence of ligand coupling and intramolecular rearrangements, the 1,2-diacylbenzene derivatives.[274] In the case of bisacylresorcinol acylhydrazones derivatives, the reaction with iodobenzene diacetate led to tetraacylbenzene derivatives.[275] These multistep reactions were explained by the formation of acetoxy azo compounds as key intermediates. Then, a series of rearrangements and internal displacements led to the final products.[276]

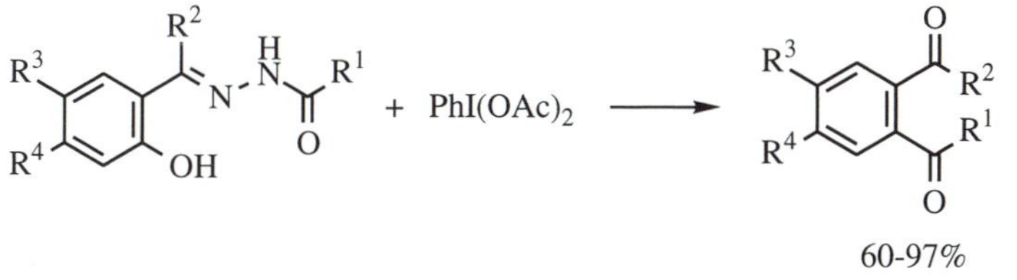

ref. 274

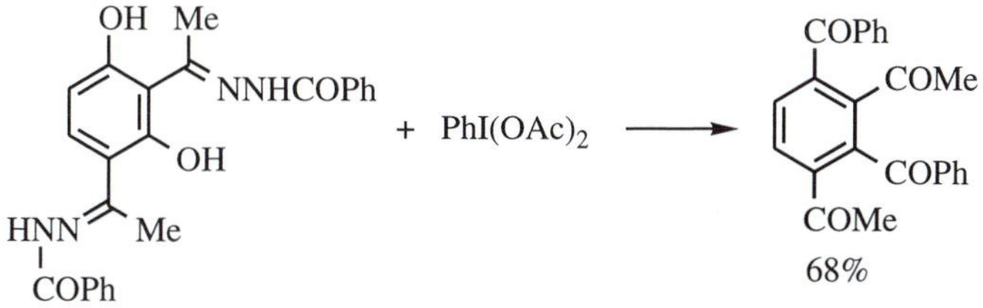

60-97%

ref. 275

68%

5.6 References

1. Banks, D.F. *Chem. Rev.* **1966**, *66*, 243-266.
2. Varvoglis, A. *Chem. Soc. Rev.* **1981**, *10*, 377-407.
3. Umemoto, T. *Yuki Gosei Kagaku Kyokai Shi* **1983**, *41*, 251-265.
4. Koser, G.F. *Hypervalent Halogen Compounds* in *The Chemistry of Functional Groups, Supplement D*; Patai, S.; Rappoport, Z., Eds.; J. Wiley & Sons: New York, **1983**; Ch. 18, pp. 721-811.
5. Koser, G.F. *Halonium Ions* in *The Chemistry of Functional Groups, Supplement D*; Patai, S.; Rappoport, Z., Eds.; J. Wiley & Sons: New York, **1983**; Ch. 25, pp. 1265-1351.
6. Varvoglis, A. *Synthesis* **1984**, 709-726.
7. Moriarty, R.M.; Prakash, O. *Acc. Chem. Res.* **1986**, *19*, 244-250.
8. Ochiai, M.; Nagao, Y. *Yuki Gosei Kagaku Kyokai Shi* **1986**, *44*, 660-673.
9. Merkushev, E.B. *Russ. Chem. Rev. (Engl. Transl.)* **1987**, *56*, 826-845.
10. Ochiai, M. *Rev. Heteroatom. Chem.* **1989**, *2*, 92-111.
11. Maletina, I.I.; Orda, V.V.; Yagupol'skii, L.M. *Russ. Chem. Rev. (Engl. Transl.)* **1989**, *58*, 544-558.
12. Moriarty, R.M.; Vaid, R.K. *Synthesis* **1990**, 431-447.
13. Moriarty, R.M.; Vaid, R.K.; Koser, G.F. *Synlett* **1990**, 365-383.
14. Stang, P.J. *Angew. Chem., Int. Ed. Engl.* **1992**, *31*, 274-285.
15. Varvoglis, A. *The Organic Chemistry of Polycoordinated Iodine*; VCH: New York, **1992**.
16. Grushin, V.V. *Acc. Chem. Res.* **1992**, *25*, 529-536.
17. Stang, P.J. in *The Chemistry of Triple-bonded Functional Groups, Supplement C2*; Patai, S., Ed.; J. Wiley & Sons: Chichester, **1994**; Ch. 20, pp. 1164-1182.
18. Prakash, O.; Saini, N.; Sharma, P.K. *Synlett* **1994**, 221-227.
19. Prakash, O.; Saini, N.; Sharma, P.K. *Heterocycles* **1994**, *38*, 409-431.
20. Prakash, O.; Singh, S.P. *Aldrichimica Acta* **1994**, *27*, 15-23.

21. Stang, P.J. *Alkynyliodonium Salts: Electrophilic Acetylene Equivalents*, in *Modern Acetylene Chemistry*; Stang, P.J.; Diederich, F., Eds.; VCH: Weinheim, **1995**; Ch. 3, pp. 67-98.
22. Koser, G.F. *Halonium Ions* in *The Chemistry of Halides, Pseudo-halides and Azides, Supplement D2*; Patai, S.; Rappoport, Z., Eds.; J. Wiley & Sons: Chichester, **1995**, Ch. 21, pp. 1173-1274.
23. Kitamura, T. *Yuki Gosei Kagaku Kyokai Shi* **1995**, *53*, 893-905.
24. Stang, P.J.; Zhdankin, V.V. *Chem. Rev.* **1996**, *96*, 1123-1178.
25. Varvoglis, A., *Hypervalent Iodine in Organic Synthesis*; Academic Press: San Diego, **1997**.
26. Sandin, R.B.; Kulka, M.; McCready, R. *J. Am. Chem. Soc.* **1937**, *59*, 2014-2015.
27. Beringer, F.M.; Forgione, P.S.; Yudis, M.D. *Tetrahedron* **1960**, *8*, 49-63.
28. Beringer, F.M.; Galton, S.A.; Huang, S.J. *J. Am. Chem. Soc.* **1962**, *84*, 2819-2823.
29. Beringer, F.M.; Forgione, P.S. *Tetrahedron* **1963**, *19*, 739-748.
30. Beringer, F.M.; Falk, R.A. *J. Chem. Soc.* **1964**, 4442-4451.
31. Tanner, D.D.; Reed, D.W.; Setiloane, B.P. *J. Am. Chem. Soc.* **1982**, *104*, 3917-3923.
32. Singh, P.R.; Khanna, R.K. *Tetrahedron Lett.* **1982**, *23*, 5355-5358.
33. Barton, D.H.R.; Finet, J.-P.; Giannotti, C.; Halley, F. *J. Chem. Soc., Perkin Trans. 1* **1987**, 241-249.
34. Banks, J.T.; Garcia, H.; Miranda, M.A.; Pérez-Prieto, J.; Scaiano, J.C. *J. Am. Chem. Soc.* **1995**, *117*, 5049-5054.
35. Budylin, V.A.; Ermolenko, M.S.; Chugtai, F.A.: Kost, A.N. *Khim. Geterotsikl. Soedin.* **1981**, 1494-1496; *Chem. Abstr.* **1982**, *96*, 142617n.
36. Irving, H.; Reid, R.W. *J. Chem. Soc.* **1960**, 2078-2081.
37. Sandin, R.B. *J. Org. Chem.* **1969**, *34*, 456-457.
38. Sato, T.; Shimizu, K.; Moriya, H. *J. Chem. Soc., Perkin Trans. 1* **1974**, 1537-1539.
39. Lucas, H.J.; Kennedy, E.R.; Wilmot, C.A. *J. Am. Chem. Soc.* **1936**, *58*, 157-160.
40. Beringer, F.M.; Brierley, A.; Drexler, M.; Gindler, E.M.; Lumpkin, C.C. *J. Am. Chem. Soc.* **1953**, *75*, 2708-2712.
41. Beringer, F.M.; Geering, E.J.; Kuntz, I.; Mausner, M. *J. Phys. Chem.* **1956**, *60*, 141-150.
42. Beringer, F.M.; Mausner, M. *J. Am. Chem. Soc.* **1958**, *80*, 4535-4536.
43. Beringer, F.M.; Gindler, E.M.; Rapoport, M.; Taylor, R.J. *J. Am. Chem. Soc.* **1959**, *81*, 351-361.
44. Le Count, D.J.; Reid, J.A.W. *J. Chem. Soc. (C)* **1967**, 1298-1301.
45. Yamada, Y.; Okawara, M. *Bull. Chem. Soc. Jpn.* **1972**, *45*, 1860-1863.
46. Yamada, Y.; Kashima, K.; Okawara, M. *Bull. Chem. Soc. Jpn.* **1974**, *47*, 3179-3180.
47. Lancer, K.M.; Wiegand, G.H. *J. Org. Chem.* **1976**, *41*, 3360-3364.
48. McEwen, W.E.; DeMassa, J.W. *Heteroatom Chem.* **1996**, *7*, 349-354.
49. Grushin, V.V.; Shcherbina, T.M.; Tolstaya, T.P. *J. Organomet. Chem.* **1985**, *292*, 105-117.
50. Grushin, V.V.; Demkina, I.I.; Tolstaya, T.P. *J. Chem. Soc., Perkin Trans. 2* **1992**, 505-511.
51. Lewis, E.S.; Stout, C.A. *J. Am. Chem. Soc.* **1954**, *76*, 4619-4621.
52. Caserio, M.C.; Glusker, D.L.; Roberts, J.D. *J. Am. Chem. Soc.* **1959**, *81*, 336-342.
53. Ptitsyna, O.A.; Lyatiev, G.G.; Reutov, O.A. *Dokl. Akad. Nauk SSSR* **1968**, *181*, 895-898; *Chem. Abstr.* **1968**, *69*, 96095h.
54. Gronowitz, S.; Holm, B. *Synth. Commun.* **1974**, *4*, 63-69.
55. Gronowitz, S.; Ander, I. *Chem. Scripta* **1980**, *15*, 135-144.
56. Yamada, Y.; Okawara, M. *Makromol. Chem.* **1972**, *152*, 163-176.
57. Gronowitz, S.; Holm, B. *Tetrahedron* **1977**, *33*, 557-561.
58. Lubinkowski, J.J.; Gomez, M.; Calderon, J.L.; McEwen, W.E. *J. Org. Chem.* **1978**, *43*, 2432-2435.
59. Grushin, V.V.; Kantor, M.M.; Tolstaya, T.P.; Shcherbina, T.M. *Izv. Akad. Nauk SSSR, Ser. Khim.* **1984**, 2332-2338.

60. Gronowitz, S.; Holm, B. *J. Heterocycl. Chem.* **1977**, *14*, 281-288.
61. Gronowitz, S.; Holm, B. *Chem. Scripta* **1974**, *6*, 133-136.
62. Van der Puy, M. *J. Fluorine Chem.* **1982**, *21*, 385-392.
63. Pike, V.W.; Aigbirhio, F.I. *J. Chem. Soc., Chem. Commun.* **1995**, 2215-2216.
64. You, J.-Z.; Chen, Z.-C. *Synth. Commun.* **1992**, *22*, 1441-1444.
65. Liu, D.-D.; Chen, D.-w.; Chen, Z.-C. *Synth. Commun.* **1992**, *22*, 2903-2908.
66. Liu, Z.-D.; Chen, Z.-C. *Synth. Commun.* **1993**, *23*, 2673-2676.
67. Liu, Z.-D.; Chen, Z.-C. *Synthesis* **1993**, 373-374.
68. Wittig, G.; Clauss, K. *Liebigs Ann. Chem.* **1949**, *562*, 187-192.
69. Clauss, K. *Chem. Ber.* **1955**, *88*, 268-270.
70. Beringer, F.M.; Chang, L.L. *J. Org. Chem.* **1971**, *36*, 4055-4060.
71. Beringer, F.M.; Chang, L.L. *J. Org. Chem.* **1972**, *37*, 1516-1519.
72. Reich, H.J.; Cooperman, C.S. *J. Am. Chem. Soc.* **1973**, *95*, 5077-5078.
73. Hepworth, H. *J. Chem. Soc.* **1921**, *119*, 1244-1249.
74. Wittig, G.; Clauss, K. *Liebigs Ann. Chem.* **1952**, *578*, 136-146.
75. Beringer, F.M.; Dehn, J.W., Jr.; Winicov, M. *J. Am. Chem. Soc.* **1960**, *82*, 2948-2952.
76. Barton, D.H.R.; Jaszberenyi, J.Cs.; Leβmann, K.; Timar, T. *Tetrahedron* **1992**, *48*, 8881-8890.
77. Moriarty, R.M.; Ku, Y.Y.; Sultana, M.; Tuncay, A. *Tetrahedron Lett.* **1987**, *28*, 3071-3074.
78. Cookson, R.C.; Farquharson, B.K. *Tetrahedron Lett.* **1979**, 1255-1256.
79. Beringer, F.M.; Daniel, W.J.; Galton, S.A.; Rubin, G. *J. Org. Chem.* **1966**, *31*, 4315-4318.
80. Rossi, R.A.; Bunnett, J.F. *J. Am. Chem. Soc.* **1974**, *96*, 112-117.
81. Gao, P.; Portoghese, P.S. *J. Org. Chem.* **1995**, *60*, 2276-2278.
82. Chen, K.; Koser, G.F. *J. Org. Chem.* **1991**, *56*, 5764-5767.
83. Kurts, A.L.; Davydov, D.V.; Bundel, Yu.G. *Vestn. Mosk. Univ., Ser. 2: Khim.* **1984**, *25*, 68-74; *Chem. Abstr.* **1984**, *100*, 209307h.
84. Chen, Z.-C.; Jin, Y.-Y.; Stang, P.J. *J. Org. Chem.* **1987**, *52*, 4115-4117.
85. Beringer, F.M.; Forgione, P.S. *J. Org. Chem.* **1963**, *28*, 714-717.
86. Tomari, K.; Machiya, K.; Ichimoto, I.; Ueda, H. *Agr. Biol. Chem.* **1980**, *44*, 2135-2138.
87. Copp, F.C.; Franzmann, K.W.; Gilmore, J.; Whalley, W.B. *J. Chem. Soc., Perkin Trans. 1* **1983**, 909-914.
88. Beringer, F.M.; Galton, S.A. *J. Org. Chem.* **1963**, *28*, 3417-3421.
89. Neilands, O.; Vanags, G.; Gudriniece, E. *Zh. Obshch. Khim.* **1958**, *28*, 1201-1205; *Chem. Abstr.* **1958**, *52*, 19988b.
90. Hampton, K.G.; Harris, T.M.; Hauser, C.R. *J. Org. Chem.* **1964**, *29*, 3511-3514.
91. Hampton, K.G.; Harris, T.M.; Hauser, C.R. *Organic Syntheses* **1971**, *51*, 128-132.
92. Kita, Y.; Okunaka, R.; Kondo, M.; Tohma, H.; Inagaki, M.; Hatanaka, K. *J. Chem. Soc., Chem. Commun.* **1992**, 429-430.
93. Kornblum, N.; Taylor, H.J. *J. Org. Chem.* **1963**, *28*, 1424-1425.
94. Park, K.P.; Clapp, L.B. *J. Org. Chem.* **1964**, *29*, 2108.
95. Kuehne, M.E. *J. Am. Chem. Soc.* **1962**, *84*, 837-847.
96. Pongratz, E.; Kappe, T. *Monatsh. Chem.* **1984**, *115*, 231-242.
97. Davidson, J.M.; Dyer, G. *J. Chem. Soc. (A)* **1968**, 1616-1617.
98. McEwen, W.E.; Lubinkowski, J.J.; Knapczyk, J.W. *Tetrahedron Lett.* **1972**, 3301-3304.
99. Lubinkowski, J.J.; Knapczyk, J.W.; Calderon, J.L.; Petit, L.R.; McEwen, W.E. *J. Org. Chem.* **1975**, *40*, 3010-3014.
100. Lubinkowski, J.J.; Arrieche, C.G.; McEwen, W.E. *J. Org. Chem.* **1980**, *45*, 2076-2079.
101. Levit, A.F.; Kiprianova, L.A.; Gragerov, I.P. *Teor. Eksper. Khim.* **1980**, *16*, 254-257; *Chem. Abstr.* **1980**, *93*, 70420.
102. Fuson, R.C.; Albright, R.L. *J. Am. Chem. Soc.* **1959**, *81*, 487-490.
103. Beringer, F.M.; Gindler, E.M. *J. Am. Chem. Soc.* **1955**, *77*, 3203-3207.

104. Crowder, J.R.; Glover, E.E.; Grundon, M.F.; Kaempfen, H.X. *J. Chem. Soc.* **1963**, 4578-4585.
105. Dibbo, A.; Stephenson, L.; Walker, T.; Warburton, W.K. *J. Chem. Soc.* **1961**, 2645-2651.
106. Humppi, T. *Synthesis* **1985**, 919-924.
107. Nevalainen, T.; Rissanen, K. *J. Chem. Soc., Perkin Trans. 2* **1994**, 271-279.
108. Nicholson, J.S.; Peak, D.A. *Chem. Ind. (London)* **1962**, 1244.
109. Taylor, E.C.; Kienzle, F. *J. Org. Chem.* **1971**, *36*, 233-235.
110. Laus, G.; Stadlwieser, J.; Klötzer, W. *Synthesis* **1989**, 773-775.
111. Abramovitch, R.A.; Alvernhe, G.; Bartnik, R.; Dassanayake, N.L.; Inbasekaran, M.N.; Kato, S. *J. Am. Chem. Soc.* **1981**, *103*, 4558-4565.
112. Abramovitch, R.A.; Inbasekaran, M.N. *Tetrahedron Lett.* **1977**, 1109-1112.
113. Iijima, H.; Endo, Y.; Shudo, K.; Okamoto, T. *Tetrahedron* **1984**, *40*, 4981-4985.
114. Cadogan, J.I.G.; Rowley, A.G. *Synth. Commun.* **1977**, *7*, 365-366.
115. Kost, D.; Berman, E. *Tetrahedron Lett.* **1980**, *21*, 1065-1068.
116. Taylor, E.C.; Inbasekaran, M. *Heterocycles* **1978**, *10*, 37-43.
117. Grubbs, E.J.; Milligan, R.J; Goodrow, M.H. *J. Org. Chem.* **1971**, *36*, 1780-1785.
118. Wentrup, C.; Gerecht, B.; Laqua, D.; Briehl, H.; Winter, H.-W.; Reisenauer, H.P.; Winnewisser, M. *J. Org. Chem.* **1981**, *46*, 1046-1048.
119. Briehl, H.; Lukosch, A.; Wentrup, C. *J. Org. Chem.* **1984**, *49*, 2772-2779.
120. Bátori, S.; Messmer, A. *J. Heterocycl. Chem.* **1988**, *25*, 437-444.
121. Sakai, S.-i.; Nakajima, K.; Ihida, A.; Ishida, I.; Saito, M. *Yakugaku Zasshi* **1962**, *82*, 1532-1537; *Chem. Abstr.* **1963**, *58*, 13912e.
122. Ptitsyna, O.A.; Lyatiev, G.G.; Reutov, O.A. *Dokl. Akad. Nauk SSSR* **1968**, *182*, 119-121; *Chem. Abstr.* **1969**, *70*, 2900n.
123. Nesmeyanov, A.N.; Makarova, L.G.; Tolstaya, T.P. *Tetrahedron* **1957**, *1*, 145-157.
124. Makarova, L.G. *Izv. Akad. Nauk SSSR, Otdel. Khim. Nauk* **1951**, 741-744; *Chem. Abstr.* **1952**, *46*, 7532d.
125. Sandin, R.B.; Brown, R.K. *J. Am. Chem. Soc.* **1947**, *69*, 2253-2254.
126. Nesmeyanov, A.N.; Makarova, L.G. *Uch. Zap. Mosk. Univ.* **1950**, *132*, 109-116; *Chem. Abstr.* **1955**, *49*, 3903a.
127. Derappe, C.; Rips, R. *C. R. Acad. Sci., Ser. C* **1975**, *281*, 789-792.
128. Sheradsky, T.; Nov, E. *J. Chem. Soc., Perkin Trans. 1* **1980**, 2781-2786.
129. Akiyama, T.; Imasaki, Y.; Kawanisi, M. *Chem. Lett.* **1974**, 229-230.
130. Chen, W.-Y.; Gilman, N.W. *J. Heterocycl. Chem.* **1983**, *20*, 663-666.
131. Messmer, A.; Hajos, G.; Fleischer, J.; Czugler, M. *Monatsh. Chem.* **1985**, *116*, 1227-1231.
132. McKillop, A.; Kobylecki, R.J. *J. Org. Chem.* **1974**, *39*, 2710-2714.
133. Kotali, E.; Varvoglis, A. *J. Chem. Soc., Perkin Trans. 1* **1987**, 2759-2763.
134. Kotali, E.; Varvoglis, A. *J. Chem. Res. (S)* **1989**, 142-143.
135. Bentrude, W.G.; Martin, J.C. *J. Am. Chem. Soc.* **1962**, *84*, 1561-1571.
136. Greidanus, J.W.; Rebel, W.J.; Sandin, R.B. *J. Am. Chem. Soc.* **1962**, *84*, 1504-1505.
137. Klose, W.; Schwartz, K. *J. Heterocycl. Chem.* **1982**, *19*, 1165-1167.
138. Drozd, V.N.; Bogomolova, G.S.; Udachin, Y.M. *Zh. Org. Khim.* **1979**, *15*, 1069-1073.
139. Makarova, L.G.; Nesmeyanov, A.N. *Izv. Akad. Nauk SSSR, Otdel. Khim. Nauk* **1945**, 617-625; *Chem. Abstr.* **1946**, *40*, 4686⁶.
140. Nesmeyanov, A.N.; Tolstaya, T.P.; Grib, A.V.; Kirgizbaeva, S.R. *Izv. Akad. Nauk SSSR, Ser. Khim.* **1973**, 1678-1679; *Chem. Abstr.* **1973**, *79*, 105167u.
141. Wu, S.K.; Fouassier, J.P.; Burr, D.; Crivello, J.V., *Polym. Bull. (Berlin)* **1988**, *19*, 457-460.
142. Crivello, J.V.; Lam, J.H.W. *J. Org. Chem.* **1978**, *43*, 3055-3058.
143. Hori, M.; Kataoka, T.; Shimizu, H.; Tomoto, A. *Tetrahedron Lett.* **1981**, *22*, 3629-3632.
144. Kataoka, T.; Tomoto, A.; Shimizu, H.; Hori, M. *J. Chem. Soc., Perkin Trans. 1* **1983**, 2913-2919.

145. Crivello, J.V.; Lee, J.L.; Conlon, D.A. *J. Polym. Sci., Part A, Polym. Chem.* **1987**, *25*, 3293-3309.
146. Kitamura, T.; Yamane, M.; Furuki, R.; Taniguchi, H.; Shiro, M. *Chem. Lett.* **1993**, 1703-1706.
147. Crivello, J.V.; Lam, J.H.W. *Synth. Commun.* **1979**, *9*, 151-156.
148. Liu, Z.-D.; Zeng, H.; Chen, Z.-C. *Synth. Commun.* **1994**, *24*, 475-479.
149. Ptitsyna, O.A.; Pudeeva, M.E.; Reutov, O.A. *Dokl. Akad. Nauk SSSR* **1965**, *165*, 582-585; *Chem. Abstr.* **1966**, *64*, 19660h.
150. Ptitsyna, O.A.; Pudeeva, M.E.; Reutov, O.A. *Dokl. Akad. Nauk SSSR* **1965**, *165*, 838-841; *Chem. Abstr.* **1966**, *64*, 5129a.
151. Kang, J.; Ku, B.C.S. *Bull. Korean Chem. Soc.* **1985**, *6*, 375-376.
152. You, J.-Z.; Chen, Z.-C. *Synthesis* **1992**, 521-522.
153. Chen, Z.-C.; Jin, Y.-Y.; Yang, R.-Y. *Synthesis* **1988**, 723-724.
154. Chen, D.-W.; Zhang, Y.-D.; Chen, Z.-C. *Synth. Commun.* **1995**, *25*, 1627-1631.
155. Chen, Z.-C.; Jin, Y.-Y.; Stang, P.J. *J. Org. Chem.* **1987**, *52*, 4117-4118.
156. You, J.-Z.; Chen, Z.-C. *Synthesis* **1992**, 633-634.
157. Lockhart, T.P. *J. Am. Chem. Soc.* **1983**, *105*, 1940-1946.
158. Block, P., Jr; Coy, D.H. *J. Chem. Soc., Perkin Trans. 1* **1972**, 633-634.
159. Humora, M.J.; Seitz, D.E.; Quick, J. *Tetrahedron Lett.* **1980**, *21*, 3971-3974.
160. Hickey, D.M.B.; Leeson, P.D.; Novelli, R.; Shah, V.P.; Burpitt, B.E.; Crawford, L.P.; Davies, B.J.; Mitchell, M.B.; Pancholi, K.D.; Tuddenham, D.; Lewis, N.J.; O'Farrell, C. *J. Chem. Soc., Perkin Trans. 1* **1988**, 3103-3111.
161. Scherrer, R.A.; Beatty, H.R. *J. Org. Chem.* **1980**, *45*, 2127-2131.
162. Rewcastle, G.W.; Denny, W.A. *Synthesis* **1985**, 220-222.
163. Le Goff, E. *J. Am. Chem. Soc.* **1962**, *84*, 3786.
164. Beringer, F.M.; Huang, S.J. *J. Org. Chem.* **1964**, *29*, 445-448.
165. Beringer, F.M.; Huang, S.J. *J. Org. Chem.* **1964**, *29*, 1637-1638.
166. Kang, S.-K.; Yamaguchi, T.; Kim, T.-H.; Ho, P.-S. *J. Org. Chem.* **1996**, *61*, 9082-9083.
167. Okuyama, T.; Takino, T.; Sueda, T.; Ochiai, M. *J. Am. Chem. Soc.* **1995**, *117*, 3360-3367.
168. Nesmeyanov, A.N.; Tolstaya, T.P.; Petrakov, A.V. *Dokl. Akad. Nauk SSSR* **1971**, *197*, 1337-1340; *Chem. Abstr.* **1971**, *75*, 140392c.
169. Nesmeyanov, A.N.; Tolstaya, T.P.; Petrakov, A.V.; Leshcheva, I.F. *Dokl. Akad. Nauk SSSR* **1978**, *238*, 1109-1112; *Chem. Abstr.* **1978**, *89*, 6030u.
170. Ochiai, M.; Oshima, K.; Masaki, Y. *J. Am. Chem. Soc.* **1991**, *113*, 7059-7061.
171. Ochiai, M.; Oshima, K.; Masaki, Y. *Chem. Lett.* **1994**, 871-874.
172. Ochiai, M.; Oshima, K.; Masaki, Y. *Tetrahedron Lett.* **1991**, *32*, 7711-7714.
173. Ochiai, M.; Sumi, K.; Nagao, Y.; Fujita, E. *Tetrahedron Lett.* **1985**, *26*, 2351-2354.
174. Ochiai, M.; Sumi, K.; Takaoka, Y.; Kunishima, M.; Nagao, Y.; Shiro, M.; Fujita, E. *Tetrahedron* **1988**, *44*, 4095-4112.
175. Zefirov, N.S.; Koz'min, A.S.; Kasumov, T.; Potekhin, K.A.; Sorokin, V.D.; Brel, V.K.; Abramkin, E.V.; Struchkov, Yu.T.; Zhdankin, V.V.; Stang, P.J. *J. Org. Chem.* **1992**, *57*, 2433-2437.
176. Stang, P.J.; Schwarz, A.; Blume, T.; Zhdankin, V.V. *Tetrahedron Lett.* **1992**, *33*, 6759-6762.
177. Beringer, F.M.; Galton, S.A. *J. Org. Chem.* **1965**, *30*, 1930-1934.
178. Beringer, F.M.; Nathan, R.A. *J. Org. Chem.* **1969**, *34*, 685-689.
179. Beringer, F.M.; Nathan, R.A. *J. Org. Chem.* **1970**, *35*, 2095-2096.
180. Stang, P.J.; Olenyuk, B.; Chen, K. *Synthesis* **1995**, 937-938.
181. Kitamura, T.; Kotani, M.; Fujiwara, Y. *Tetrahedron Lett.* **1996**, *37*, 3721-3722.
182. Stang, P.J.; Blume, T.; Zhdankin, V.V. *Synthesis* **1993**, 35-36.
183. Ishikura, M.; Terashima, M. *Heterocycles* **1988**, *27*, 2619-2625.
184. Ishikura, M.; Terashima, M. *J. Chem. Soc., Chem. Commun.* **1989**, 727-728.

185. Stang, P.J.; Ullmann, R.A. *Angew. Chem., Int. Ed. Engl.* **1991**, *30*, 1469-1470.
186. Beringer, F.M.; Ganis, P.; Avitabile, G.; Jaffe, H. *J. Org. Chem.* **1972**, *37*, 879-886.
187. Ochiai, M.; Takaoka, Y.; Masaki, Y.; Inenaga, M.; Nagao, Y. *Tetrahedron Lett.* **1989**, *30*, 6701-6704.
188. Nesmeyanov, A.N.; Tolstaya, T.P.; Sokolova, N.F.; Varfolomeeva, V.N.; Petrakov, A.V. *Dokl. Akad. Nauk SSSR* **1971**, *198*, 115-117; *Chem. Abstr.* **1971**, *75*, 48576.
189. Ochiai, M.; Oshima, K.; Masaki, Y.; Kunishima, M.; Tani, S. *Tetrahedron Lett.* **1993**, *34*, 4829-4830.
190. Ochiai, M.; Kitagawa, Y.; Toyonari, M.; Uemura, K. *Tetrahedron Lett.* **1994**, *35*, 9407-9408.
191. Ochiai, M.; Kunishima, M.; Nagao, Y.; Fuji, K.; Shiro, M.; Fujita, E. *J. Am. Chem. Soc.* **1986**, *108*, 8281-8283.
192. Stang, P.J.; Surber, B.W. *J. Am. Chem. Soc.* **1985**, *107*, 1452-1453.
193. Stang, P.J.; Surber, B.W., Chen, Z.-C.; Roberts, K.A.; Anderson, A.G. *J. Am. Chem. Soc.* **1987**, *109*, 228-235.
194. Tykwinski, R.R.; Stang, P.J. *Tetrahedron* **1993**, *49*, 3043-3052.
195. Stang, P.J.; Boehshar, M.; Lin, J. *J. Am. Chem. Soc.* **1986**, *108*, 7832-7834.
196. Stang, P.J.; Boehshar, M.; Wingert, H.; Kitamura, T. *J. Am. Chem. Soc.* **1988**, *110*, 3272-3278.
197. Stang, P.J.; Kitamura, T.; Boehshar, M.; Wingert, H. *J. Am. Chem. Soc.* **1989**, *111*, 2225-2230.
198. Nagaoka, T.; Sueda, T.; Ochiai, M. *Tetrahedron Lett.* **1995**, *36*, 261-264.
199. Margida, A.J.; Koser, G.F. *J. Org. Chem.* **1984**, *49*, 4703-4706.
200. Lermontov, S.A.; Velikohat'ko, T.N.; Zefirov, N.S. *Russ. Chem. Bull.* **1996**, *45*, 977-978.
201. Stang, P.J.; Kitamura, T. *J. Am. Chem. Soc.* **1987**, *109*, 7561-7563.
202. Kitamura, T.; Tanaka, T.; Taniguchi, H.; Stang, P.J. *J. Chem. Soc., Perkin Trans. 1* **1991**, 2892-2893.
203. Kitamura, T.; Lee, C.H.; Taniguchi, H.; Matsumoto, M.; Sano, Y. *J. Org. Chem.* **1994**, *59*, 8053-8057.
204. Kitamura, T.; Lee, C.H.; Taniguchi, H.; Fujiwara, Y.; Matsumoto, M.; Sano, Y. *Mol. Cryst. Liq. Cryst.* **1996**, *287*, 93-100.
205. Yagupol'skii, L.M. *J. Fluorine Chem.* **1987**, *36*, 1-28.
206. Umemoto, T. *Chem. Rev.* **1996**, *96*, 1757-1777.
207. Yagupolskii, L.M.; Maletina, I.I.; Kondratenko, N.V.; Orda, V.V. *Synthesis* **1978**, 835-837.
208. Umemoto, T.; Kuriu, Y.; Shuyama, H. *Chem. Lett.* **1981**, 1663-1666.
209. Yagupol'skii, L.M.; Mironova, A.A.; Maletina, I.I. *Zh. Org. Khim.* **1980**, *16*, 232-233.
210. Umemoto, T.; Kuriu, Y. *Tetrahedron Lett.* **1981**, *22*, 5197-5200.
211. Umemoto, T.; Gotoh, Y. *Bull. Chem. Soc. Jpn.* **1986**, *59*, 439-445.
212. Umemoto, T.; Kuriu, Y.; Nakayama, S.-i.; Miyano, O. *Tetrahedron Lett.* **1982**, *23*, 1471-1474.
213. Umemoto, T.; Kuriu, Y.; Miyano, O. *Tetrahedron Lett.* **1982**, *23*, 3579-3582.
214. Umemoto, T.; Kuriu, Y.; Nakayama, S.-i. *Tetrahedron Lett.* **1982**, *23*, 1169-1172.
215. Umemoto, T.; Nakamura, H. *Chem. Lett.* **1984**, 983-984.
216. Umemoto, T.; Kuriu, Y. *Chem. Lett.* **1982**, 65-66.
217. Umemoto, T.; Miyano, O. *Bull. Chem. Soc. Jpn.* **1984**, *57*, 3361-3362.
218. Klyuchinskii, S.A.; Grishnyakov, S.B.; Demchenko, L.V.; Zavgorodnii, V.S. *Zh. Obshch. Khim.* **1993**, *63*, 2394-2395; *Russ. J. Gen. Chem.* **1993**, *63*, 1661-1662.
219. Umemoto, T.; Gotoh, Y. *J. Fluorine Chem.* **1986**, *31*, 231-236.
220. Umemoto, T.; Gotoh, Y. *Bull. Chem. Soc. Jpn.* **1987**, *60*, 3307-3313.
221. Umemoto, T.; Gotoh, Y. *Bull. Chem. Soc. Jpn.* **1991**, *64*, 2008-2010.
222. Montanari, V.; Resnati, G. *Tetrahedron Lett.* **1994**, *35*, 8015-8018.
223. Manfredi, A.; Montanari, V.; Quici, S.; Resnati, G. *Abstracts of the 12th Winter Fluorine Conference* Florida, **1995**, p. 62.

224. Mironova, A.A.; Soloshonok, I.V.; Maletina, I.I.; Orda, V.V.; Yagupol'skii, L.M. *Zh. Org. Khim.* **1988**, *24*, 593-598; *J. Org. Chem. USSR* **1988**, 530-535.

225. Umemoto, T.; Gotoh, Y. *Bull. Chem. Soc. Jpn.* **1987**, *60*, 3823-3825.

226. Kashin, A.N.; Tul'chinskii, M.L.; Bumagin, N.A.; Beletskaya, I.P.; Reutov, O.A. *Zh. Org. Khim.* **1982**, *18*, 1588-1595; *J. Org. Chem. USSR* **1982**, *18*, 1390-1395.

227. Moriarty, R.M.; Prakash, O.; Duncan, M.P. *J. Chem. Soc., Chem. Commun.* **1985**, 420.

228. Moriarty, R.M.; Prakash, O.; Duncan, M.P. *J. Chem. Soc., Perkin Trans. 1* **1987**, 559-561.

229. Moriarty, R.M.; Prakash, O.; Duncan, M.P. *Synth. Commun.* **1985**, *15*, 789-795.

230. Zhdankin, V.V.; Tykwinski, R.; Caple, R.; Berglund, B.; Koz'min, A.S.; Zefirov, N.S. *Tetrahedron Lett.* **1988**, *29,* 3717-3720.

231. Zhdankin, V.V.; Mullikin, M.; Tykwinski, R.; Berglund, B.; Caple, R.; Zefirov, N.S.; Koz'min, A.S. *J. Org. Chem.* **1989**, *54*, 2609-2612.

232. Zhdankin, V.V.; Tykwinski, R.; Caple, R.; Berglund, B.; Koz'min, A.S.; Zefirov, N.S. *Tetrahedron Lett.* **1988**, *29,* 3703-3704.

233. Zhdankin, V.V.; Mullikin, M.; Tykwinski, R.; Berglund, B.; Caple, R.; Zefirov, N.S.; Koz'min, A.S. *J. Org. Chem.* **1989**, *54*, 2605-2608.

234. Moriarty, R.M.; Vaid, R.K.; Ravikumar, V.T.; Vaid, B.K.; Hopkins, T.E. *Tetrahedron* **1988**, *44,* 1603-1607.

235. Bregant, N.; Matijević, J.; Širola, I.; Balenović, K. *Bull. Sci. Cons. Acad. Sci. Arts RSF Yougosl. Sect. A.* **1972**, *17*, 148-150; *Chem. Abstr.* **1973**, *78*, 4047.

236. Ochiai, M.; Fujita, E.; Arimoto, M.; Yamaguchi, H. *Chem. Pharm. Bull.* **1985**, *33*, 41-47.

237. Lee, K.; Kim, D.Y.; Oh, D.Y. *Tetrahedron Lett.* **1988**, *29*, 667-668.

238. Ochiai, M.; Fujita, E.; Arimoto, M.; Yamaguchi, H. *J. Chem. Soc., Chem. Commun.* **1982**, 1108-1109.

239. Koser, G.F.; Relenyi, A.G.; Kalos, A.N.; Rebrovic, L.; Wettach, R.H. *J. Org. Chem.* **1982**, *47,* 2487-2489.

240. Koser, G.F.; Rebrovic, L.; Wettach, R.H. *J. Org. Chem.* **1981**, *46,* 4324-4326.

241. Rebrovic, L.; Koser, G.F. *J. Org. Chem.* **1984**, *49*, 2462-2472.

242. Zefirov, N.S.; Zhdankin, V.V.; Dan'kov, Yu.V.; Koz'min, A.S. *Zh. Org. Khim.* **1984**, *20*, 446-447; *J. Org. Chem. USSR* **1984**, *20,* 401-402.

243. Hembre, R.T.; Scott, C.P.; Norton, J.R. *J. Org. Chem.* **1987**, *52*, 3650-3654.

244. Zefirov, N.S.; Zhdankin, V.V.; Dan'kov, Yu.V.; Sorokin, V.D.; Semerikov, V.N.; Koz'min, A.S.; Caple, R.; Berglund, B. *Tetrahedron Lett.* **1986**, *27,* 3971-3974.

245. Potekhin, K.A.; Yanovskii, A.I.; Struchkov, Yu.T.; Sorokin, V.D.; Zhdankin, V.V.; Koz'min, A.S.; Zefirov, N.S. *Dokl. Akad. Nauk SSSR* **1988**, *301,* 119-123.

246. Moriarty, R.M.; Prakash, O.; Duncan, M.P.; Vaid, R.K.; Rani, N. *J. Chem. Res. (S)* **1996**, 432-433.

247. Zefirov, N.S.; Zhdankin, V.V.; Dan'kov, Yu.V.; Samoshin, V.V.; Koz'min, A.S. *Zh. Org. Khim.* **1984**, *20*, 444-445; *J. Org. Chem. USSR* **1984**, *20,* 400.

248. Mizukami, F.; Ando, M.; Tanaka, T.; Imamura, J. *Bull. Chem. Soc. Jpn.* **1978**, *51*, 335-336.

249. Lodaya, J.S.; Koser, G.F. *J. Org. Chem.* **1988**, *53,* 210-212.

250. Hatzigrigoriou, E.; Varvoglis, A.; Bakola-Christianopoulou, M. *J. Org. Chem.* **1990**, *55*, 315-318.

251. Zefirov, N.S.; Zhdankin, V.V.; Dan'kov, Yu.V.; Koz'min, A.S.; Chizhov, O.S. *Zh. Org. Khim.* **1985**, *21*, 2461-2462; *J. Org. Chem. USSR* **1985**, *21*, 2252-2253.

252. Tuncay, A.; Dustman, J.A.; Fisher, G.; Tuncay, C.I.; Suslick, K.S. *Tetrahedron Lett.* **1992**, *33,* 7647-7650.

253. Moriarty, R.M.; Vaid, R.K.; Hopkins, T.E.; Vaid, B.K.; Tuncay, A. *Tetrahedron Lett.* **1989**, *30,* 3019-3022.

254. Moriarty, R.M.; Prakash, O.; Duncan, M.P. *Synthesis* **1985**, 943-944.

255. Moriarty, R.M.; Prakash, O.; Duncan, M.P. *J. Chem. Soc., Perkin Trans. 1* **1987**, 1781-1784.
256. Cox, P.J.; Simpkins, N.S. *Synlett* **1991**, 321.
257. Moriarty, R.M.; Prakash, O.; Duncan, M.P.; Vaid, B.K.; Musallam, H.A. *J. Org. Chem.* **1987**, *52,* 150-153.
258. Moriarty, R.M.; Epa, W.R.; Penmasta, R.; Awasthi, A.K. *Tetrahedron Lett.* **1989**, *30,* 667-670.
259. Moriarty, R.M.; Penmasta, R.; Awasthi, A.K.; Epa, W.R.; Prakash, I. *J. Org. Chem.* **1989**, *54,* 1101-1104.
260. Moriarty, R.M.; Penmasta, R.; Prakash, I.; Awasthi, A.K. *J. Org. Chem.* **1988**, *53,* 1022-1025.
261. Lim, B.-W.; Ahn, K.-H. *Synth. Commun.* **1996**, *26,* 3407-3412.
262. Evans, D.E.; Faul, M.M.; Bilodeau, M.T. *J. Org. Chem.* **1991**, *56,* 6744-6746.
263. Evans, D.E.; Faul, M.M.; Bilodeau, M.T.; Anderson, B.A.; Barnes, D.M. *J. Am. Chem. Soc.* **1993**, *115,* 5328-5329.
264. Li, Z.; Conser, K.R.; Jacobsen, E.N. *J. Am. Chem. Soc.* **1993**, *115,* 5326-5327.
265. Evans, D.E.; Faul, M.M.; Bilodeau, M.T. *J. Am. Chem. Soc.* **1994**, *116,* 2742-2753.
266. Ratner, V.G.; Pashkevich, K.I. *Izv. Akad. Nauk* **1994**, 541-542.
267. Weiss, R.; Seubert, J.; Hampel, F. *Angew. Chem., Int. Ed. Engl.* **1994**, *33,* 1952-1953.
268. Zefirov, N.S.; Surmina, L.S.; Sadovaya, N.K.; Blokhin, A.V.; Tyurekhodzhaeva, M.A.; Bubnov, Yu.N.; Lavrinovich, L.I.; Ignatenko, A.V.; Grishin, Yu.K.; Zelenkina, O.A.; Kolotyrkina, N.G.; Kudrevitch, S.V.; Koz'min, A.S. *Zh. Org. Khim.* **1990**, *26,* 2317-2333; *J. Org. Chem. USSR* **1990**, *26,* 2002-2014.
269. Gallos, J.; Varvoglis, A. *J. Chem. Soc., Perkin Trans. 1* **1983**, 1999-2002.
270. see reference 15, p. 97.
271. see reference 15, p. 98.
272. Barton, D.H.R.; Jaszberenyi, J.Cs.; Shinada, T. *Tetrahedron Lett.* **1993**, *34,* 7191-7194.
273. Barton, D.H.R.; Jaszberenyi, J.Cs.; Liu, W.; Shinada, T. *Tetrahedron* **1996**, *52,* 14673-14688.
274. Moriarty, R.M.; Berglund, B.A.; Rao, M.S.C. *Synthesis* **1993**, 318-321.
275. Kotali, A. *Tetrahedron Lett.* **1994**, *35,* 6753-6754.
276. Katritzky, A.R.; Harris, P.A.; Kotali, A. *J. Org. Chem.* **1991**, *56,* 5049-5051.

Chapter 6

Ligand Coupling Involving Organobismuth Compounds

The chemistry of organic compounds of bismuth is rich and has a long history, as organobismuth compounds have been known for more than a century.[1,2] However, the use of organobismuth derivatives in organic chemistry is relatively recent, as the first applications appeared in the mid 1970s. Since then, their use in organic synthesis has become more frequent.[3-8] The chemistry of organobismuth compounds is for the main part based on the existence of two states of oxidation: Bi(III) and Bi(V). Pentavalent organobismuth compounds constitute efficient selective oxidation reagents.[9,10] The weakness of the carbon-bismuth bond has led to the development of a number of reactions of ligand transfer, and the ligand coupling mechanism is frequently implied in these reactions, although other mechanisms can also play a significant role in some instances.

The synthesis of organobismuth compounds has been extensively reviewed.[1,2] For the synthesis of arylbismuth derivatives, the most general and high-yielding method of preparation involves the reaction of Grignard reagents with bismuth trihalides. The triarylbismuthanes are then oxidized to the corresponding pentavalent derivatives, which can be treated with aryllithium derivatives, affording pentaarylbismuth. Ligand exchange reactions lead selectively to different types of pentavalent triarylbismuth derivatives: carbonate, diacylate, difluoride or disulfonates... (Scheme 6.1)

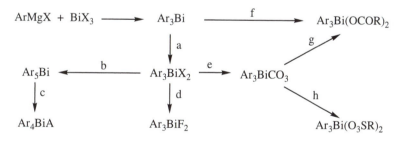

a) X_2; b) ArLi; c) HA; d) KF; e) K_2CO_3; f) $tert$-BuOOH, RCO_2H, or RCO_3H, or $NaBO_3 + RCO_2H$ (ref. 11); g) RCO_2H; h) RSO_3H

Scheme 6.1: Synthesis of organobismuth compounds

6.1 ARYLATION OF INORGANIC SUBSTRATES

Triarylbismuthanes react with halogens such as chlorine or bromine to afford the stable pentavalent dihalotriarylbismuth.[12-15] These compounds can decompose with carbon-bismuth bond cleavage to lead to the aryl-halogen derivatives by a ligand coupling reaction. Reaction of triphenylbismuthane (**1**) with iodine leads only to phenyl iodides, the pentavalent intermediates undergoing directly the ligand coupling reaction.[13,16] On the other hand, thermal decomposition of triphenylbismuth difluoride (2) led to a moderate yield of biphenyl, the presence of fluorobenzene being not reported.[17]

$$Ph_3Bi \ + \ ICl \ \longrightarrow \ PhI \ + \ Ph_2BiCl$$
$$(1)$$

$$Ph_3BiF_2 \ \xrightarrow{\ PhH \ / \ reflux\ } \ Ph\text{-}Ph \quad (40\%)$$
$$(2)$$

Similarly, triphenylbismuthane reacts with halogen halides and halogen cyanides to give Ar_2BiX and ArY, in which X is the most electronegative fragment of the halogen derivative XY.[12,13,16,18]

$$Ph_3Bi \ + \ ICl \ \longrightarrow \ PhI \ + \ Ph_2BiCl$$

$$Ph_3Bi \ + \ BrCN \ \longrightarrow \ PhBr \ + \ Ph_2BiCN$$

Triphenylbismuth dichloride (3) reacts with sodium azide and sodium cyanide to afford triphenyl-bismuth diazide and triphenylbismuth dicyanide. Thermolysis of these compounds leads to phenyl azide or to benzonitrile, respectively.[19]

$$Ph_3BiCl_2 \ + \ NaN_3 \ \longrightarrow \ Ph_3Bi(N_3)_2 \ \xrightarrow{\ \Delta\ } \ PhN_3 \ + \ Ph_2BiN_3 \ + \ Ph_3Bi$$
$$(3)$$

$$Ph_3BiCl_2 \ + \ NaCN \ \longrightarrow \ Ph_3Bi(CN)_2 \ \xrightarrow{\ \Delta\ } \ PhCN \ + \ Ph_3Bi$$

Reaction of sodium thiocyanate with triphenylbismuth dichloride (3) leads directly to phenyl thiocyanate. Although no intermediate was detected, the reaction is likely to proceed by ligand coupling of a pentavalent thiocyanato derivative.[14]

$$Ph_3BiCl_2 \ + \ NaSCN \ \longrightarrow \ PhSCN \ + \ Ph_2BiSCN$$

The tetraphenylbismuthonium salts behave similarly with various inorganic substrates. However, the reagent itself, tetraphenylbismuthonium chloride, is not stable and decomposes in the solid state at room temperature to afford chlorobenzene.[20]

$$Ph_4BiCl \ \xrightarrow{\ RT\ } \ PhCl \ + \ Ph_3Bi$$

$$Ph_4BiCl \ + \ KBr \ \longrightarrow \ PhBr \ + \ Ph_3Bi$$

$$Ph_4BiCl \ + \ KCN \ \longrightarrow \ PhCN \ + \ Ph_3Bi$$

$$Ph_4BiCl \ + \ NaNO_2 \ \longrightarrow \ PhNO_2$$

Tetraphenylbismuthonium salts react also with triphenylphosphane to afford the corresponding tetraphenylphosphonium salts. Under photolytic conditions, a mechanism involving single electron transfer takes place and the tetraphenylphosphonium salt is obtained in 90% yield. But under thermal conditions, the reaction is much slower, and the phenyl group transfer may be explained by a ligand coupling involving a covalent bismuth-phosphonium salt.[21]

$$Ph_4Bi^{\oplus} BF_4^{\ominus} \ + \ PPh_3 \ \longrightarrow \ Ph_4P^{\oplus} BF_4^{\ominus} \ + \ Ph_3Bi$$

The reactions of pentaphenylbismuth (4) with bromine[20] and sulfur dioxide[22] are formally ligand coupling reactions. However, they should involve as an intermediate or transition state an hypervalent hexa or heptacoordinate bismuth derivative, which undergoes a ligand coupling-type reaction.

$$Ph_5Bi + Br_2 \longrightarrow [PhBr + Ph_4BiBr] \longrightarrow Ph_3Bi + 2PhBr$$
$$(4)$$

$$Ph_5Bi + SO_2 \longrightarrow Ph_2SO_2$$

6.2 C-ARYLATION WITH PENTAVALENT ORGANOBISMUTH

The first studies on the application of pentavalent organobismuth derivatives in organic synthesis focused on oxidation reactions.[9,10] In the course of these studies, when Barton *et al.* treated quinine with triphenylbismuth carbonate (5), the reaction led to a rather modest yield of the expected quininone (34%). Using an excess of the reagent (5) afforded a good yield of a mixture of diastereo-isomeric α-phenylated ketones (75%).[23]

In subsequent studies, a number of enolisable substrates, such as phenols and enols as well as various anions, were found to be arylated under mild conditions by triarylbismuth carbonates, as well as by other pentavalent triarylbismuth or tetraphenylbismuth compounds and by pentaphenylbismuth.

6.2.1 C-Arylation of phenols

The interaction of a phenolic derivative with a pentavalent organobismuth reagent can lead more or less selectively, in yields ranging from moderate to frequently excellent, to one of the two types of aryl derivatives: the 2-hydroxybiphenyl substituted derivative or the O-aryl ether. The outcome of the reaction is dependent on the nature of the organobismuth reagent in close relation to the reaction conditions. But, for a given set of reaction conditions and nature of the organobismuth reagent, the electronic properties of the phenolic substrate play also a determinant role in the regioselectivity.

6.2.1.1 Influence of the nature of the organobismuth reagent

Three types of pentavalent arylbismuth derivatives can be used for the arylation of phenols. They derive from the triaryl, the tetraaryl and the pentaaryl series. The C-arylation of phenolic compounds can be performed by triarylbismuth carbonate, the presence of a base being not mandatory. It can also be realized by a variety of pentavalent triarylbismuth compounds and tetraarylbismuthonium salts, both in the presence of a base, in various solvents (methylene chloride, benzene, toluene or tetrahydrofuran).[24] The nature of the base (NaH, TMG = N,N,N',N'-tetramethylguanidine or BTMG = N-*tert*-butyl-N',N',N'',N''-tetramethylguanidine) does not change the yield of arylated product. Similar C-phenylation of phenols can also be performed by reaction of the phenol with pentaphenylbismuth under neutral conditions.[25] The yield of the C-phenylated derivatives is not very dependent upon the choice of the reagent. However, the tetraphenylbismuthonium salts require generally milder conditions with shorter reaction times. They are also less prone to side reactions such as oxidative dimerisation, which are relatively common with triarylbismuth carbonate. But, only phenylation has been performed with these reagents as tetraphenylbismuthonium salts are the only tetraaryl compounds which have been prepared.

A variety of pentavalent substituted triarylbismuth derivatives are known and electron-withdrawing or electron-donating substituted phenyl groups can be introduced by arylation with these reagents. The electronic properties of the substituents of the arylbismuth do not influence significantly the overall yield of arylated products (arylation of 2-naphthol: 84% with the 4-nitrophenyl derivative, 76% with the unsubstituted phenyl derivative and 69% with 4-methoxyphenyl derivative).

Table 6.1: Arylation of 2-naphthol by organobismuth (V) reagents.

Reagent	Reaction conditions	O-Aryl (%)	C-Aryl (%)	Starter (%)	Ref
Ph$_3$BiCO$_3$ (**5**)	TMG, CH$_2$Cl$_2$, RT, 12 h	-	76	-	24
Ph$_3$BiCO$_3$ (**5**)	CH$_2$Cl$_2$, RT, 24 h	-	76	-	26
(4-MeOC$_6$H$_4$)$_3$BiCO$_3$	CH$_2$Cl$_2$, RT, 24 h	-	69	-	26
(4-MeC$_6$H$_4$)$_3$BiCO$_3$	CH$_2$Cl$_2$, RT, 24 h	-	71	-	26
(4-NO$_2$C$_6$H$_4$)$_3$BiCO$_3$	CH$_2$Cl$_2$, RT, 24 h	-	84	-	26
Ph$_3$BiCl$_2$ (**3**)	BTMG, PhH, RT, 4.5 h	-	90	-	24
Ph$_3$BiCl$_2$ (**3**)	NaH, THF, RT, 3 h	-	86	-	24
Ph$_3$Bi(ONO$_2$)$_2$	TMG, CH$_2$Cl$_2$, RT, 6.5 h	-	60	-	24
Ph$_3$Bi(OTs)$_2$	TMG, CH$_2$Cl$_2$, RT, 7.5 h	-	59	-	24
Ph$_4$BiOAc	PhH, 80°C, 24 h	25	25	19	24
Ph$_4$BiOAc	BTMG, PhH, RT, 1 h	-	95	-	24
Ph$_4$BiOCOCF$_3$ (**6**)	PhH, RT, 24 h	50	-	36	27
Ph$_4$BiOCOCF$_3$ (**6**)	PhH, 80°C, 15 h	82	-	-	28
Ph$_4$BiOCOCF$_3$ (**6**)	PhH, Cl$_3$CCO$_2$H (0.6 mol equiv.), 60°C, 24 h	91	-	-	24
Ph$_4$BiOCOCF$_3$ (**6**)	BTMG, PhH, RT, 1 h	-	94	-	24
Ph$_4$BiOTs (**7**)	PhH, 80°C, 24 h	42	-	58	24
Ph$_4$BiOTs (**7**)	BTMG, PhH, RT, 2 h	-	90	-	24
Ph$_4$BiOSO$_2$CF$_3$	BTMG, PhH, RT, 3 h	-	86	-	24
Ph$_5$Bi (**4**)	PhH, RT, 2.5 h	-	90	-	29

6.2.1.2 Influence of the nature of the phenolic substrate

The nature of the substituents on the phenolic substrate appeared to govern more or less efficiently the regioselectivity of the arylation reactions. This was clearly demonstrated by a study of the phenylation reaction of a variety of substituted phenols by triphenylbismuth dichloride under basic conditions.[30]

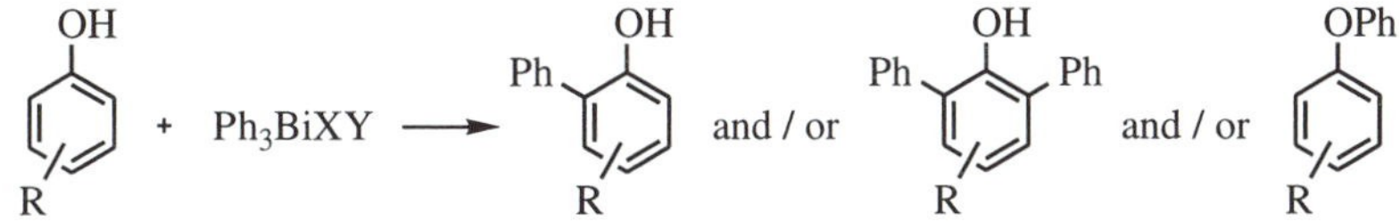

Phenols bearing electron-donating substituents are essentially *ortho-C*-phenylated. A small amount of the *O*-phenyl ether is found in some instances. When the two *ortho*-positions are substituted by an alkyl group, good yields of the 6-arylcyclohexa-2,4-dienone are obtained. Phenols bearing *para* electron-withdrawing substituents give exclusively *O*-phenyl ethers after thermal decomposition of the aryloxybismuth intermediate (see section 6.3). Unactivated phenols or phenols bearing *meta* electron-withdrawing substituents afford complex mixtures of *O*-phenyl- and *ortho-C*-phenyl-phenols. Phenol itself **(8)** and pentaphenylbismuth **(4)** gave diphenylether **(9)** in 42% yield.[31] But the reaction of phenol **(8)** with triphenylbismuth dichloride **(3)** under basic conditions (BTMG) led to a more complex mixture, in which the formation of diphenylether **(9)** (8%), 2-phenylphenol **(10)** (30%), 2,6-diphenylphenol **(11)** (7%) and biphenyl **(12)** (8%) was observed with a recovery of unreacted phenol **(8)** (40%).

(8) **(3)** **(9)** **(10)** **(11)** **(12)** **(8)**

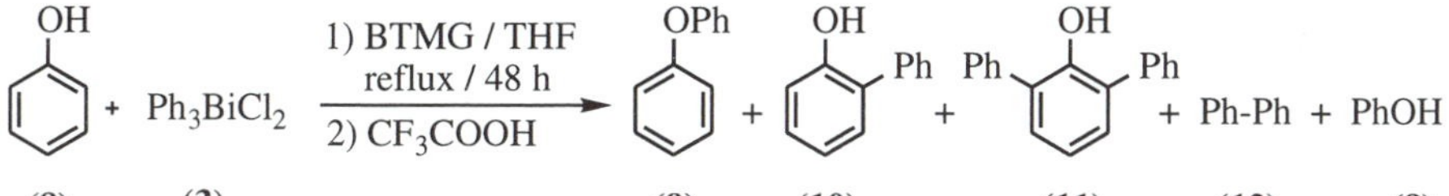

Mono or diarylation is realized for most of the phenols substituted with electron-donating groups. The *C*-phenylation is nearly always *ortho*-selective: no trace of the 3- or 4-phenyl isomers was detected.[30] The presence of small amounts of the *O*-phenyl ether is seldom observed, and mostly in the case of phenols substituted on the 3- or on the 3- and 5-positions, as in the case of phenol **(16)**. The only known exception, where *para*-phenylation occurs, is the case of 2,6-di-*tert*-butylphenol derivatives (see p. 165).

(13)	1 equiv.	NaH	**(14)**, 26%	**(15)**, 1-2%
	2 equiv.	TMG	-	77%

(16) 45% 30% 10%

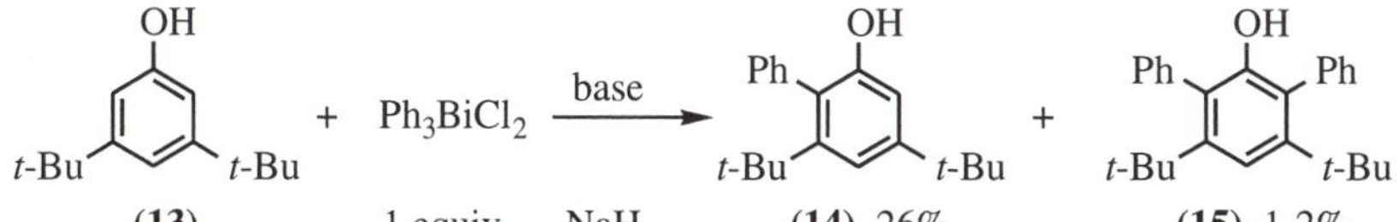

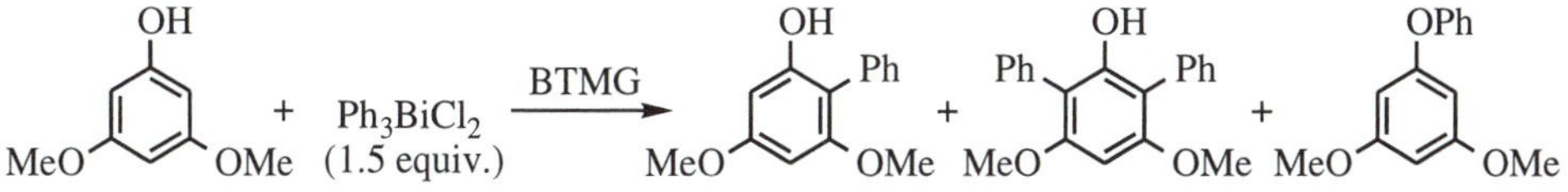

The steric influence of the *ortho*-substituent does not interfere strongly with the overall yield of the *C*-arylation: 2,4-di-*tert*-butylphenol (**17**) afforded the 6-phenyl derivative (**18**) in 65% yield by reaction with pentaphenylbismuth and in 81% yield with triphenylbismuth bis(trifluoroacetate) and BTMG (*N-tert*-butyl-*N',N',N'',N''*-tetramethylguanidine).

Ph₅Bi   65%
Ph₃Bi(OCOCF₃)₂ / BTMG   81%

In the case of 2,6-disubstituted phenols, the nature of the bismuth reagent, the nature of the alkyl substituents and the reaction conditions determined the outcome of the reactions.[32] Thus, in the reactions of 2,6-dimethylphenol (**19**) with pentaphenylbismuth (**4**) or with tetraphenylbismuthonium derivatives under basic conditions, *ortho C*-phenylation resulted in the formation of 6-phenylcyclo-hexadienone (**20**) in good yield.[25] On the other hand, oxidative dimerisation took place in the reaction of 2,6-dimethylphenol with triphenylbismuth carbonate to afford the diphenoquinone (**21**) quantitatively.[25]

2-Alkyl-6-phenylphenol derivatives can be phenylated to afford the corresponding 2-alkyl-2,6-diphenylcyclohexadienone. The reaction of 2,4-dimethylphenol (**22**) with pentaphenylbismuth (**4**) gave a mixture of the expected 2,4-dimethyl-6-phenylphenol (**23**) together with the corresponding 6-phenyl-cyclohexadienone (**24**). However, the *ortho* position bearing already one aryl group does not undergo a second arylation. Indeed, 2,6-diphenylphenol did not react with any organobismuth reagent to lead to 2,2,6-triphenylcyclohexadienone.[25]

The behaviour of the most sterically hindered 2,6-di-*tert*-butylphenol (**25**) and its derivatives was very dependent upon the reaction conditions. In the presence of a base favouring electron-transfer (such as BTMG), oxidation reactions took place with triphenylbismuth dichloride or triphenylbismuth carbonate leading to the diphenoquinone (**26**). *Para*-phenylation with formation of (**27**) was observed for the first time in the reaction of the potassium phenolate of 2,6-di-*tert*-butylphenol with triphenylbismuth dichloride and in the reaction of the phenol (**25**) with tetraphenylbismuthonium tosylate in the presence of BTMG. Even *ortho*- and *para*-phenylation with concomitant de-*tert*-butylation occurred in the reaction of the potassium salt of 2,4,6-tri-*tert*-butylphenol with triphenylbismuth dichloride.[32]

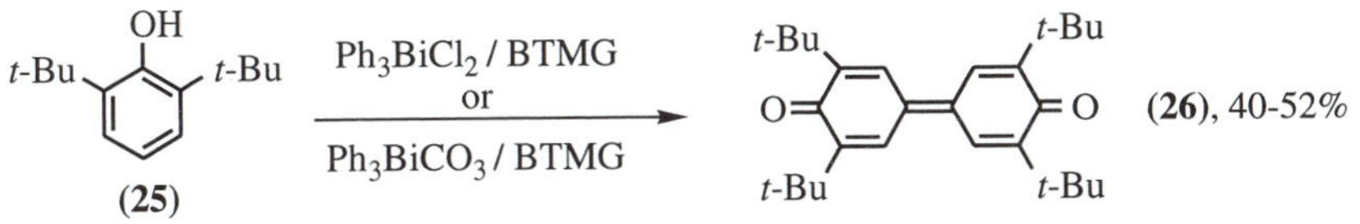

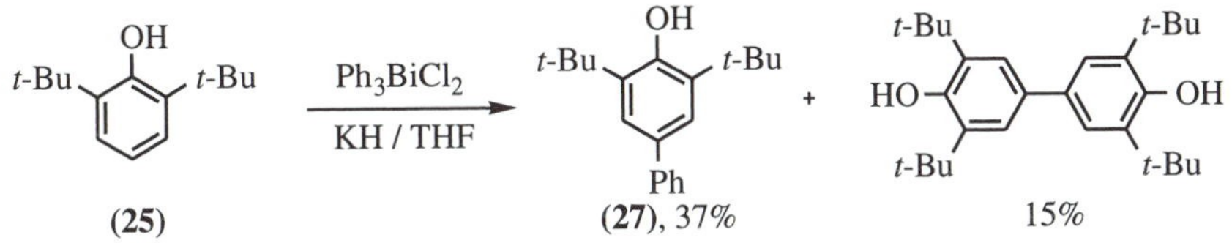

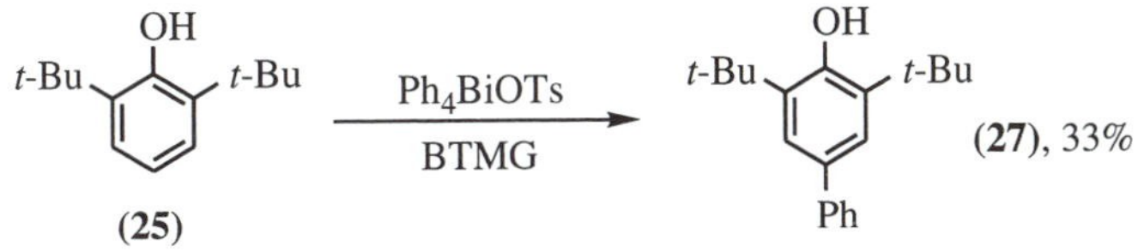

In the case of polyhydroxylic phenols, perphenylation occurred. Thus, phloroglucinol (**28**) reacted with triphenylbismuth carbonate to give a mixture of 2,4,6-triphenylphloroglucinol (**29**) and 2,2,4,5-tetraphenylcyclopent-4-en-1,3-dione (**30**).[24] The dione (**30**) is formed from (**29**) by phenylation, decarboxylation and oxidation. When a large excess of triphenylbismuth carbonate (5 mol equiv.) was used, the dione (**30**) was the only isolated product in 60% yield. In contrast, the reaction of phloroglucinol with pentaphenylbismuth led to a complex mixture.[25]

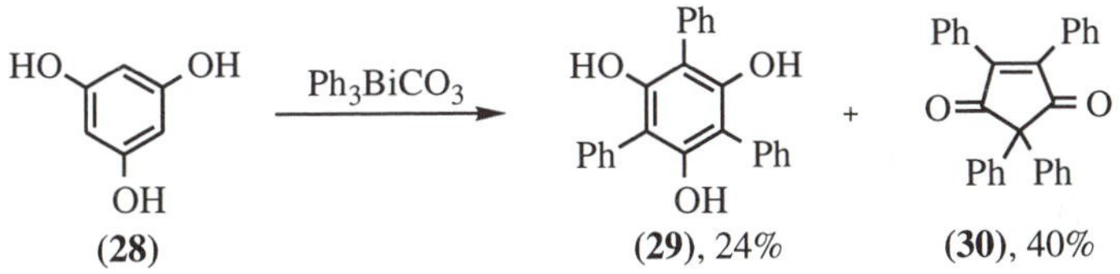

6.2.2 *C*-Arylation of enols and enolate anions

6.2.2.1 1,3-Dicarbonyl compounds

α-*C*-arylation of 1,3-dicarbonyl compounds such as β-diketones, β-ketoesters or malonic esters is easily performed with a variety of pentavalent organobismuth reagents under neutral or basic conditions. When the α-carbon bears two hydrogen atoms, monoarylation is difficult to obtain selectively in most cases. Good yields of the 2,2-diarylated 1,3-dicarbonyl compounds are generally obtained with triphenylbismuth dichloride and tetraphenylbismuthonium derivatives under basic conditions and with triphenylbismuth carbonate, triphenylbismuth diacetate and pentaphenylbismuth under neutral conditions. Such was the case for acetylacetone (**31**), ethyl acetoacetate (**32**), diethyl malonate (**33**), ethyl cyclohexanone-2-carboxylate (**34**) and ethyl cyclopentanone-2-carboxylate.[28,33] In the case of acetylacetone (**31**), the reaction with triphenylbismuth dichloride (2 equiv.) in the presence of tetramethylguanidine (TMG) as a base led to 3,3-diphenylacetylacetone (**36**) (74%), whereas tetraphenylbismuthonium trifluoroacetate (1 equiv.) in benzene under reflux in the absence of a base gave a moderate yield of the 3-monophenyl derivative (**35**) (34%).[33]

(31) → (35) + (36)

	(35)	(36)
Ph$_3$BiCO$_3$, CH$_2$Cl$_2$, reflux:	33%	40%
Ph$_3$BiCl$_2$, TMG, PhH, reflux:	-	74%
Ph$_4$BiOCOCF$_3$, PhH, no base, reflux:	34%	-

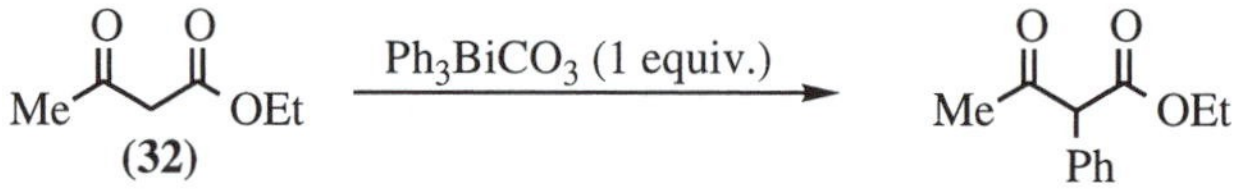

(32) $\xrightarrow{\text{Ph}_3\text{BiCO}_3 \text{ (1 equiv.)}}$ 59%

(32) →

		Ph$_2$CHCO$_2$Et
Ph$_3$BiCO$_3$ (excess)	55%	21%
Ph$_3$BiCl$_2$ / BTMG	82%	-

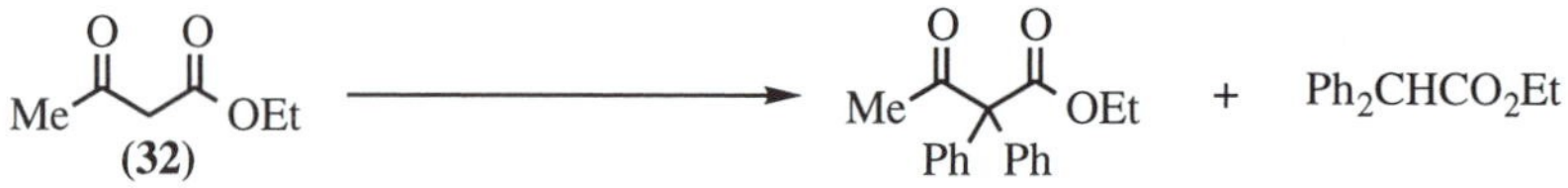

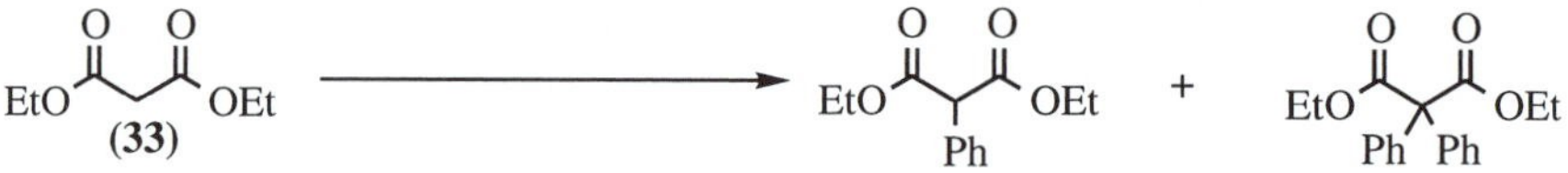

(33) →

	Ph	Ph Ph
Ph$_3$BiCO$_3$ (1 equiv.)	53%	13%
Ph$_3$BiCO$_3$ (2 equiv.)	18%	77%

(34) →

Ph$_3$BiCl$_2$/BTMG	75%
Ph$_4$BiOCOCF$_3$/BTMG	91%
Ph$_5$Bi	57%

By contrast, the sodium salt of acetylacetone (31) was reported to react with triphenylbismuth dichloride in THF to lead to an unstable bismuthonium ylide (37). It was formed nearly quantitatively after 1 hour at 0-5°C, but was not isolated. Addition of triethylamine and methanesulfonyl chloride led to the isolation of a 1,3-oxathiole-3,3-dioxide derivative (38) in 65% yield. On the other hand, this bismuthonium ylide (37) afforded the mono-*C*-phenylated product, 3-phenylacetylacetone (35), on standing at room temperature. However, the precise experimental conditions and the yields of the latter reaction were not reported.[34]

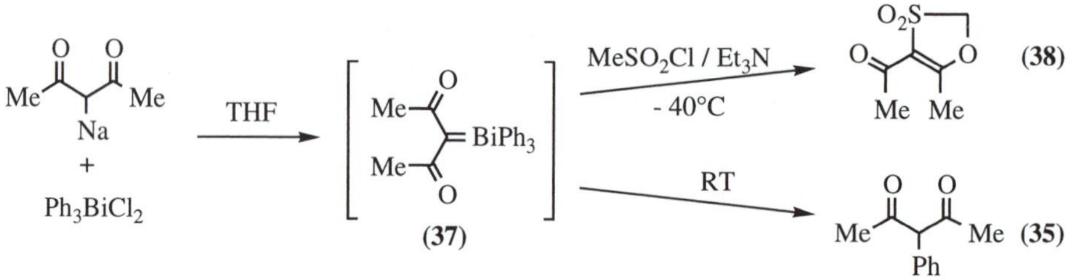

(31) Na + Ph$_3$BiCl$_2$ $\xrightarrow{\text{THF}}$ (37) $\xrightarrow[\text{- 40°C}]{\text{MeSO}_2\text{Cl / Et}_3\text{N}}$ (38)

(37) $\xrightarrow{\text{RT}}$ (35)

The reaction of dimedone (39) with tetraphenylbismuthonium derivatives and BTMG gave the α,α-diphenyl derivative (40).[33] But when dimedone (39) was treated with triphenylbismuth carbonate, an ylide (41) was obtained.[33] This ylide was later isolated as a stable crystalline compound.[35] This ylide (41) can also be prepared by reaction of the sodium salt of dimedone either with triphenylbismuth dichloride or with triphenylbismuth oxide.[34] Similarly, Meldrum's acid gave the corresponding bismuthonium ylide with triphenylbismuth carbonate and with triphenylbismuth dichloride.[34,36] Such ylides can also be made by decomposition of the appropriate dicarbonyl diazonium derivative in the presence of triphenylbismuthane catalysed by bis(hexafluoroacetylacetonato)copper (II).[37] These ylides react with aldehydes to give cyclopropanes, dihydrofurans and α,β-unsaturated carbonyl compounds.[34,38]

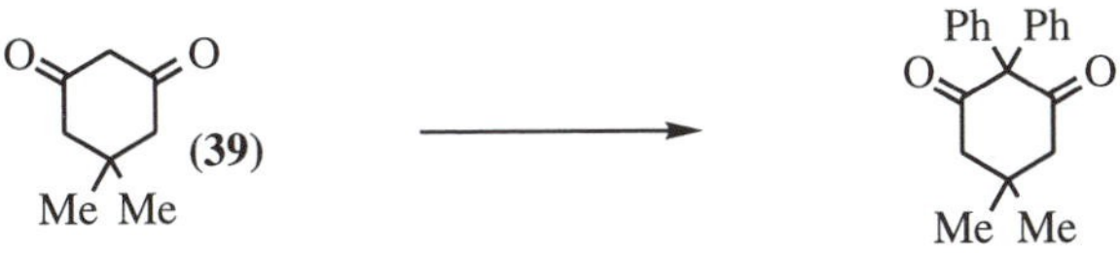

Ph$_4$BiOCOCF$_3$, BTMG	75%	ref. 33
Ph$_4$BiOTs, BTMG	81%	ref. 33

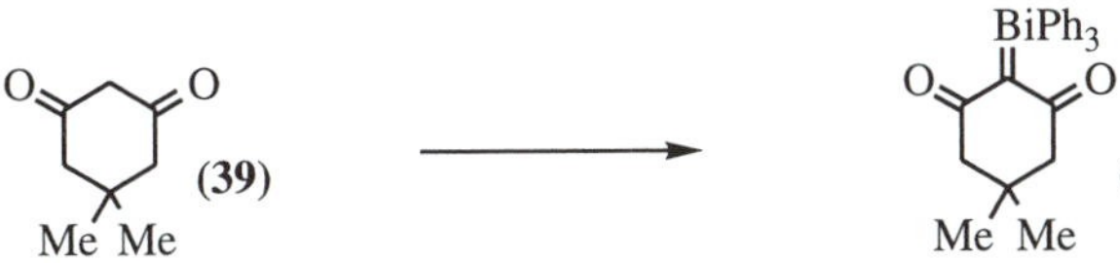

Ph$_3$BiCO$_3$, CH$_2$Cl$_2$, reflux	75-83%	ref. 33,35
Ph$_3$BiCl$_2$, NaH, THF	52%	ref. 34
Ph$_3$BiO, NaH, THF	84%	ref. 34

The arylation reaction of 1,3-dicarbonyl compounds was extended to the use of variously substituted triarylbismuth carbonates. The electronic properties of the substituents of the arylbismuth do not influence significantly the overall yield of arylated products. Arylation of ethyl 2-oxo-cyclohexanecarboxylate (34) with substituted triarylbismuth carbonate derivatives afforded the products in yields ranging from 81 to 90%.[26]

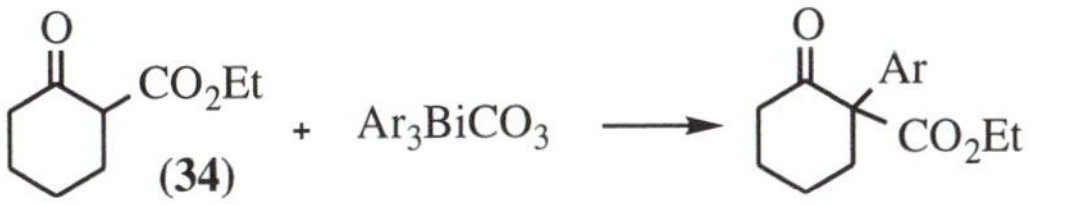

Ar = 4-NO$_2$C$_6$H$_4$	90%
C$_6$H$_5$	86%
4-MeC$_6$H$_4$	81%
4-MeOC$_6$H$_4$	86%

With caprolactam-type malonic derivatives, completely different results were observed depending on the structure of the lactams. Thus, 3-cyano-ε-caprolactam (42) reacted with pentaphenylbismuth to give 74% of the expected 3-cyano-3-phenyl-ε-caprolactam (43). But 3-methoxycarbonyl-ε-capro-lactams (44) and (45) did not react with pentaphenylbismuth. The N-phenylcaprolactam derivative (46) was obtained through reaction of (44) with triphenylbismuth dichloride and base.[39]

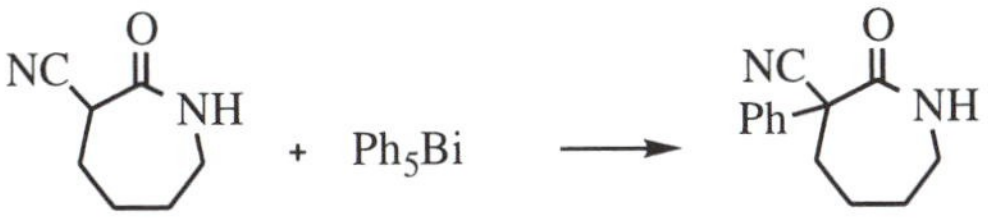

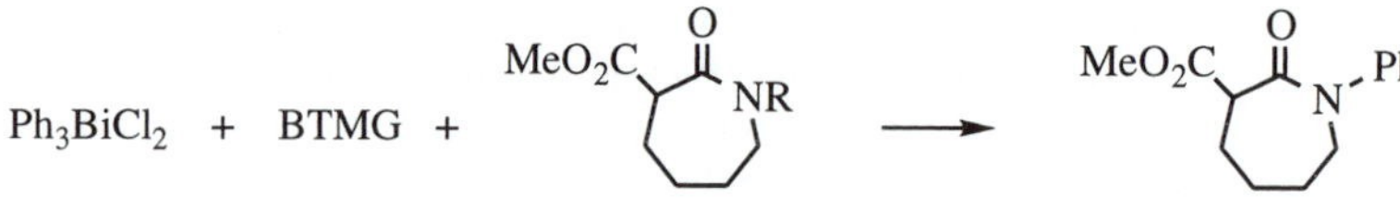

Ph_3BiCl_2 + BTMG +

(44) R = H   - - - - ➤ (46), 62%
(45) R = COO*t*-Bu - - - - ➤ no reaction

The reaction of arylation of 1,3-dicarbonyl derivatives was also used in a new approach towards the synthesis of isoflavanones and 3-aryl-4-hydroxycoumarins.[40] Phenylation of activated chroman-4-ones to isoflavanones was obtained in moderate to good yields. Selective monophenylation was performed in high yield in a low-temperature reaction (- 23°C) of the activated chroman-4-one with pentaphenylbismuth and mixtures of mono- and di-aryl derivatives were obtained in modest yields with other types of arylbismuth reagents.

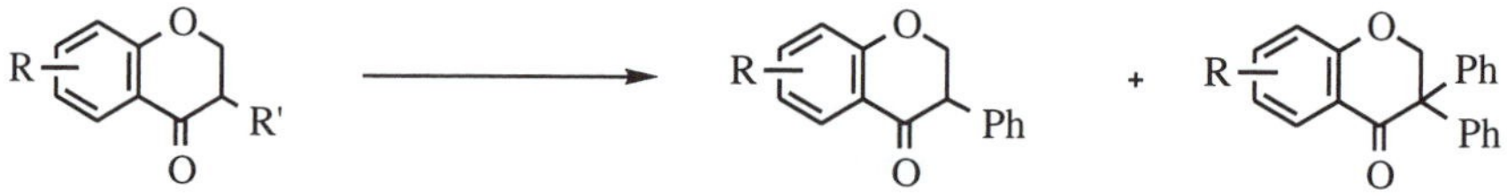

R' = H	Ph_3BiXY	0-22%	17-38%
R' = CHO	Ph_3BiXY	10-32%	13-40%
R' = CHO	Ph_5Bi / - 23°C	84%	-
R' = $COCO_2Et$	Ph_5Bi / - 23°C	88%	3%

Various 3-(3'- and 4'-substituted phenyl)-4-hydroxycoumarins were easily accessible by reaction of 4-hydroxycoumarins with triarylbismuth dichloride and triarylbismuth dinitrate under basic conditions or with triphenylbismuth diacetate (47) under neutral conditions. In the latter reaction, the best yields were obtained when the reaction was performed in methylene dichloride in the dark.[40]

$Ph_3Bi(OAc)_2$ (47), BTMG, CH_2Cl_2, RT   81%
$Ph_3Bi(OAc)_2$ (47), no base, CH_2Cl_2, reflux, dark   92%

The heterocyclic 10,10-dichloro-10-(4'-methylphenyl)phenothia-$10\lambda^5$-bismine-5,5-dioxide (48) is a relatively stable substance, which slowly decomposes at room temperature to afford the ligand coupling product, 4-chlorotoluene. However, treatment of this bismuth compound with the sodium salt of dibenzoylmethane resulted only in the transfer of the tolyl group. The monoaryl derivative (49) and the bismuth containing fragment (50) were recovered in good yields (76% and 74% respectively).[41]

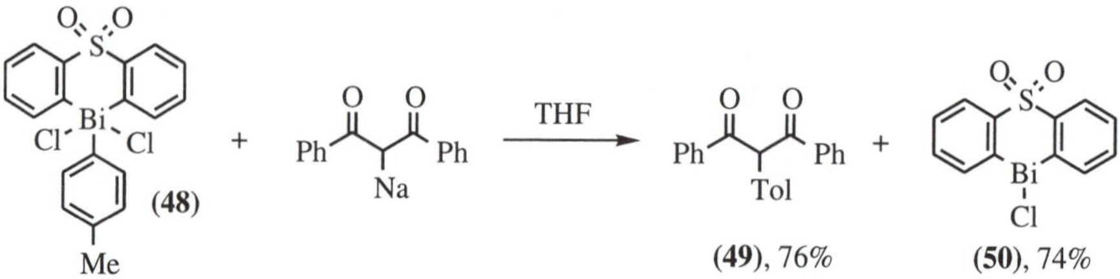

(48)   Na   Tol

(49), 76%   (50), 74%

Other enolised compounds were also easily arylated. Such was the case of β-ketosulfones, which reacted easily with triphenylbismuth carbonate (5) to afford the corresponding α-phenyl-β-ketosulfones. This reaction was put to good use in an efficient synthesis of isoflavanone and isoflavone derivatives.[42]

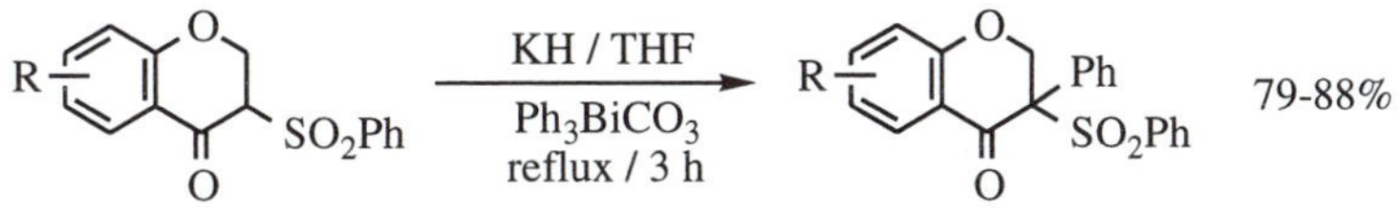

α-Diketones led to the formation of α'-aryl derivatives on the most substituted carbon, in modest yields.[33] However, treatment of tropolone (51) with tri(p-tolyl)bismuth diacetate in methanol under reflux led to stable crystals of the pentavalent di(O,O-tropolonato)tri(p-tolyl)bismuth (V) (52) resulting from ligand exchange.[43]

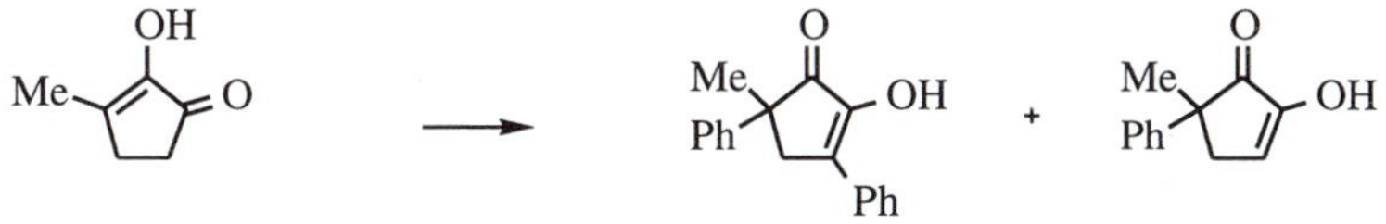

Ph₃BiCO₃	25%	-	ref. 33
Ph₅Bi	30%	-	ref. 33
Ph₄BiOTs, NaH	44%	10%	ref. 33

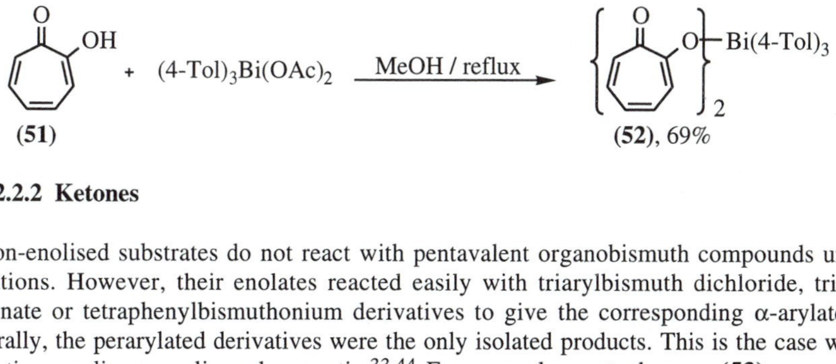

(51)   **(52), 69%**

6.2.2.2 Ketones

Non-enolised substrates do not react with pentavalent organobismuth compounds under neutral conditions. However, their enolates reacted easily with triarylbismuth dichloride, triarylbismuth carbonate or tetraphenylbismuthonium derivatives to give the corresponding α-arylated products. Generally, the perarylated derivatives were the only isolated products. This is the case with ketones, aliphatic, acyclic or cyclic and aromatic.[33,44] For example, acetophenone (53) was converted by reaction with triphenylbismuth carbonate into phenyl(triphenylmethyl)ketone (54) in 69% yield. The unknown hindered hexaphenylacetone remained inaccessible by phenylation of 1,3-diphenylacetone (55): only pentaphenylacetone (56) was isolated from the reaction of the potassium enolate of 1,3-diphenylacetone with triphenylbismuth carbonate or with tetraphenylbismuthonium derivatives.[33] Similarly, cyclohexanone (59) afforded 2,2,6,6-tetraphenylcyclohexanone (60) in 80% yield by reaction with tetraphenylbismuthonium tosylate and potassium hydride and 4,4-dimethylcholest-5-en-3-one (61) gave the 2,2-diphenyl analogue (62) in 80% yield. Similarly, arylation of chroman-4-ones yielded only the 3,3-diaryl derivatives, moreover in modest yields (34-38%).[40] Attempted synthesis of α,α'-diphenylcyclanones from α,α'-dibromocyclanones by reaction with lithium diphenylcuprate (LiCuPh₂) followed by treatment with triphenylbismuth carbonate led only to poor yields of the diphenylcyclanones.[45]

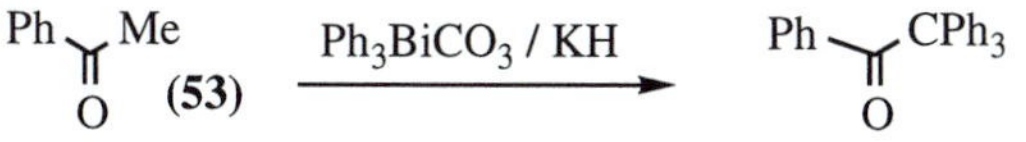

(54), 69%   ref. 33

Ph₃BiCO₃ / KH	60%	ref. 33
Ph₄BiOTs / BTMG	88%	ref. 33

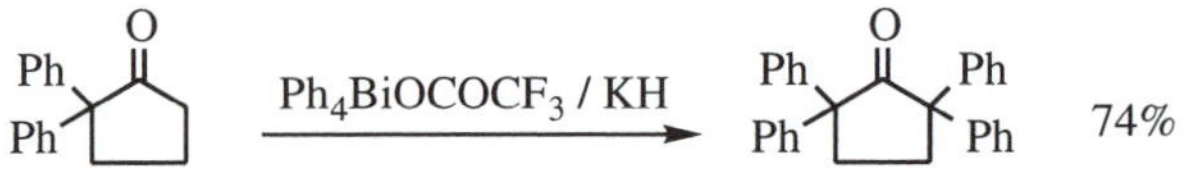

Ph₄BiOCOCF₃ / KH   74%   ref. 44

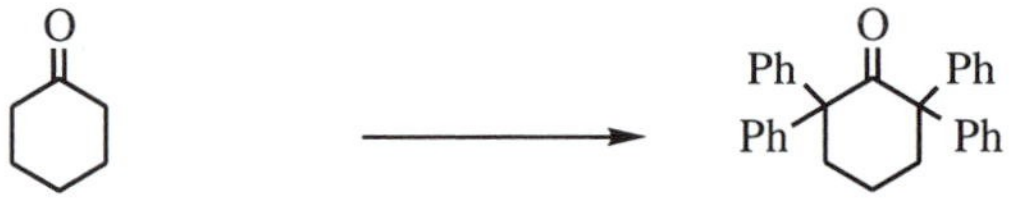

Ph₃BiCO₃ / KH	93%	ref. 44
Ph₄BiOTs / KH	80%	ref. 44

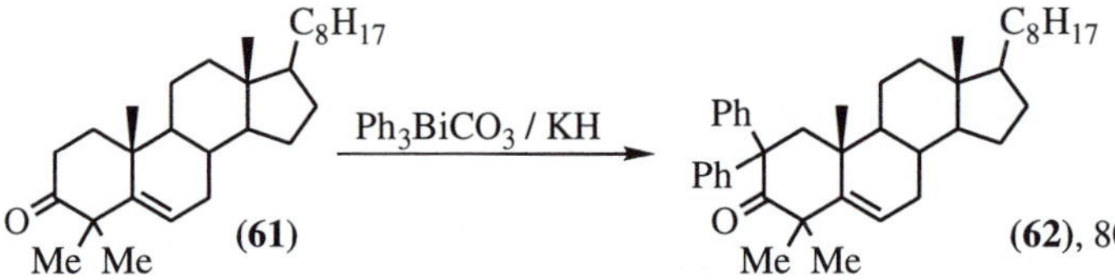

Ph₃BiCO₃ / KH   ref. 33

(62), 80%

6.2.3 Arylation of various anions and nucleophiles

α-Phenylglycine derivatives were obtained by the reaction of arylidene imines of glycine ethyl esters with triphenylbismuth carbonate under neutral conditions.[46] When the arylidene group was derived from *p*-chlorobenzaldehyde, only a modest yield of the diphenylated product was obtained, but selective monophenylation was realized with the benzophenone imine, due to the lowered acidity of the monophenyl derivative resulting from the greater steric hindrance of the diphenylmethylene group.[47] This arylation procedure was also extended to the synthesis of α-substituted-α-phenylglycines.[48]

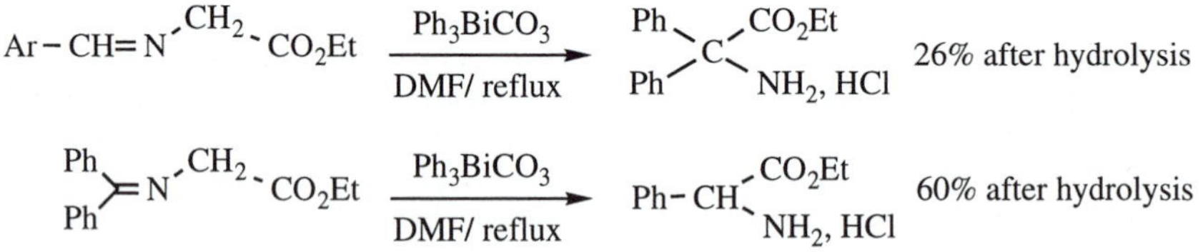

26% after hydrolysis

60% after hydrolysis

Other stabilized anions that were successfully *C*-arylated include the anions of nitroalkanes, α-methyl-α-nitrocarboxylic acid derivatives, esters, triphenylmethane, indole and scatole.[33] Nitroalkanes gave high yields of the α-*C*-aryl derivatives. In the presence of BTMG, 2-nitropropane reacts with triphenylbismuth dichloride or tetraphenylbismuthonium *p*-toluenesulfonate to lead to α-nitrocumene (86 or 77% respectively).[28] The reaction using triphenylbismuth dichloride and TMG was applied to the synthesis of α-methyl-α-phenylglycine *via* arylation of α-nitropropionic acid esters.[49]

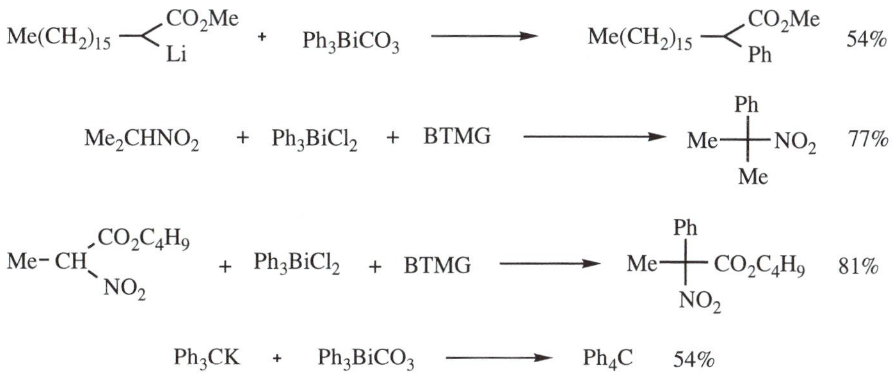

Me(CH$_2$)$_{15}$—C(CO$_2$Me)(Li) + Ph$_3$BiCO$_3$ ⟶ Me(CH$_2$)$_{15}$—C(CO$_2$Me)(Ph) 54%

Me$_2$CHNO$_2$ + Ph$_3$BiCl$_2$ + BTMG ⟶ Me—C(Ph)(Me)—NO$_2$ 77%

Me—CH(CO$_2$C$_4$H$_9$)(NO$_2$) + Ph$_3$BiCl$_2$ + BTMG ⟶ Me—C(Ph)(NO$_2$)—CO$_2$C$_4$H$_9$ 81%

Ph$_3$CK + Ph$_3$BiCO$_3$ ⟶ Ph$_4$C 54%

In the case of indole derivatives, 3-phenylindole (**64**) was formed in the reaction of indole (**63**) with tetraphenylbismuthonium trifluoroacetate or tetraphenylbismuthonium *p*-toluenesulfonate in benzene under reflux under neutral conditions. But 3,3-diphenyl-*3H*-indole (**65**) was obtained by treatment of the sodium salt of indole with tetraphenylbismuthonium *p*-toluenesulfonate. A related example is the 2-phenylation of *N,N,N′,N′*-tetramethyl-*p*-phenylenediamine which was realized with triphenylbismuth dichloride in the presence of BTMG or tetraphenylbismuthonium *p*-toluenesulfonate in the presence of BTMG and with triphenylbismuth carbonate alone. Phenylation was not obtained in the reaction with tetraphenylbismuthonium tetrafluoroborate.[21,28,33]

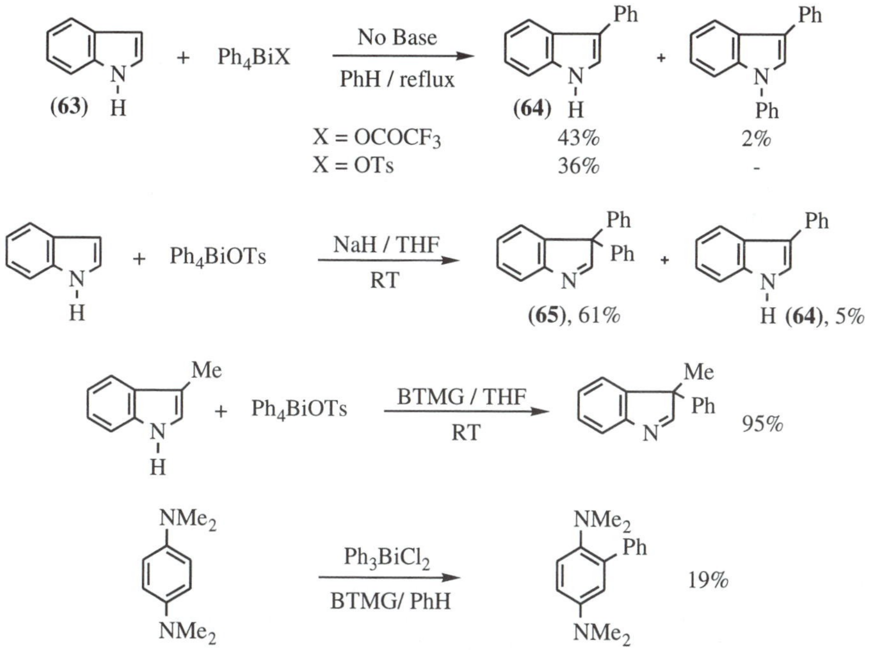

(63) H + Ph$_4$BiX —(No Base / PhH / reflux)→ (64) H + Ph

X = OCOCF$_3$ 43%   2%
X = OTs   36%   -

+ Ph$_4$BiOTs —(NaH / THF, RT)→ (65), 61% + H (64), 5%

+ Ph$_4$BiOTs —(BTMG / THF, RT)→ 95%

+ Ph$_3$BiCl$_2$ —(BTMG/ PhH)→ 19%

Triphenylbismuth difluoride (**2**) reacted with terminal acetylenes in the presence of a catalytic amount of cuprous chloride to afford moderate to good yields of the phenylacetylene derivative. The reaction was claimed to involve the formation of a pentaorganyl intermediate, which undergoes a ligand coupling process. Better yields were obtained when the reaction was performed with catalytic amounts of the copper salt than with the stoichiometric copper acetylide.[17]

$$Ph-C\equiv CH \ + \ \underset{(2)}{Ph_3BiF_2} \quad \xrightarrow[\text{reflux}]{CuCl \, / \, PhH} \quad \underset{87\%}{Ph-C\equiv C-Ph} \ + \ \underset{\sim 1\%}{Ph-Ph}$$

$$PhCH_2OCH_2-C\equiv CH \ + \ \underset{(2)}{Ph_3BiF_2} \quad \xrightarrow[\text{reflux}]{CuCl \, / \, toluene} \quad \underset{55\%}{PhCH_2OCH_2-C\equiv C-Ph} \ + \ \underset{\sim 3\%}{Ph-Ph}$$

$$Ph-C\equiv C-Cu \ + \ \underset{(2)}{Ph_3BiF_2} \quad \xrightarrow[\text{reflux}]{PhH} \quad \underset{17\%}{Ph-C\equiv C-Ph} \ + \ \underset{40\%}{Ph-Ph}$$

6.3 *O*-ARYLATION WITH PENTAVALENT ORGANOBISMUTH

6.3.1 Phenol and enol derivatives

The first example of *O*-arylation of a phenolic compound was reported by Razuvaev *et al.* who showed that the decomposition of pentaphenylbismuth (**4**) in the presence of phenol (**8**) led to diphenyl ether (**9**).[31] An intermediate phenoxytetraphenylbismuth (**66**) was suggested to decompose by two different pathways, the first one affording diphenyl ether (**9**) in 46% and the second one leading to phenol in 48% yield and to benzyne, eventually leading to tarry products.

$$\underset{(8)}{PhOH} \ + \ \underset{(4)}{Ph_5Bi} \ \longrightarrow \ \underset{\sim 100\%}{PhH} \ + \ \underset{(66)}{\left[PhO\text{-}BiPh_4\right]} \ \longrightarrow \ \underset{(9),\,46\%}{Ph\text{-}O\text{-}Ph} \ + \ \underset{48\%}{PhOH} \ + \ \underset{90\%}{Ph_3Bi} \ + \ tars$$

However, when a solution of pentaphenylbismuth (**4**) in carbon tetrachloride was heated under reflux in the presence of 2,3,4,5-tetraphenylcyclopentadienone (**67**), an efficient benzyne trap,[50] the benzyne adduct (**68**), 1,2,3,4-tetraphenylnaphthalene, was indeed obtained in 50% yield together with triphenylbismuth (76%).[51] Thus the tarry products may more likely originate directly from the decomposition of pentaphenylbismuth.

If this *O*-phenylation reaction of a phenol involves a covalent aryloxytetraarylbismuth intermediate formed by displacement of a phenyl ligand by the incoming phenol, its occurrence could be ascertained if its synthesis by an alternate way led to the same result, *O*-phenylation of the phenol. Indeed, when the postulated intermediate, phenoxytetraphenylbismuth (**66**), was prepared by an alternate way, such as by base-catalysed reaction of phenol with tetraphenylbismuthonium trifluoroacetate, a similar yield of the *O*-phenyl ether was obtained, and *ortho*-*C*-phenylation products were also formed.[30]

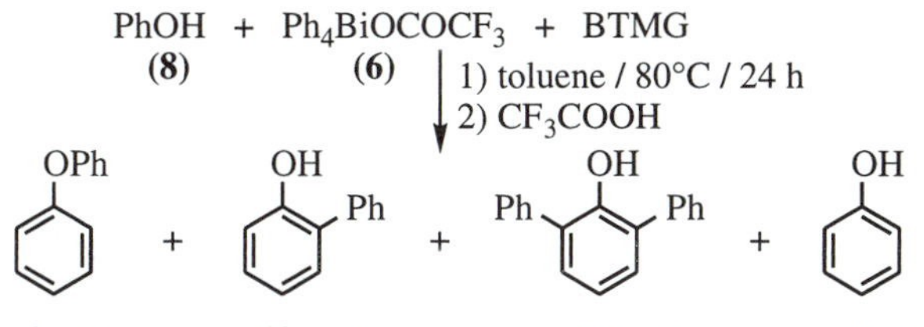

PhOH + Ph$_4$BiOCOCF$_3$ + BTMG
(8)   (6) | 1) toluene / 80°C / 24 h
                    2) CF$_3$COOH

(9), 42%   (10), 29%   (11), 3%   (8), 22%

When 4-nitrophenol (69) was treated with pentaphenylbismuth (4) or with tetraphenylbismuthonium tosylate (7) in the presence of a base, the stable covalent (4-nitrophenoxy)tetraphenylbismuth (70) was isolated. Its thermal decomposition, by refluxing a toluene solution for 4 hours, led to 4-nitrodiphenyl ether (71) in 98% yield.[23-25]

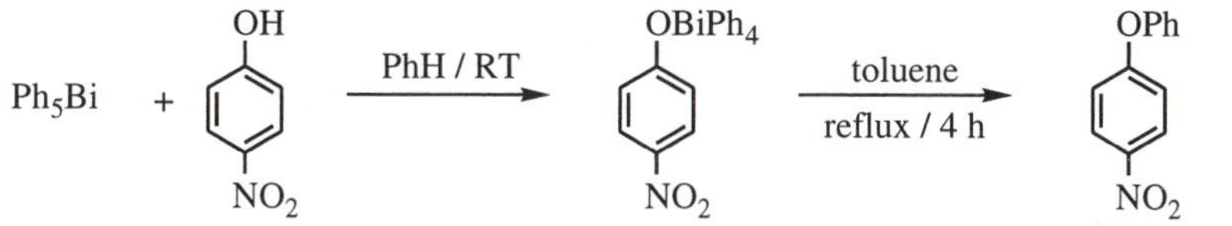

Ph$_5$Bi + (69) → PhH / RT → (70), 88% → toluene reflux / 4 h → (71), 98%

(4)

This observation led Barton *et al.* to consider that the presence of an electron-withdrawing ligand may influence the regioselectivity of the arylation towards *O*-arylation, whereas the presence of an electron-donating ligand would favour the *C*-arylation. Therefore they decided to study the reaction of electron-rich phenols with pentavalent organobismuth reagents bearing one electron-withdrawing group and four phenyl groups.[25] A range of phenols were treated with tetraphenylbismuthonium acylates (Ph$_4$BiOCOR, with R = Me or CF$_3$) and with tetraphenylbismuthonium tosylates under neutral conditions in benzene under reflux. The *O*-phenyl ethers were obtained in moderate to good yields, together with variable amounts of the *C*-phenylation product. In the reaction of β-naphthol with tetraphenylbismuthonium acetate, the *O*- and *C*-phenyl products were obtained both in 25% yield. The best yields of *O*-phenyl ethers were obtained in the reactions with tetraphenylbismuthonium trifluoroacetate. This latter reaction was susceptible to acid catalysis (Cl$_3$CCO$_2$H, 0.6 equiv.), which improved significantly the yields (Tables 6.1 and 6.2).[24]

Table 6.2: Phenylation of phenols with Ph$_4$BiOCOCF$_3$ under neutral conditions

ArOH + Ph$_4$BiOCOCF$_3$ → PhH / 80°C → Ar-O-Ph

Phenol	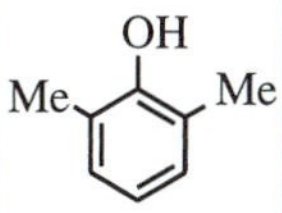 Me ⌬ Me (2,6-dimethylphenol)	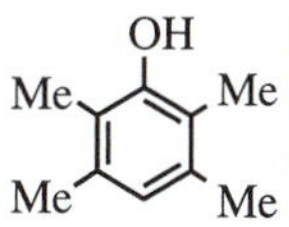 Me ⌬ Me / Me ⌬ Me	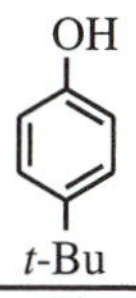 ⌬ t-Bu	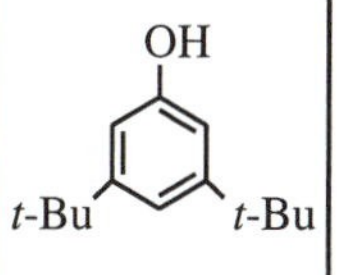 t-Bu ⌬ t-Bu	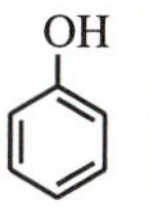 ⌬ (phenol)	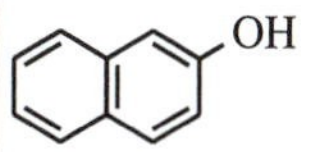 ⌬⌬ (naphthol)	⌬⌬ OH
Reaction time	72 h	24 h	4 h	4 h	140 h	15 h	15 h
Yield	58%	57%	53%	68%	100%	82%	85%[a]
Ref	24	24	24	24	24	28	28

a) reaction performed in the presence of 1,1-diphenylethylene (72) [Ph$_2$C=CH$_2$] (2 equiv.)

O-Phenylation of 3,5-di-*tert*-butylphenol (**13**) was also obtained by reaction with triphenylbismuth diacetate (**47**) in methylene dichloride under reflux, under neutral conditions.[27] However, as this reaction is greatly improved by copper catalysis (see section 6.7), the scope of the non-catalysed system was not studied.

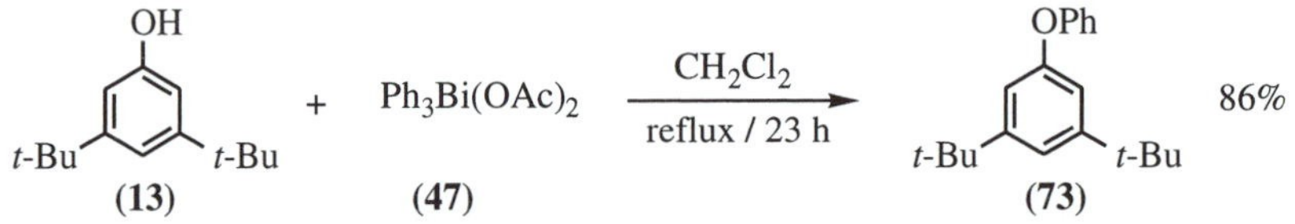

These two systems of phenol *O*-phenylation [*O*-phenylation with tetraphenylbismuthonium trifluoroacetate (**6**) or with triphenylbismuth diacetate (**47**)] may be ligand coupling reactions involving hexavalent transient intermediates, but no proof of such structures was ever detected. The former reaction does not involve free radicals, as addition of 1,1-diphenylethylene (**72**) (DPE), did not affect the overall yield. These reactions were explained by a direct aromatic S_N2 displacement, facilitated by the partial charge on the carbon bonded to the bismuth atom.[24] (see section 6.6)

A second type of *O*-arylation was developed from the original isolation of a covalent pentavalent stable aryloxybismuth intermediate. When 4-nitrophenol (**69**) reacts with triphenylbismuth derivatives (Ph_3BiX_2, with X = Cl or $OCOCF_3$) under basic conditions, again a covalent stable aryloxybismuth derivative is obtained. But their thermal decomposition did not lead to the *O*-phenyl ether (**71**). However, when the THF solution of other phenols containing electron-withdrawing groups in the *para* position was treated by triphenylbismuth dichloride and BTMG at room temperature, a coloured solution was quickly formed. When the solution was heated under reflux, the coloration of the solution faded and eventually the corresponding *O*-phenyl ethers (**74**) - (**76**) were obtained in good to excellent yields (70-91%).[30]

In the case of phenols bearing the electron-attracting groups in the *meta* position, the reaction led to mixtures of predominantly the *O*-phenyl ether together with minor amounts of *C*-phenyl products. The *O*-phenyl ethers were obtained in 50-60% yields.

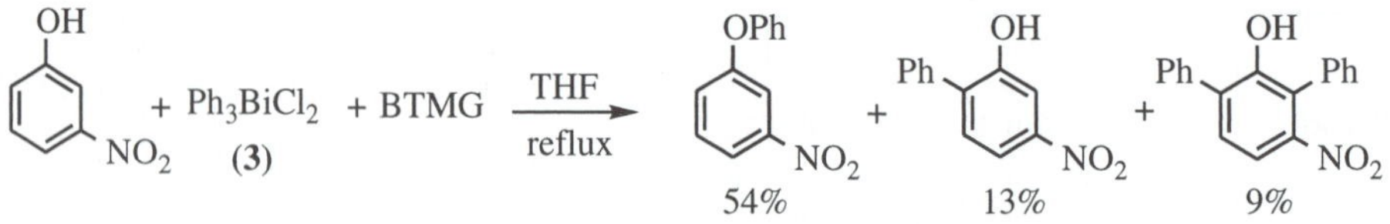

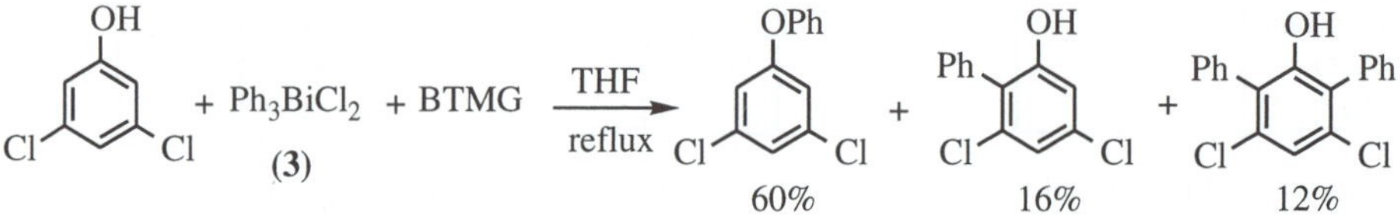

The *O*-phenylation reaction with tetraphenylbismuthonium trifluoroacetate (**6**) was also extended to enols. A series of enolised dicarbonyl compounds were *O*-phenylated in modest to moderate yields with this reagent under neutral or acidic conditions, the best yields being again obtained under acidic conditions.[33]

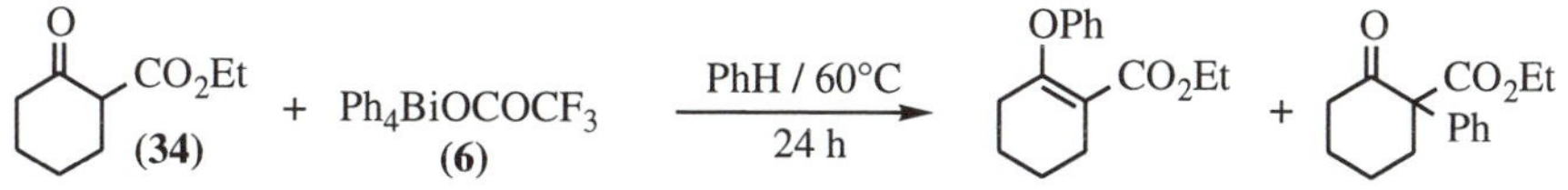

	Neutral conditions	30%	5%
	CCl_3COOH (0.6 equiv.)	57%	-

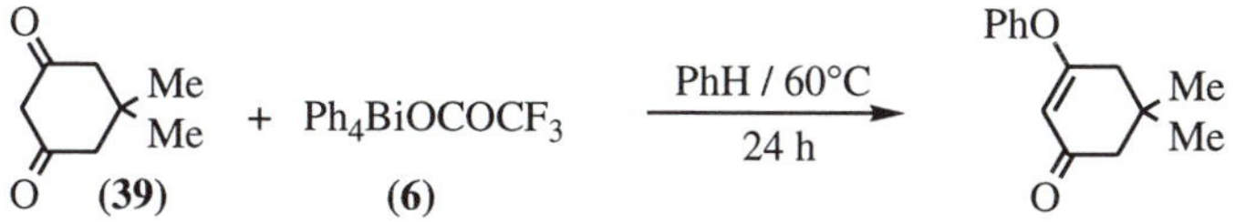

	Neutral conditions	56%
	CCl_3COOH (0.6 equiv.)	88%

6.3.2 Alcohol derivatives

As with phenols and enols, the reaction of alcoholic substrates with tetraphenylbismuthonium trifluoroacetate appeared to be pH dependent. The reactions of primary and secondary alcohols with tetraphenylbismuthonium trifluoroacetate or related tetraphenylbismuthonium reagents under basic conditions resulted in oxidation to the corresponding carbonyl derivatives. Under neutral or acidic conditions, the *O*-phenyl ethers of primary and secondary alcohols were obtained in modest yields (secondary alcohols) to fairly good yields (primary alcohols) by reaction of the alcohol with tetraphenylbismuthonium trifluoroacetate (**6**) in benzene or in toluene under reflux. Tertiary alcohols gave intractable mixtures.[51]

$$R\text{-}OH \quad + \quad Ph_4BiOCOCF_3 \quad \xrightarrow[\text{reflux}]{\text{PhH or toluene}} \quad R\text{-}O\text{-}Ph$$
$$\text{(6)}$$

Primary alcohols	50-70%
Secondary alcohols	20-30%
Allylic alcohols	50-60%
Diols	45-60%

The important drop in the yields from primary to secondary alcohols was attributed in part to a competing trifluoroacetylation induced by the presence of phenyl trifluoroacetate, a decomposition product of the reagent (**6**). Reaction of 3β-cholestanol (**77**) with (**6**) led the *O*-phenyl ether (**78**) in 29% yield, together with 3β-cholestanyl trifluoroacetate (**79**) in 65% yield.

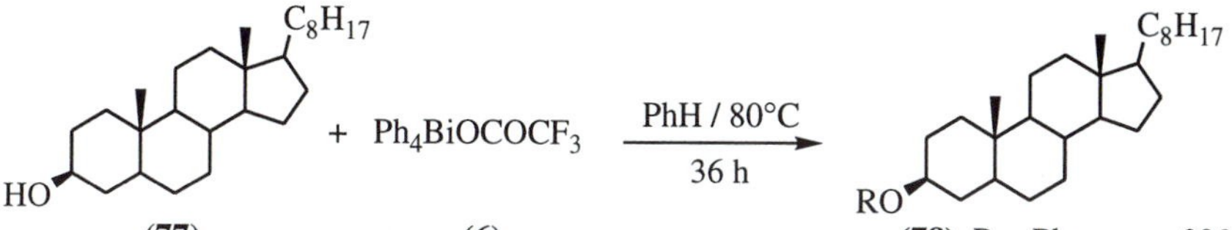

(**78**), R = Ph   29%
(**79**), R = $COCF_3$ 65%

O-Phenylation of tertiary alcohols can be realized through reaction with pentaphenylbismuth (**4**), a benzyne intermediate being involved. *tert*-Butyl phenyl ether (**80**) was also obtained in rather low yields in the reaction of triphenylbismuth diacetate with *tert*-butyl alcohol used as solvent in the presence of potassium carbonate or by reaction of triphenylbismuth dichloride with lithium *tert*-butoxide.[31,52,53]

$$Ph_5Bi \ + \ t\text{-BuOH} \longrightarrow t\text{-BuO-Ph} \underset{\diagdown}{\overset{\diagup}{}} \quad \begin{array}{l} Ph_3BiCl_2 \ + \ t\text{-BuOLi} \\[1em] Ph_3Bi(OAc)_2 \ + \ t\text{-BuOH} \ + \ K_2CO_3 \end{array}$$

$$\textbf{(4)} \qquad\qquad\qquad \textbf{(80)}$$

The reaction of triphenylbismuth diacetate (**47**) with secondary alcohols under neutral conditions resulted in poor to moderate yields of the corresponding *O*-phenyl ethers: 2-propanol gave about 20% of the *O*-phenyl ether with a large excess of the alcohol, cyclohexanol less than 3% and 3β-cholestanol gave 36% of the *O*-phenyl ether. Moreover, no reaction was observed between triphenylbismuth diacetate and primary alcohols.[53-55]

$$R_2CH\text{-OH} \ + \ Ph_3Bi(OAc)_2 \xrightarrow[\text{reflux}]{CH_2Cl_2} R_2CH\text{-O-Ph} \quad 3\text{-}40\%$$
$$\textbf{(47)}$$

The reactivity of tetraphenylbismuthonium trifluoroacetate (**6**) and triphenylbismuth diacetate (**47**) towards hydroxylic functions was greatly improved by the presence of a neighbouring group. Whereas reaction of cyclohexanol with tetraphenylbismuthonium trifluoroacetate (**6**) in benzene under reflux afforded only 22% of the *O*-phenyl ether after 18h, *cis*-cyclohexane-1,2-diol gave 52% of the mono-*O*-phenyl ether and 13% of the bis-*O*-phenyl ether under the same conditions.[51] A selective mono-*O*-phenylation of glycols by triphenylbismuth diacetate (**47**) was discovered by David and Thieffry during a comparative study of the action of various oxidants towards α-glycols. Good to excellent yields were obtained in the reaction of (**47**) with a series of combinations of primary and secondary glycols and also for a secondary-tertiary glycol. Even ditertiary glycols afforded the mono-ether, although in a very modest yield. The distance between the two hydroxyl functions is not determinant as only 1,5- and 1,6-glycols showed a decrease in the average yields. An axial preference was also noted in conformationally rigid glycols.[54,56]

$$\begin{array}{c} \text{OH} \\ \left(\right. \\ \text{OH} \end{array} + \ Ph_3Bi(OAc)_2 \xrightarrow[\text{reflux}]{CH_2Cl_2} \begin{array}{c} \text{O-Ph} \\ \left(\right. \\ \text{OH} \end{array}$$
$$\textbf{(47)}$$

Selected examples of mono- **O**-*phenylation of glycols with* **triphenylbismuth diacetate** (**47**)

HO−(CH₂)ₙ−OH (Ph)

n = 2	85%
3	87%
4	80%
5	50%
6	40%

OH (Ph) / Me, Me / OH   86%

OH (Ph) / Me Me / OH   84%

OH (Ph) / Me Me / OH   15%

OH (Ph) / Ph, Ph / OH   37%

OH (Ph) / Me Ph, Ph / OH   50%

OH / Ph Ph / OH   0%

7h, 60%

60h, R^1 = Ph, R^2 = H : 5%
      R^1 = H, R^2 = Ph : 5%

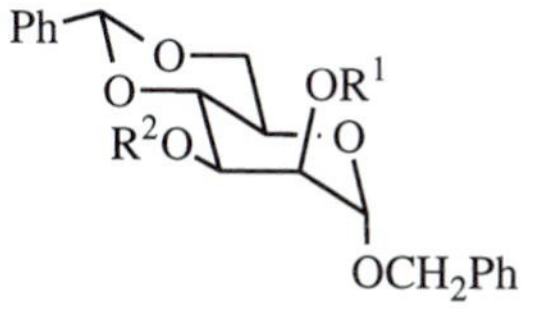

OCH$_2$Ph
R^1 = Ph, R^2 = H : 42%
R^1 = H, R^2 = Ph : 11%

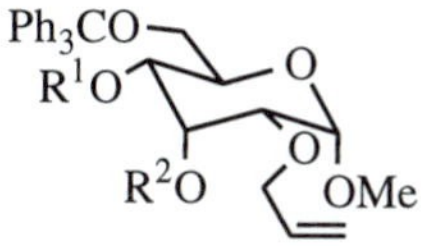

R^1 = Ph, R^2 = H : 3%
R^1 = H, R^2 = Ph : 27%

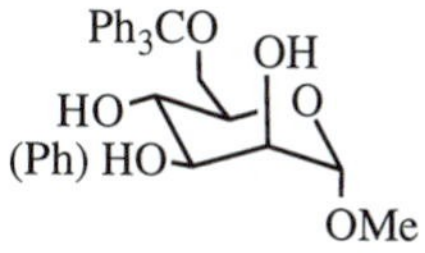

32%

It was subsequently found that this reaction is not restricted to glycols. Ether-alcohols, amino-alcohols and benzoin were also prone to the phenylation reaction with triphenylbismuth diacetate (**47**).[55] In the case of the amino-alcohol, the amino group is preferentially phenylated to afford the *N*-phenyl derivative, which can undergo subsequently a second arylation. This second arylation occurs both on the nitrogen and oxygen atoms, preferentially on the oxygen rather than on the aniline-type nitrogen.

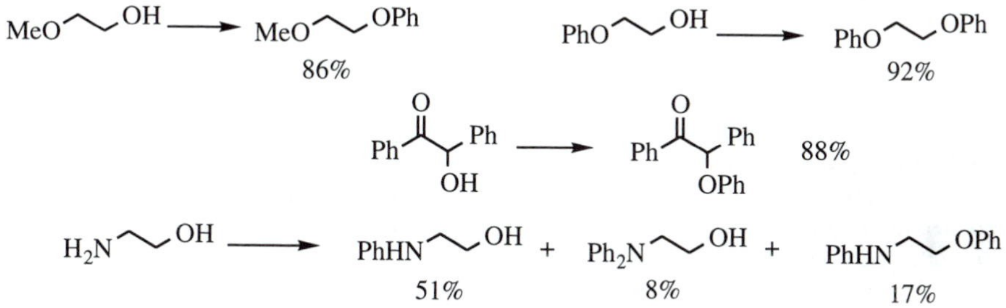

6.4 *S*- AND *N*-ARYLATION WITH PENTAVALENT ORGANOBISMUTH

The number of *S*- and *N*-arylations with arylbismuth reagents is relatively limited. The reaction of thiols with pentavalent organobismuth reagents leads mostly to oxidation products and, in some instances, to *S*-phenylation. In the case of the nitrogen derivatives, the most efficient arylation by arylbismuth reagents are reactions which are copper-catalysed, and these will be reviewed in a later part of this chapter (see section 6.7).

6.4.1 Sulfur derivatives

Arylthiols were converted to mixed arylphenylsulfides by reaction with pentaphenylbismuth (**4**) or with tetraphenylbismuthonium trifluoroacetate (**6**). Oxidation products were also formed in minor amounts (<15%). On the other hand, thiols were readily converted to the corresponding disulfides by treatment either with triphenylbismuth dichloride (**3**) or with triphenylbismuth carbonate (**5**) in the presence of sodium hydride.[25,33]

$$\text{Ar-SH} + \underset{(4)}{\text{Ph}_5\text{Bi}} \xrightarrow[\text{reflux}]{\text{PhH or toluene}} \underset{32\text{-}65\%}{\text{Ar-S-Ph} + \text{Ar-S-S-Ar}}$$

$$\text{Ar-SH} + \underset{(6)}{\text{Ph}_4\text{BiOCOCF}_3} \xrightarrow[\text{reflux}]{\text{PhH}} \underset{70\text{-}80\% \qquad 0\text{-}15\%}{\text{Ar-S-Ph} + \text{Ar-S-S-Ar}}$$

$$\text{Ar-SH} + \underset{(3)}{\text{Ph}_3\text{BiCl}_2} + \text{NaH} \longrightarrow \underset{95\text{-}100\%}{\text{Ar-S-S-Ar}}$$

p-Toluenesulfinic acid reacted with pentaphenylbismuth (4) or with tetraphenylbismuthonium trifluoroacetate (6) under neutral or basic conditions to afford phenyl *p*-tolylsulfone (81) in high yields. A similar result was also obtained with triphenylbismuth carbonate (5).[33,57]

Me—⟨⟩—SO$_2$H + Ph$_5$Bi ⟶ Me—⟨⟩—SO$_2$—⟨⟩   (81), 87%
(4)

Me—⟨⟩—SO$_2$Na + Ph$_4$BiOCOCF$_3$ ⟶ Me—⟨⟩—SO$_2$—⟨⟩   (81), 86%
(6)

6.4.2 Nitrogen derivatives

Triphenylbismuth carbonate generally does not react with amines, although phenyl transfer to the amino functionality was first observed during an attempt of oxidation of the hydroxyl function of an aminoalcohol by triphenylbismuth carbonate in chloroform.[58] It also reacted with *N*-phenylhydroxylamine under neutral or basic conditions to give diphenylnitroxyl and *N,N*-diphenylacetamide after reductive acetylation (29-33%).[28] No reaction was noted between triphenylbismuth carbonate and nitrosobenzene. High-yielding *N*-arylation was found in the reaction of amino-alcohols with triarylbismuth diacetate in methylene dichloride under reflux.[55]

$$\text{PhNHOH} + \text{Ar}_3\text{BiCO}_3 \longrightarrow \underset{\text{Ar}'}{\overset{\text{Ph}}{>}}\text{N-O}^\bullet \xrightarrow[\text{Ac}_2\text{O}]{\text{Fe, AcOH}} \underset{\underset{29\text{-}33\%}{\text{Ar}}}{\text{Ph-N-Ac}}$$

Ph$_3$Bi(OAc)$_2$ + H$_2$N⌒⌒OH ⟶ PhHN⌒⌒OH + Ph$_2$N⌒⌒OH + PhHN⌒⌒OPh
                               51%   8%   17%

Amides are generally not phenylated by triphenylbismuth carbonate and only in poor yields with tetraphenylbismuthonium derivatives under neutral conditions.[25,51] However, amides can be *N*-phenylated by triphenylbismuth dichloride in the presence of BTMG, although only one example was reported (see also section 6.2.2.1).[39]

MeO$_2$C—C(=O)—NH (ring) + Ph$_3$BiCl$_2$ + BTMG ⟶ MeO$_2$C—C(=O)—N-Ph (ring)
(44)   (3)   (46), 62%

Oximes were cleaved by tetraphenylbismuthonium trifluoroacetate under basic conditions to regenerate the ketone,[51] but stable *O*-triphenylbismuth bisoximates were obtained from the reaction of an oxime with triphenylbismuth dichloride under basic conditions.[59] Imides were phenylated in moderate yields by tetraphenylbismuthonium trifluoroacetate.[51] In the reaction of indole with tetraphenylbismuthonium trifluoroacetate under neutral conditions, 3,*N*-diphenylindole was isolated as a very minor side product (2%).[33]

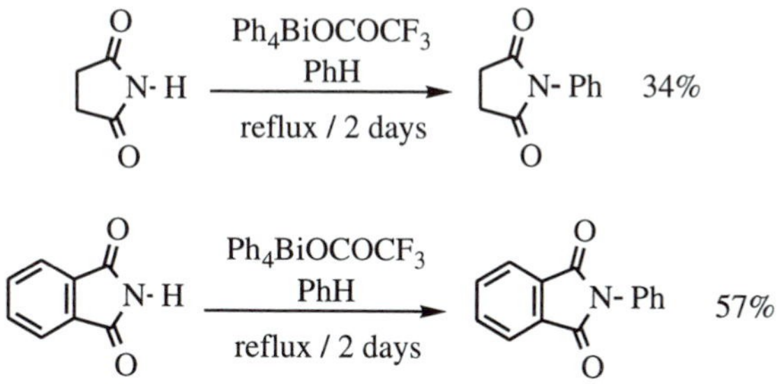

6.5 LIGAND COUPLING WITH ORGANOBISMUTHONIUM SALTS

Three different types of bismuthonium salts are known: the tetraaryl-, the tetraalkyl- and the mixed alkylaryl-bismuthonium salts. The reactivity of the tetraarylbismuthonium compounds parallels that of the triaryl- and pentaaryl-bismuth (V) derivatives. A nucleophile attacks their bismuth atom to lead to pentacoordinated tetraaryl compounds which subsequently undergo reductive processes: ligand coupling or oxidation of the nucleophile. Very few tetraalkyl compounds are known, in part due to the difficulties associated with their preparation.[60] Functionalised alkyl- and alkenyltriphenylbismuth-onium salts have been studied by Suzuki *et al.* in recent years.[8] Efficient ways of synthesis have been developed and their chemical behaviour can be mostly explained by invoking three main pathways: ligand coupling, S_N2 reaction on the α-carbon of the alkyl chain, and ylid chemistry leading to Darzens-type products. Free carbene intermediates were also sometimes observed.

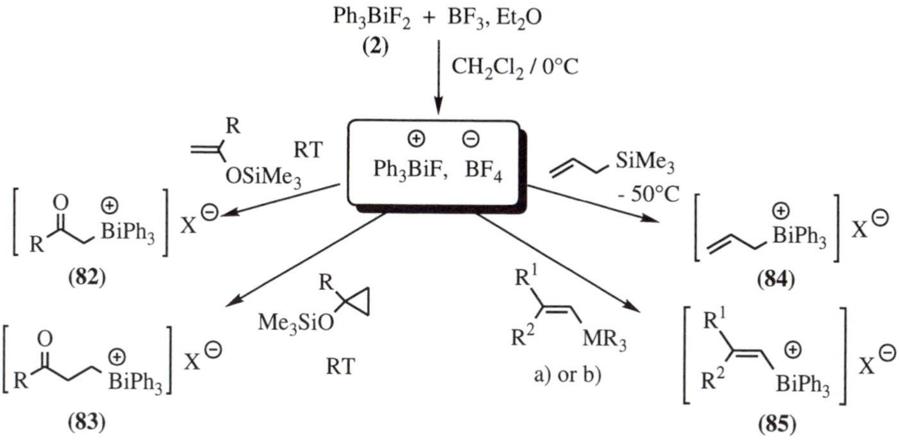

a) $MR_3 = SiMe_3$: reflux, 12-25h b) *i*- Me_3SiCN, RT, 1h, *ii*- $MR_3 = SnMe_3$ or $SnBu_3$, RT, 12h

Scheme 6.2: Synthesis of alkyltriphenylbismuthonium salts

Reaction of triphenylbismuth difluoride (2) with a Lewis acid, such as boron trifluoride diethyl etherate, in methylene dichloride at 0°C gives an association complex, probably fluorotriphenyl-bismuthonium tetrafluoroborate. Addition of a silyl enol ether led to the formation of (2-oxoalkyl) triphenylbismuthonium salts (82) in good yields.[61] The corresponding reaction using silyloxycyclo-propane afforded (3-oxoalkyl)triphenylbismuthonium salts (83) in moderate yields.[62] Under the same conditions, reaction of the activated triphenylbismuth difluoride with allyltrimethylsilane at low temperature led to the unstable allyltriphenylbismuthonium salts (84).[63] Treatment of the activated triphenylbismuth difluoride with either alkenyltrimethylsilane or with cyanotrimethylsilane-alkenyltrialkylstannane gave the stable alkenyltriphenylbismuthonium salts (85) in good yields.[64] (Scheme 6.2)

The stability of these bismuthonium salts (82) - (85) depends on the nature of the anionic counterpart and of the fourth ligand. The allyl derivatives (84) are too unstable to be isolated. On the other hand, the tetrafluoroborate salts of the derivatives of the three other types of onium compounds are thermally stable. Ligand coupling- type products are formed with more nucleophilic anions, for example during the attempted exchange of anion with bromide or iodide, or during the slow decomposition of a solution of bismuthonium tosylates in chloroform.[61,62]

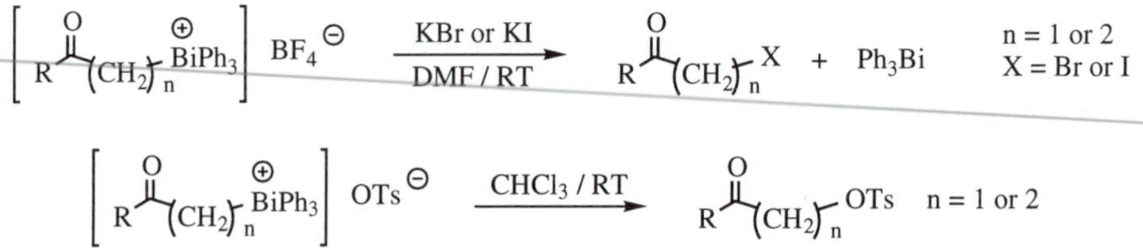

2-Oxoalkyltriphenylbismuthonium salts (82) react with nucleophiles to afford α-substituted carbonyl compounds. The 2-oxoalkyl moiety is transferred as an acylmethyl cation equivalent. Reaction with piperidine and various sodium salts gave moderate to good yields of the neutral α-substituted carbonyl compounds. Pentavalent covalent species were suggested as possible intermediates, although they were not detected.[61] (Scheme 6.3)

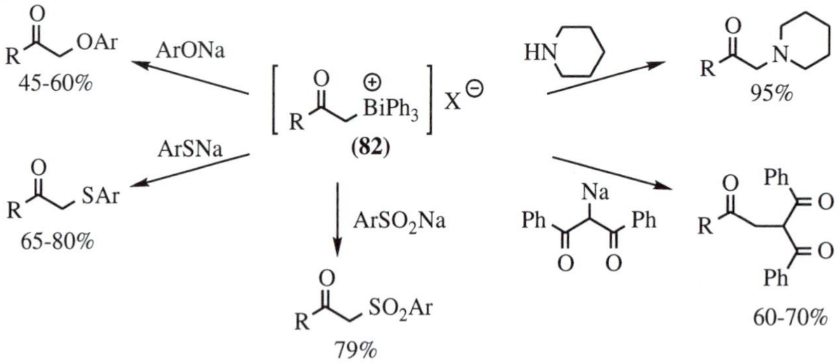

Scheme 6.3: Reactions of 2-oxoalkyltriphenylbismuthonium salts (82)

In the reaction of 2-oxoalkyltriphenylbismuthonium salts (82) with triphenylphosphine and with dimethylsulfide, onium exchange takes place with quantitative formation of triphenylbismuthane.[61] (Scheme 6.4)

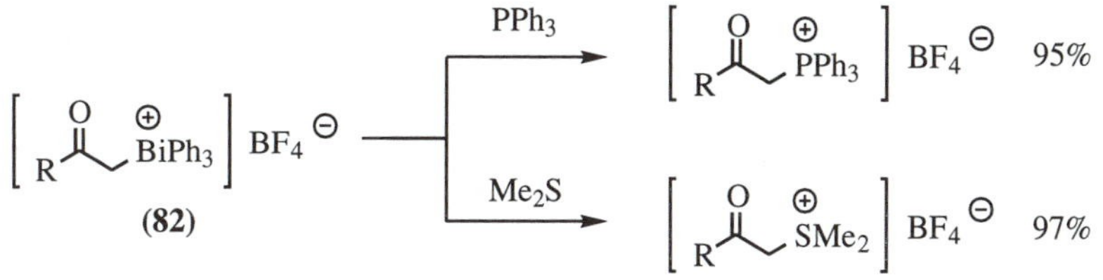

Scheme 6.4

Reaction of 2-oxoalkyltriphenylbismuthonium salts (**82**) with a base such as potassium *tert*-butoxide led to triphenylbismuthonium 2-oxoalkylides (**86**). These compounds showed the behaviour of moderately stabilized ylides towards various nucleophiles.[65-67]

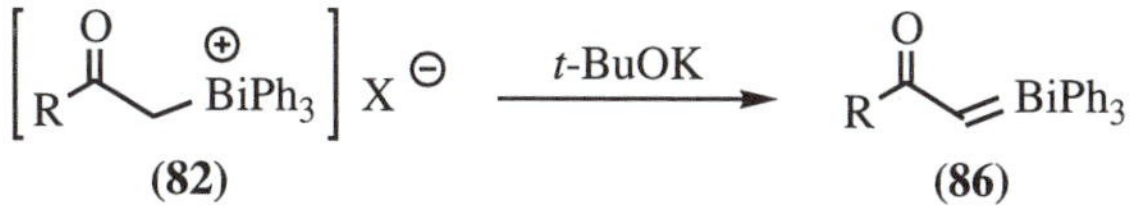

3-Oxoalkyltriphenylbismuthonium salts (**83**) reacted with some of the same nucleophiles as the 2-oxoalkyl derivatives, to afford the corresponding β-substituted carbonyl compounds. Moreover they also react with protic and aprotic solvents to yield mixtures of β-substituted carbonyl compounds and α,β-unsaturated carbonyl compounds, together with stoichiometric recovery of triphenylbismuthane.[62] (Scheme 6.5)

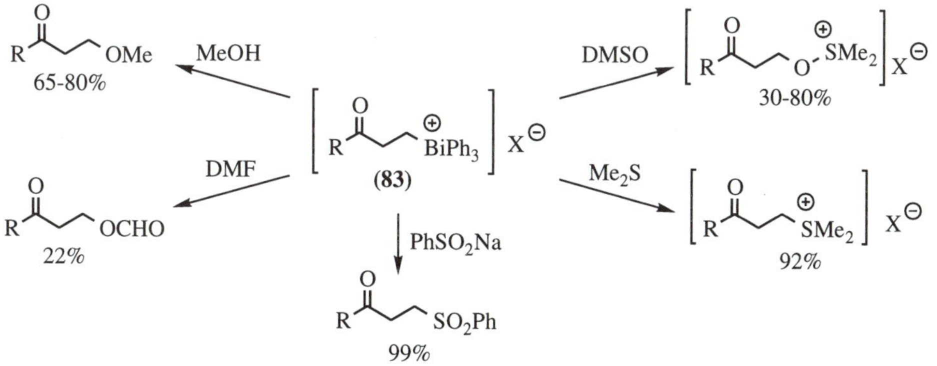

Scheme 6.5: Reactions of 3-oxoalkyltriphenylbismuthonium salts (**83**)

Reaction of 3-oxoalkyltriphenylbismuthonium salts (**83**) with potassium *tert*-butoxide led only to the β-elimination product.

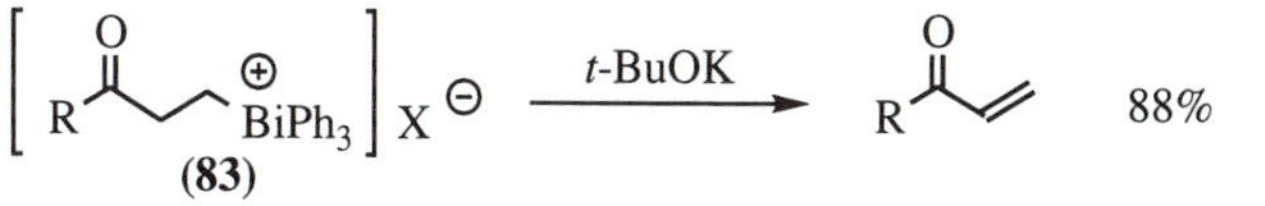

Allyltriphenylbismuthonium salt (**84**), generated *in situ*, decomposes to give allylbenzene. As the *ipso* substitution product was predominant in the reaction of tris(*ortho*-tolyl)- and tris(*para*-tolyl)-

bismuth difluoride with allyltrimethylsilane, an intramolecular reductive coupling was suggested as the major pathway. In the presence of electron-rich arenes, allyl transfer takes place to yield a mixture of mono- and di-allyl derivatives, the monoallyl being usually predominant. Other nucleophiles such as triphenylphosphine, dimethylsulfide, and *p*-toluenesulfinate led to the substitution products in excellent yields. Reaction with thiophenol led eventually to the product of double allylation in high yield (86%).[63] (Scheme 6.6)

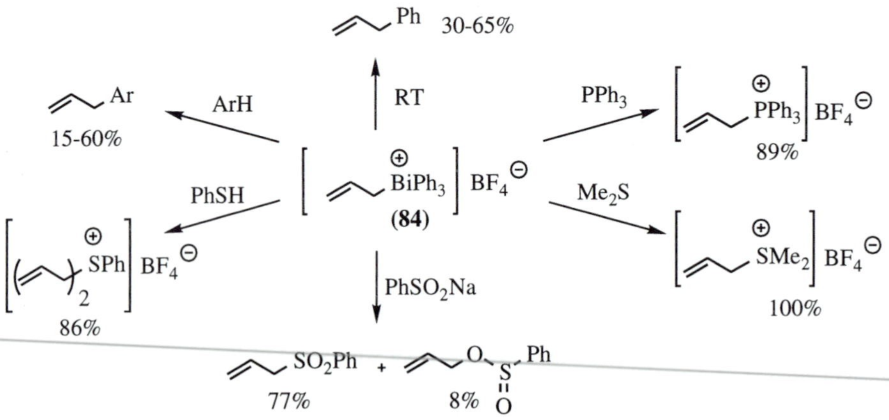

Scheme 6.6: Reactions of allyltriphenylbismuthonium salt (84)

The greater stability of 1-alkenyltriphenylbismuthonium salts (85) compared to the other types of alkylbismuthonium salts is reflected in their reactivity towards nucleophiles.[64] They do not react with triphenylphosphine, dimethylsulfide or piperidine. Sodium sulfinates and thiolates afforded the corresponding sulfones and sulfides, respectively. Both the alkenyl and the phenyl group can be transferred by ligand coupling. In the reaction with sodium sulfinates, the alkenyl is preferentially transferred to the nucleophile at the α-position with retention of the olefinic geometry affording the arylvinylsulfones. However, in the reaction with sodium thiolates, nearly equivalent amounts of the two sulfides, phenyl-vinylsulfide and phenyl-phenylsulfide, were obtained. The ligand coupling taking place on the covalent pentacoordinate intermediate, possessing four carbon-bismuth bonds and one sulfur-bismuth bond, is not chimioselective.

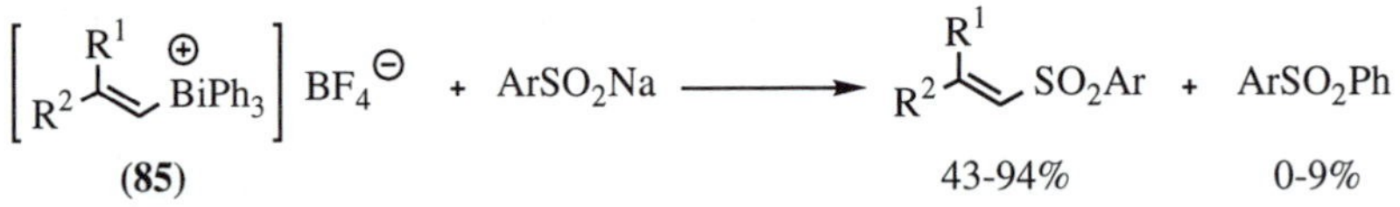

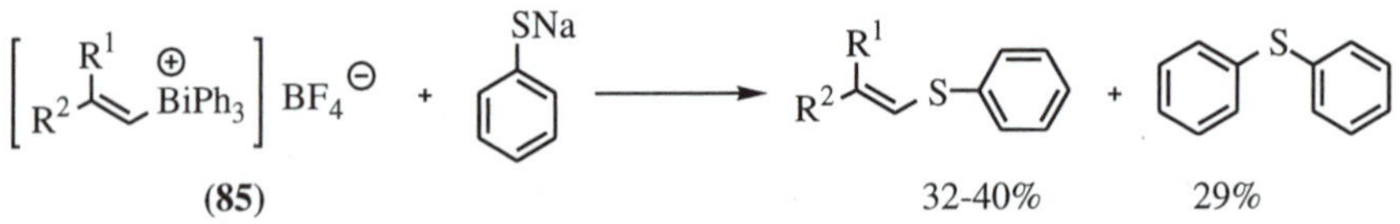

Reaction of 1-alkenyltriphenylbismuthonium salts (85) with potassium *tert*-butoxide led to the carbene derivatives, 1-alken-1-ylidenes, which can either be trapped by styrene to give cyclopropanes or isomerise to give the corresponding acetylenes.[64]

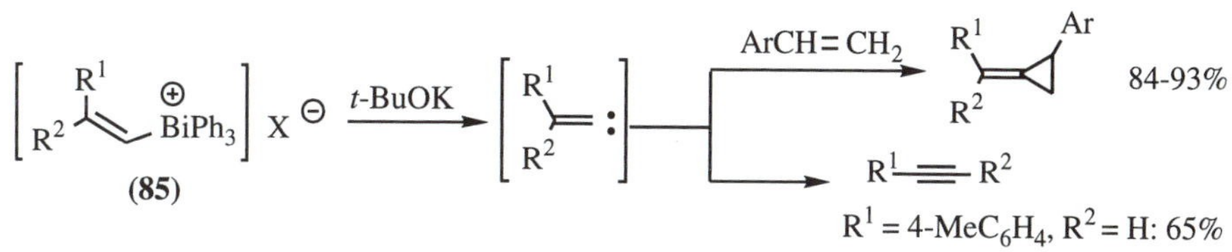

$$\left[\underset{R^2}{\overset{R^1}{\diagdown}}\hspace{-0.5em}\underset{}{\overset{\oplus}{\diagup}}\hspace{-0.5em}BiPh_3 \right] X^{\ominus} \xrightarrow{\textit{t-}BuOK} \left[\underset{R^2}{\overset{R^1}{\diagdown}}\hspace{-0.5em}=: \right]$$

(85)

$ArCH=CH_2 \longrightarrow \underset{R^2}{\overset{R^1}{\diagdown}}\hspace{-0.3em}\overset{Ar}{\triangle}$ 84-93%

$R^1 \!\!\equiv\!\! R^2$

$R^1 = 4\text{-MeC}_6H_4,\ R^2 = H\text{: }65\%$

6.6 MECHANISTIC STUDIES

Pentavalent arylbismuth derivatives act as regioselective reagents for the arylation of a wide range of substrates under mild conditions. The mechanisms of all these reactions can be explained by one of four possible mechanistic pathways: ligand coupling, aromatic S_N2 nucleophilic substitution, free radical or benzyne intermediates. The occurrence of benzyne intermediates was clearly demonstrated by a trapping experiment using 2,3,4,5-tetraphenylcyclopentadienone only in the case of the decomposition of pentaphenylbismuth. The facile ligand exchange which is observed with pentavalent organobismuth derivatives led to postulate the formation of a covalent pentacoordinate intermediate (87) as the most likely pathway. In a second step, its decomposition is governed by the nature of the substrate, the nature of the substituents on the organobismuth reagent and the reaction conditions. (Scheme 6.7)

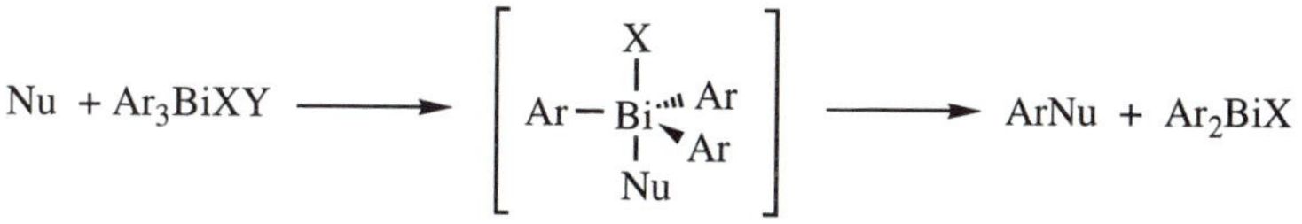

$$Nu + Ar_3BiXY \longrightarrow \left[\begin{array}{c} X \\ | \\ Ar-Bi^{\text{\tiny\hspace{0.1em}}}\hspace{-0.3em}Ar \\ | \hspace{0.3em}\searrow Ar \\ Nu \end{array} \right] \longrightarrow ArNu + Ar_2BiX$$

(87)

Scheme 6.7: General mechanism of ligand transfer with organobismuth derivatives

6.6.1 Formation of covalent intermediates in *C*- and *O*-arylations

The formation of a covalent bismuth-substrate intermediate such as (87) was first suggested in order to explain the oxidation reactions of alcohols by pentavalent organobismuth compounds. The decomposition of carvyloxy(tris-*p*-anisyl)bismuth semicarbonates, (88) (X = OCO_2^-), formed by interaction of deuterocarveol with tris(*p*-anisyl)bismuth carbonate afforded a mixture of products resulting from two types of ligand interactions.[68] *p*-Deuterioanisol and tris(*p*-anisyl)bismuth were isolated from the reaction mixture. (Scheme 6.8)

The formation of a covalent Bi-O intermediate was later demonstrated by a series of ^{1}H NMR experiments involving the reaction of triphenylbismuth dichloride with 2,2-dimethylpropan-1-ol. This type of intermediate was also detected by ^{1}H NMR monitoring of the ligand exchange reaction between 2-isopropoxytributylstannane and triphenylbismuth dichloride, leading to chloro(2-isopropoxy)triphenylbismuth. An alternate approach involving the low-temperature synthesis of 2-isopropyl-hypobromite (generated *in situ* by reaction of the isopropoxytributylstannane with bromine), followed by reaction with triphenylbismuthane led to the observation of the same sets of NMR signals.[51]

By analogy, similar types of intermediate were postulated to intervene in the reactions of *C*- and *O*-phenylation.[25,69]

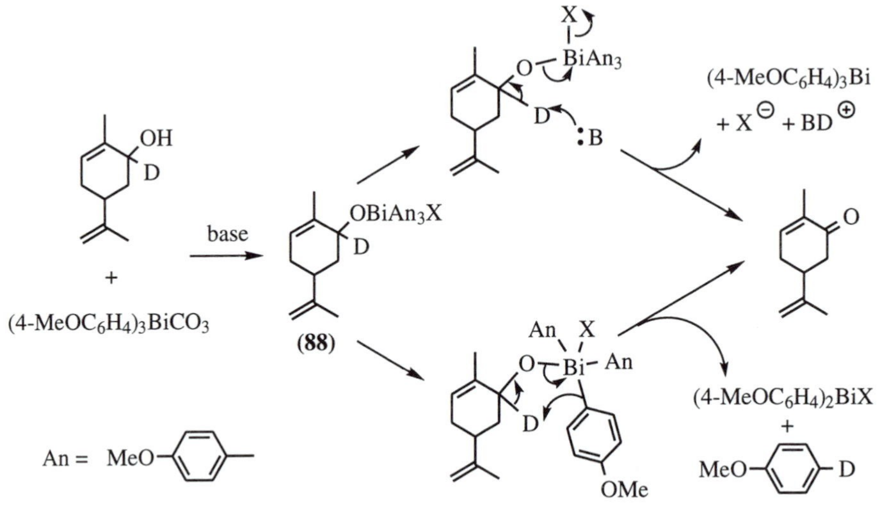

Scheme 6.8: Mechanism of the oxidation of carveol by organobismuth derivatives

In support of the formation of a more or less stable covalent intermediate, was the observation and isolation of (4-nitrophenoxy)tetraphenylbismuth (**70**) resulting from the treatment of 4-nitrophenol (**69**) with pentaphenylbismuth (**5**).[24] Thermal decomposition of this product led to the *O*-phenyl ether (**71**) in quantitative yield. (Scheme 6.9)

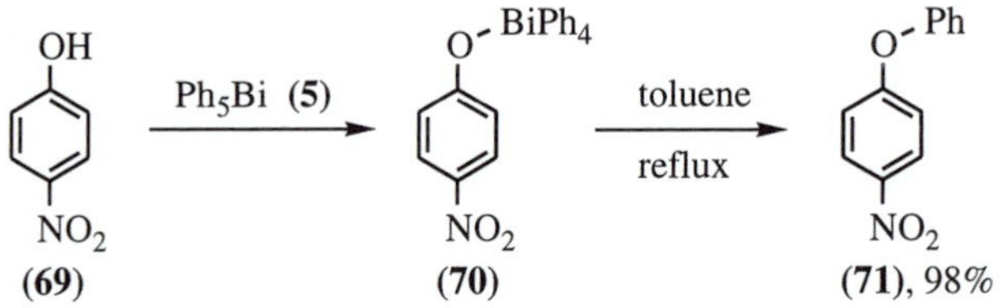

Scheme 6.9: Isolation of a stable covalent organobismuth intermediate in an *O*-phenylation reaction

This reaction led Barton *et al.* to consider that the breakdown of the pentavalent covalent bismuth intermediate was controlled by the nature of the fifth ligand.[57,69] An electron-withdrawing group would orient towards *O*-phenylation (Scheme 6.10), and an electron-donating group would favour *C*-phenylation (Scheme 6.11). The *O*-phenylation of phenols and enols by tetraphenylbismuthonium trifluoroacetate under neutral or acidic conditions was a supporting argument. Under basic conditions, the same phenols and enolic substrates led to products of *C*-phenylation with tetraphenylbismuthonium trifluoroacetate.

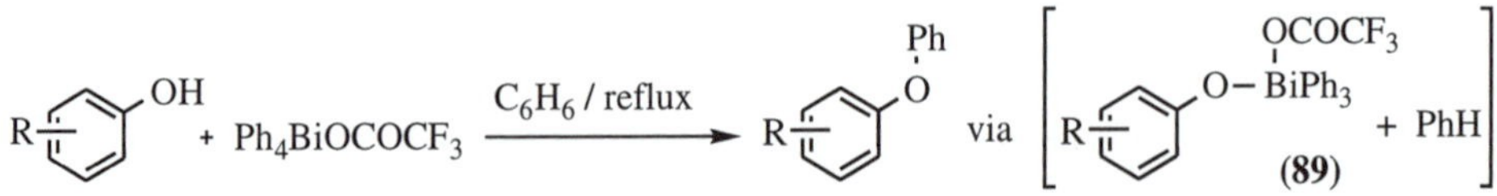

Scheme 6.10: *O*-Phenylation with an electron-withdrawing group as fifth ligand

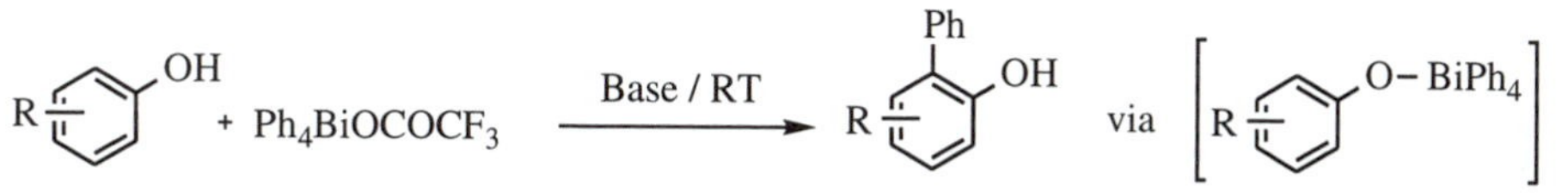

Scheme 6.11: *C*-Phenylation with an electron-donating group as fifth ligand

If these hypotheses are true, synthesis of the postulated intermediate (89) should be possible by different routes. A good example of the extreme regioselectivity was the case of β-naphthol, which led to high yields of the *O*- or *C*-phenylation products, as a function of the reaction conditions.

In the case of β-naphthol:	*C*-phenyl	*O*-phenyl
Ph$_5$Bi / C$_6$H$_6$:	90%	-
Ph$_4$BiOCOCF$_3$ / C$_6$H$_6$:	-	82-91%
Ph$_4$BiOCOCF$_3$ + Base / C$_6$H$_6$:	94%	-

Synthesis of compound (90) (case of β-naphthol) corresponding to the postulated intermediate in the *O*-phenylation reaction was performed by the base-catalysed reaction of β-naphthol with triphenylbismuth bistrifluoroacetate. However, after decomposition, this reaction led only to the products of *C*-phenylation.[24,70] (Scheme 6.12) Moreover, the *C*-phenylation is a fast reaction which can be completed at room temperature, or below, when the *O*-phenylation under neutral or acidic conditions requires long periods of reflux of the benzene solutions.

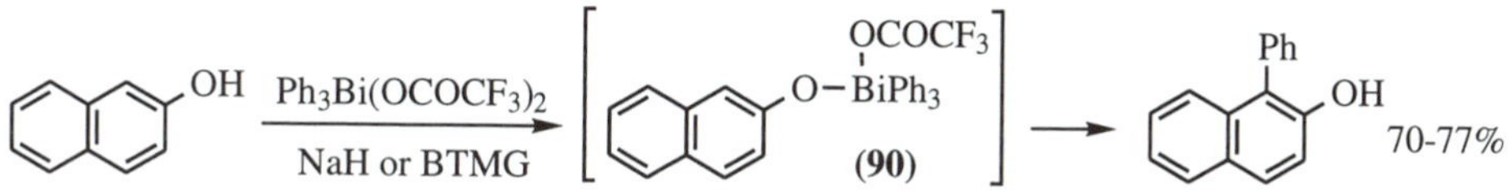

Scheme 6.12: Synthesis and evolution of the intermediate (90)

The same type of reactivity patterns was observed in the case of the *O*-phenylation of alcohols by tetraphenylbismuthonium trifluoroacetate under neutral or acidic conditions.[51] These observations led to the conclusion that these reactions of *O*-phenylation of alcohols and phenols are better explained by an aromatic S$_N$2 substitution, involving nucleophilic attack of the bismuth-bearing aromatic carbon by the hydroxyl function of the substrate. The electron-withdrawing trifluoroacetoxy substituent induces a partial positive charge on the *ipso* carbon, which facilitates the substitution.[24,51]

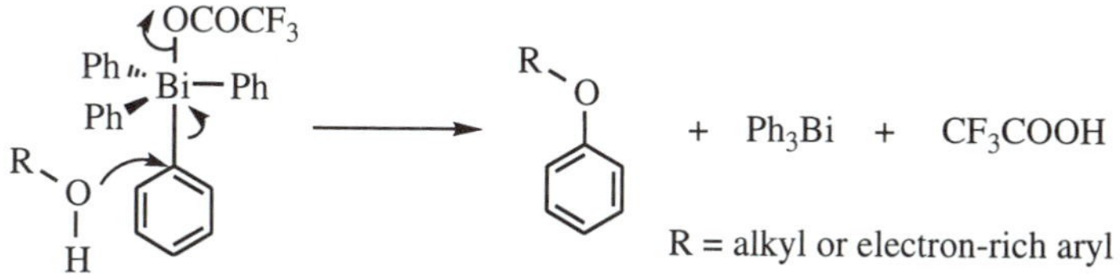

R = alkyl or electron-rich aryl

Although unlikely, hexa- or hepta-coordinated intermediates, such as (91) and (92), would be compatible with the observed bimolecular kinetics.[71] However, this type of covalent intermediate is more usually obtained under fairly basic conditions. The reaction of pentaphenylbismuth with phenyllithium was proposed to involve an hexacoordinate lithium hexaphenylbismuthate.[72-75]

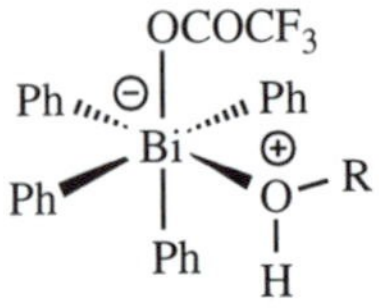

(91) 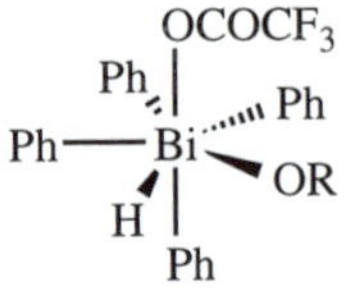   (92)

In the reaction of alcohols with the various types of pentavalent organobismuth compounds, O-phenylation is usually disfavoured, oxidation taking place preferentially. However, in the case of the reaction of glycols with triphenylbismuth diacetate, an apical-equatorial chelation (93) was suggested to explain the facile mono-O-phenylation which is observed.[56] By contrast, diapical chelation (94) takes place with monofunctional alcohols, and the ensuing reductive process leads to oxidation.

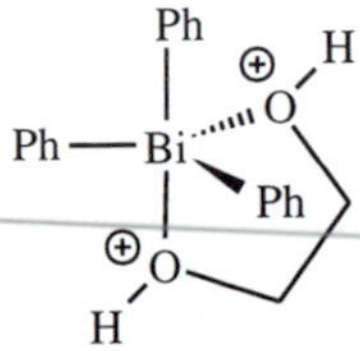

(93) 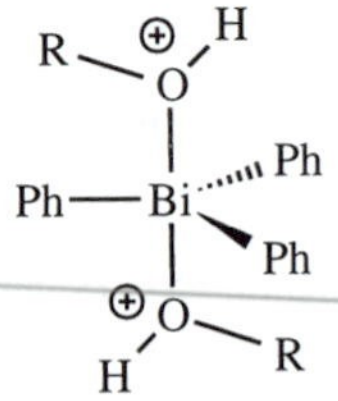   (94)

X-ray crystallographic studies have later shown that the structure of bis(acyloxy)triphenyl bismuth compounds is not a trigonal bipyramid, but rather like a distorted pentagonal bipyramid (95), in which the four oxygen atoms and one C-phenyl atoms can be viewed as forming the equatorial plane.[76,77]

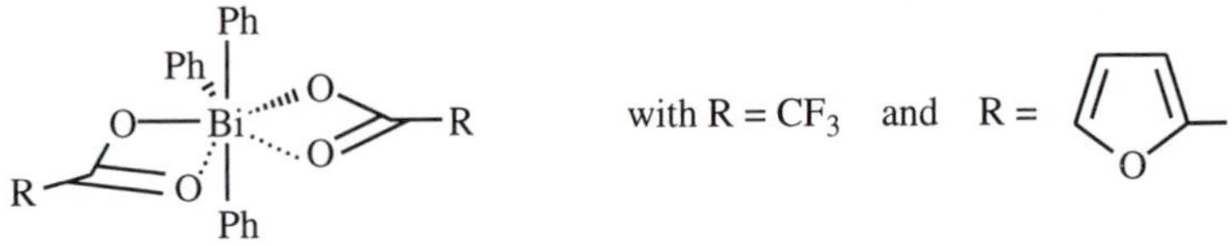

(95)   with R = CF_3 and   R =

As the two weakly bonded oxygen atoms are *cis*, the O-phenylation reaction of glycols could be better explained by the formation of a transition state with a pentagonal bipyramid structure (96), in which the two heteroatoms of the substrate have first displaced the weakly bonded oxygen atoms of the organobismuth reagent. A ligand coupling-like process then realizes the phenyl transfer to the more available heteroatom.

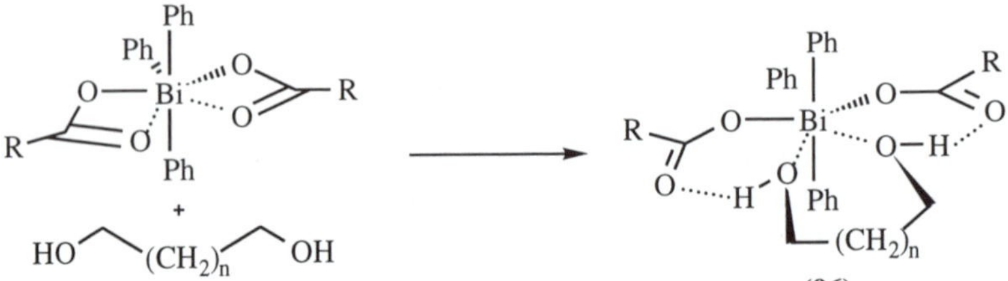

(96)

It is not unlikely that a similar type of transition state may be also involved in the *O*-phenylation of phenols and alcohols by tetraphenylbismuthonium trifluoroacetate under neutral or acidic conditions. The existence of an hexacoordinate structure for tetraphenylbismuthonium trifluoroacetate is possible, but attempted crystallization towards the isolation of crystals suitable for X-ray studies resulted instead in the formation of crystalline tetraphenylbismuthonium diphenylbis(trifluoroacetoxy) bismuthate.[78] However, hexacoordinate structures were observed in the case of chloro(8-quinolinato) triphenylbismuth[78] and its 2-methyl-8-quinolinato analogue,[79] which showed a Bi-O bond length of 2.175(7) Å and 2.19(2) Å respectively, together with a weak nitrogen-bismuth bond [2.807(10) Å and 2.71(2) Å respectively].

In the *C*-phenylation process, the postulated covalent bismuth-substrate intermediates were first detected by physical methods and later isolated. Reaction of 3,5-di-*tert*-butylphenol (**13**) with various phenylbismuth reagents afforded high yields of the 2,6-diphenyl derivative (**15**). Advantage was taken of the simplicity of its ^{1}H NMR pattern to study the formation of covalent intermediates. ^{1}H NMR monitoring of the CDCl$_3$ solutions of 3,5-di-*tert*-butylphenol with pentaphenylbismuth and with triphenylbismuth derivatives in the presence of *N,N,N',N'*-tetramethylguanidine (TMG) has given evidence of the presence of covalent intermediates.[24]

^{1}**H NMR Study of the Formation of Covalent Intermediates using Phenol (13).**
(Spectra taken in CDCl$_3$ solutions)[24]

	δ (H-*ortho*)	δ (H-*ortho*)
Phenol (**13**)	6.66	6.97
(**13**) + TMG	6.67	6.82
(**13**) + TMG + Ph$_3$BiCl$_2$	6.10	6.50
(**13**) + TMG + Ph$_4$Bi(OCOCF$_3$)$_2$	6.13	6.46
(**13**) + Ph$_5$Bi	6.15	6.47

Such intermediates were subsequently isolated, and their controlled thermal degradation led to the *ortho* mono-*C*-phenyl derivative (**14**).[24]

A possible alternative would be the formation of a covalent carbon-bismuth intermediate. Such an intermediate (**97**) may be involved in the *ortho* and *para*-phenylation of the sterically hindered 2,6-di-*tert*-butylphenol derivatives.[32] (Scheme 6.13) However, its occurrence has still not received any conclusive experimental support.

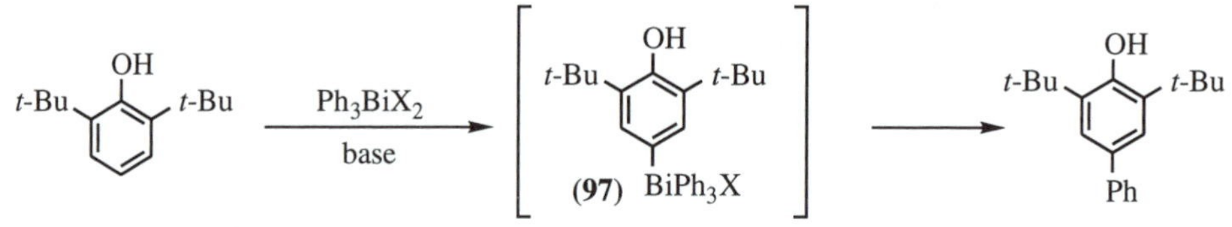

Scheme 6.13: Possible mechanism of 4-phenylation of 2,6-di-*tert*-butylphenol via intermediate (**97**)

Moreover, the synthesis and good stability of the covalent carbon-bismuth compound in the case of acetone, 2-oxopropyltriphenylbismuthonium perchlorate, excluded this type of intermediate as a common pathway,[80] although this view was questioned by Suzuki *et al.*[34] The reaction of the sodium salt of acetylacetone with triphenylbismuth dichloride led to a bismuthonium salt (**37**) decomposing slowly at room temperature to give eventually 3-phenylpentane-2,4-dione (**35**).[34]

6.6.2 Mechanism of the ligand coupling step

The second step of the arylation process involves a coupling reaction between two of the ligands linked to the bismuth atom. Various mechanistic pathways were considered to be possible. Study of the relative migratory aptitudes of aryl groups indicated that the *C*-phenylation reaction with either a phenol or a β-dicarbonyl does not follow an ionic pathway. Although these migratory aptitudes were of the same order as a free-radical type, the relative ratios are more consistent with a non-synchronous concerted mechanism.[26]

Table 6.3: Average relative migratory aptitudes[26]

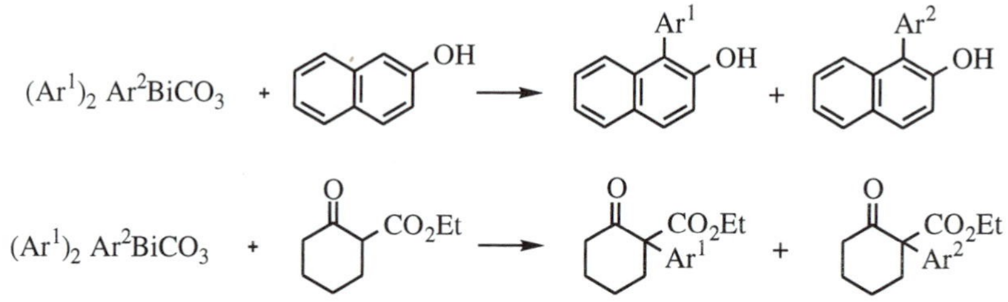

	4-NO$_2$-C$_6$H$_4$	C$_6$H$_5$	4-CH$_3$-C$_6$H$_4$	4-CH$_3$O-C$_6$H$_4$
2-Naphthol	3.55	1	0.45	0.22
Ethyl Cyclohexanonecarboxylate	3.55	1	0.60	0.25
Pinacol (Cationic), ref. 81	0.1	1	15.7	500
1,2-aryl radical migration, ref. 82	31	1	0.72	0.35

From all these observations, it was deduced that the mechanism of the ligand coupling process is a concerted mechanism which requires the favourable overlap between the π-systems of the aryl group and either the aryloxy or the vinyloxy partner. The change of the oxidation state from Bi(V) in the reagents to the more stable Bi(III) in the organobismuth by-products appears to be the driving force of the overall system. (Scheme 6.14)

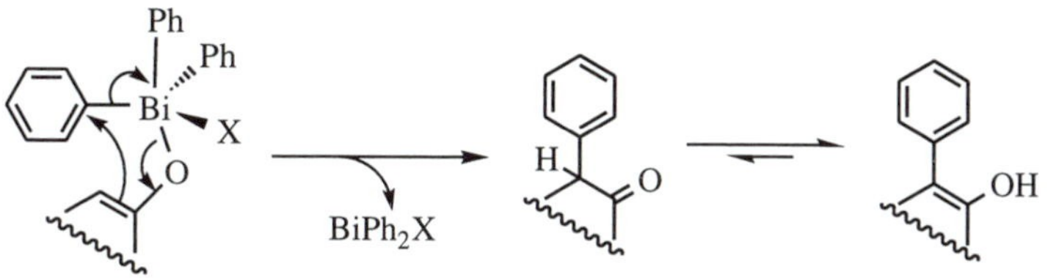

Scheme 6.14: Mechanism of the ligand coupling step

The reaction of phenols with triphenylbismuth dichloride in the presence of a base leads to an unstable covalent (aryloxy)bismuth compound (**98**). (Scheme 6.15) Depending on the electronic nature of the substituents present on the aryloxy ligand, topological transformations by either Berry pseudorotation (BPR) or turnstile rotation (TR) can take place. In the case of electron-rich phenols, the first formed intermediate with the apical aryloxy ligand will evolve into a topoisomeric intermediate having the aryloxy ligand in equatorial position. Eventually, this will lead to the *C*-phenylation products. If the two ligands in the apical positions are too strongly electron-withdrawing, interconversion does not happen, and the ligand coupling will imply two equatorial ligands, hence the good yields of biphenyl. In the intermediate cases, the relative weight of the different topoisomers will result in a more complex mixture, with *O*- and *C*-arylation products being obtained.[30]

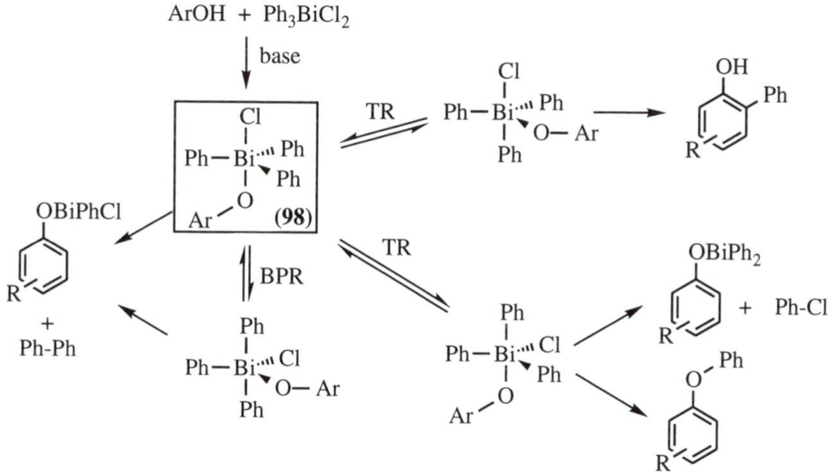

BPR = Berry Pseudorotation; TR = Turnstile Rotation

Scheme 6.15: Topological transformations in the ligand coupling involving phenolic compounds

6.6.3 Occurrence of free radicals

The phenylation of 1,1,3,3-tetraphenylacetone (**99**) affording pentaphenylacetone (**56**) in a high yield raised doubts on the generality of the formation of a covalent intermediate (**100**). Intervention of free radicals appeared as a plausible alternative which could be more compatible with the formation of highly hindered phenylated products. (Scheme 6.16)

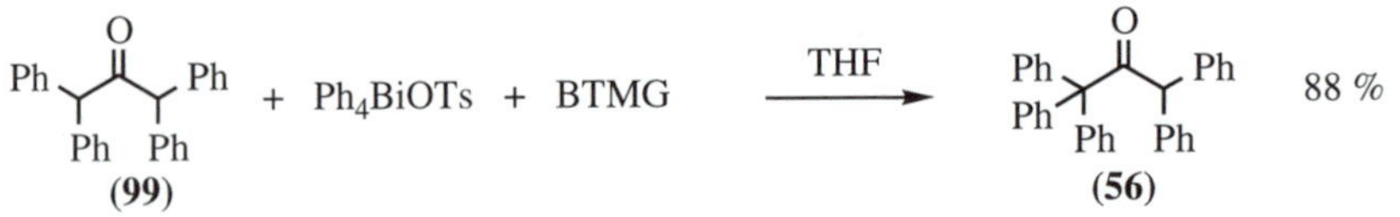

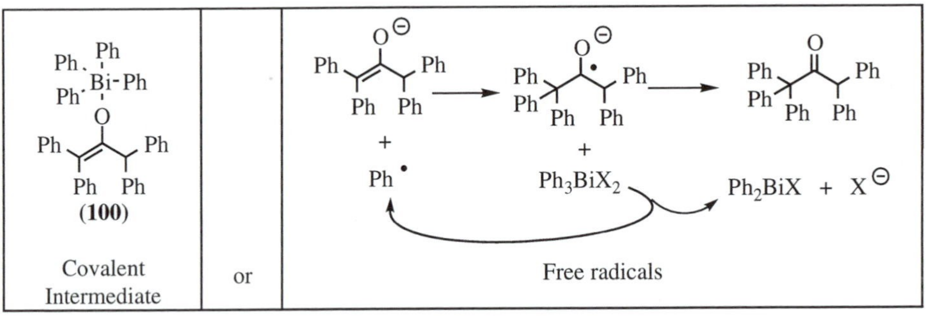

Scheme 6.16: Possible alternative mechanisms

Such a possibility was quickly supported by qualitative ESR studies, which revealed the presence of phenyl free radicals in the *C*-phenylation of phenols and enols. For example, when the reaction of tetraphenylacetone with tetraphenylbismuthonium tosylate and BTMG as a base was performed in the presence of a spin-trap, phenyl-*tert*-butylnitrone, (PBN), the phenyl free radical adduct (**101**) was detected by ESR. The same free radical adduct (**101**) was seen in the absence of tetraphenylacetone.[28] (Scheme 6.17)

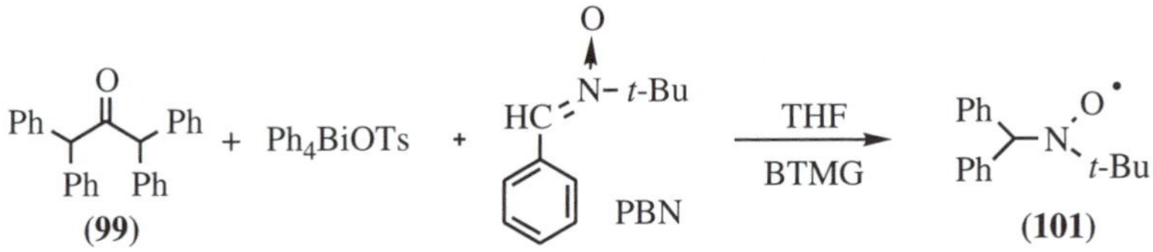

Blank experiments:

> THF + PBN
> Tetraphenylacetone + PBN in THF
> Ph₄BiOTs + PBN in THF ⟩ ⟶ No ESR Signal
> BTMG + PBN in THF

but:
Ph₄BiOTs + BTMG + PBN in THF ⟶ ESR Signal of (**101**)

Scheme 6.17: Qualitative ESR observations

To check the real weight of the free radical pathway in the overall arylation reactions mechanisms, quantitative chemical trapping experiments were then performed on a variety of arylation reaction systems. When nitrosobenzene was used as a trapping agent for phenyl free radicals, the yields of the *C*-arylation products were not affected. After reductive acetylation to *N,N*-diphenyl-acetamide, the measured amount of diphenylnitroxide was never exceeding a few percent. (Scheme 6.18) A second stoichiometric free-radical trapping agent was also used, 1,1-diphenylethylene (**72**)

(DPE). *C*- and *O*-phenylation reactions were not affected by its addition to the reaction medium. This stoichiometric chemical trapping method was also used in the study of other related reactions such as the *C*-phenylation of nitroalkanes and indoles, and in the thermal decomposition of (4-nitrophenoxy) tetraphenylbismuth derivatives. No free radicals were involved although, again, qualitative ESR studies had revealed their presence.[28] (Scheme 6.19)

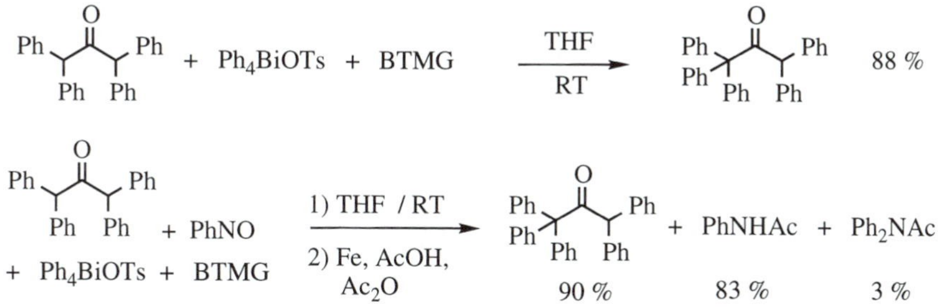

Scheme 6.18: Quantitative trapping with nitrosobenzene, PhNO

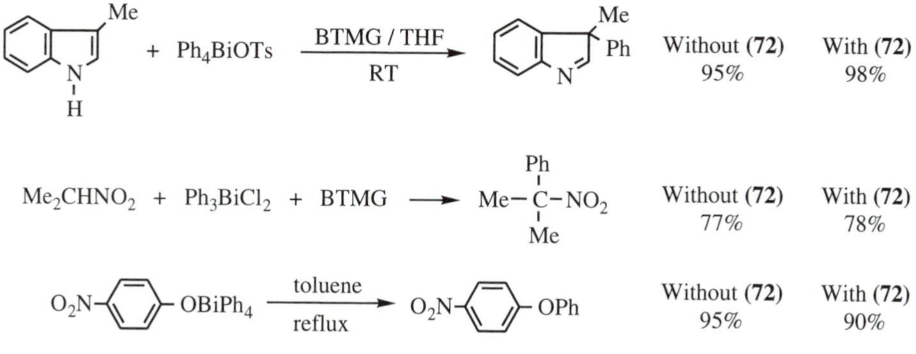

Scheme 6.19: Influence of the addition of 1,1-diphenylethylene (**72**) (DPE) on the yield of arylated product in arylation reactions

Therefore, the observation of ESR signals due to phenyl free-radicals resulted from a minor competing decomposition pathway of the bismuth reagent, and the arylation reaction itself does not proceed by a free-radical pathway.

Similarly, phenyl radicals were detected by ESR in the *O*-phenylation of *tert*-butyl alcohol. However, no chemical trapping has been performed.[52] A one-electron transfer was invoked in the reaction of tetraphenylbismuthonium tetrafluoroborate with *N,N,N',N'*-tetramethyl-*para*-phenylene diamine.[21] Oxidative dimerisation and oxidation of sterically hindered 2,6-disubstituted phenolic substrates by triphenylbismuth dichloride and triphenylbismuth carbonate in the presence of BTMG are likely to occur *via* aryloxyl radicals generated by electron transfer between BTMG and the bismuth reagent followed by reaction of the intermediate radical species with the phenol.[32] Dimerisation of 2,6-dimethylphenol by reaction with triphenylbismuth carbonate was shown to be purely ionic with formation of a covalent bismuth-phenol intermediate.

The possibility of a cage-mechanism involving a biradical intermediate (102) was excluded on the basis of the reaction of pentaphenylbismuth with a phenol bearing an internal radical trap, which did not afford the expected product from an internal trapping of free radicals.[4] (Scheme 6.20)

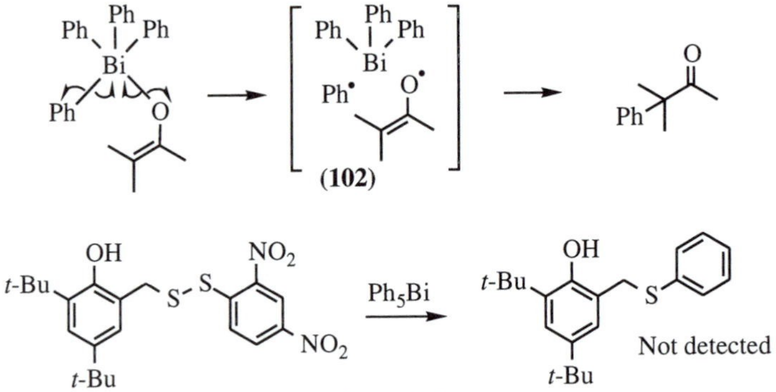

Scheme 6.20

6.7 COPPER-CATALYSED REACTIONS

The observation of a catalytic effect of copper salts on the decomposition of triphenylbismuth diacylates[53] led to the development of a copper-catalysed arylation reaction by triarylbismuth derivatives in the presence of copper or its metallic salts.

6.7.1 *O*-Arylation

The *O*-arylation of hydroxyl groups is obtained by reaction either with pentavalent organobismuth and catalytic copper or by reaction with trivalent organobismuth and stoichiometric copper diacetate, acting as an oxidant.

6.7.1.1 Triarylbismuth diacetate - copper catalysed arylation

The arylation of lower aliphatic alcohols was observed when the copper-catalysed decomposition of triarylbismuth diacetate was carried out in simple alcohols, used as solvent (20 mL per mmole of bismuth reagent). The yields of alkyl aryl ethers (based on the bismuth reagent) ranged from 60 to 95% for primary and secondary alcohols, but only 9% were obtained in the case of *tert*-butyl alcohol.[83-85] Under stoichiometric conditions, the *O*-phenylation of 3-β-cholestanol by triphenyl-bismuth diacetate (1 equiv.) was not significantly improved upon addition of copper diacetate.[55] Different copper compounds, such as $Cu(OAc)_2$, $CuCl_2$, CuCl or metallic copper, can be used as effective catalysts.

$$ROH + Ph_3Bi(OAc)_2 + CuX_n \longrightarrow R\text{-}O\text{-}Ph$$

The mono-*O*-phenylation reaction of glycols discovered by David and Thieffry presents some characteristic features which are not compatible with the postulated covalent bismuth intermediate.[56] (see section 6.3.2) The reaction is solvent-selective (methylene dichloride under reflux), light-catalysed (no reaction in the dark), and presents an induction period (2 hours in the case of the phenylation of 2,2-dimethylpropan-1,3-diol). Addition of small amounts of copper diacetate had a marked effect on the reaction. The reaction became fast: reaction times can be as short as 15 minutes. Moreover, the copper-catalysed reaction is no more solvent selective nor light-catalysed.[55]

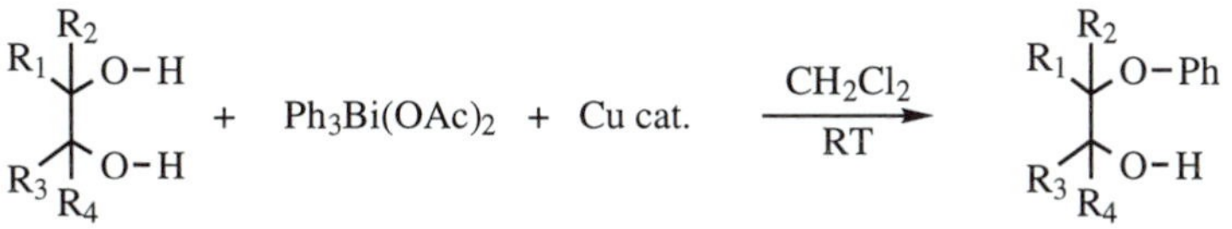

Optical inductions were observed when the glycol O-phenylation reaction was performed in the presence of a chiral pyridinyloxazoline ligand. Enantiomeric excesses up to 50% were reached in the presence of the ligand (103), [R^4 = CHMe$_2$ and R^1, R^2, R^3 = H). However, the yields were lower under these conditions: 35-45% instead of 87% in absence of catalyst for the O-phenylation of *cis*-cyclohexane-1,2-diol.[86-88]

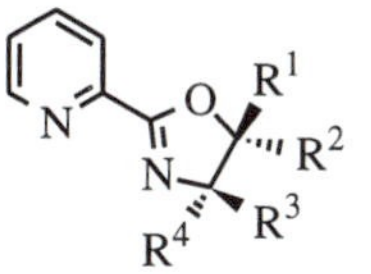

The reaction is not limited to phenylation, as the analogous arylation of hydroxylic substrates was performed with various substituted triarylbismuth diacylates in the case of the 3-hydroxymethyl benzofuran derivative.[89]

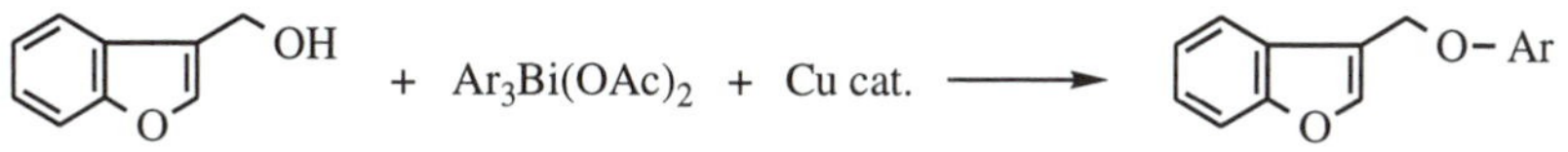

$$Ar = 3\text{-}NO_2C_6H_4, \ 4\text{-}CH_3C_6H_4, \ 4\text{-}CH_3OC_6H_4$$

The copper-catalysed arylation reaction was recently applied to the selective functionalisation of relatively complex natural products, such as the immunosuppressive macrolides ascomycine, (104), $R_{(31)}$ = Me and L-683,742, (104), $R_{(31)}$ = H.[90,91] A variety of triarylbismuth derivatives substituted with electron-donating or with electron-withdrawing groups has been prepared and used in the O-arylation reaction. With ascomycine, yields of the C-32 O-aryl ethers (105) ranged from modest to relatively good. The groups introduced on the C-32 hydroxyl included: 3,5-bis(trifluoromethyl)phenyl, 46%; 4-trifluoromethylphenyl, 51%; 4-tolyl, 42%; 4-anisyl, 18%; 4-dimethylaminophenyl, 17%; 4-(*tert*-butyldimethylsilyloxy)phenyl, 65%. The 5-indolyl and N-methyl-5-indolyl groups were also introduced in 10 and 50% respectively. In the case of L-683,742, a 1:1 mixture of the C-31 and C-32 O-aryl ethers was obtained with all the bismuth derivatives.

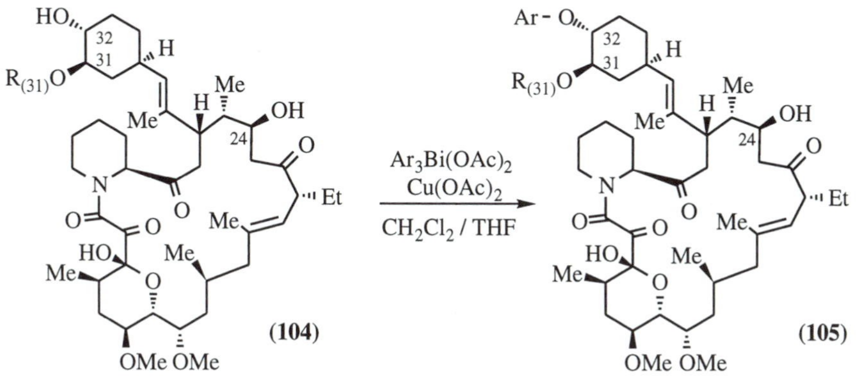

This copper-catalysed phenylation reaction with triphenylbismuth diacetate is not limited to glycols as it has been extended to the arylation of other hydroxylic substrates such as phenols and enols.[27] Variously substituted phenolic substrates have been selectively O-phenylated under very mild neutral conditions (1 to 24 hours at room temperature in methylene dichloride). The best yields were obtained with metallic copper used as the catalyst. The electronic nature of the substituents of the phenol did not influence the yields (4-nitro, 97%; 3,5-dimethoxy, 90%). Only steric hindrance of the 2- and 6-substituents interfered with the reactivity. Thus only 26% of the O-phenyl ether was obtained in the phenylation of 2,4-di-*tert*-butylphenol, and no phenylation took place in the case of 2,4,6-tri-*tert*-butylphenol.[27,84,92,93]

$$ArOH + Ph_3Bi(OAc)_2 + Cu\ cat. \xrightarrow[RT]{CH_2Cl_2} Ar\text{-}O\text{-}Ph$$

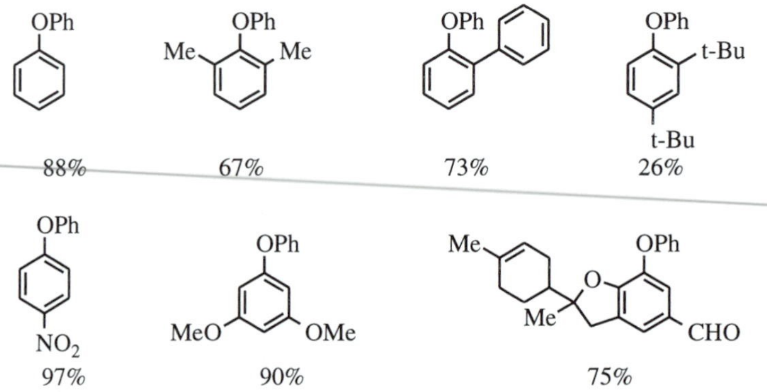

88%	67%	73%	26%

97%	90%	75%

Optically active (S)-6,6'-diphenoxy-2,2'-biphenyldiols (**107**) were obtained by diphenylation of a menthyl-protected 2,2',6,6'-biphenyltetrol (**106**) by treatment with an excess of triphenylbismuth diacetate (4 molar equiv.) in benzene in the presence of metallic copper. The 2,2'-diphenoxy derivative was obtained in 48% yield and quantitatively deprotected to give (**107**).[94]

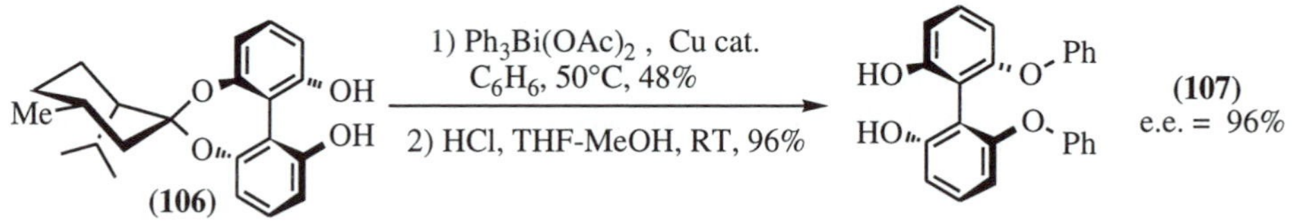

The O-phenylation reaction of enols by tetraphenylbismuthonium trifluoroacetate (**6**) was marginally improved by copper catalysis. Enolised diketones are O-phenylated by triphenylbismuth diacylates in moderate to good yields. With the α-diketone (**108**), the best yields were obtained with triphenylbismuth bis(trifluoroacetate) in the presence of metallic copper. β-Diketones reacted only with triphenylbismuth bis(trifluoroacetate) in the presence of a copper catalyst.[27]

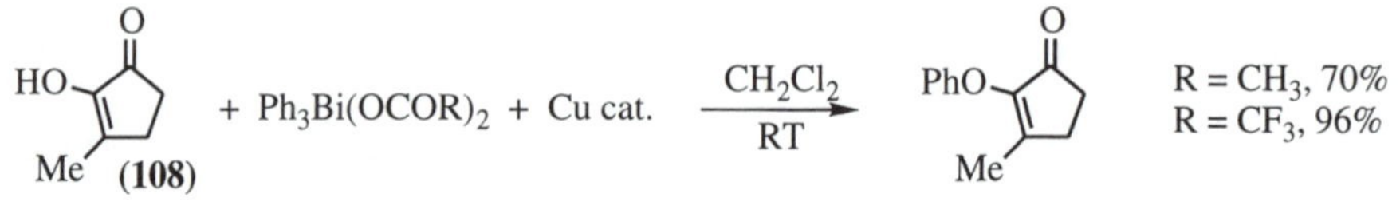

$$R = CH_3,\ 70\%$$
$$R = CF_3,\ 96\%$$

6.7.1.2 Triarylbismuthane - copper diacetate arylation

Although there is no reaction between triphenylbismuthane and copper diacetate,[95] triphenyl-bismuthane can transfer a phenyl group to alcohols and phenols when a stoichiometric amount of copper diacylate is used. When primary or secondary alcohols, used in large excess, were treated with triphenylbismuthane in the presence of copper diacetate in the ratio $Ph_3Bi : Cu(OAc)_2 = 1 : 2$, without solvent in sealed ampoules, the *O*-phenyl ethers were formed in 43-91% yields (based on the bismuth reagent), at a very slow rate (several days at room temperature). No reaction with phenol was described under these conditions.[96-98]

$$ROH + Ph_3Bi + Cu(OAc)_2 \xrightarrow{\text{neat}} R\text{-}O\text{-}Ph$$

When a solution of 3,5-di-*tert*-butylphenol in methylene dichloride was treated with triphenylbismuthane (1.2 molar equiv.) in the presence of various amounts of copper diacetate (up to 2 equiv.), no reaction was observed. However, when triethylamine (6 molar equiv.) was added to the mixture of 3,5-di-*tert*-butylphenol, triphenylbismuthane (1.2 molar equiv.) and copper diacetate (2 molar equiv.), the *O*-phenyl ether was now obtained in a moderate 44% yield, after 10 hours at room temperature.

These reactions of *O*-phenylation are preferably performed with the copper catalysed-triphenylbismuth diacetate system which is more general and gives higher yields, particularly in the case of phenol *O*-phenylation.[95]

6.7.2 *N*-Arylation

The *N*-arylation of different types of amino groups, aliphatic, heterocyclic or aromatic, is obtained by treatment either with pentavalent organobismuth and catalytic copper or by reaction with trivalent organobismuth and stoichiometric copper diacetate, acting as an oxidant.

6.7.2.1 Triarylbismuth diacetate - copper catalysed arylation

The reaction of simple aliphatic amines with triphenylbismuth diacetate and copper diacetate (the ratio used being: 5-20 : 1 : 0.01-0.02) in tetrahydrofuran led to good yields of the derived arylamine (60-85% based on the bismuth reagent). In the case of the phenylation of *N*,*N*-diphenylamine, a very poor yield was (< 3%) of triphenylamine was realized.[84,98,99]

$$RR'NH + Ph_3Bi(OAc)_2 + Cu(OAc)_2 \xrightarrow[60-180 \text{ h}]{\text{THF}} RR'N\text{-}Ph$$

10-20 equiv.   1 equiv.

R = H, R' = *i*-Pr, *i*-Bu, *t*-Bu, Ph

R, R' = Et or Bu

However, when the reaction is performed in methylene dichloride in the presence of catalytic amounts of metallic copper, the *N*-monophenylated amine derivatives are obtained in preparatively useful yields. Phenylation of aliphatic and aromatic amines with triphenylbismuth diacetate and metallic copper (ratio : 1 : 1.1 : 0.1) in methylene dichloride at room temperature led to good to high yields of the derived anilines (up to 96% based on the amine).[100] In the case of less reactive substrates, use of copper diacetate instead of metallic copper as catalyst constitutes a more reactive system which allows the preparation of the *N*-phenyl derivatives in moderate yields. These yields can also be improved by performing the reaction at a higher temperature.[101]

Primary amines:

$$RNH_2 + Ph_3Bi(OAc)_2 + Cu\ cat. \xrightarrow{CH_2Cl_2} RNH\text{-}Ph + RNPh_2$$

Amine	N-Phenyl (%)	N-Diphenyl (%)	Reaction conditions
n-Butylamine	60	20	$Ph_3Bi(OAc)_2$ (1.1eq.), Cu, RT, 4 h
n-Butylamine	-	70	$Ph_3Bi(OAc)_2$ (2.2eq.), Cu, RT, 3 h
t-Butylamine	15		$Ph_3Bi(OAc)_2$ (1.1eq.), $Cu(OAc)_2$, RT, 24 h
t-Butylamine	50		$Ph_3Bi(OAc)_2$ (1.1eq.), $Cu(OAc)_2$, reflux, 24 h
Cyclohexylamine	90		$Ph_3Bi(OAc)_2$ (1.1eq.), Cu, RT, 4 h

Secondary amines:

$$RR'NH + Ph_3Bi(OAc)_2 + Cu\ cat. \xrightarrow{CH_2Cl_2} RR'N\text{-}Ph$$

Amine	N-Phenyl (%)	Reaction conditions
Piperidine	58	$Ph_3Bi(OAc)_2$ (1.1eq.), Cu, RT, 4 h
Diethylamine	22	$Ph_3Bi(OAc)_2$ (1.1eq.), $Cu(OAc)_2$, RT, 24 h
Diethylamine	42	$Ph_3Bi(OAc)_2$ (1.1eq.), $Cu(OAc)_2$, reflux, 24 h
Ph_2N-NHPh	10, (Ph_3N)	$Ph_3Bi(OAc)_2$ (1.1eq.), $Cu(OAc)_2$, RT, 24 h

Anilines:

$$ArNH_2 + Ph_3Bi(OAc)_2 + Cu \xrightarrow[RT]{CH_2Cl_2} ArNH\text{-}Ph$$

Aniline, Ar =	N-Phenyl (%)	Reaction conditions
Phenyl	96	$Ph_3Bi(OAc)_2$ (1.1eq.), 2 h
p-Tolyl	97	$Ph_3Bi(OAc)_2$ (1eq.), 0.75 h
p-Anisyl	91	$Ph_3Bi(OAc)_2$ (1eq.), 0.25 h
p-Nitrophenyl	90	$Ph_3Bi(OAc)_2$ (2.2eq.), 16 h
Mesityl	92	$Ph_3Bi(OAc)_2$ (2.2eq.), 24 h

N-Phenylanilines:

$$ArNH\text{-}Ph + Ph_3Bi(OAc)_2 + Cu \xrightarrow[RT]{CH_2Cl_2} ArNPh_2$$

Aniline, Ar =	N-Phenyl (%)	Reaction conditions
Phenyl	23	$Ph_3Bi(OAc)_2$ (1.1eq.), 48 h
p-Anisyl	78	$Ph_3Bi(OAc)_2$ (2.2eq.), 72 h

In the phenylation reaction of substituted anilines, the electronic nature of the substituents plays a role only on the reaction rate, not on the overall yield (4-methoxyphenyl: 91% after 15 minutes and 4-nitrophenyl: 90% after 16 hours). The steric hindrance is also a factor which influences the reaction in slowing down the reaction rate. To obtain good to excellent yields, an excess of bismuth reagent is then required with hindered substrates. For example, mesitylamine required 2.2 equivalents of triphenylbismuth diacetate to afford the N-phenyl derivative in 92% after 24 hours. Moreover, triphenylamine was obtained in 23% yield after 48 hours by arylation of N,N-diphenylamine. A variety of aliphatic, alicyclic, heterocyclic and aromatic amines as well as hydrazines were N-arylated by this system.[100] No reaction took place with α-amino acids but their esters were mono-N-phenylated under mild conditions.[102]

$$\underset{R}{\overset{H}{H_2N{+}COOR'}} + Ph_3Bi(OAc)_2 + Cu \xrightarrow[RT]{CH_2Cl_2} \underset{R}{\overset{H}{PhNH{+}COOR'}}$$

R = H, R' = Et : 81%; R = Bzl , R' = Bzl : 80%; R = BzlO$_2$CCH$_2$, R' = Bzl : 50%;
R = BzlO$_2$CCH$_2$, R' = Bzl : 50%; R = BzlO$_2$C(CH$_2$)$_2$, R' = Bzl : 58%;
R = 3-Indolylmethyl, R' = Me : 66%; Methyl prolinate : 61%

The most conveniently prepared reagent is triphenylbismuth diacetate, but it is less reactive than triphenylbismuth bis(trifluoroacetate). Faster and higher yielding reactions were realized with the bis(trifluoroacetate) derivative in the presence of metallic copper as catalyst. Indeed, after 45 minutes, mesitylamine gave 40% of *N*-phenylmesitylamine with the diacetate, but 95% with the bis(trifluoro-acetate). Similarly, 4-nitroaniline gave 26% of 4-nitrodiphenylamine with the diacetate, and 98% with the bis(trifluoroacetate).[100] This reagent was the only one to be efficient for the *N*- and *C*-3 phenylation of indole derivatives. Preferential *C*-3 phenylation took place in high yields when the *C*-3 carbon of the substrate was not substituted. In this case even *N*,2-dimethylindole was phenylated in high yield (84%). With *C*-3 substituted indoles, modest yields of the *N*-phenyl derivatives were obtained (15-30%).[103]

C-Phenylation

50%   60-95%   84-94%

N-Phenylation

30%   21%   58-84%   15%

6.7.2.2 Triarylbismuthane - copper diacetate arylation

Triphenylbismuthane acts also as a phenylating agent towards a variety of amines in the presence of a stoichiometric amount of copper diacetate. Amines reacted smoothly with triphenylbismuthane (1.2 molar equiv.) and copper diacetate (0.5 molar equiv.) to give high yields of *N*-mono or *N*,*N*-diphenylated amines. The reaction was usually performed by stirring the mixture in methylene dichloride at room temperature under inert atmosphere for 18 to 24 hours. Generally monophenylated compounds were obtained with primary amines, in poor to excellent yields depending upon the basicity and steric hindrance of the substrate (6% for 4-nitroaniline, 25% for mesitylamine, but 82% for 4-methoxyaniline). With *n*-butylamine, the *N*-monophenyl (60%) and *N*,*N*-diphenyl (38%) derivatives were obtained.[95]

$$RNH_2 + Ph_3Bi + Cu(OAc)_2 \xrightarrow[RT]{CH_2Cl_2} RNH{-}Ph$$

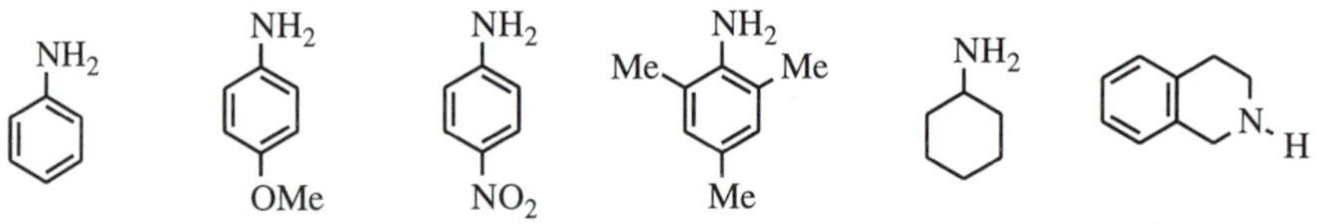

| 48% | 82% | 6% | 25% | 76% | 90% |

The poor yield of the phenylation reaction of 4-nitroaniline became an excellent yield upon addition of triethylamine. When only 0.1 equivalent of triethylamine was used, the diarylamine was obtained in 94% yield.[104]

A variety of substituted *N*-arylpiperidines **(110)** were obtained by arylation of 4-oxopiperidines **(109)** with various tris(3-substituted phenyl)bismuthane and copper diacetate.[105]

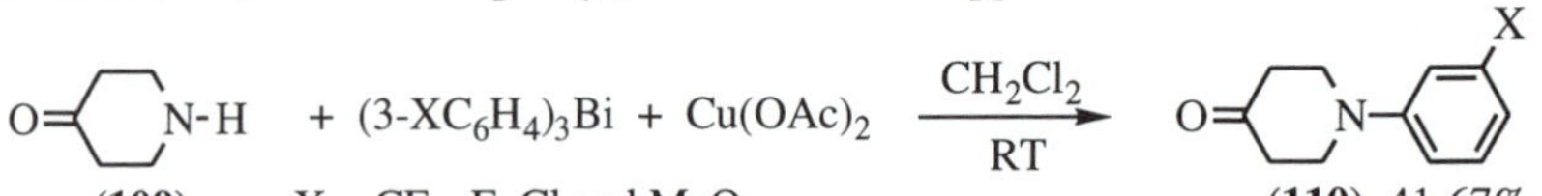

A number of amides, imides, ureas, carbamates and sulfonamides have also been efficiently arylated by the triphenylbismuthane-copper diacetate system, when a tertiary amine promoter, such as triethylamine or pyridine, was added.[104]

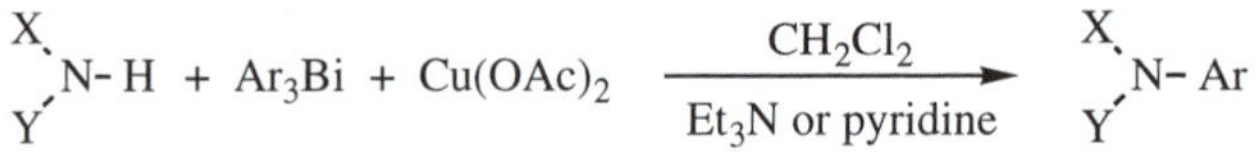

X, Y = COR, CO$_2$R, CONR$_2$, SO$_2$R, R, H (with R = alkyl or aryl)

Some examples: (the amine promoter being indicated with the yields).

| O
‖
Ph – C – N – Me
          H
Et$_3$N, 81% | Et$_3$N, 100% | N – H
Et$_3$N, 93% | AcO—⬠=O
Et$_3$N, 75% |

| N – H
pyridine, 99% | Me$_2$N⏜NHPh
Et$_3$N, 77% | pyridine, 64% | NH
pyridine, 82% |

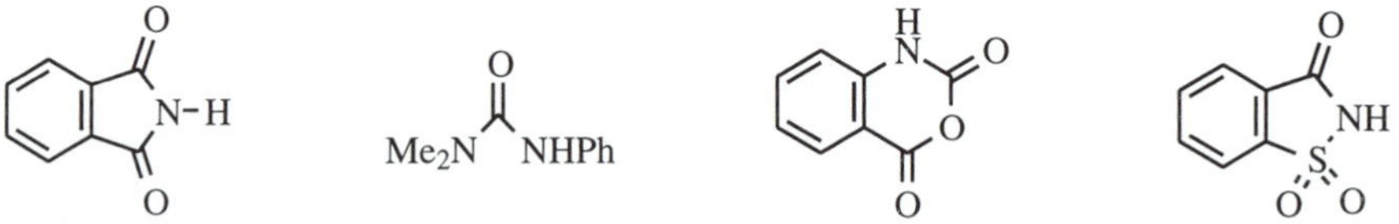

6.7.3 Alkylation reactions

Reaction of amines with alkyldiphenylbismuthane in the presence of a stoichiometric amount of copper diacetate afforded mixtures of the corresponding mono- and di-substituted amines, the alkyl group being first transferred, followed by the phenyl group. The transfer of two different substituents was explained by a disproportionation reaction of the alkyldiphenylbismuthane into trialkylbismuthane and triphenylbismuthane.[106]

$$RNH_2 + Ph_2BiR' + Cu(OAc)_2 \xrightarrow{CH_2Cl_2} R\text{-}NH\text{-}R' + R\text{-}N\underset{Ph}{\overset{R'}{\diagdown}}$$

(1 equiv.) (1.2 equiv.) (1 equiv.)

Modest to moderately good yields of the alkylated amines were obtained upon treatment with the symmetrical trialkylbismuth derivatives. With primary amines, trimethylbismuthane led preferentially to the disubstituted product, but tris(2-phenylethyl)bismuthane afforded only the product of monosubstitution.[106]

$$RNH_2 + Me_3Bi + Cu(OAc)_2 \xrightarrow{CH_2Cl_2} R\text{-}NMe_2 \qquad\qquad 31\text{-}51\%$$

$$RNH_2 + (PhCH_2CH_2)_3Bi + Cu(OAc)_2 \xrightarrow{CH_2Cl_2} R\text{-}NH\text{-}CH_2CH_2Ph \qquad 24\text{-}32\%$$

$$\underset{R'}{\overset{R}{\diagdown}}NH + Me_3Bi + Cu(OAc)_2 \xrightarrow{CH_2Cl_2} \underset{R'}{\overset{R}{\diagdown}}N\text{-}Me \qquad\qquad 47\text{-}52\%$$

$$\underset{R'}{\overset{R}{\diagdown}}NH + (PhCH_2CH_2)_3Bi + Cu(OAc)_2 \xrightarrow{CH_2Cl_2} \underset{R'}{\overset{R}{\diagdown}}N\text{-}CH_2CH_2Ph \qquad 49\text{-}53\%$$

The reaction of tributylbismuthane or tribenzylbismuthane with alcohols or phenol in the presence of copper diacetate led to very poor yields of the alkylbutyl or alkylbenzyl ethers (10-20% based on the bismuth reagent).[107]

$$RO\text{-}H + Bu_3Bi + Cu(OAc)_2 \xrightarrow{CH_2Cl_2} R\text{-}O\text{-}Bu$$

6.7.4 Mechanism of the copper - catalysed reactions

Two different mechanisms have been postulated to explain the copper-catalysed arylation reactions. Barton *et al.* suggested a mechanism involving the formation of a complex between the substrate and the copper species.[95] Oxidative addition on this copper (I) species leads to a copper (III) intermediate, such as (111) in the case of glycols (Scheme 6.21) or (112) in the case of amines (Scheme 6.22). These copper (III) intermediates bear an aryl ligand together with the substrate in the coordination sphere. A reductive elimination then affords the various products, depending on the substrates. In the triphenylbismuthane reaction, copper diacetate acts first as an oxidant of bismuth (III) to bismuth (V) and then as a catalyst for the transfer of ligands, in a manner similar to the bismuth diacetate system. Various types of copper species are involved in all these pathways. In order for the oxidation of triphenylbismuthane to proceed, the presence of an amine ligand on the copper center is required.

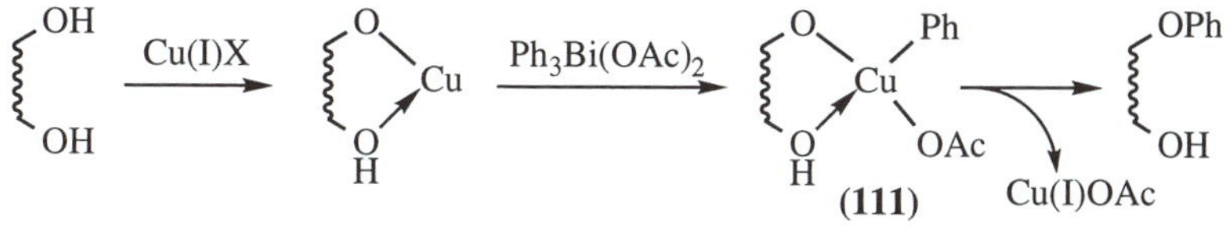

Scheme 6.21: Mechanism of glycol *O*-phenylation

In the case of the arylation of phenols or amides, no or very low-yielding arylation reactions took place in the absence of a tertiary amine. The mechanism of these two systems does not involve free radicals, as addition of 1,1-diphenylethylene **(72)** did not affect the yield of the *N*-arylation product in the triphenylbismuth diacetate-catalytic copper system as well as in the triphenylbismuthane-stoichiometric copper diacetate system.

Scheme 6.22: Mechanisms of amine *N*-phenylation as suggested by Barton *et al.*[95]

A- Triphenylbismuth diacetate and catalytic amount of a copper salt

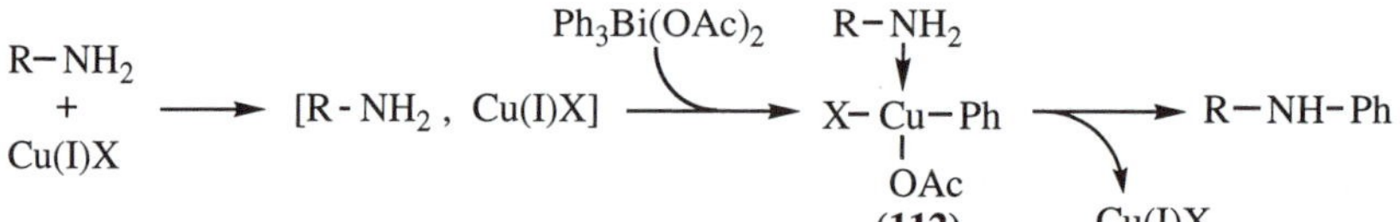

$$
\begin{array}{c}
R-NH_2 \\
+ \\
Cu(I)X
\end{array}
\longrightarrow
[R-NH_2,\ Cu(I)X]
\xrightarrow{\ Ph_3Bi(OAc)_2\ }
\underset{\substack{\big| \\ OAc \\ (112)}}{X-\underset{\big|}{Cu}-Ph}
\xrightarrow{\substack{R-NH_2}}
R-NH-Ph
$$

B- Triphenylbismuthane and stoichiometric amount of copper diacetate

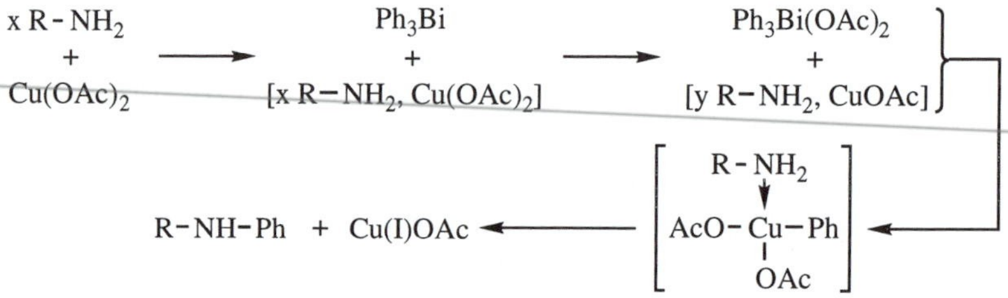

$$
\begin{array}{c}
x\,R-NH_2 \\
+ \\
Cu(OAc)_2
\end{array}
\longrightarrow
[x\,R-NH_2,\ Cu(OAc)_2]
\xrightarrow{\ Ph_3Bi\ }
\left.\begin{array}{c}
Ph_3Bi(OAc)_2 \\
+ \\
[y\,R-NH_2,\ CuOAc]
\end{array}\right\}
$$

$$
R-NH-Ph\ +\ Cu(I)OAc \longleftarrow
\left[\,AcO-\underset{\substack{\big| \\ OAc}}{Cu}-Ph\,\right]
\longleftarrow
$$

When the copper-catalysed reactions (type A and type B) were performed in the cavity of an ESR spectrometer, the presence of phenyl radicals was detected by way of their spin adducts with 2-methyl-2-nitrosopropane and with 2,4,6-tribromonitrosobenzene.[98] In view of these observations, Dodonov *et al.* suggested that a free-radical mechanism was involved in both reaction types, A and B, and explained the formation of the hypervalent copper (III) intermediate. (Scheme 6.23) The copper (III) intermediate is formed by two consecutive one-electron oxido-reduction elementary steps. The copper (I) catalytic species is first oxidised to a copper(II) species which is then oxidised by a phenyl radical to the active copper (III) intermediate. This hypervalent species then undergoes a ligand coupling reaction with the substrate, either hydroxylic or an amino derivative. In the type B reaction, the *in situ* generated phenylcopper (III) diacetate reacts with the substrate to eventually afford the *O*- or the *N*-phenyl derivative.

Scheme 6.23: Mechanism of generation of the hypervalent copper species involving free radicals as suggested by Dodonov *et al.*[98]

A- Triphenylbismuth diacetate and catalytic amount of a copper salt

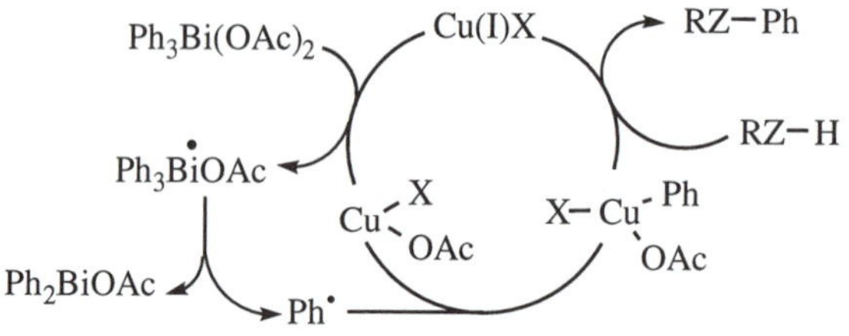

with
Z = O, NH or NR'

B- *Triphenylbismuthane and stoichiometric amount of copper diacetate*

1 - Ph$_3$Bi + Cu(OAc)$_2$ $\longrightarrow$ Ph$_2$BiOAc + PhCuOAc $\longrightarrow$ Ph$^•$ + CuOAc

2 - Ph$^•$ + Cu(OAc)$_2$ $\longrightarrow$ PhCu(OAc)$_2$

6.8 REFERENCES

1. Wieber, M. *Gmelin Handbuch der Anorganische Chemie*; Springer-Verlag: Berlin, **1977**; Band 47, Bismut-Organische Verbindungen.
2. Freedman, L.D.; Doak, G.O. *Chem. Rev.* **1982**, *82*, 15-57.
3. Barton, D.H.R.; Finet, J.-P. *Pure Appl. Chem.* **1987**, *59*, 937-946.
4. Abramovitch, R.A.; Barton, D.H.R.; Finet, J.-P. *Tetrahedron* **1988**, *44*, 3039-3071.
5. Finet, J.-P. *Chem. Rev.* **1989**, *89*, 1487-1501.
6. Freedman, L.D.; Doak, G.O. in *The Chemistry of the Metal-Carbon Bond*; Hartley, F.R., Ed.; Vol. 5; J. Wiley & Sons: New York, **1989**; Ch. 9, pp. 397-433.
7. Dodonov, V.A.; Gushchin, A.V. *Russ. Chem. Bull.* **1993**, *42*, 1955-1959.
8. Matano, Y.; Suzuki, H. *Bull. Chem. Soc. Jpn.* **1996**, *69*, 2673-2681.
9. Kitchin, J.P. Bismuth Salt Oxidations, in *Organic Syntheses by Oxidation with Metal Compounds*; Mijs, W.J.; De Jonge, C.R.H.I., Eds.; Plenum Press: New York, **1986**; pp. 817-837.
10. Postel, M.; Duñach, E. *Coord. Chem. Rev.* **1996**, *155*, 127-144.
11. Combes, S.; Finet, J.-P. *Synth. Commun.* **1996**, *26*, 4569-4575.
12. Challenger, F.; Allpress, C.F. *J. Chem. Soc.* **1921**, *119*, 913-926.
13. Challenger, F.; Allpress, C.F. *Proc. Chem. Soc., London* **1914**, *30*, 292.
14. Challenger, F.; Wilkinson, J.F. *J. Chem. Soc.* **1922**, *121*, 91-104.
15. Challenger, F. *J. Chem. Soc.* **1914**, *105*, 2210-2218.
16. Challenger, F.; Wilkinson, J.F. *J. Chem. Soc.* **1924**, *125*, 854-864.
17. Lermontov, S.A.; Rakov, I.M.; Zefirov, N.S.; Stang, P.J. *Tetrahedron Lett.* **1996**, *37*, 4051-4054.
18. Bhattacharya, S.N.; Singh, M. *Indian J. Chem., Sect. A* **1978**, *16*, 778-781.
19. Goel, R.G.; Prasad, H.S. *J. Organomet. Chem.* **1973**, *50*, 129-134.
20. Wittig, G.; Clauss, K. *Liebigs Ann. Chem.* **1952**, *578*, 136-146.
21. Ptitsyna, O.A.; Gurskii, M.E.; Reutov, O.A. *Izv. Akad. Nauk SSSR, Ser. Khim.* **1973**, 229-230; *Chem. Abstr.* **1973**, *78*, 159764j.
22. Sharutin, V.V.; Ermoshkin, A.E. *Izv. Akad. Nauk SSSR, Ser. Khim.* **1987**, 2598-2599.
23. Barton, D.H.R.; Lester, D.J.; Motherwell, W.B.; Barros Papoula, M.T. *J. Chem. Soc., Chem. Commun.* **1980**, 246-247.
24. Barton, D.H.R.; Bhatnagar, N.Y.; Blazejewski, J.-C.; Charpiot, B.; Finet, J.-P.; Lester, D.J.; Motherwell, W.B.; Papoula, M.T.B.; Stanforth, S.P. *J. Chem. Soc., Perkin Trans. 1* **1985**, 2657-2665.
25. Barton, D.H.R.; Blazejewski, J.-C.; Charpiot, B.; Lester, D.J.; Motherwell, W.B.; Papoula, M.T.B. *J. Chem. Soc., Chem. Commun.* **1980**, 827-829.
26. Barton, D.H.R.; Bhatnagar, N.Y.; Finet, J.-P.; Motherwell, W.B. *Tetrahedron* **1986**, *42*, 3111-3122.
27. Barton, D.H.R.; Finet, J.-P.; Khamsi, J.; Pichon, C. *Tetrahedron Lett.* **1986**, *27*, 3619-3622.
28. Barton, D.H.R.; Finet, J.-P.; Giannotti, C.; Halley, F. *J. Chem. Soc., Perkin Trans. 1* **1987**, 241-249.
29. Charpiot, B. *PhD thesis*, Université de Paris-Sud, Orsay, 1983.

30. Barton, D.H.R.; Yadav-Bhatnagar, N.; Finet, J.-P.; Khamsi, J.; Motherwell, W.B.; Stanforth, S.P. *Tetrahedron* **1987**, *43*, 323-332.
31. Razuvaev, G.A.; Osanova, N.A.; Sharutin, V.V. *Dokl. Akad. Nauk SSSR* **1975**, *225*, 581-582; *Chem. Abstr.* **1976**, *84*, 105704v.
32. Barton, D.H.R.; Finet, J.-P.; Giannotti, C.; Halley, F. *Tetrahedron* **1988**, *44*, 4483-4494.
33. Barton, D.H.R.; Blazejewski, J.-C.; Charpiot, B.; Finet, J.-P.; Motherwell, W.B.; Papoula, M.T.B.; Stanforth, S.P. *J. Chem. Soc., Perkin Trans. 1* **1985**, 2667-2675.
34. Ogawa, T.; Murafuji, T.; Suzuki, H. *J. Chem. Soc., Chem. Commun.* **1989**, 1749-1751.
35. Yasui, M.; Kikuchi, T.; Iwasaki, F.; Suzuki, H.; Murafuji, T.; Ogawa, T. *J. Chem. Soc., Perkin Trans. 1* **1990**, 3367-3368.
36. Suzuki, H.; Murafuji, T.; Ogawa, T. *Chem. Lett.* **1988**, 847-848.
37. Glidewell, C.; Lloyd, D.; Metcalfe, S. *Synthesis* **1988**, 319-321.
38. Ogawa, T.; Murafuji, T.; Suzuki, H. *Chem. Lett.* **1988**, 849-852.
39. Akhtar, M.S.; Brouillette, W.J.; Waterhous, D.V. *J. Org. Chem.* **1990**, *55*, 5222-5225.
40. Barton, D.H.R.; Donnelly, D.M.X.; Finet, J.-P.; Stenson, P.H. *Tetrahedron* **1988**, *44*, 6387-6396.
41. Suzuki, H.; Murafuji, T.; Azuma, N. *J. Chem. Soc., Perkin Trans. 1* **1992**, 1593-1600.
42. Santhosh, K.C.; Balasubramanian, K.K. *J. Chem. Soc., Chem. Commun.* **1992**, 224-225.
43. Dittes, U.; Keppler, K.; Nuber, B. *Angew. Chem., Int. Ed. Engl.* **1996**, *35*, 67-68.
44. Barton, D.H.R.; Charpiot, B.; Ingold, K.U.; Johnston, L.J.; Motherwell, W.B.; Scaiano, J.C.; Stanforth, S.P. *J. Am. Chem. Soc.* **1985**, *107*, 3607-3611.
45. Lei, X.; Doubleday, C., Jr; Turro, N.J. *Tetrahedron Lett.* **1986**, *27*, 4671-4674.
46. O'Donnell, M.J.; Bennett, W.D.; Jacobsen, W.N.; Ma, Y.; Huffman, J.C. *Tetrahedron Lett.* **1989**, *30*, 3909-3912.
47. O'Donnell, M.J.; Bennett, W.D.; Bruder, W.A.; Jacobsen, W.N.; Knuth, K.; LeClef, B.; Polt, R.L.; Bordwell, F.G.; Mrozack, S.R.; Cripe, T.A. *J. Am. Chem. Soc.* **1988**, *110*, 8520-8525.
48. O'Donnell, M.J.; Bennett, W.D., Jacobsen, W.N.; Ma, Y. *Tetrahedron Lett.* **1989**, *30*, 3913-3914.
49. Lalonde, J.J.; Bergbreiter, D.E.; Wong, C.-H. *J. Org. Chem.* **1988**, *53*, 2323-2327.
50. Le Goff, E. *J. Am. Chem. Soc.* **1962**, *84*, 3786.
51. Barton, D.H.R.; Finet, J.-P.; Motherwell, W.B.; Pichon, C. *J. Chem. Soc., Perkin Trans. 1* **1987**, 251-259.
52. Dodonov, V.A.; Gushchin, A.V.; Grishin, D.F.; Brilkina, T.G. *Zh. Obshch. Khim.* **1984**, *54*, 100-103.
53. Dodonov, V.A.; Gushchin, A.V.; Brilkina, T.G. *Zh. Obshch. Khim.* **1985**, *55*, 73-80.
54. David, S.; Thieffry, A. *Tetrahedron Lett.* **1981**, *22*, 5063-5066.
55. Barton, D.H.R.; Finet, J.-P.; Pichon, C. *J. Chem. Soc., Chem. Commun.* **1986**, 65-66.
56. David, S.; Thieffry, A., *J. Org. Chem.*, **1983**, 48, 441-447.
57. Barton, D.H.R.; Blazejewski, J.-C.; Charpiot, B.; Lester, D.J.; Motherwell, W.B. *J. Chem. Soc., Chem. Commun.* **1981**, 503-504.
58. Walts, A.E.; Roush, W.R. *Tetrahedron* **1985**, *41*, 3463-3478.
59. Singhal, K.; Raj, P.; Jee, F. *Synth. React. Inorg. Met.-Org. Chem.* **1987**, *17*, 559-566.
60. Wallenhauer, S.; Seppelt, K. *Angew. Chem., Int. Ed. Eng.* **1994**, *33*, 976-978.
61. Matano, Y.; Azuma, N.; Suzuki, H. *J. Chem. Soc., Perkin Trans. 1* **1994**, 1739-1747.
62. Matano, Y.; Azuma, N.; Suzuki, H. *J. Chem. Soc., Perkin Trans. 1* **1995**, 2543-2549.
63. Matano, Y.; Yoshimune, M.; Suzuki, H. *Tetrahedron Lett.* **1995**, *36*, 7475-7478.
64. Matano, Y.; Yoshimune, M.; Azuma, N.; Suzuki, H. *J. Chem. Soc., Perkin Trans. 1* **1996**, 1971-1977.
65. Matano, Y. *J. Chem. Soc., Perkin Trans. 1* **1994**, 2703-2709.
66. Matano, Y.; Yoshimune, M.; Suzuki, H. *J. Org. Chem.* **1995**, *60*, 4663-4665.

67. Matano, Y.; Suzuki, H. *J. Chem. Soc., Chem. Commun.* **1996**, *69*, 2697-2698.
68. Barton, D.H.R.; Kitchin, J.P.; Lester, D.J.; Motherwell, W.B.; Papoula, M.T.B. *Tetrahedron* **1981**, *37*, Suppl. 1, 73-79.
69. Barton, D.H.R.; Charpiot, B.; Motherwell, W.B. *Tetrahedron Lett.* **1982**, *23*, 3365-3368.
70. Barton, D.H.R.; Barros Papoula, M.T.; Guilhem, J.; Motherwell, W.B.; Pascard, C.; Tran Huu Dau, E. *J. Chem. Soc., Chem. Commun.* **1982**, 732-734.
71. Halley, F. *PhD thesis*, Université de Paris-Sud, Orsay, December 1986.
72. Hellwinkel, D.; Kilthau, G. *Liebigs Ann. Chem.* **1967**, *705*, 66-75.
73. Hellwinkel, D.; Bach, U. *Liebigs Ann. Chem.* **1968**, *720*, 198-200.
74. Yamamoto, Y.; Ohdoi, K.; Chen, X.; Kitano, M.; Akiba, K.-y. *Organometallics* **1993**, *12*, 3297-3303.
75. Wallenhauer, S.; Leopold, D.; Seppelt, K. *Inorg. Chem.* **1993**, *32*, 3948-3951.
76. Domagala, M.; Preut, H.; Huber, F. *Acta Cryst., Ser. C* **1988**, *C44*, 830-832.
77. Ferguson, G.; Kaitner, B.; Glidewell, C.; Smith, S. *J. Organomet. Chem.* **1991**, *419*, 283-291.
78. Barton, D.H.R.; Charpiot, B.; Dau, E.T.H.; Motherwell, W.B.; Pascard, C.; Pichon, C. *Helv. Chim. Acta* **1984**, *67*, 586-599.
79. Faraglia, G.; Graziani, R.; Volponi, L.; Casellato, U. *J. Organomet. Chem.* **1983**, *253*, 317-327.
80. Goel, R.G.; Prasad, H.S. *J. Chem. Soc. A* **1971**, 562-563.
81. *Molecular Rearrangements*; de Mayo, P., Ed.; Interscience: New York, **1964**; Vol. 1, p. 22.
82. a) *Molecular Rearrangements*; de Mayo, P., Ed.; Interscience: New York, **1964**; Vol. 1, p. 429. b) Ruechardt, C. *Chem. Ber.* **1961**, *94*, 2609-2623.
83. Dodonov, V.A.; Gushchin, A.V.; Brilkina, T.G. *Zh. Obshch. Khim.* **1984**, *54*, 2157-2158.
84. Dodonov, V.A.; Gushchin, A.V.; Brilkina, T.G. *Zh. Obshch. Khim.* **1985**, *55*, 2514-2519.
85. Dodonov, V.A.; Starostina, T.I.; Kuznetsova, Yu.L.; Gushchin, A.V. *Iz. Akad. Nauk, Ser. Khim.* **1995**, 156-158; *Russ. Chem. Bull.* **1995**, *44*, 151-153.
86. Brunner, H.; Obermann, U.; Wimmer, P. *J. Organomet. Chem.* **1986**, *316*, C1-C3;
87. Brunner, H.; Obermann, U.; Wimmer, P. *Organometallics* **1989**, *8*, 821-826.
88. Brunner, H.; Chuard, T. *Monatsh. Chem.* **1994**, *125*, 1293-1300.
89. Busteed, R.M.M. *PhD Thesis*, University College Dublin, **1988**.
90. Sinclair, P.J.; Wong, F.; Wyvratt, M.; Staruch, M.J.; Dumont, F. *Bioorg. Med. Chem. Lett.* **1995**, 5, 1035-1038.
91. Sinclair, P.J.; Wong, F.; Staruch, M.J.; Wiederrecht, G.; Parsons, W.H.; Dumont, F.; Wyvratt, M. *Bioorg. Med. Chem. Lett.* **1996**, 6, 2193-2196.
92. Srivastava, R.P.; Zhu, X.; Walker, L.A.; Sindelar, R.D. *Bioorg. Med. Chem. Lett.* **1995**, 5, 1751-1755.
93. Srivastava, R.P.; Zhu, X.; Walker, L.A.; Sindelar, R.D. *Bioorg. Med. Chem. Lett.* **1995**, 5, 2429-2434.
94. Harada, T.; Ueda, S.; Yoshida, T.; Inoue, A.; Takeuchi, M.; Ogawa, N.; Oku, A.; Shiro, M. *J. Org. Chem.* **1994**, *59*, 7575-7576.
95. Barton, D.H.R.; Finet, J.-P.; Khamsi, J. *Tetrahedron Lett.* **1987**, *28*, 887-890.
96. Gushchin, A.V.; Brilkina, T.G.; Dodonov, V.A. *Zh. Obshch. Khim.* **1985**, *55*, 2630.
97. Dodonov, V.A.; Gushchin, A.V.; Brilkina, T.G.; Muratova, L.V. *Zh. Obshch. Khim.* **1986**, *56*, 2714-2721.
98. Dodonov, V.A.; Gushchin, A.V. *Metalloorg. Khim.* **1990**, *56*, 2714-2721.
99. Dodonov, V.A.; Gushchin, A.V.; Brilkina, T.G. *Zh. Obshch. Khim.* **1985**, *55*, 466-467.
100. Barton, D.H.R.; Finet, J.-P.; Khamsi, J. *Tetrahedron Lett.* **1986**, *27*, 3615-3618.
101. Khamsi, J. *PhD Thesis*, Université de Paris-Sud, Orsay, September 1987.
102. Barton, D.H.R.; Finet, J.-P.; Khamsi, J. *Tetrahedron Lett.* **1989**, *30*, 937-940.
103. Barton, D.H.R.; Finet, J.-P.; Khamsi, J. *Tetrahedron Lett.* **1988**, *29*, 1115-1118.
104. Chan, D.M.T. *Tetrahedron Lett.* **1996**, *37*, 9013-9016.

105. Banfi, A.; Bartoletti, M.; Bellora, E.; Bignotti, M.; Turconi, M. *Synthesis* **1994**, 775-776.
106. Barton, D.H.R.; Ozbalik, N.; Ramesh, M. *Tetrahedron Lett.* **1988**, *29*, 857-860.
107. Dodonov, V.A.; Starostina, T.I.; Belukhina, E.V.; Vorobeva, N.V. *Russ. Chem. Bull.* **1993**, *42*, 2023-2025.

Chapter 7

Ligand Coupling Involving Organolead Compounds

Since its discovery[1] and its first use as an oxidant,[2] lead tetraacetate has been widely used in organic chemistry. Depending on the nature of the substrate and the reaction conditions, it can react as a radical or ionic oxidant, or take part in substitution, elimination, addition and fragmentation reactions. A large number of papers and review articles have been devoted to its chemistry.[3-6] On the other hand, organolead compounds have also been known for a long time, in part due to the commercial importance of tetraethyllead.[7-8] However, it is only in the last twenty years that their interest as reagents for organic synthesis has been developed. Depending on the nature of the ligands, radical, anionic or cationic behaviours have been observed.[9-12] The occurrence of mechanisms of the ligand coupling type have been suggested for acetoxylation reactions with lead tetraacetate nearly forty years ago, and this mechanism is frequently implied in the chemistry of monoorganolead reagents of the type $RPbX_3$.

7.1 ACYLOXYLATION REACTIONS, A BRIEF OVERVIEW

The substitutive acyloxylation reaction can take place with a wide variety of substrates and different types of metal acylates can be used.[3] Among them, lead tetraacetate shows a wide range of possibilities. Depending upon the nature of the organic substrates, a number of mechanistic pathways have been invoked to rationalize the observations.[6] The formation of a covalent substrate-lead intermediate was suggested by Corey.[13] The ligand coupling mechanism was postulated by Henbest[14] and Marshall[15] to explain the α-acetoxylation of ketones. This ligand coupling mechanism is most frequently involved in the reactions of lead tetraacetate with compounds presenting either keto-enol or imine-enamine tautomerism.

7.1.1 Keto-enol tautomers

Compounds presenting the possibility of keto-enol tautomerism react easily with lead tetraacetate to afford products of α-acetoxylation. This reaction is found in the case of ketones and phenols, although it is frequently accompanied by other products derived from alternate mechanistic pathways.

7.1.1.1 Phenols

Oxidation of phenols with lead tetraacetate was extensively studied by the group of Wessely.[16-21] The nature of the products is dependent upon a number of factors, such as the nature and position of the substituents, the nature of the solvent and the ratio lead tetraacetate : substrate. In acetic acid, quinol acetates, *ortho*-quinone diacetates and quinones are formed. In non-polar solvents, mostly dimeric products are obtained.

It is now generally accepted that the first step involves the formation of aryloxy-lead (IV) triacetates (**1**). These can decompose homolytically, leading to dimeric products *via* the aryloxyl radicals (**2**). They can also decompose heterolytically to cationic aryloxy species, which are then trapped by external nucleophiles. This has been exemplified by the reaction of phenols with lead tetraacetate in the presence of acetic acid or methanol. In the latter case, *ortho*-methoxy derivatives were formed. However, the preferential formation of 6-acetoxy-2,4-dienone derivatives is more likely explained by an intramolecular ligand coupling reaction (**3**) rather than by occurrence of a cationic species.[22,23]

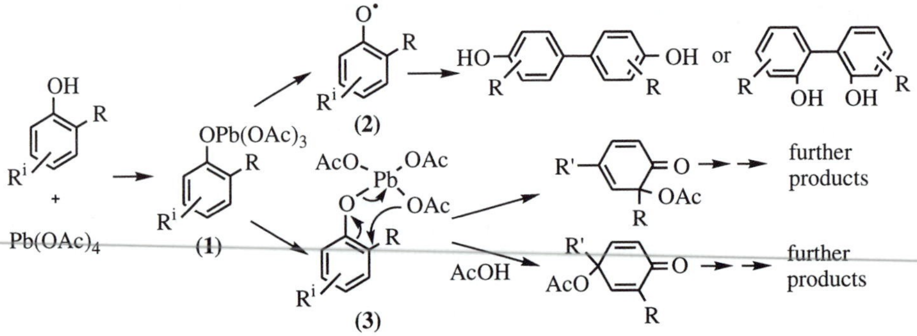

Scheme 7.1

Reaction of a 2,6-dimethylphenol (**4**) with lead tetraacetate in the presence of an excess of acrylic acid leads to *ortho*-quinol acrylates (**5**) which then can lead to tricyclic lactones (**6**) by intramolecular Diels-Alder cyclisation.[24,25]

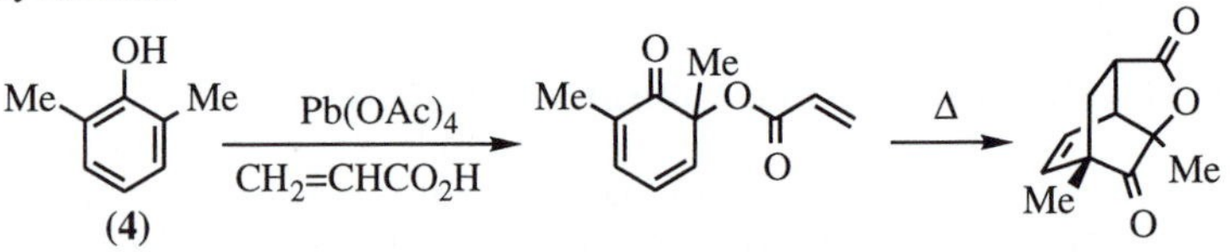

An elegant application of the oxidative coupling of phenolic compounds through reaction with lead tetraacetate was described by Feldman in the synthesis of ellagitannins by intramolecular coupling of galloyl esters, which could imply a ligand coupling mechanism in the crucial step.[26-30]

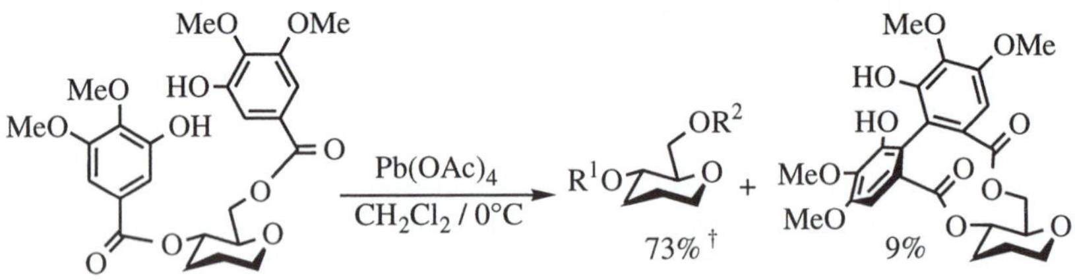

†: as a mixture of quinone ketals with R^1, R^2 =

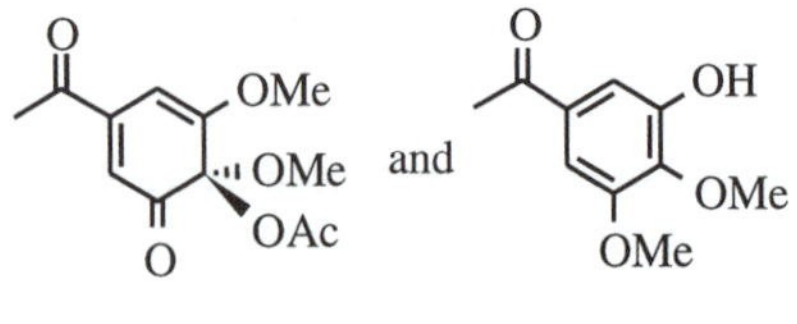

But, with a different phenol protecting group, the coupling product became the major one:

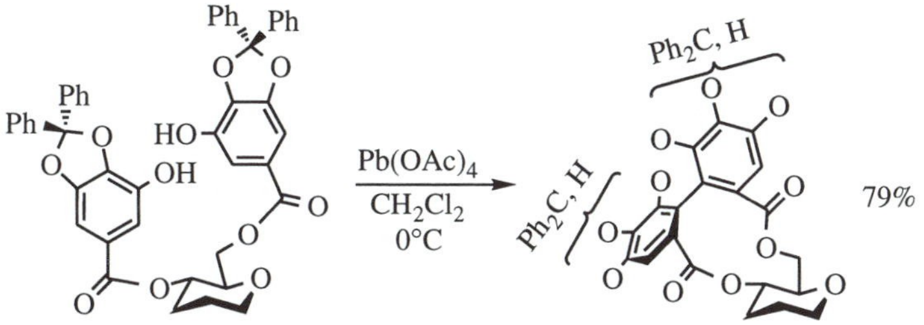

79%

7.1.1.2 Ketones, β-dicarbonyl derivatives and enol silyl ethers

Carbonyl compounds possessing at least one hydrogen on the α-carbon react with lead tetraacetate in benzene or in acetic acid to give the α-acetoxy derivatives in moderate yields in most cases.[3,4,6] Best yields are generally obtained with ketones. When two electron-withdrawing groups are attached to a methylene group, this substrate becomes very reactive and the reaction occurs even at room temperature. This is the case of β-dicarbonyl, β-ketoesters and malonic esters.[31] High reactivity is also observed with completely enolised carbonyl compounds, such as 2,2-diarylacetaldehyde, which gives the α-acetoxy derivative under very mild conditions.[32]

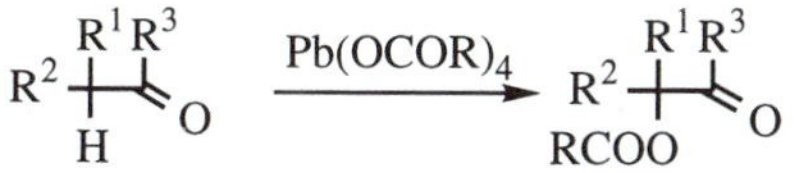

Two mechanisms have been suggested. A radical mechanism was first proposed and its involvement is supported by the presence of dimeric products.[31] However, the ligand coupling mechanism is now generally accepted.[13,14] An enol-lead (IV) triacetate intermediate (**7**) is first formed by reaction of lead tetraacetate with the enol form. Its formation is accelerated by catalysis by boron trifluoride.[14,33] Treatment of the preformed enolate with lead tetraacetate performs α-acetoxylation at lower temperature and more rapidly than in the reaction with the corresponding enol.[34] Ligand coupling then takes place on this intermediate to lead to the α-acetoxycarbonyl derivative.

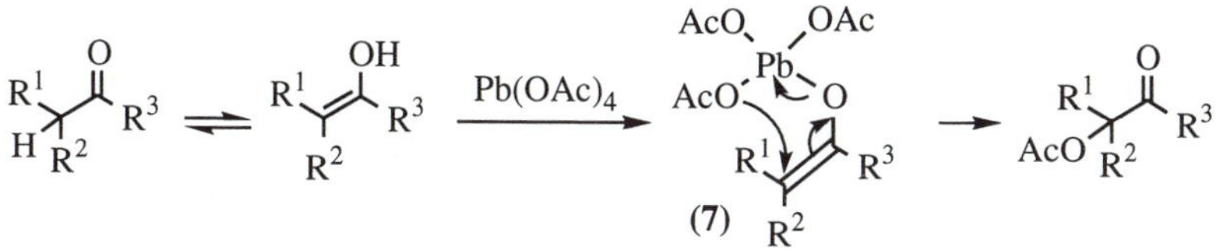

In the case of the reaction of α,β-unsaturated ketones with lead tetraacetate, the product of α'-oxidation is generally formed, as in the following example.[35]

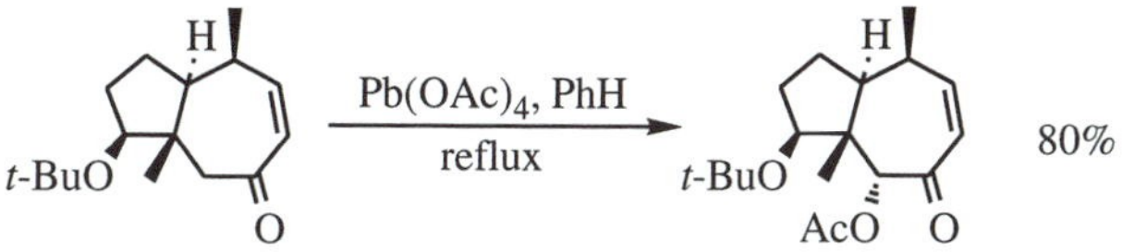

80%

α-Acetoxyketones have also been obtained by lead tetraacetate oxidation of enol acetates, the geometry of the enol acetate being generally transferred to the product.[36-38]

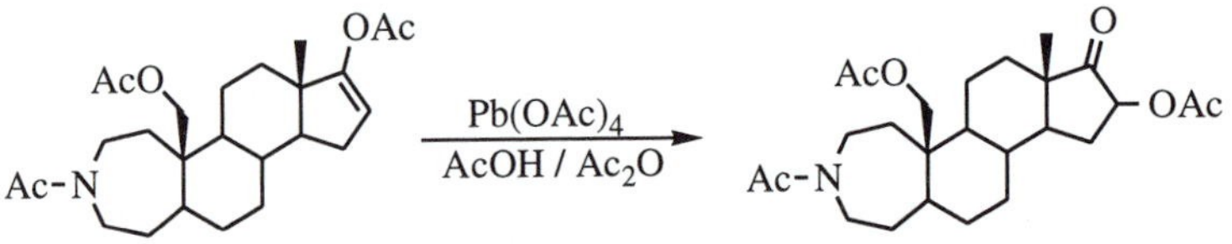

Trimethylsilyl enol ethers of acetophenone derivatives (**8**) react with lead tetraacetate in benzene at room temperature to give unstable phenacyllead triacetate intermediates (**9**), which decompose to the α-acetoxyacetophenones (**10**) in high yields (90-95%), when a 1:1 ratio of the enol ether to lead tetraacetate is used.[39] However when a 2:1 ratio is used and the reaction performed in CH$_2$Cl$_2$ or in THF at - 78°C, only small amounts of the α-acetoxyacetophenone derivatives (**11**) are formed, the main product being the 1,4-diketone dimers of acetophenone (40-60%).[40]

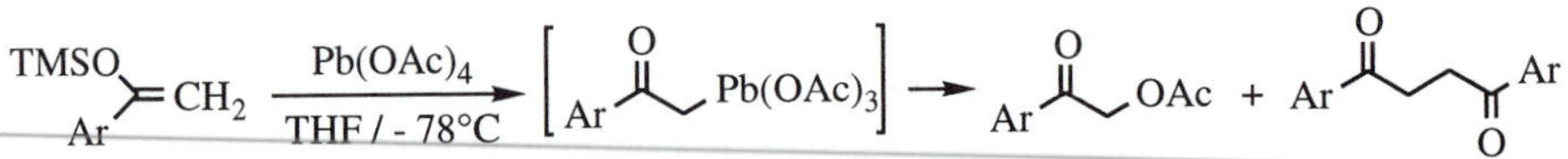

	(8)	(9)	(10)	(11)
1 equiv.			90-95%	
2 equiv.			< 10%	40-60%

Reaction of enol silyl ethers with lead tetrabenzoate followed by treatment with triethylammonium fluoride leads to the α-benzoyloxyketones.[41] In both cases, the sequence involves bisoxygenation of the double bond followed by hydrolysis. α-Acetoxylation is also possible, but the benzoate is the reagent of choice for that system.[41] In the cyclohexanone series, reaction of lead tetraacetate with cyclohexanone trimethylsilyl enol ether afforded also α-acetoxycyclohexanone after hydrolysis.[42] When the same reaction was performed on the triethyltin enol ether, the main product was now α-acetoxycyclohexanone with a small amount of dimeric product. This latter result could be explained by a transmetallation followed by ligand coupling, in the major pathway.

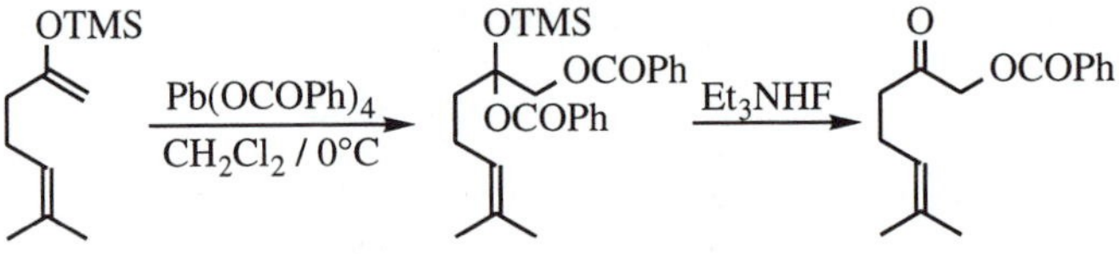

The stable α-metalloketones ArCOCH$_2$Tl(*p*-tolyl)(OCOCF$_3$) (**12**) or ArCOCH$_2$Pb(*p*-An)(OAc)$_2$ (**13**) are generated by reaction of the enol silyl ethers with respectively 4-MeC$_6$H$_4$Tl(OCOCF$_3$)$_2$ (**14**) and 4-MeOC$_6$H$_4$Pb(OAc)$_3$ (**15**).[43,44] They react with lead tetraacetate to afford good yields of the α-acetoxy derivatives ArCOCH$_2$OAc (**10**). The reaction was postulated to involve metal-metal exchange, followed by ligand coupling.[43]

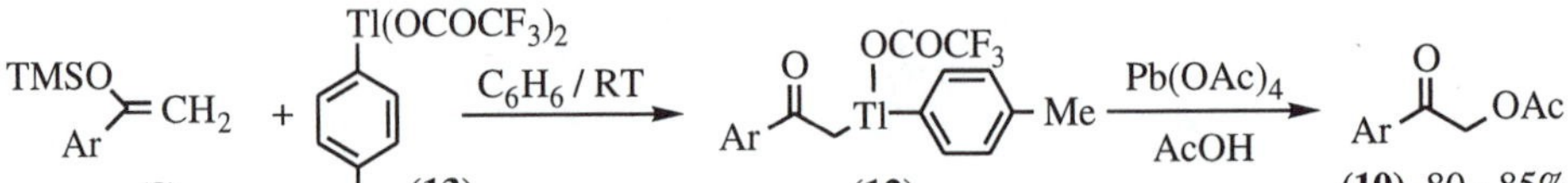

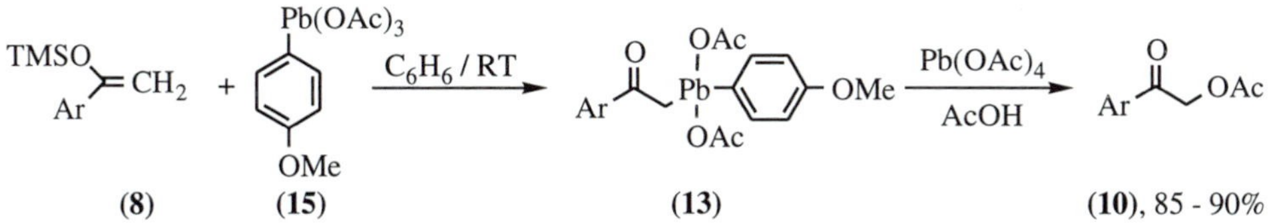

(8)   (15)   (13)   (10), 85 - 90%

7.1.2 Nitrogen derivatives: imine-enamine tautomers

Nitrogen containing compounds which are prone to imine-enamine tautomerism react with lead tetraacetate to give ligand coupling compounds with transfer of an acetyl group.[5]

7.1.2.1 Anilines

Aromatic primary amines react with lead tetraacetate to give symmetrical azo compounds in varying yields, via hydrazo intermediates.[45-48] However, in the case of 2,4,6-tri-tert-butylaniline, reaction with lead tetraacetate in benzene at 5°C led to a mixture of three products, the formation of which can be explained by a ligand coupling process.[49] (Scheme 7.2)

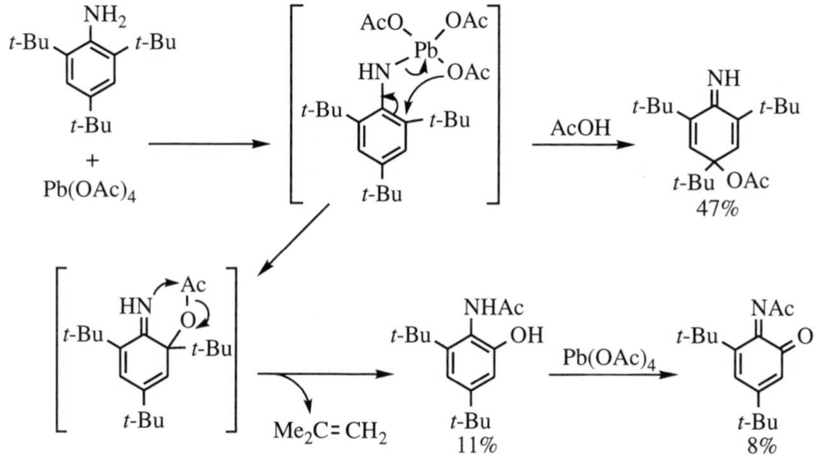

Scheme 7.2

In a synthetic approach towards mitomycin C, a similar reaction of α-oxidation was observed in the treatment of an aniline derivative with lead tetraacetate.[50]

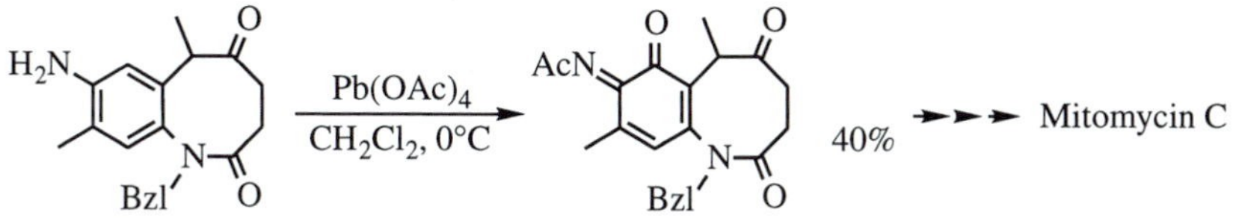

Reaction of *N*-sulfonylanilines with lead tetraacetate proceeds by ligand coupling to give the *ortho*-acetoxy *N*-sulfonylcyclohexadienonimine.[51-53]

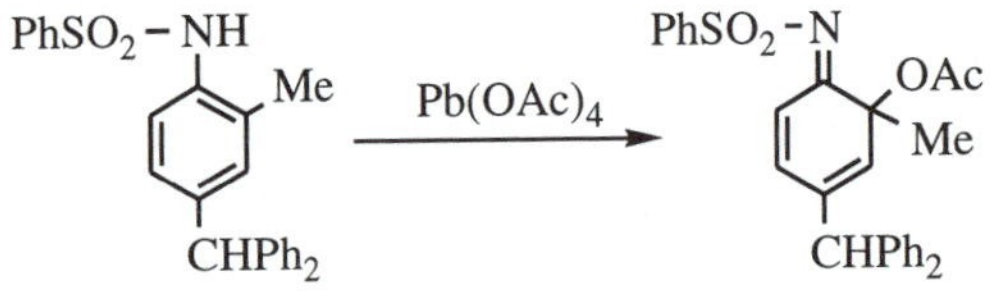

7.1.2.2 Enamines

Enamines having a tertiary nitrogen react with lead tetraacetate in benzene to afford complex mixtures in which products of ligand coupling are present.[54] Although a diacetoxy derivative was originally postulated as the key intermediate, a ligand coupling process was later invoked.[5] From a synthetic point of view, this reaction does not compete favorably with the reaction with thallium salts. Enamines react with thallium triacetate in acetic acid[55] to afford good yields of the α-acetoxyketones, and α-methoxy ketones are obtained by reaction with thallium trinitrate in methanol.[56,57] These thallium reactions involve intermolecular S_N2 reaction with the solvent, acetic acid or methanol.

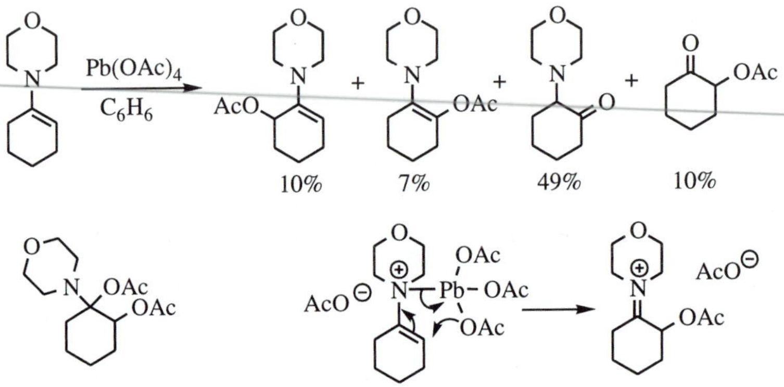

Scheme 7.3

In the case of enamines with a secondary nitrogen, a clean and high yielding ligand coupling took place in the case of the following steroidal enamide (16) affording the 17α-acetoxyenamide (17). A second reaction with lead tetraacetate led to the 17α,21-diacetoxy compound.[58,59]

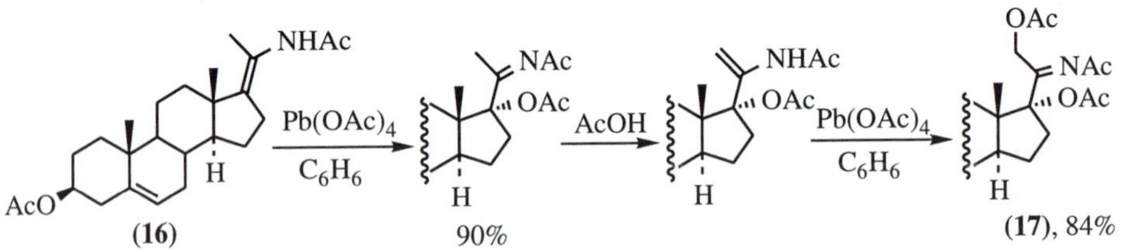

Scheme 7.4

Oxidation of *N*-alkyl or *N*-aryl enamines, with a secondary nitrogen, can lead to poor to modest yields of ligand coupling products or their products of imine hydrolysis.[60] Mixtures of dimeric products are obtained in the reactions of β-aminoacrylic acid derivatives.[61-64] Lead tetraacetate oxidation of α-oxoketene-S,N-phenyl or N-benzyl acetals [(**18**), R = Ph or PhCH$_2$] gave the iminoacetates and α-acetoxy-S,N-acetal by ligand coupling.[65]

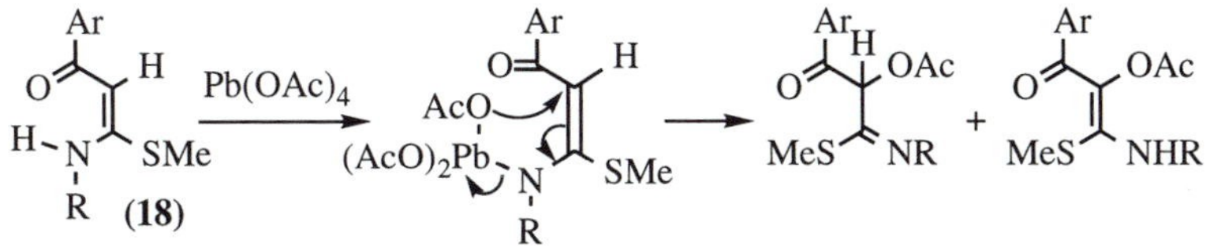

7.1.2.3 Nitroalkanes

Nitroalkanes react with lead tetraacetate to afford α-acetoxynitroalkanes, but no yields were reported. The reaction is accelerated by addition of base, but not affected by radical initiators.[66] These α-acetoxynitroalkanes can also be obtained by oxidation of the corresponding α-acetoxynitroso alkanes by hydrogen peroxide-sodium nitrite (see section 7.1.2.5).[67]

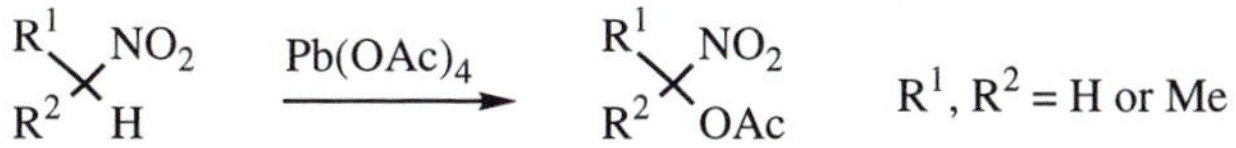

7.1.2.4 Hydrazones

The outcome of the reaction of hydrazones with lead tetraacetate is dependent upon the nature of the hydrazone and the reaction conditions. Mechanistically, the reaction was first suggested to imply the formation of a pair of free radicals, Pb(OAc)$_3$• and the hydrazonyl radical.[68] However, ESR studies failed to detect the presence of radicals in solutions of hydrazones and lead tetraacetate. The reaction was explained by a two step mechanism, involving first the formation of a covalent hydrazone-lead intermediate followed by an intramolecular rearrangement leading to the α-acetoxy products which can then evolve into different derivatives.[69,70]

Unsubstituted hydrazones undergo a dehydrogenation to give diazoalkanes, presumably via an hydrazone-lead intermediate. Ligand coupling leads to the acetoxyazo compound which eventually loses acetic acid.[71,72] Only the more stable diazo compounds can be isolated, such as those containing two electron-withdrawing substituents at the α-carbon.[73,74] Generally non-stabilized diazo compounds react further either with lead tetraacetate to give 1,1-diacetoxy derivatives or with acetic acid to give the monoacetoxy derivatives.[71,72]

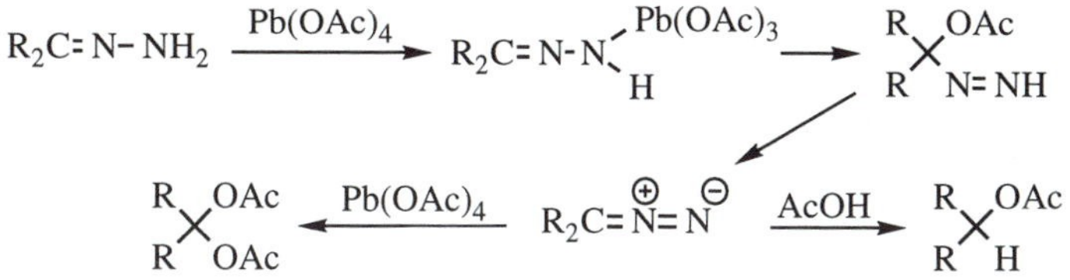

Scheme 7.5

Monosubstituted hydrazones of aldehydes (**19**) are oxidised by lead tetraacetate to afford N-acetyl N'-acylhydrazones.[75-77] The reaction proceeds via a N-hydrazone intermediate which undergoes a ligand coupling reaction to give an hydrazonyl acetate (**20**) which then rearranges to the final product. Monosubstituted hydrazones of ketones (**22**) react with lead tetraacetate to give α-azoacetates (**23**).[68]

$$R-CH=NNHR' \xrightarrow{Pb(OAc)_4} R-CH=N-N\overset{Pb(OAc)_3}{\underset{R'}{\diagdown}} \xrightarrow{} \overset{OAc}{\underset{R-C=NNHR'}{}} \xrightarrow{} R-\overset{O}{\overset{\|}{C}}-\overset{}{\underset{H}{N}}-N\overset{Ac}{\underset{R'}{\diagdown}}$$

(19)   (20)

$$R_2C=NNHR' \xrightarrow{Pb(OAc)_4} R_2C=N-N\overset{Pb(OAc)_3}{\underset{R'}{\diagdown}} \xrightarrow{} \overset{OAc}{\underset{R_2C-N=NR'}{}}$$

(22)   (23)

When these reactions are performed in alcoholic solvent, ethers are also obtained. Hydrazones of aldehydes (19) give a mixture of hydrazonyl acetate (20) and hydrazonyl ether (21). Hydrazones of ketones (22) give similarly mixtures of α-azoacetates (23) and α-azoethers (24) although in relatively poor yields.[69,76,78]

$$R-CH=N-NHR' \xrightarrow[R''OH]{Pb(OAc)_4} \overset{AcO}{\underset{R}{\diagdown}}C=N-NHR' + \overset{R''O}{\underset{R}{\diagdown}}C=N-NHR'$$

(19)   (20)   (21)

$$R_2C=N-NHR' \xrightarrow[R''OH]{Pb(OAc)_4} \overset{OAc}{\underset{R_2C-N=NR'}{}} + \overset{OR''}{\underset{R_2C-N=NR'}{}}$$

(22)   (23)   (24)

Lead tetraacetate oxidation of *N*-aroylhydrazones (25) gives highly reactive azoacetates (26), which then undergo cyclisation to 1,3,4-oxadiazolines (27).[79] The lead tetraacetate oxidation of acylhydrazones of *ortho*-hydroxyarylketones gives 1,2-diacylbenzenes, *via* ligand coupling followed by intramolecular rearrangements, for which two possible pathways were suggested.[80-82]

$$\overset{R^1}{\underset{R^2}{\diagup}}=N-NH-COAr \xrightarrow{Pb(OAc)_4} \overset{R^1}{\underset{R^2}{}}\overset{N=N-COAr}{\underset{OAc}{\times}} \xrightarrow{} \overset{R^1}{\underset{R^2}{}}\overset{N=N}{\underset{O}{\times}}\overset{}{\underset{OAc}{Ar}}$$

(25)   (26)   (27)

(structures for salicyl-derived compounds)

7.1.2.5 Oximes

A number of competing pathways takes place in the reaction of oximes with lead tetraacetate. The outcome of the reaction depends on various factors such as the structure of the oxime, the ratio lead tetraacetate:substrate, the temperature, the nature of the solvent and the presence of nitric oxide. A common intermediate, an *O*-triacetoxyplumbyl oxime ester (28), is formed and its decomposition leads to the ligand coupling products, the α-acetoxynitroso compounds (29) (from aliphatic and alicyclic ketoximes at low temperature),[67,83-86] to the iminoxyl radicals (30) (at higher temperatures)[87] or to the nitrile oxides (31) (from aliphatic *syn*-aldoximes at - 78°C).[88] In the case of aliphatic *anti*-aldoximes, ligand coupling occurs followed by dimerisation to α-nitrosoacetate dimers.[89] When the reaction is performed in acetic acid, decomposition of the nitrosoacetates leads to regeneration of the parent carbonyl compounds.[83,84]

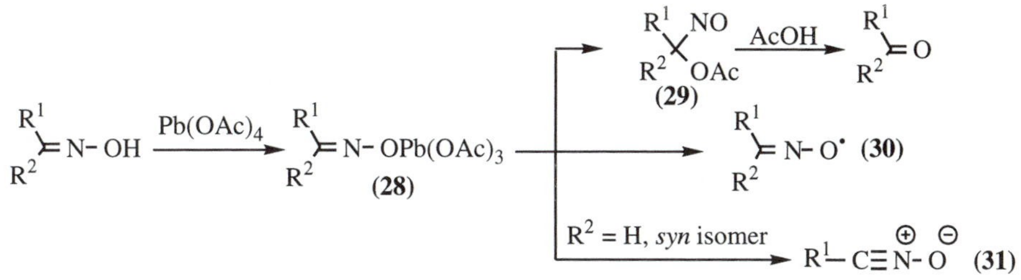

$$R^1R^2C=N-OH \xrightarrow{Pb(OAc)_4} R^1R^2C=N-OPb(OAc)_3 \quad (28)$$

R¹R²C(NO)(OAc) (29) $\xrightarrow{AcOH}$ R¹R²C=O

R¹R²C=N-O· (30)

R² = H, *syn* isomer

$$R^1-C\equiv \overset{\oplus}{N}-\overset{\ominus}{O} \quad (31)$$

Scheme 7.6

7.1.3 Reaction of alkenes and arenes

The reaction of lead tetraacetate with alkenes and arenes involve first an electrophilic addition step. The intermediate organolead derivative then can follow various pathways to yield oxidation products.

7.1.3.1 Reaction of alkenes with lead tetraacetate

Lead tetraacetate reacts with alkenes to give generally complex mixtures of products resulting either from the addition of oxygen functional groups to the double bond or from allylic oxidation.[90,91] Moreover, other types of reactions, such as skeletal rearrangements, double bond migration or C-C bond cleavage can take place, depending on the structure of the alkene. The formation of β-acetoxy-alkyllead triacetates has been proposed as intermediates in these oxidation reactions.[92,93] Generally highly unstable, they undergo a change of the oxidation state from Pb^{+4} to Pb^{+2}, with either formation of a carbocation, or inter- or intramolecular displacement.

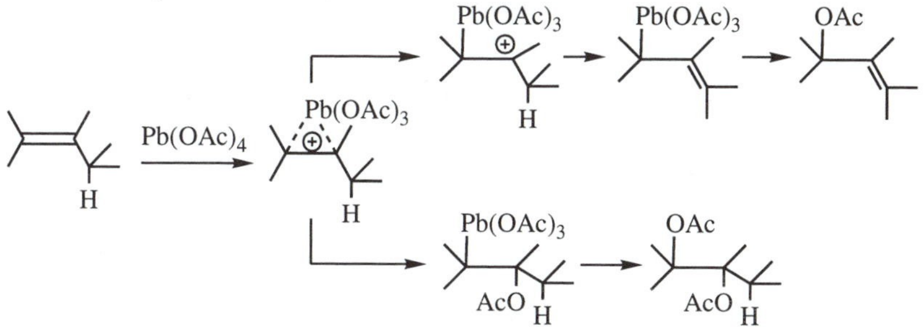

Scheme 7.7

The occurrence of a ligand coupling mechanism could be an alternative mechanistic pathway taking part in the evolution of the alkyllead intermediates, although it was excluded in the decomposition of the only known case of an alkyllead diacetate derivative (32).[94]

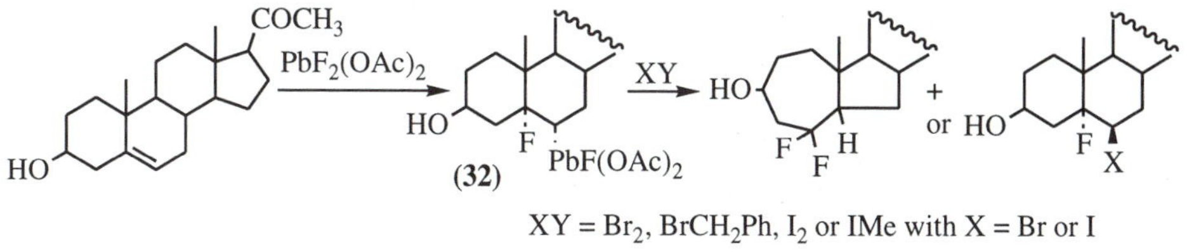

$XY = Br_2, BrCH_2Ph, I_2$ or IMe with X = Br or I

7.1.3.2 Reaction of arenes with lead tetraacetate

Lead tetraacetate reacts with arenes to lead either to aromatic nucleus substitution or to substitution on the benzylic position of the side chain. Substitution on the nucleus involves electrophilic attack of $(AcO)_3Pb^+$ to give aryllead tricarboxylates. Subsequently, these aryl species react with acid to afford eventually the corresponding aryl esters.[95,96]

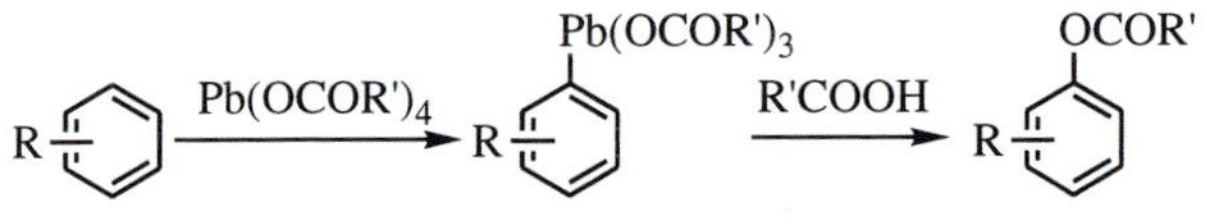

Scheme 7.8

Benzene is relatively stable to lead tetraacetate oxidation and is used as solvent in oxidation of more reactive substrates. But it reacts with lead tetrakis(trifluoroacetate) to afford the corresponding trifluoroacetoxy derivatives.[97,98] Oxidation of aromatic rings bearing electron donating substituents with lead tetraacetate in trifluoroacetic acid at low temperature leads usually to the corresponding aryl trifluoroacetates.[99] The reaction is likely to involve acid-catalysed heterolytic cleavage of the C-Pb bond in an intermediately formed aryllead tris(trifluoroacetate). This one can result either from the reaction of the arene with lead tetrakis(trifluoroacetate) or from ligand exchange when an aryllead triacetate is treated with an excess of trifluoroacetic acid. The intermediate aryl cation, free or incipient, then reacts with the solvent.[100] Sometimes various amounts of side products such as dimers or diarylmethanes are observed.[100-103] The dimerisation products have been postulated to arise from an electron transfer and radical-cation mechanism.[103]

Aryltrimethylsilanes react also with lead tetraacetate or lead tetrakis(trifluoroacetate), in trifluoroacetic acid, to yield the corresponding aryllead tris(trifluoroacetate), which again can further react with TFA to lead to the aryltrifluoroacetates.[97,104,105]

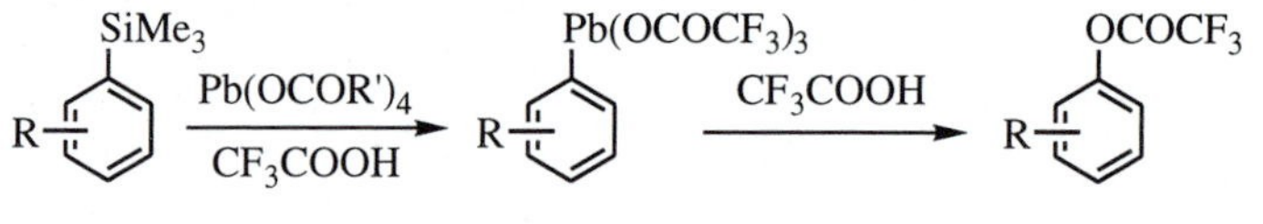

Scheme 7.9

The reaction of aromatic compounds with lead tetraacetate can be catalysed by boron trifluoride to give mixtures of dimerisation and acetoxylation products with the dimeric products being usually predominant.[106]

7.1.3.3 Preparation of organolead triacetates

Direct electrophilic plumbation of aromatic compounds can be used to prepare a small range of aryllead tricarboxylates.[10,11,105] It is limited to substrates which are more electron rich than halobenzenes. Plumbation of halobenzenes can be conveniently performed by treatment with lead tetraacetate in the presence of trichloroacetic or trifluoroacetic acid. For the plumbation of compounds of intermediate reactivity between the halobenzenes and 1,3-dimethoxybenzene, the use of haloacetic acid (mono-, di- or trichloroacetic acid) is required to optimize the yield of aryllead triacetates. For 1,3-dimethoxybenzene and the more electron-rich aromatics, the reaction can be suitably performed in acetic acid.

The most general route to organolead tricarboxylates, such as aryllead, vinyllead or alkynyllead tricarboxylates, is the metal-metal exchange which can involve mercury-lead, thallium-lead, silicon-lead, tin-lead, boron-lead or zinc-lead transmetallation.

The most useful methods for the synthesis of aryllead tricarboxylates are either the tin-lead[107-109] or the boron-lead[110] exchange. These are generally performed in high yield by reacting the appropriate aryltributylstannane or arylboronic acid with lead tetraacetate in the presence of a catalytic amount of mercury (II) salts. Ultrasound activation can be used as an alternative to catalyse the exchange.[111] Although the tributyltin derivative is generally preferred to the trimethyltin analogue, the exchange can be sometimes efficiently performed with the trimethyltin compound in the case of relatively hindered aryl substrates.[112]

$$R\text{-}M\text{-}X_n \quad + \quad Pb(OCOR')_4 \quad \xrightarrow{\text{cat.}} \quad R\text{-}Pb(OCOR_4)_3$$

$$M = Sn, X = Bu \text{ or } Me, n = 3; \; M = B, X = OH, n = 2$$

$$\text{cat.} = Hg(II) \text{ salt or sonication}$$

The metal-lead exchange can also be successfully applied to the synthesis of some heteroaryllead triacetates, although the range of known compounds is limited to furan and thiophene.[113] Moreover, the intermediate organolead triacetate can experience overoxidation in the presence of an excess of lead tetraacetate.[114,115]

Vinyllead and alkynyllead tricarboxylates can only be prepared by metal-metal exchange, such as mercury-lead,[116,117] tin-lead,[116,117] boron-lead[118] or zinc-lead.[119] In the case of tin-lead exchange, cleavage of the methyl-tin bond is much slower than cleavage of the vinyl-tin or alkynyl-tin bond. Therefore (trimethyl)vinyltin derivatives are preferred to the corresponding (tributyl)vinyltin analogues for the synthesis of the unstable vinyllead and alkynyllead compounds.[117]

7.1.3.4 Reaction of arenes with aryllead triacetate

Reaction of arenes with lead tetraacetate in TFA (trifluoroacetic acid) can afford symmetrical dimeric compounds in moderate to good yields.[103] (see section 7.1.3.2) Unsymmetrical biaryls can be produced by reaction of aryllead triacetates substituted with electron-donating as well as electron-withdrawing groups with *p*-xylene or more highly methylated benzenes.[100,120] For example, in the arylation with 4-fluorophenyllead triacetate the yields vary from 0% for nitrobenzene, 2% for benzene and toluene to 80% for pentamethylbenzene and 88% for mesitylene.

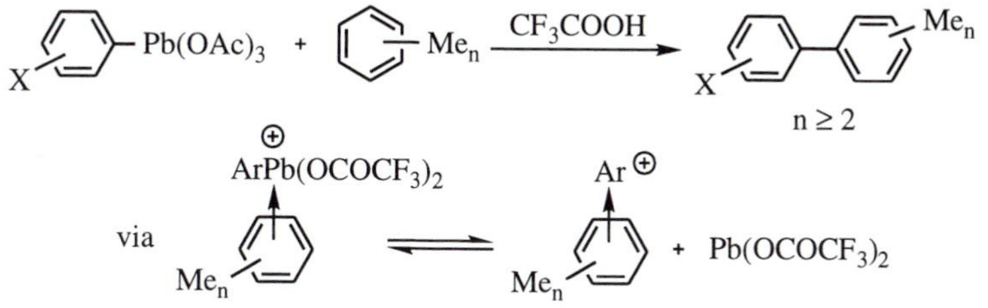

Scheme 7.10

It is assumed that, in the reaction of arenes with aryllead triacetates, the arylation takes place *via* the corresponding cationic π-complexes.[100] In the reaction with lead tetraacetate, an electron transfer-radical cation mechanism was postulated.[103] In agreement with this assumption, 4,6,8-trimethylazulene (33) was arylated by treatment with 4-methoxyphenyllead triacetate (15) in TFA to afford only one isomer, the 1-arylazulene (34), although in modest yield (27%). A minor by-product was also formed (4%), the 1,1'-bis-azulene (35), which could result from one-electron oxidation of the primary product, 1-arylazulene, by the 4-methoxyphenyl cation in the corresponding π-complex.[121]

(33)   (34), 27%   (35), 4%

7.2 ARYLATION REACTIONS

In a series of outstanding papers, Pinhey *et al.* have shown that aryllead tricarboxylates react with soft nucleophiles to afford *C*-arylation products. These aryllead derivatives behave as aryl cation equivalents in reactions which involve a ligand coupling mechanism (see section 7.5).[9-12] In most cases, the reactions proceed in chloroform at 40-60°C in the presence of pyridine as a base with a ratio of substrate to organolead derivative to pyridine of 1 : 1 : 3. The substrates which easily undergo *C*-arylation include phenols, β-dicarbonyl compounds and their vinylogues, α-cyanoesters, α-hetero-substituted ketones, enamines and nitroalkanes. A very limited number of non-carbon nucleophiles has also been reported to react.

7.2.1 Phenols

Aryllead (IV) triacetates react with phenols to give mainly products of *ortho*-*C*-arylation, formed by ligand coupling mechanisms.[122-124] In an attempt to extend the arylation of polymethylbenzenes to phenolic substrates, Pinhey *et al.* treated mesitol (36) with *p*-methoxyphenyllead triacetate (15) in CHCl₃. The reaction afforded a mixture of the *C*-arylated products (37) and (38) together with minor amounts of the *C*-acetoxy product (39) and *O*-aryl ether (40).[123]

(36)   (37), 48%   (38), 13%   (39), 2%   (40), 2%

with An = 4-MeO-C₆H₄

When the reactions of phenols with aryllead triacetates were performed in the presence of pyridine, only the *C*-arylated dienones were formed in a nearly quantitative yield (90%; ratio *ortho* : *para* = 4:1). Under the classical conditions (phenol : organolead triacetate : base in a ratio 1:1:3), the reaction of methylated phenols only proceeds in high yield when both *ortho* positions are substituted. There is a preference for attack *ipso* to a methoxyl group compared to a methyl group. As in the Wessely acetoxylation, there is a marked preference for *ortho*-arylation.

Table 7.1: Arylation of phenols with 4-methoxyphenyllead triacetate **(15)** and pyridine.[123]

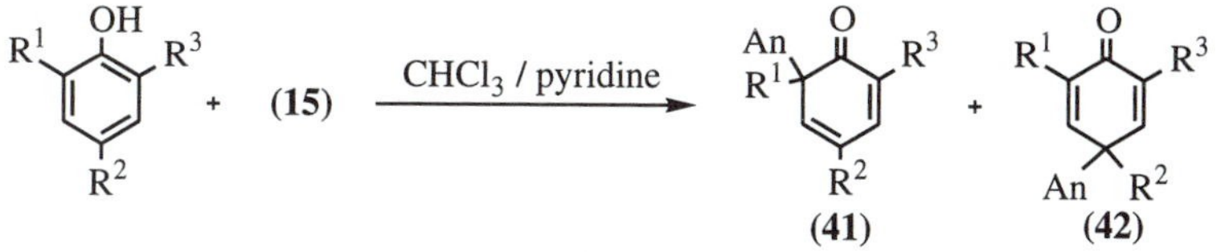

Phenol			*ortho*-Aryl dienone **(41)** (%)	*para*-Aryl dienone **(42)** (%)	Anisole (%)	Other products, (%)
R^1	R^2	R^3				
H	Me	H	-	-	12	4-MeOC$_6$H$_4$OH, 1
Me	H	H	5-10	-	5	
Me	H	Me	75-90	-	3	**(43)**, 1-2
Me	Me	Me	75	20	1-3	
Cl	H	Cl	-	-	-	No reaction
MeO	H	H	48	-	6	
H	MeO	H	-	-	29	**(44)**, 34

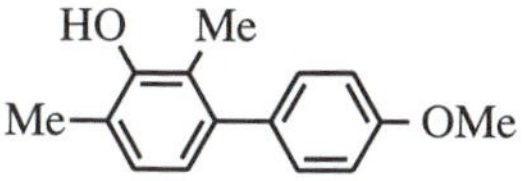

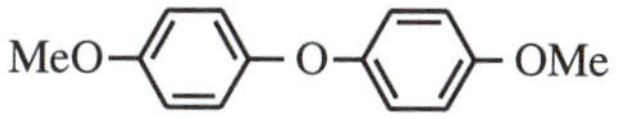

The presence of two or more alkyl groups on the phenolic substrate is required but they do not necessarily have to be on either of the *ortho* positions. Indeed, 3,4,5-trimethylphenol **(45)** has been reported to yield a modest amount of 2-(4-methoxyphenyl)-3,4,5-trimethylphenol in the reaction with 4-methoxyphenyllead triacetate **(15)**.[123]

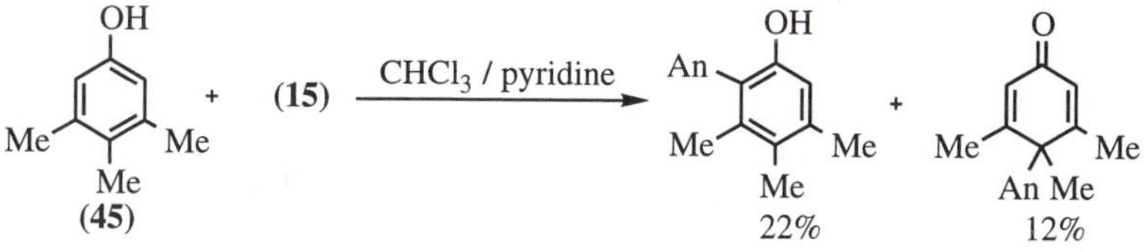

However, under more forcing conditions, high yields of the 2,6-diaryl derivatives **(46)** can be obtained, and particularly with more reactive aryllead derivatives, such as 2,4,6-trimethoxy-phenyllead triacetate **(48)**.[12] Highly hindered structures of the 2'-hydroxy-1',3'-terphenyl type can be obtained. For example, 3,5-di-*tert*-butylphenol **(47)** reacts with 2,4,6-trimethoxyphenyllead triacetate **(48)** to afford the very sterically hindered 2,6-diarylphenol **(49)** in 87% yield.[125,126]

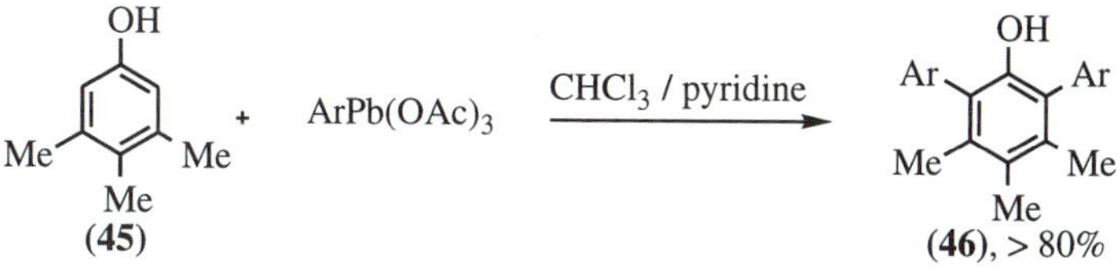

(47) (48) (49), 87% ref. 125

In general, the rate of the reaction of arylation of phenols by aryllead triacetate increases with the electron density of the phenolic substrate. Thus, when pyridine is used as a base, no reaction is generally observed with electron-poor phenols, such as 2,6-dichlorophenol or 2,6-dichloro-4-nitrophenol.[123] However, the reaction of the sodium salt of perfluorophenol (50) with phenyllead triacetate under more forcing conditions led to the product of *ortho*-arylation, the 6-aryl-2,4-cyclohexadienone (51) together with minor amounts of the product of *para*-arylation (52) and the unsymmetrical diaryl ether (53).[127]

(50) (51), 49% (52), 16% (53), 4%

As a synthetic method, the reaction of aryllead triacetates with phenols is useful for the production of two structural types: the 6-aryl-2,4-cyclohexadienones and the sterically hindered 2,6-diarylphenols substituted with alkyl groups on the C-3 to C-5 positions. The steric compression in the putative aryloxylead intermediate seems to play a decisive role in the success of the ligand coupling step. This is particularly true for the arylation of hindered phenols with the more hindered aryllead reagents, which takes place at room temperature. In contrast to the reactions of phenols with arylbismuth reagents (chapter 6), *O*-arylation products are exceptional and generally obtained in very small yields.

7.2.2 Dicarbonyl compounds and derivatives

A number of enolised compounds reacts with aryllead triacetate to afford the corresponding α-arylketone derivatives. Due to the lower reactivity of aryllead (IV) triacetate compared to lead (IV) tetraacetate, the range of compounds is narrower and, for the most part, restricted to substrates with pK_A values similar or less than that of phenol. The reaction conditions are closely similar to the reactions with phenols, that is a ratio of substrate to organolead derivative to pyridine of 1:1:3 in $CHCl_3$ solution at 40-60°C. Although pyridine is usually preferred, it can be advantageously replaced by bases more strongly coordinating to lead, such as 2,2'-bipyridine, 1,10-phenanthroline or 4-dimethylaminopyridine (DMAP). The substrates which easily react with aryllead triacetates include β-diketones, β-ketoaldehydes, α- and β-ketoesters, β-ketoester vinylogues, malonic acid derivatives such as their esters, amides or nitriles. β-Dicarbonyl compounds with one α-hydrogen give the α-aryl derivatives in good to high yields. On the other side, β-dicarbonyl compounds with two α-hydrogen give variable mixtures of mono- and di-arylated derivatives, in modest to moderate yield.

7.2.2.1 β-Diketones

Whether linear or cyclic, β-diketones with only one α-hydrogen react with aryllead triacetates to afford consistently high yields of arylated product.[128] However, in the case of 2-methyl-1,3-cyclopentanedione (54), the monoarylketone was obtained in 49%, together with products derived from oxidation of the dione to the enedione.

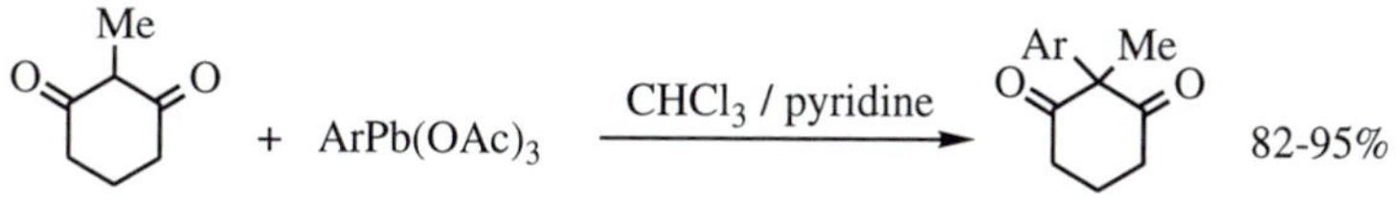

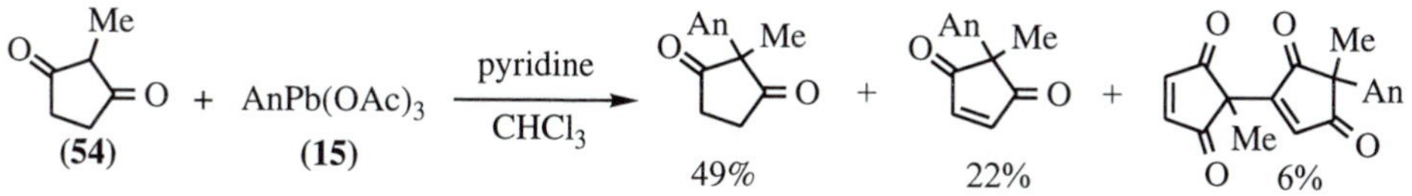

The reactivity of β-diketones with two α-hydrogen is dependent on the structure of the substrate. For cyclic substrates such as dimedone (55), only the diaryl product was observed in high yield. The second arylation takes place faster than the first one, so that even with 1 equivalent of aryllead triacetate, the monoaryl product was not detected.[128]

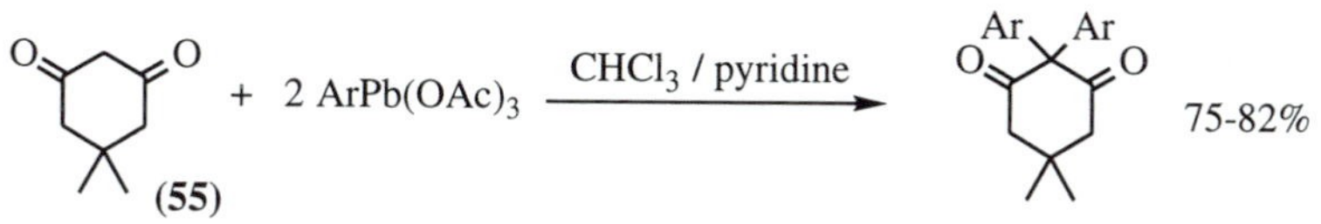

Acyclic substrates afforded modest to moderate amounts of monoaryl products, for example in the case of 2,4-pentanedione (56). Longer reaction times were required to obtain poor yields of the α,α-diaryl β-diketone.[128-132]

Table 7.2: Arylation of 2,4-pentanedione (56) with aryllead triacetates

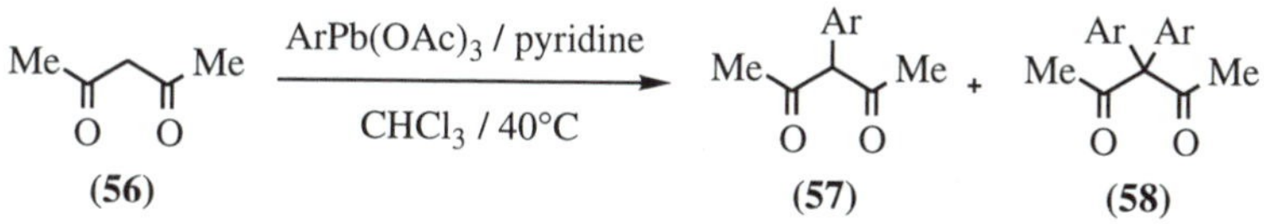

Ar	t (h)	(57) (%)	(58) (%)	Ref
4-MeOC$_6$H$_4$	24	19	2.5	128,129
4-MeC$_6$H$_4$	24	32, 50	-	128, 130
4-MeC$_6$H$_4$	42	-	17	128
3,4-(MeO)$_2$C$_6$H$_3$	48	56	-	131
3,4,5-Me$_3$C$_6$H$_2$	24	63	-	132

The related α-hydroxymethylene ketones behaved similarly. However, when the reactions are performed in CHCl$_3$, loss of the formyl group followed by a second arylation results in the formation of the α,α-diarylketone.[133] The deformylation of the α-aryl-α-formylketone was suppressed by performing the reaction in THF instead of CHCl$_3$. Good yields of the mono α-arylketones were then obtained.[133,134]

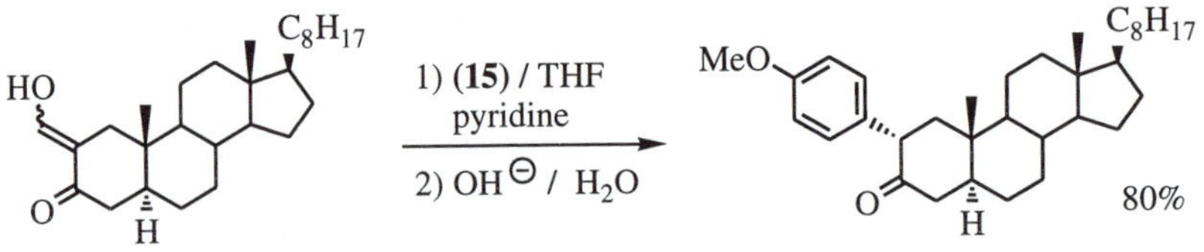

7.2.2.2 β-Ketoesters

β-Ketoesters behave similarly to β-diketones in their reactions with aryllead triacetate. The acyclic β-ketoesters possessing two α-hydrogen again afforded mixtures of mono and diaryl derivatives.[135]

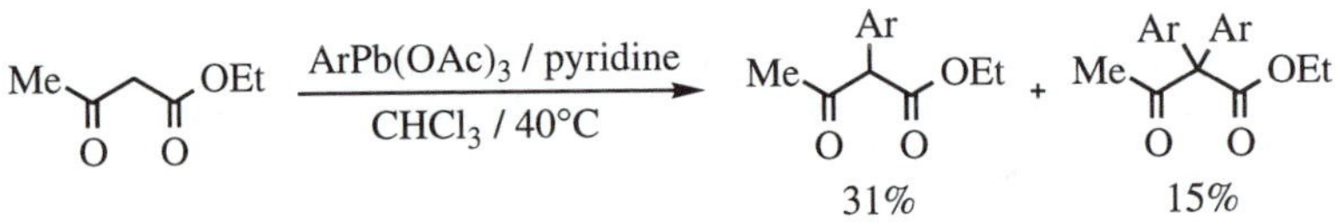

Consistently good to high yields (60-95%) are obtained with β-ketoesters possessing only one α-hydrogen.[135,136]

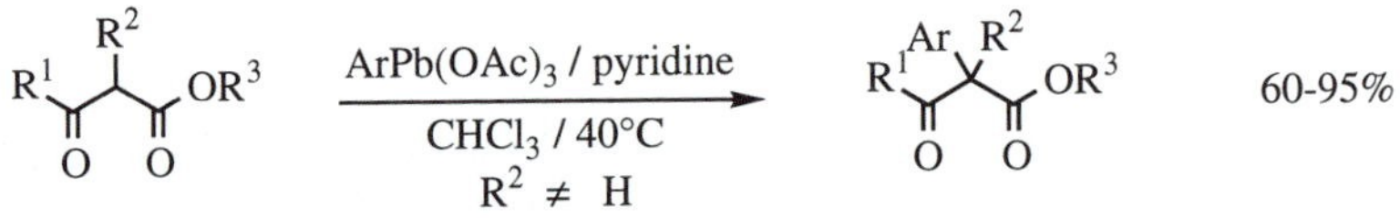

It requires rather hindered β-ketoesters to observe a significant drop in the yield of the arylation reaction.[135]

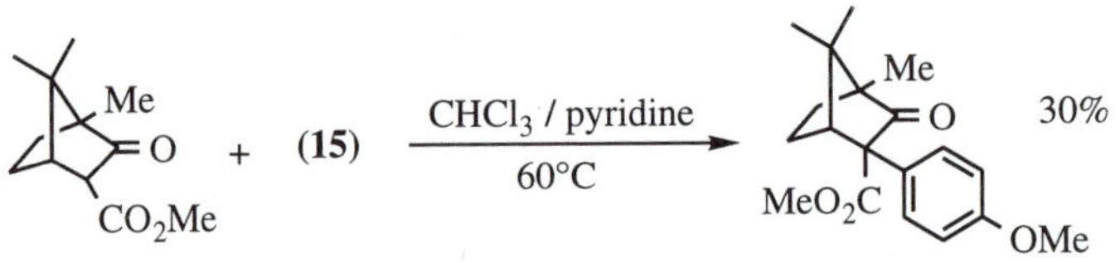

The aryllead triacetate reagent is generally used as the purified reagent. However, *in situ* generated reagents can also be conveniently used as efficient alternatives, particularly in the case of relatively unstable reagents. The *in situ* synthesis of the aryllead triacetate can be done either by mercury-lead exchange[137] or more preferably by boron-lead exchange.[110] Combined with the Krapcho decarboxylation,[138] the arylation of methyl β-ketoesters constitute an efficient method for the synthesis of α-arylketones.[133] However, difficulties are sometimes encountered with sterically rigid or sterically demanding tetrasubstituted α-aryl β-ketoesters.[139] Prolonged heating at 180°C can result in low yields either by decomposition of the product or by oxidation to α,β-unsaturated ketones.[133] This drawback can be avoided by using a modified β-ketoester, such as the benzyl or allyl esters.[111,140] The reaction of arylation of β-ketoesters by aryllead triacetates has been used as a key step in the synthesis of a number of natural products or analogues. Carbocyclic β-ketoesters were involved in the synthesis of various lignan structures. Substrates bearing two β-ketoester moieties, such as dibenzyl 3,7-dioxobicyclo [3.3.0]-octane-2,6-dicarboxylate (**59**), can be bisarylated with aryllead triacetate.[111,141]

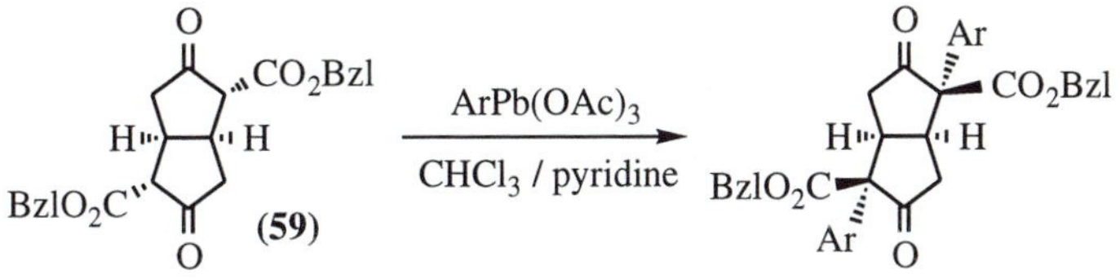

Elaboration of the carbon skeleton led to the total synthesis of natural products possessing an aryltetrahydrofuran structure: (±) sesamin,[111,141] (±) eudesmin,[111] (±) yangambin.[111] Mixed arylation was used to obtain unsymmetrically diarylated compounds. One-pot treatment of the dibenzyl 3,7-dioxobicyclo [3.3.0]-octane-2,6-dicarboxylate (59) with a mixture of two aryllead triacetates (60) and (61) afforded a mixture of the unsymmetrical diaryl derivative (63) with the symmetrical diaryl compounds (62) and (64). When the arylation was performed with 1.1 equivalent of the aryllead reagents (60) and (61), a mixture of (62), (63) and (64) in a ratio 0.46:1:0.72 was obtained. This ratio became 0.43:1:0.49, when 0.86 equivalent of (60) and 1.33 equivalent of (61) were used. (Scheme 7.11) This compound (63) isolated in a 33% yield was eventually elaborated to complete a total synthesis of (±) methyl piperitol.[142]

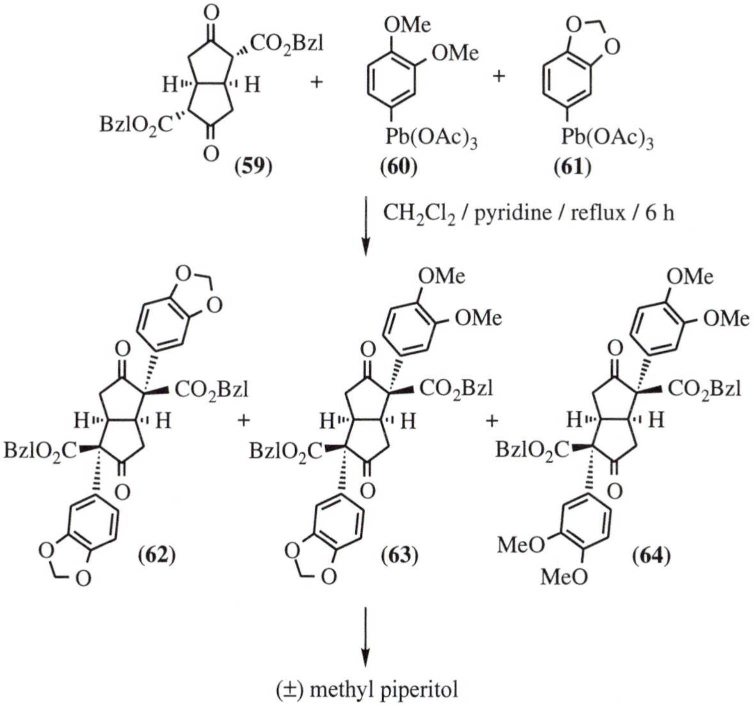

(±) methyl piperitol

Scheme 7.11

Arylation reactions of heterocyclic β-ketoesters were employed in the synthesis of a number of 2-arylbenzofurans,[140,143] isoflavanones and isoflavones,[144,145] as well as for the synthesis of 2-aryl-(2H)-indole derivatives.[146]

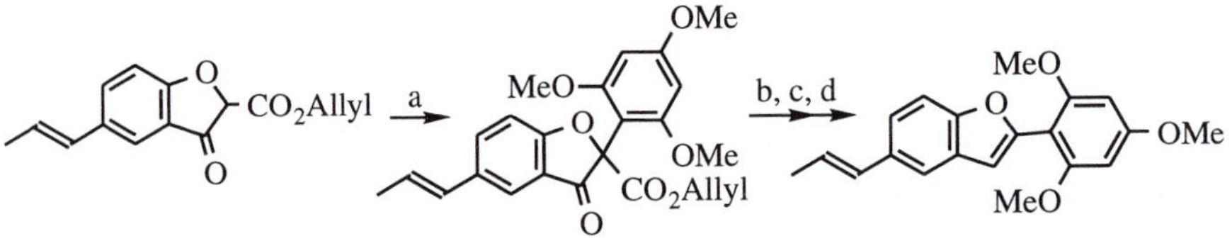

a) ArPb(OAc)$_3$ (**48**), pyridine, CHCl$_3$, 60°C, 85% b) Pd(OAc)$_2$, PPh$_3$, HCOOH, Et$_3$N, THF, RT c) NaBH$_4$, EtOH d) 10% aqueous HCl

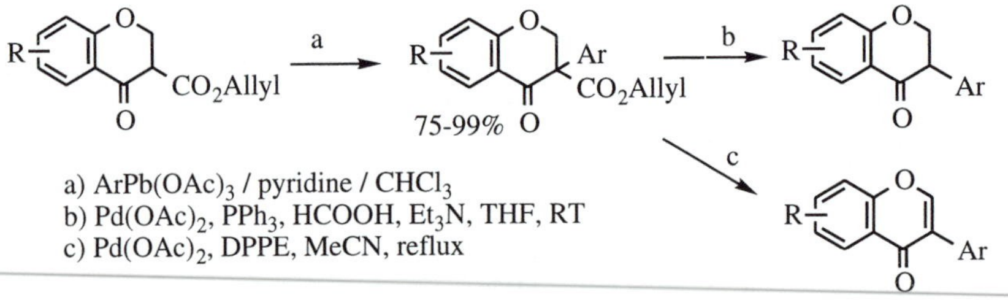

a) ArPb(OAc)$_3$ / pyridine / CHCl$_3$
b) Pd(OAc)$_2$, PPh$_3$, HCOOH, Et$_3$N, THF, RT
c) Pd(OAc)$_2$, DPPE, MeCN, reflux

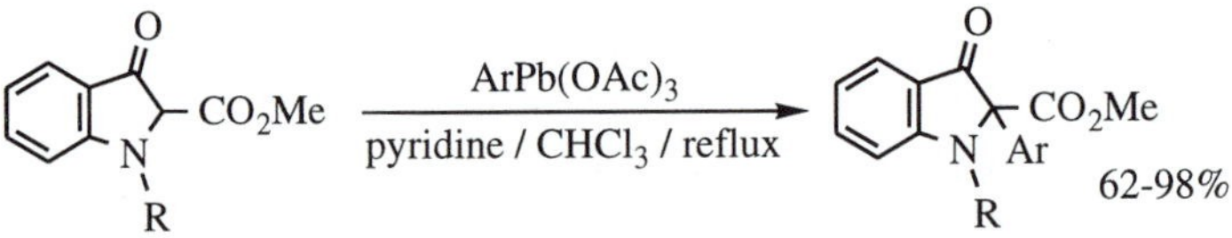

ref. 144

ref. 146

α-Methylene α-arylketones can be easily and selectively obtained by arylation of allyl β-ketoesters which are eventually deprotected by the Tsuji's procedures.[147] Deallyloxycarbonylation is performed by treatment of the allyl α-aryl-β-ketoesters with catalytic amounts of palladium (II) acetate, triethylammonium formate and triphenylphosphine in THF at room temperature and affords the α-arylketones in 75-97% yield.[144] Deallyloxycarbonylation-dehydrogenation can be realized with the same allyl esters by treatment with catalytic amounts of palladium (II) acetate and 1,2-bis(diphenyl-phosphino)ethane (DPPE) in acetonitrile under reflux and affords the α-aryl α,β-unsaturated ketones in 60-90% yield.[144] In particular, this reaction afforded a direct synthesis of 2'-hydroxyisoflavones involving arylation of the appropriate allyl β-ketoester with the MOM-protected 2-methoxymethoxy-phenyllead triacetate derivative.[145] An alternative system for the synthesis for α-methylene α-aryl-ketones involves the arylation of an heterocyclic β-diketolactone. Under the conditions of the reaction, three steps (arylation, β-dicarbonyl cleavage and β-elimination) occur in one-pot to afford directly the α-aryl α-methylene ketones or esters.[148]

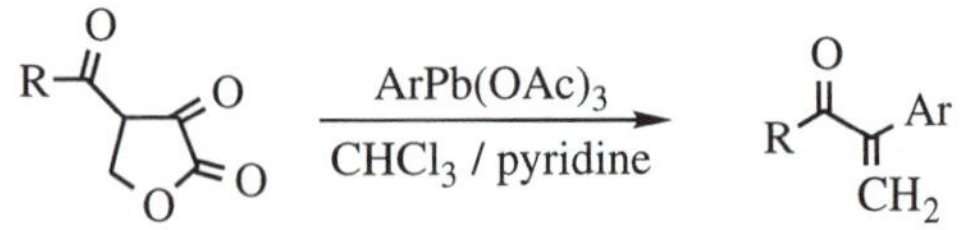

4-Hydroxycoumarins are cyclic completely enolised β-ketoesters. In their reactions with aryllead triacetates, they behave more like phenols than like β-ketoesters bearing two α-hydrogens. Indeed only monoarylation took place to provide a convenient access to 3-aryl-4-hydroxycoumarins, which belong to the group of highly oxygenated isoflavonoids.[109,149] This was applied for example to the synthesis of isorobustin (**65**), robustin (**66**) and robustic acid.[109,150,151]

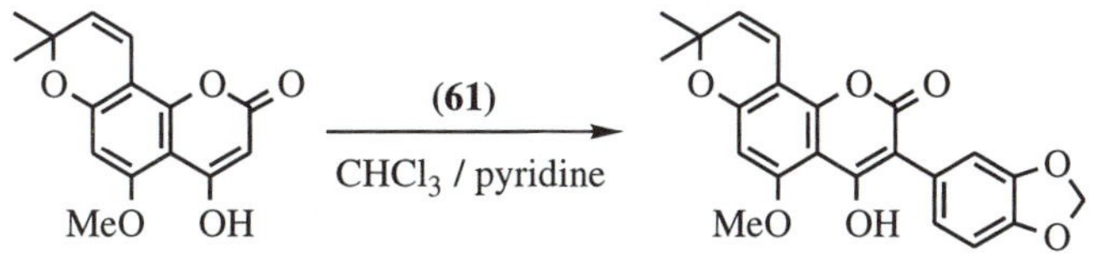

(65), 84% ref. 109

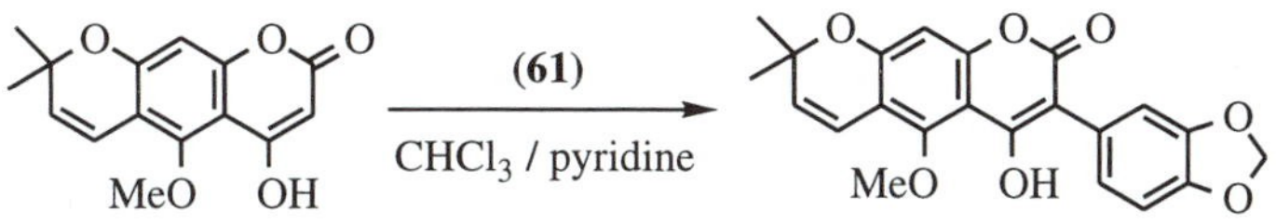

(66), 84% ref. 151

This reaction was also applied to the isomeric α-ketoester system, the 3-hydroxycoumarins. Monoarylation was observed in good to high yields, thus giving an entry into 4-aryl-3-hydroxy-coumarins, also called 3-hydroxyneoflavonoids.[152]

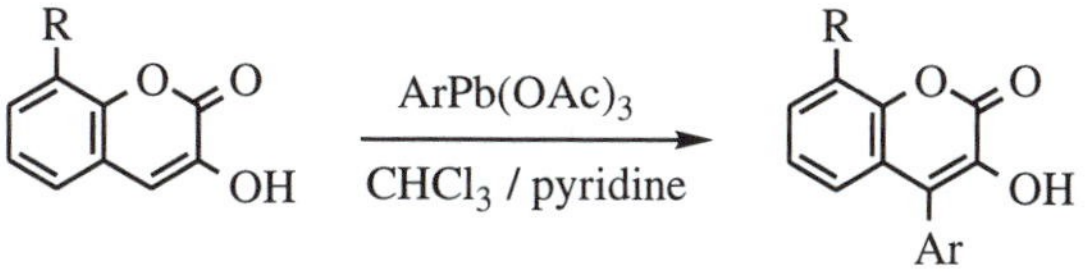

R = H, OMe
59-92%

Under the conditions suitable for the *N*-arylation of amidic nitrogen atoms, (see section 7.6.1), reaction of β-ketoanilides with 4-methylphenyllead triacetate (67) in the presence of sodium hydride and copper diacetate afforded only products of *C*-arylation (68), although in poor yields (11-33%). The α-aryl-β-ketoanilides are unstable in solution and easily oxidised by air to the α-aryl-α-ketoanilide (69), also isolated in modest yields (8-19%).[153]

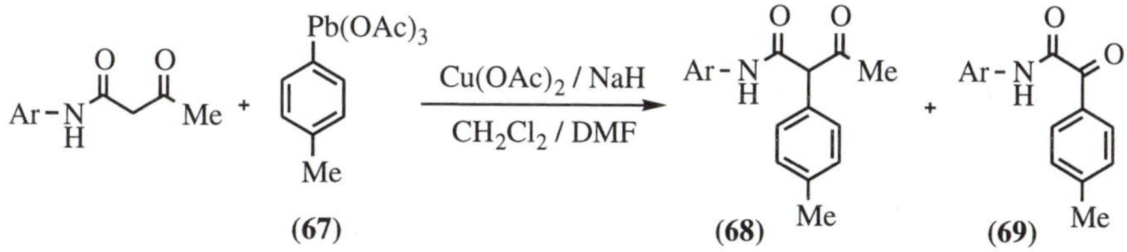

7.2.2.3 β-Dicarbonyl vinylogues

Vinylogues of β-dicarbonyl compounds, such as β-diketones or β-ketoesters react with aryllead triacetates to afford preferably the quaternary centred derivatives.[128,154-157] In the only reported example of arylation of a vinylogous β-diketone, the steroidal derivative (70), a mixture of 6-α and 6-β aryl derivatives was obtained with no trace of the 4-aryl derivative.[128]

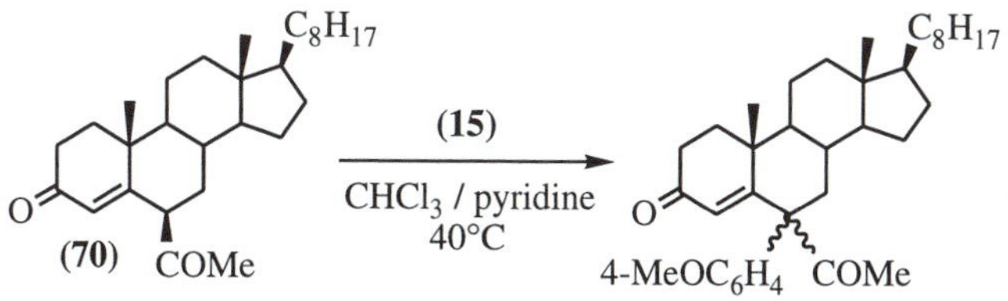

57%
6α / 6β = 1.75

Hagemann's ester (**71**) reacted similarly to afford the product of arylation on the carbon α to the ester group.[154,155]

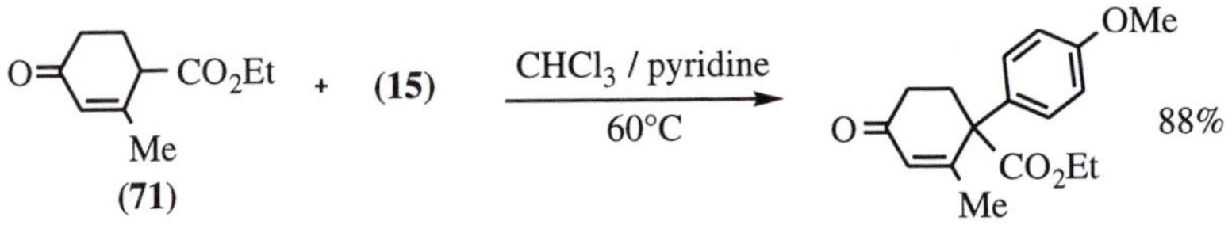

This reaction was used in a short high-yielding synthesis of the alkaloids (±)lycoramine,[154,156] (±)*O*-methyljoubertiamine[154,157] and (±)mesembrine.[154,157] All these syntheses involved in the key step the selective high-yielding arylation of alkyl 4-oxocyclohex-2,3-ene carboxylates (**72**) with organolead reagents.

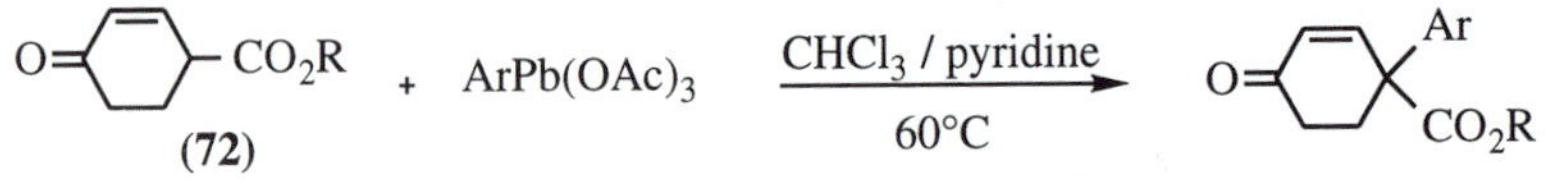

R = Et	Ar =	2,3-(MeO)$_2$C$_6$H$_3$	96%
R = Me		4-MeO-C$_6$H$_4$	90%
R = Me		3,4-(MeO)$_2$C$_6$H$_3$	90%

7.2.2.4 Malonic acid derivatives, α-cyanoesters and malononitriles

Due to the lower acidity of the α-hydrogen of malonic acid derivatives compared to the acidity of the corresponding hydrogens of β-diketones and β-ketoesters, the reactivity of malonic acid derivatives towards aryllead compounds is very dependent upon the nature of the substrate.[158,159] Under the classical conditions (pyridine, CHCl$_3$, room temperature or 40-60°C), diethyl malonate did not react with aryllead triacetates and the α-methyl and α-phenyl derivatives reacted slowly to give poor yields (~25%) of the corresponding α-aryl α-substituted malonic acid diesters. Moderate to good yields were obtained by reacting the sodium salt of substituted malonic esters with aryllead triacetates in THF in the presence of pyridine, but the sodium salt of diethyl malonate again failed to react.[159]

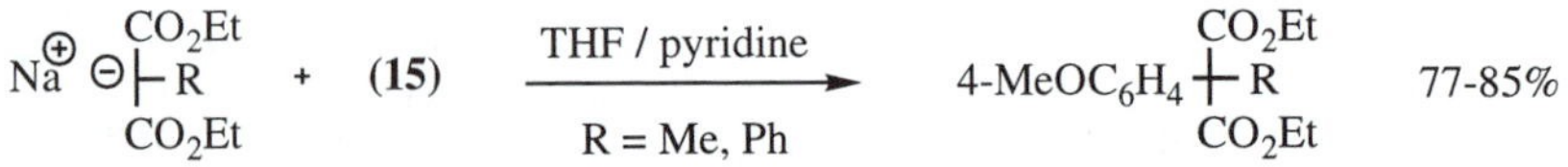

In view of the high acidity of the cyclic ester analogue, Meldrum's acid derivatives underwent rapid arylation with aryllead reagents.[159] The size of the substituent present on the α-carbon has a significant influence on the overall yield. Whereas the α-methyl and α-phenyl derivatives gave nearly

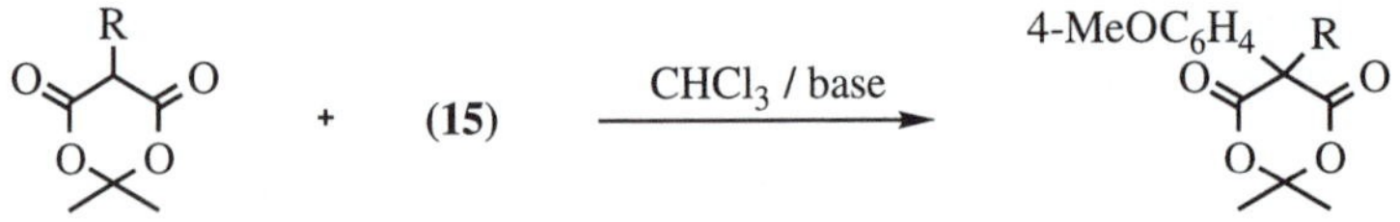

R =		base	yield
	Me	pyridine	91%
	Phe	pyridine	92%
	Et	pyridine	74%
	i-Pr	pyridine	29%
	i-Pr	Bipy	51%
	i-Pr	Phen	87%

quantitatively the aryl product (91-92%), a modest drop was noted for the α-ethyl analogue (74%), but the reaction became sluggish and low-yielding for the α-isopropyl compound (29%). In this latter case, the influence of the base appeared as determinant [pyridine: 29%; 2,2'-bipyridine (Bipy): 51%; 1,10-phenanthroline (Phen): 87%].

Phenylene bis(lead triacetate) reagents, generated *in situ* from the corresponding bis(boronic acid) derivatives and lead tetraacetate, react with the α-methyl Meldrum's acid derivative to afford the *meta*- or *para*-phenylene bis(Meldrum's acid) derivatives in *ca* 45% yield.[110] Similarly to malonic acid compounds, the unsubstituted Meldrum's acid was very slow to react and the only observed product was the α,α-diarylated product in 7-17%.

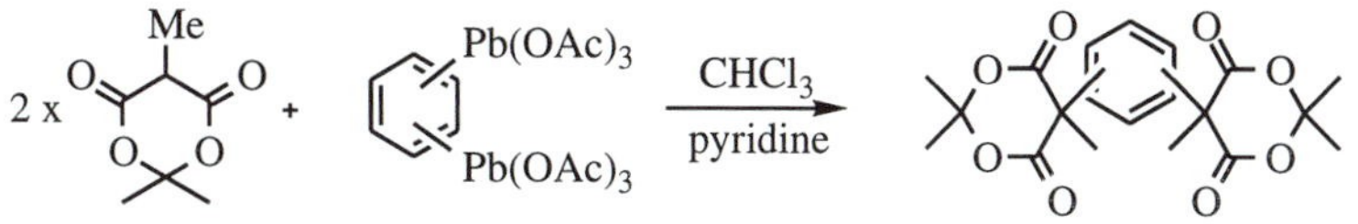

The 5-substituted derivatives of barbituric acid behaved similarly to the Meldrum's acid congeners to afford good yields of the quaternary derivatives. For example, phenobarbital [(74), R[1] = Et, Ar = Ph] was obtained in 91% by phenylation of 5-ethylbarbituric acid [(73), R = Et] with phenyllead triacetate. However, in sharp contrast, the 5-unsubstituted barbituric acid behaved like dimedone to give good yields of the product of diarylation (55-61% with 2 equivalents of aryllead triacetate).[159]

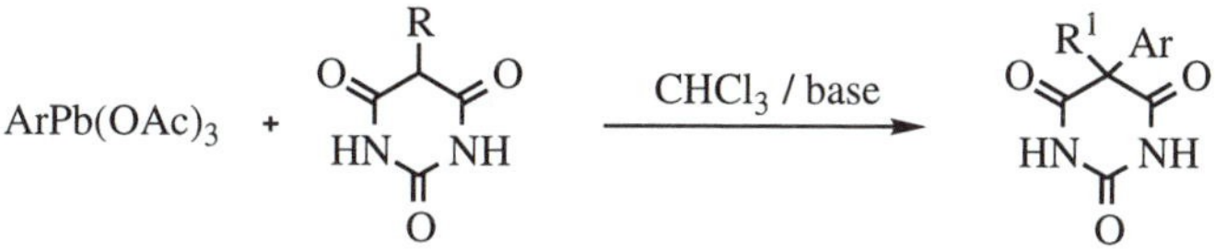

R = H	R[1] = Ar, 55-61%
R = Et	R[1] = Et, 81-91%

Ethyl α-cyanoacetate has a lower pK_a and a higher kinetic acidity than diethyl malonate.[160] However it did not react with aryllead triacetates, but its α-substituted derivatives showed a higher reactivity than the corresponding malonic acid derivatives. Good yields of the α-aryl compounds were obtained either by using better complexing bases, such as DMAP or phenanthroline in chloroform, or by performing the reaction in DMSO.[161] As DMSO is known to form strong complexes with organolead compounds,[162] the presence of pyridine or any other complexing base is not required.

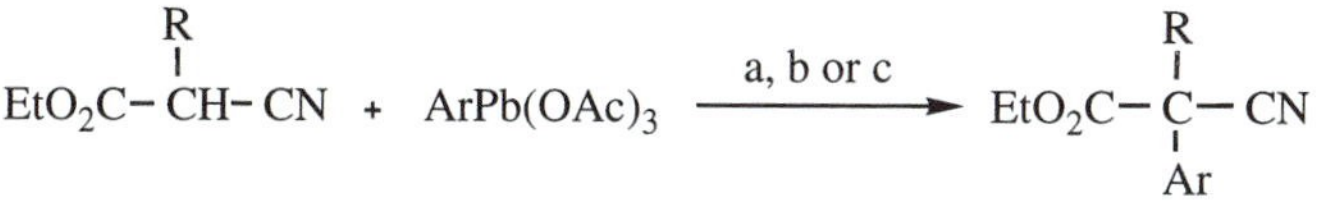

R = Et	a) DMSO	70-81%
R = Et	b) DMAP / CHCl$_3$	72%
R = *i*-Pr	c) Phen / CHCl$_3$	54%

Substituted malononitriles again behaved similarly, but more slowly, to give the α-arylmalononitriles in moderate yields.[161]

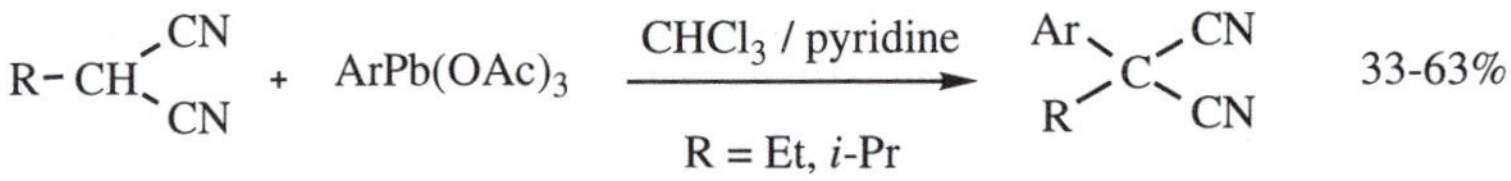

$$R-CH\begin{smallmatrix}CN\\CN\end{smallmatrix} + ArPb(OAc)_3 \xrightarrow[\text{R = Et, }i\text{-Pr}]{\text{CHCl}_3 \text{ / pyridine}} \overset{Ar}{\underset{R}{C}}\begin{smallmatrix}CN\\CN\end{smallmatrix} \quad 33\text{-}63\%$$

As α-substituted malonates are good substrates for arylation, diethyl α-acetamidomalonate could have been a convenient substrate for the synthesis of α-arylglycines. However, unlike its carbon analogues, it did not react with aryllead triacetates. The more acidic 5-oxazolone derivatives (75), existing as the enol in chloroform, underwent rapid arylation with aryllead triacetates under mild conditions to give the 4-aryloxazolone, which can be hydrolysed and decarboxylated.[163] (Scheme 7.12) Combined with the use of enzymatic resolution (with either an acylase such as PKA = porcine kidney acylase or with an esterase such as Subtilisine Carlsberg esterase), this method provides an efficient route to a wide variety of optically active D- and L-arylglycines.[164] (Scheme 7.13)

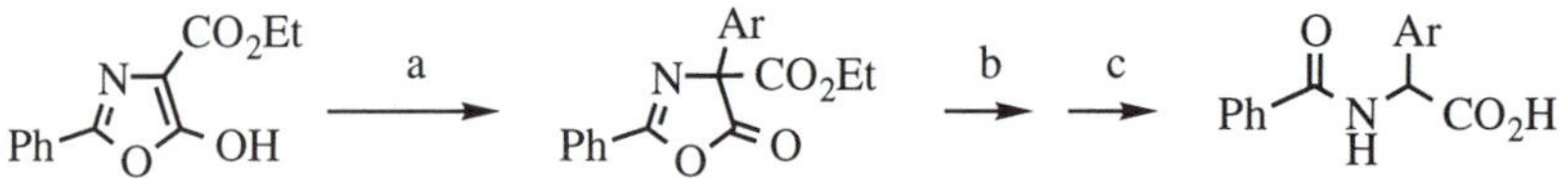

a) ArPb(OAc)₃ / CHCl₃ / pyridine / 40°C   overall yields: 70-94%
b) NaOH, EtOH, H₂O, Δ
c) H₃O⁺

Scheme 7.12

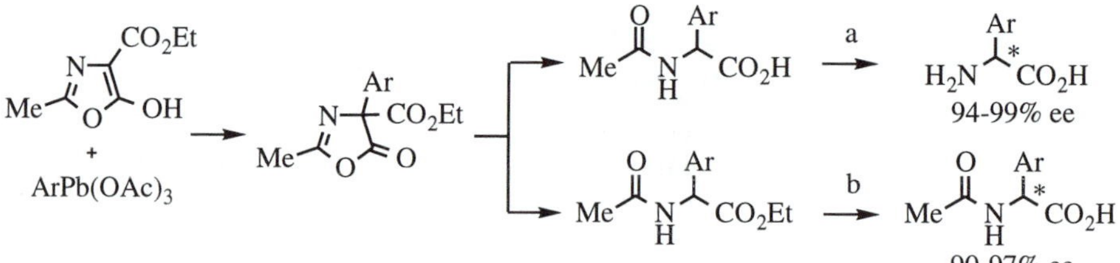

a) PKA b) Subtilisine Carlsberg

Scheme 7.13

In an approach towards the synthesis of the marine natural product cytotoxin Diazonamide A, the arylation of a lactonic β-diester (76) with a tyrosinyllead triacetate was realized and afforded (77) as a mixture of diastereomers in 85% overall yield from 3-iodotyrosine.[165]

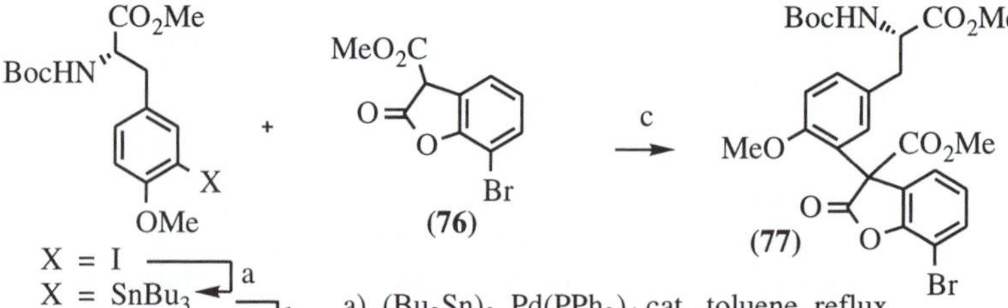

X = I
X = SnBu₃ ◄ a
X = Pb(OAc)₃ ◄ b

a) (Bu₃Sn)₂, Pd(PPh₃)₄ cat., toluene, reflux
b) Pb(OAc)₄, Hg(OAc)₂ cat., CHCl₃, 40°C
c) pyridine (3 equiv.), CHCl₃, 40°C

7.2.3 Ketones and derivatives

The reactivity of ketones towards aryllead (IV) triacetates is quite different from their reactivity towards lead (IV) tetraacetate, with which they are easily functionalised to give the α-acetoxyketones (see section 7.1.1.2). Under the usual conditions for arylation, ketones remain generally unaffected. Only ketone enolates and some specially activated ketones have been successfully arylated with aryllead reagents.

7.2.3.1 Ketones

The enolate salts of simple ketones react with aryllead reagents, but the reaction is of little practical value. Indeed, only trisubstituted α-carbon atoms are reactive. The case of the potassium enolate of cyclohexanone derivatives is significant: no reaction with cyclohexanone enolate (78), 36% with the mixture of enolates of 2-methylcyclohexanone (79), and 75% in the case of 2,6-dimethyl cyclohexanone enolate (80).[10,11]

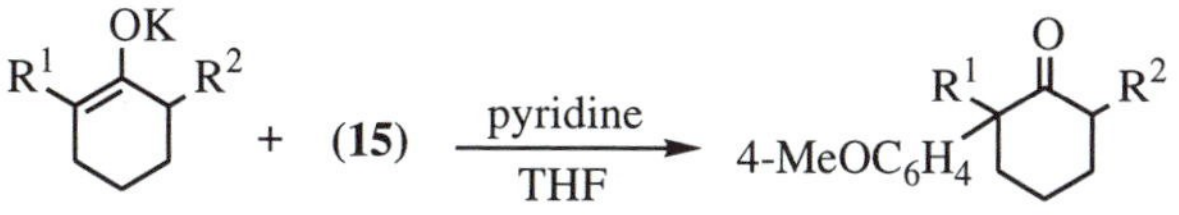

(78)	$R^1 = R^2 = H$	No reaction
(79)	$R^1 = H, R^2 = Me$ (33%) $R^1 = Me, R^2 = H$ (67%)	36%
(80)	$R^1 = R^2 = Me$	75%

The silyl enol ethers of ketones react with aryllead triacetates to afford completely different types of product, depending on the nature of the substrate. The reaction of cyclohexanone trimethylsilyl enol ether with 4-methoxyphenyllead triacetate (15) afforded a mixture of 2-*p*-anisylcyclohexanone (44%) and 2-acetoxycyclohexanone (32%).[44]

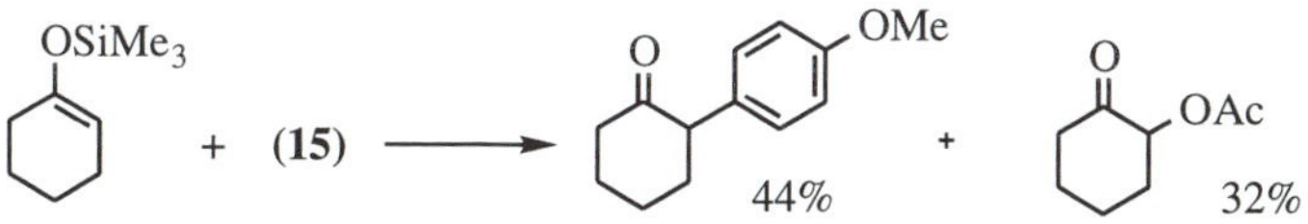

On the other hand, trimethylsilyl enol ethers of acetophenone derivatives (81) did not afford the expected arylacetophenones. Instead high yields of the corresponding arylphenacyllead (IV) diacetates (82) were isolated.[44] Originally reported to be performed under neutral conditions, this reaction in fact requires an acid catalysis, which is provided either by traces of dichloroacetic acid contained in the reagent or by addition of boron trifluoride-ether.[166]

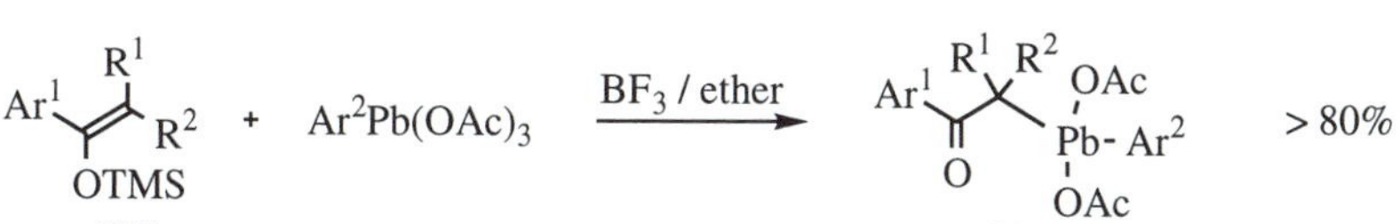

In the case of non-activated ketones bearing an aryl group on the α-carbon, a slow reaction was observed with aryllead triacetates, which afforded only poor to modest yields of arylated products.[10,11,133] However, this reaction was fast enough to compete with the arylation of α-hydroxy-methyleneketones in the usual conditions (see section 7.2.2.1).[133]

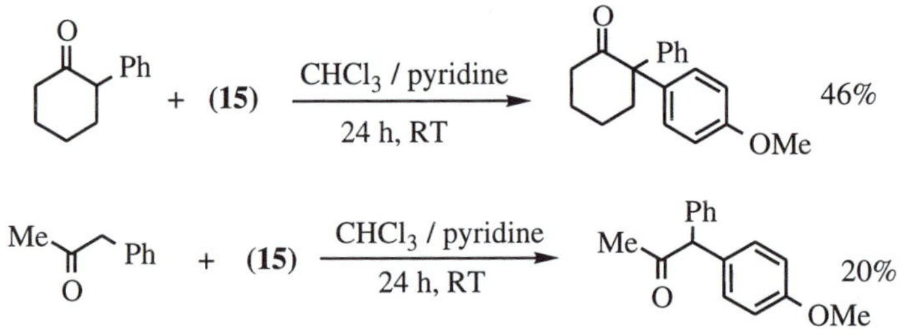

7.2.3.2 α-Substituted ketones

Variable results were obtained with a special type of ketone derivatives, the ketones bearing an arylheteroatom group on the α-carbon, which can present a high degree of enolisation. The heteroatom can be nitrogen (3-oxo-2,3-dihydroindole), oxygen (3-benzofuranone) or sulfur (arylthioflavanone, for example).

N-Substituted 3-oxo-2,3-dihydroindoles underwent arylation with aryllead triacetates in the chloroform-pyridine system in generally good yields.[146]

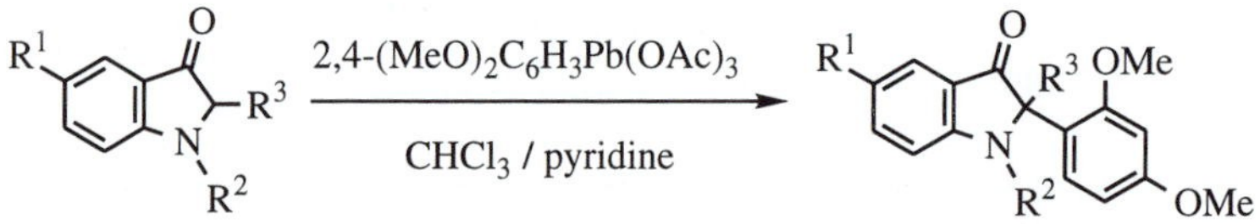

R^1 = H, Me or MeO; R^2 = Ac or PhSO$_2$ R^3 = H   26-81%
R^3 = Me   72%
R^3 = Ph   42%

An oxygen analogue, the 3-(2*H*)-benzofuranone, reacted with aryllead triacetate to afford either the mono or the diaryl derivative. In the usual system (chloroform, pyridine), only modest yields were obtained, but use of *N,N,N',N'*-tetramethylguanidine (TMG) allowed the reaction to proceed in better yields. However, the scope of this reaction is narrow, as only the diphenyl and the mono 2,4,6-trimethoxyphenyl derivatives were obtained in clean reactions in moderate to good yields.[140,167]

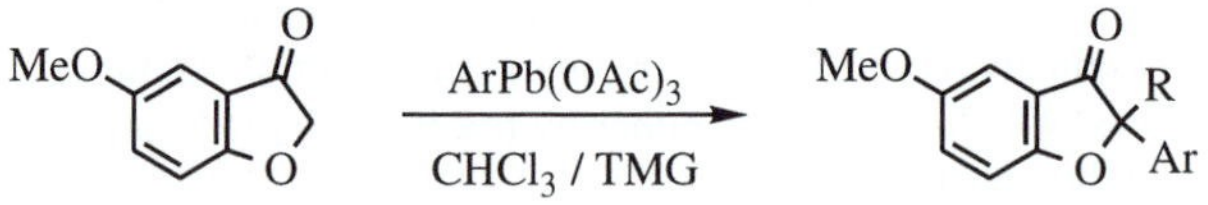

phenyllead triacetate: 1 equiv.   R = Ph; Ar = Ph   21%
phenyllead triacetate: 2.2 equiv.   R = Ph; Ar = Ph   84%
R = H; Ar = 2,4,6-(MeO)$_3$C$_6$H$_2$ 31%

The trisubstituted α-carbon of α-phenylthioketones reacted with aryllead triacetates under classical conditions to afford good to high yields of the α-aryl α-phenylthio derivatives, without competing oxidation of the sulfur of the phenylthio group. Sometimes, α-acetoxylation was a competing side reaction, particularly in the case of *ortho*-substituted aryllead derivatives; for example, a 55% yield of (**84**) was reached in the reaction of (**83**) with the 2,4,6-trimethoxyphenyllead compound (**48**).[167,168]

The activating α-phenylthio group can then be removed either by oxidation with dimethyldioxirane followed by thermolysis to afford the isoflavones or through reduction by nickel boride to give the corresponding isoflavanones.[168]

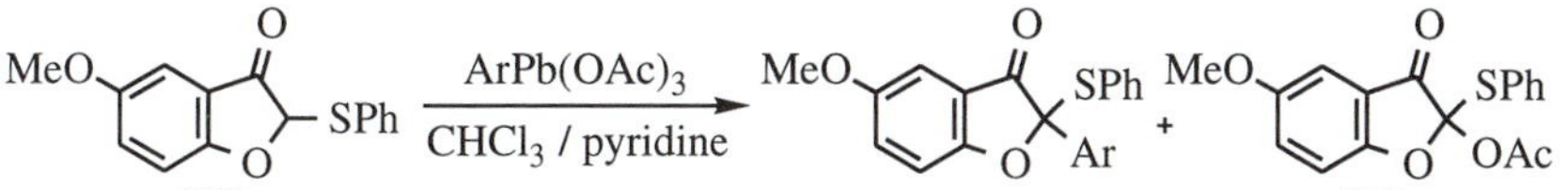

(83) → (84) ref. 167

14-92%   0-55%

ArPb(OAc)₃, CHCl₃ / pyridine, R = H, R = 4-MeO-C₆H₄, 50-93% / 82-99%, 0-10% ref. 168

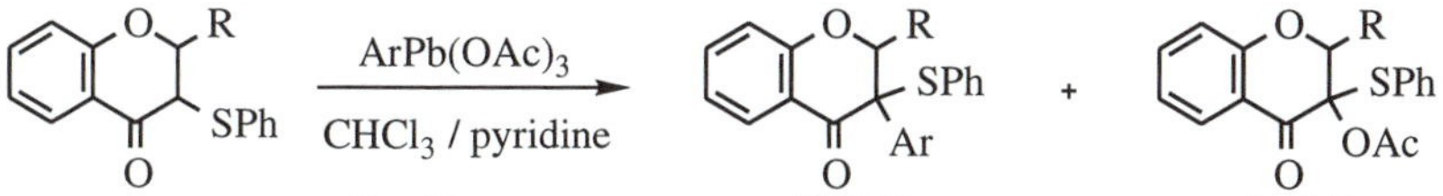

This reaction was used in the synthesis of biflavonoid structures (85), which are analogues of natural compounds isolated from *Garcinia* species, members of the Guttiferae family.[112,169]

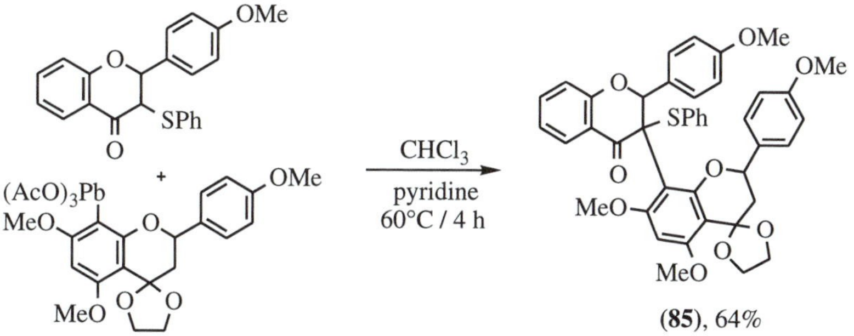

(85), 64%

The α-phenylsulfonylketone analogues reacted similarly with aryllead triacetates to yield the corresponding α-aryl derivatives in good yields (69-74%) except in the case of 2,4,6-trimethoxy-phenyllead triacetate with which no reaction took place.[169,170]

7.2.3.3 Enamines

The reaction of enamines with aryllead (IV) triacetates presents some similarities with their reaction with lead (IV) tetraacetate. The enamines of aldehydes do not react, and the enamines of ketones undergo an exothermic reaction to afford moderate to good yields of α-arylketones. When the ring size or the steric hindrance in the vicinity of the double bond increases, the yield falls significantly, and α-acetoxylation becomes a major competing pathway.[171]

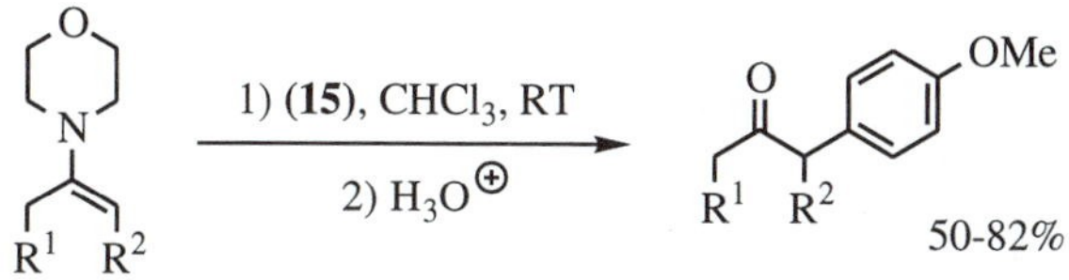

1) (15), CHCl₃, RT   2) H₃O⊕   50-82%

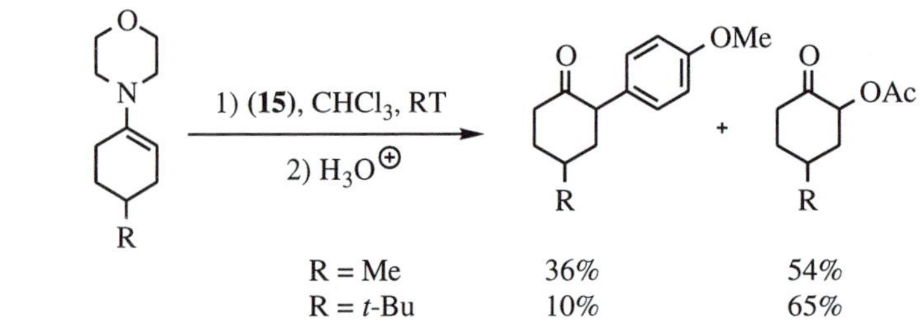

R = Me	36%	54%
R = *t*-Bu	10%	65%

7.2.4 Nitroalkanes and nitroacetic acid derivatives

Under the classical conditions (chloroform, pyridine) for the arylation reactions of β-dicarbonyl compounds, nitroalkanes react very slowly to afford modest yields of the *C*-arylated products.[172,173] This phenomenon was attributed to the lower kinetic acidity of nitroalkanes. Dipolar aprotic solvents can be used with aryllead reagents to conduct the arylations, as the deprotonation is accelerated in these solvents. The reactions are preferably performed in DMSO. As DMSO forms strong complexes with organolead compounds,[162] the presence of pyridine or any other complexing base is not required and the reaction proceeded smoothly at 40°C in 30 hours to afford good yields of the α-arylated derivatives.

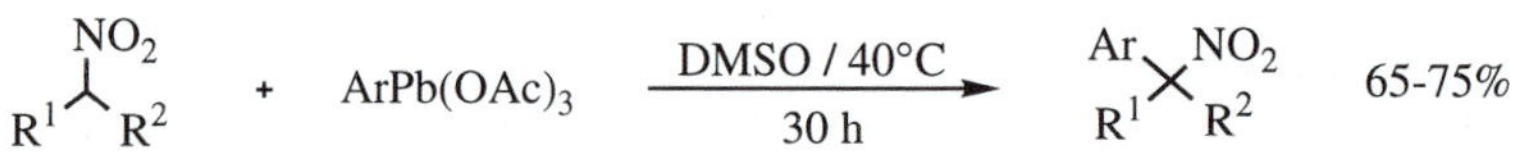

The reaction rate can be significantly accelerated by performing the reaction with the nitronate salts. It then proceeds at room temperature in 30 minutes to afford 56-71% yields of isolated products.

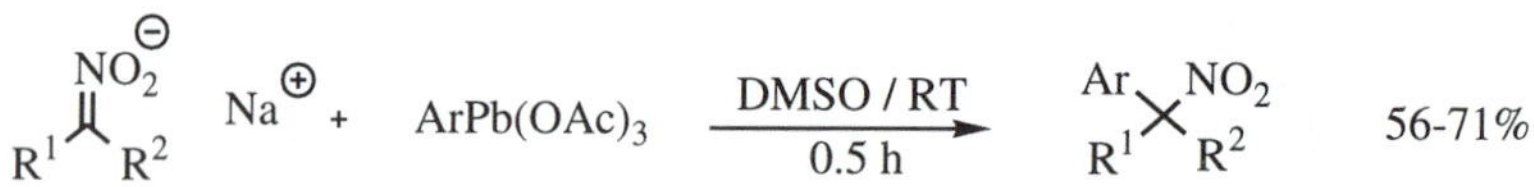

In the case of compounds bearing two α-hydrogen atoms, the mono- or the di-arylated derivatives can be selectively prepared by using 1.1 or 2.2 equivalents of the aryllead triacetate. This is in contrast with β-dicarbonyl compounds for which the diaryl derivatives were preferentially obtained.

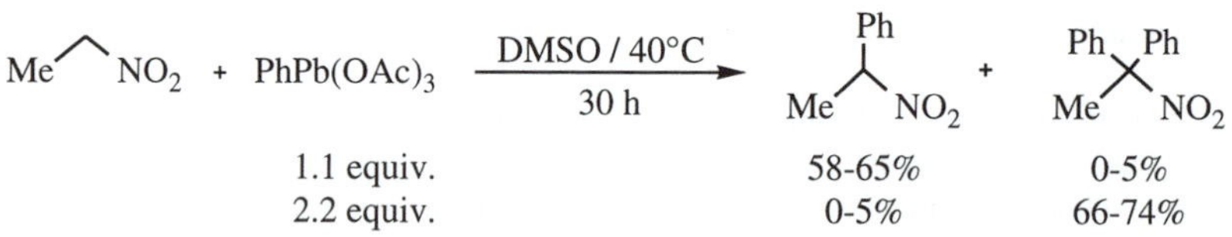

1.1 equiv.	58-65%	0-5%
2.2 equiv.	0-5%	66-74%

Arylation of methyl nitroacetate can also be controlled so as to proceed to mono or to diarylation. However, stepwise arylation is more effective to obtain the diaryl product in good yield.

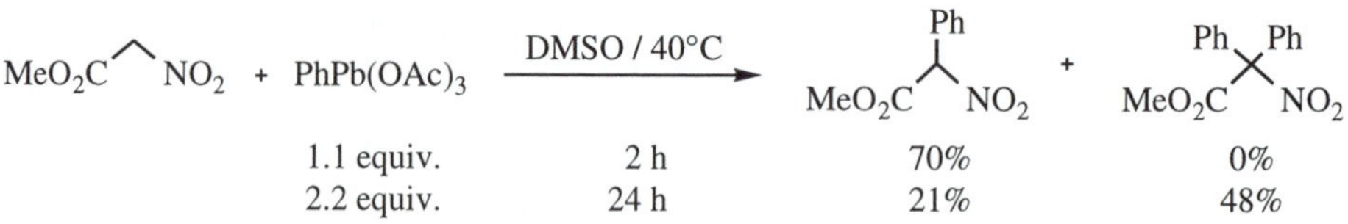

1.1 equiv.	2 h	70%	0%
2.2 equiv.	24 h	21%	48%

$$\underset{\underset{MeO_2C}{}}{\overset{Ph}{\bigwedge}}\underset{NO_2}{} \quad + \quad PhPb(OAc)_3 \quad \xrightarrow[24\ h]{DMSO\ /\ 40°C} \quad \underset{\underset{MeO_2C}{}\ NO_2}{\overset{Ph\ \ Ph}{\times}} \quad 71\%$$

7.2.5 Inorganic nucleophiles

A limited range of inorganic nucleophiles has been found to react with aryllead triacetates. Aryl iodides are easily produced by the reaction of aryllead triacetates with aqueous potassium iodide.[174]

$$ArPb(OAc)_3 \ + \ KI \quad \xrightarrow{H_2O\ /\ RT} \quad ArI \qquad >90\%$$

The reaction of sodium azide with aryllead triacetates in DMSO leads to aryl azides in high yields. A one-pot system can be used with profit to transform the easily available arylboronic acids into aryl azides by metal-metal exchange followed by reaction with sodium azide.[175]

$$ArPb(OAc)_3 \ + \ NaN_3 \quad \xrightarrow{DMSO\ /\ RT} \quad ArN_3 \qquad > 80\%$$

$$ArB(OH)_2 \quad \xrightarrow[2)\ NaN_3\ /\ DMSO]{1)\ Pb(OAc)_4} \quad ArN_3 \qquad 59\text{-}82\%$$

Although these two reactions are similar to the corresponding reactions of arenediazonium salts, aryl radicals are not involved in the reaction of the lead reagents and these reactions are better explained by a ligand coupling mechanism.

A third type of reaction with inorganic nucleophiles is the reaction of aryllead triacetates with boron trifluoride etherate, which affords the corresponding aryl fluorides in moderate to good yields (49-82%).[176,177] In the case of *ortho*-substituted aryllead reagents (*o*-fluoro, 0% and *o*-methoxy, 14%), an *ortho*-effect was invoked to explain the poor reactivity of these compounds. Triarylboroxins, electron-rich aryltrimethylsilanes and some arenes, which yield aryllead triacetate when treated with lead tetraacetate in acid catalysed reactions, are also converted into aryl fluorides when stirred with lead tetraacetate in the presence of boron trifluoride etherate. Originally believed to proceed by a ligand coupling-like mechanism,[176] these reactions of fluorodeplumbation are now considered to be acid-catalysed reactions involving the formation of aryl cation (or incipient cation) by heterolytic cleavage of the carbon-lead bond.[177]

$$ArPb(OAc)_3 \ + \ BF_3.Et_2O \quad \xrightarrow{RT} \quad \underset{0\text{-}82\%}{ArF} \ + \ \underset{0\text{-}11\%}{ArH}$$

The last system which has been briefly reported is the reaction of aryllead triacetates with copper (I) derivatives, such as copper (I) chloride, bromide and cyanide, which afford the corresponding aryl chlorides, bromides or cyanides.[10] As the yields are only moderate, these reactions do not present any synthetic interest and they are only of mechanistic interest. A catalytic effect of copper was claimed to be operative in the reductive displacement of lead (II) by Cl⁻, Br⁻ or CN⁻, and this catalytic effect of copper compounds is also involved in the reactions of amine derivatives with aryllead reagents (see section 7.6.1).

$$ArPb(OAc)_3 \ + \ CuX \quad \xrightarrow[(X = Cl,\ Br\ or\ CN)]{Me_2NCHO} \quad ArX$$

7.2.6 Heteroaromatic lead compounds

The reaction of aryllead triacetates is not limited to substituted phenyllead derivatives. It has also been extended to the use of heteroaryllead compounds, derived from furan and thiophene.[113] Due to their relative instability and moisture-sensitivity, they are best prepared by metal exchange, such as mercury-lead or tin-lead, and used *in situ* in the reactions of carbon nucleophiles.

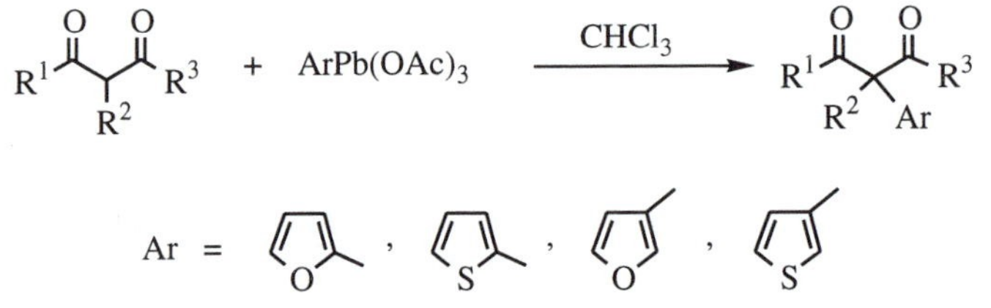

β-Dicarbonyl compounds generally reacted with heteroaryllead reagents to afford the α-aryl derivatives in synthetically useful yields. However, in the reaction of 2-thienyllead triacetate with 2-ethoxycarbonylcyclopentanone (**86**), the *C*-arylation product was observed in a low yield (9%), together with the presence of a dimer resulting from a radical oxidative coupling induced by the organolead reagent. Use of 2-thienyllead tribenzoate, a weaker oxidant, restored the classical reactivity. When the reaction was performed in pyridine as solvent, the arylation product, the 2-(2-thienyl)cyclopentanone derivative, was obtained in 76% yield.[113]

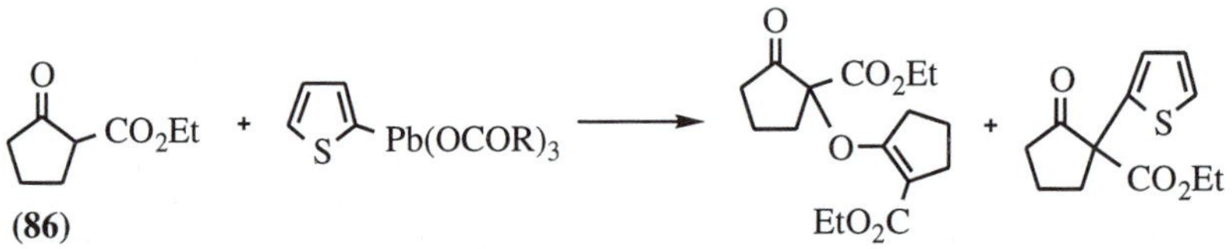

7.3 VINYLATION REACTIONS

In parallel to their work on the arylation reactions with aryllead tricarboxylates, Pinhey *et al.* have examined, although in much less detail, the reactivity of vinyllead tricarboxylates as vinylation reagents for soft nucleophiles.[10,11]

The possible involvement of vinyllead tricarboxylates has been suggested by Corey and Wollenberg[178] and by Larock *et al.*[179] as reactive intermediates. Their existence is supported by NMR studies[118,180,181] and by the isolation of cyclopent-1-enyllead triacetate.[182] Vinyllead triacetates are extremely unstable compounds which generally decompose by formation of vinyl cations and afford acetylenes or enol acetates, depending on the precursor and the substitution pattern.[178-181,183] Vinyltin derivatives react with lead tetraacetate to yield usually the alkynes,[178,183] and vinyl-mercurials react with lead tetraacetate to yield the corresponding enol acetates.[179] However, addition of mercury(II) salts to the vinyltin reactions draw the reaction towards formation of the enol acetate.[180] The involvement of an alkylidenecarbene intermediate as an alternative decomposition pathway has been excluded.[180]

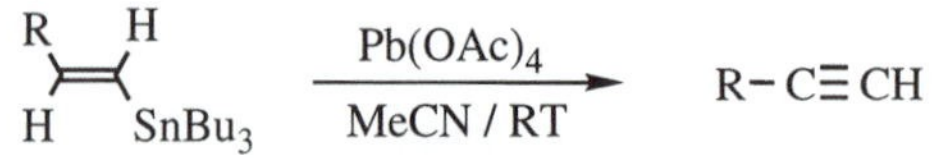

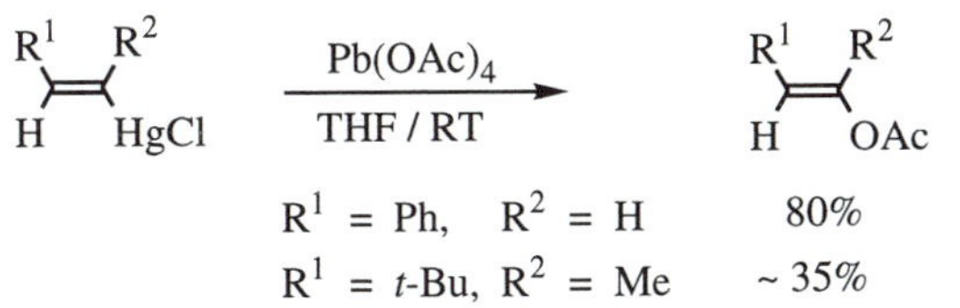

ref. 179

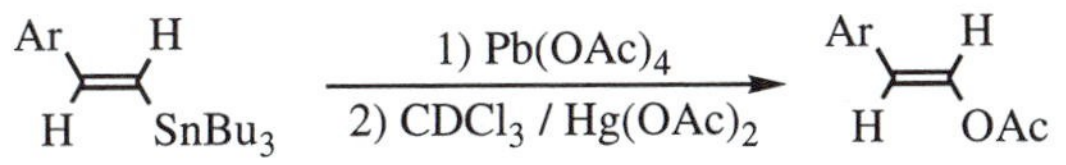

$R^1 = Ph, \quad R^2 = H \qquad 80\%$

$R^1 = t\text{-Bu}, \quad R^2 = Me \qquad \sim 35\%$

ref. 180

In the presence of a soft nucleophile, the unstable intermediate vinyllead triacetate (87) behaves as a *C*-vinylating reagent, whatever the mode of formation of the vinyllead intermediate (metal-metal exchange from divinylmercury or from vinyltin compounds).[116,182]

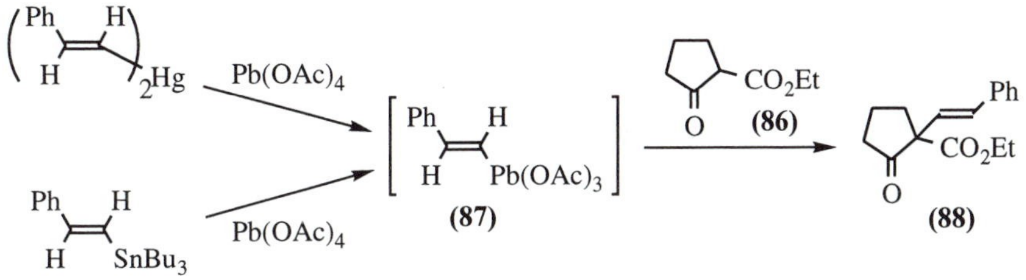

β-Diketones and β-ketoesters possessing only one α-hydrogen react smoothly with vinyllead tricarboxylates to afford the corresponding α-vinyl derivatives.[182] When the vinyllead reagent is generated *in situ* by metal-metal exchange between lead tetraacetate and divinylmercury, tributylvinyltin or vinylzinc chloride, an excess of the vinyl precursor is required. This is unsatisfactory in the case of expensive vinyl groups.[184] However when the metal-metal exchange is performed on the trimethyltin derivative, only one equivalent is required to afford good yields of coupling products with nucleophiles. The most efficient system is the metal-metal exchange between the (trimethyl)vinyltin precursor and lead tetrabenzoate, which provides significantly higher yields of coupling product.[117] (Table 7.3)

Table 7.3: Vinylation of ethyl 2-oxocyclopentanecarboxylate (86) with *in situ* generated styryllead triacylate in chloroform.

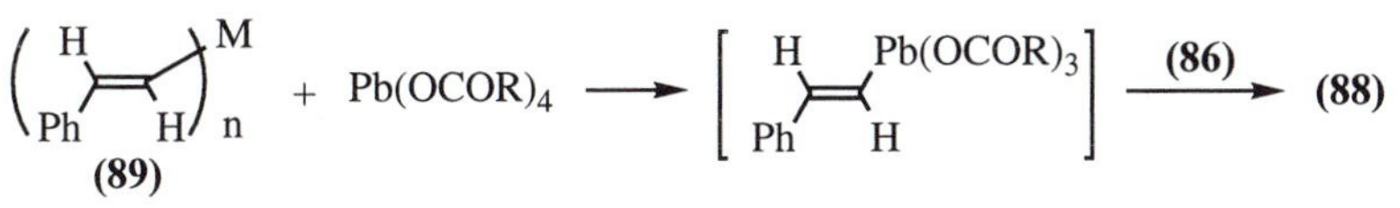

(89)		Pb(OCOR)₄	Pyridine	(88) (%)	Ref
M = Hg	n = 2	R = Me	-	65	116
M = Bu₃Sn	n = 1	R = Me	+	55	182
M = Me₃Sn	n = 1	R = Me	+	78	117
M = Me₃Sn	n = 1	R = Ph	-	84	117

An elegant application of the reaction of vinyllead triacetates with β-ketoesters was described in a convergent synthesis of (+)isocarbacyclin, an analogue of prostacyclin (PGI₂). This synthesis involved the vinylation of a benzyl β-ketoester.[119,184] The benzyl ester was later removed under conditions compatible with the other functionalities of the molecule.

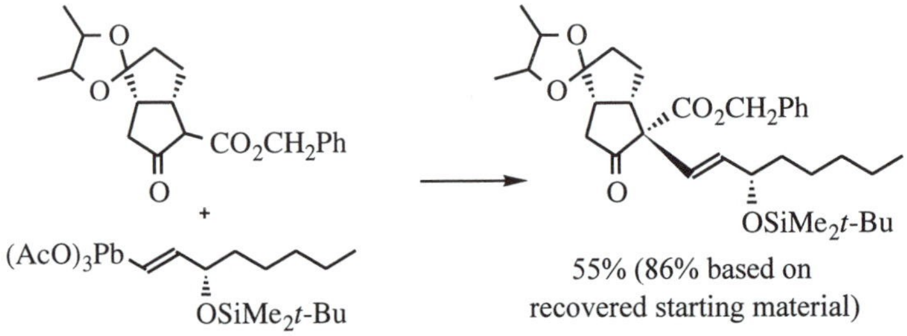

The range of dicarbonyl compounds which were successfully vinylated include aliphatic or cyclic β-diketones, ethyl or benzyl β-ketoesters and 5-ethylbarbituric acid derivatives. Vinylogous β-ketoesters reacted with vinyllead triacetates in the same way as they do with aryllead tricarboxylates to afford the products with the vinyl group linked to the carbon α to the ester group.[116,182] An heterocyclic lactonic β-diester gave also good yields of the corresponding α-vinyl aminoacid derivatives, after hydrolysis of the 5-oxazolone ring.[164]

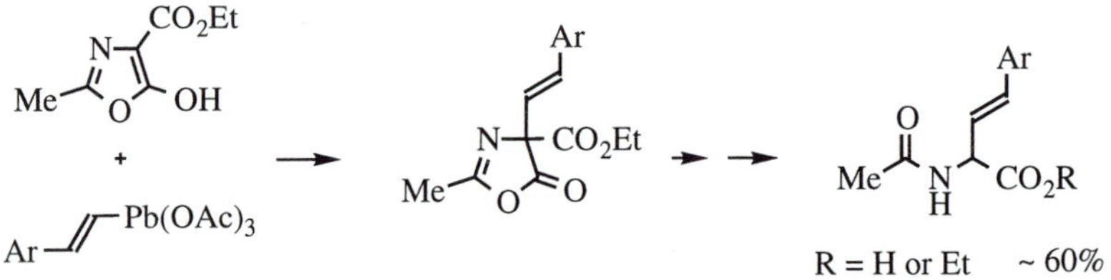

Good yields were also obtained in the reaction of vinyllead derivatives, generated from the trimethyltin derivatives, with the sodium salts of 2-nitropropane or of nitrocyclohexane.[117]

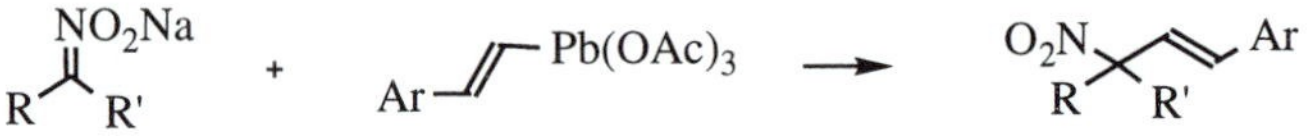

High yields of 6-(*E*)-styrylcyclohexa-2,4-dienones can be isolated from the reaction of 2,6-di-methylphenol derivatives. When the 4-position is not substituted or substituted by a methoxy group, moderate yields of the dienone or of the products of dimerisation of the dienone are obtained.[185]

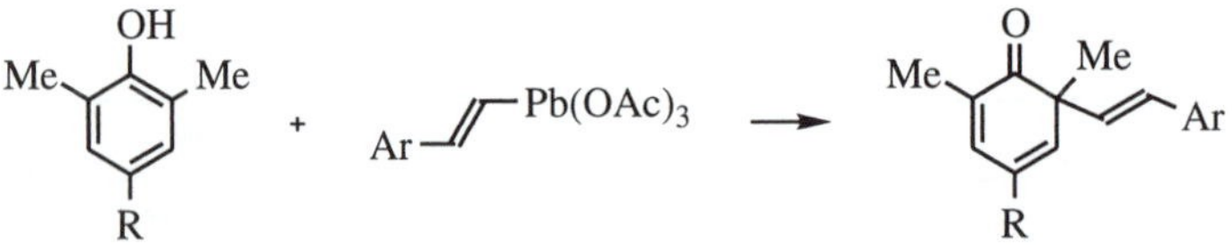

R = Me	Ar = Ph	82-83%
R = H	Ar = Ph	40%
R = H	Ar = 4-MeOC₆H₄	41%
R = Br	Ar = Ph	67-74%
R = OMe	Ar = Ph	30%

In the case of the reaction of 2,6-dimethylphenol (R = H) with (E)-p-methoxystyryllead triacetate (Ar = 4-methoxyphenyl), the 6(E)-styrylcyclohexa-2,4-dienone was not detected, and the vinylation product was isolated as a mixture of dimers (90) in a modest yield (41%). Moreover, the presence of 2,2',6,6'-tetramethyldiphenoquinone (91) was also noted.

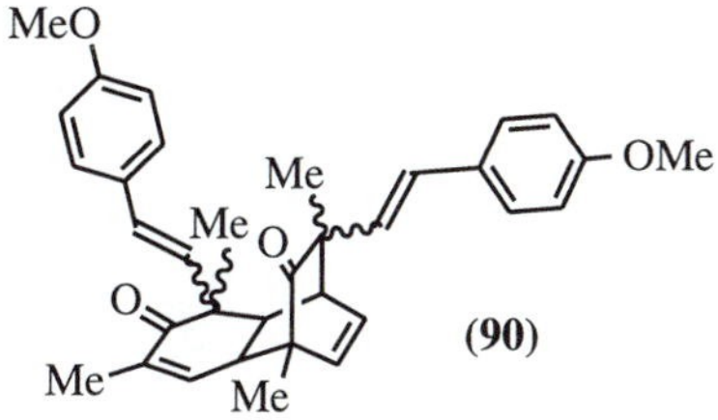

7.4 ALKYNYLATION REACTIONS

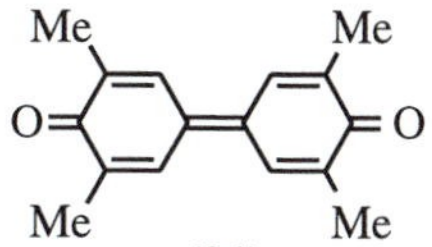

Alkynyllead triacetates can be prepared by metal-metal exchange between lead tetraacetate and either dialkynylmercury or alkynyltrimethyltin derivatives.[186,187] A straightforward and more convenient method is the reaction of alkynyllithium with lead tetraacetate with[188] or without[189] addition of zinc chloride. The chloroform solution of alk-1-ynyllead triacetate behaved in the same way towards soft carbon nucleophiles as do their aryl and vinyl analogues. The acetylenic group of a number of alkynyllead derivatives has been introduced into β-dicarbonyl compounds and nitronate salts in good to high yields.[186,187]

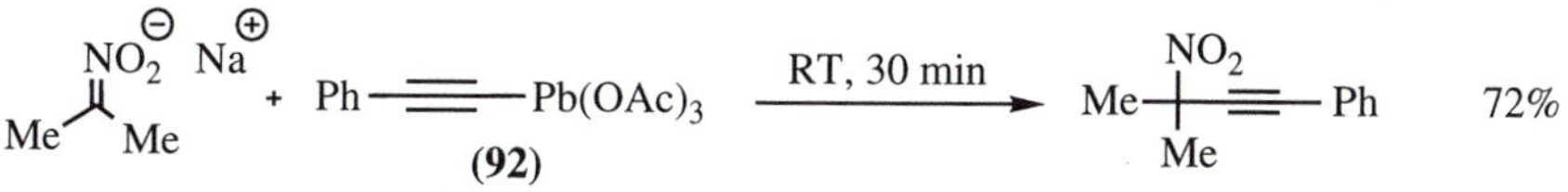

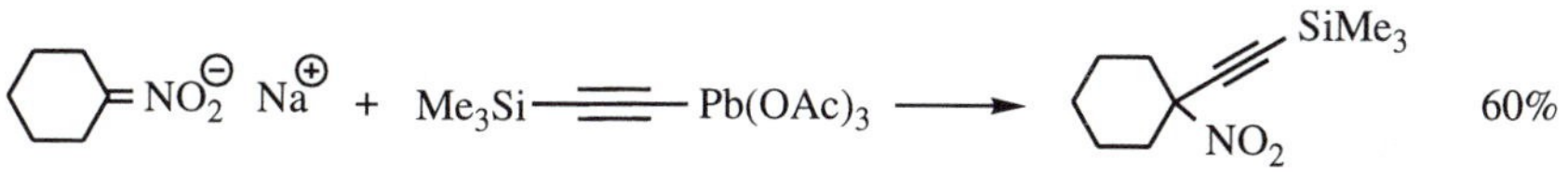

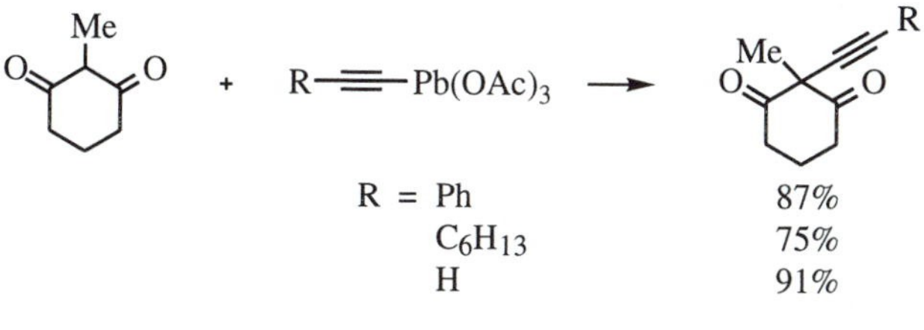

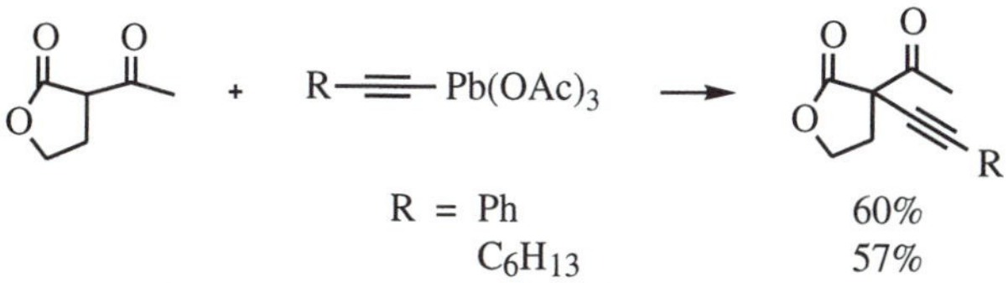

Phenylethynyllead triacetate (92) reacted with mesitol (36), in the presence of 2,2'-bipyridine acting as the base, at low temperature to afford the 2,4-cyclohexadienone product which underwent *in situ* a

[2+4] cycloaddition, and was isolated as its dimer in 78% yield.[185] In the case of the 2,6-dimethylphenol (4), the analogous dimer (93) was isolated in a very poor yield (12%), and its structure determined by single crystal X-ray analysis. The major product was the 2,2',6,6'-diphenoquinone (91) (28%), showing that phenylethynyllead triacetate is a strong enough oxidant to induce oxidative coupling of the phenol.

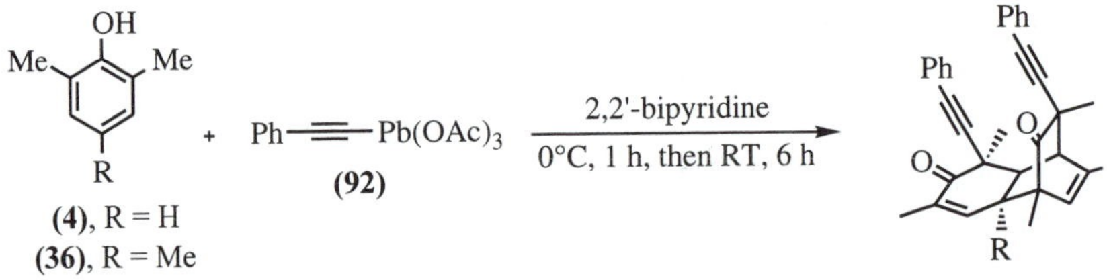

(4), R = H
(36), R = Me

(92)

2,2'-bipyridine
0°C, 1 h, then RT, 6 h

(93)

7.5 MECHANISTIC STUDIES ON THE REACTIONS OF ORGANOLEAD DERIVATIVES

The mechanistic studies on the reactions of organolead tricarboxylates have been performed only with aryllead compounds. Due to the similarity in the reactivity between aryl, vinyl and alkynyl compounds, it is likely that a similar ligand coupling mechanism is operating in the three systems. In the case of the vinyl reagents, the intervention of an alkylidenecarbene has been excluded.[180]

The general patterns of the reactivity of aryllead triacetates could be explained by the occurrence of free radicals, generated by a single-electron-transfer mechanism or via a $S_{RN}1$ process.

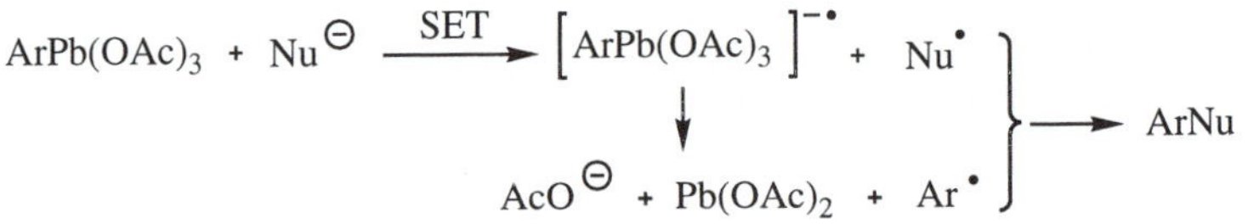

$$ArPb(OAc)_3 \ + \ Nu^{\ominus} \ \xrightarrow{\text{SET}} \ \left[ArPb(OAc)_3\right]^{-\bullet} \ + \ Nu^{\bullet}$$

$$AcO^{\ominus} \ + \ Pb(OAc)_2 \ + \ Ar^{\bullet} \Bigg\} \longrightarrow ArNu$$

Scheme 7.14

Such a mechanism was unlikely as addition of an external trap, 1,1-diphenylethylene, had no effect on the course of the arylation of β-ketoesters.[109,190] A second approach involved the use of an internal trapping system which had been successfully used in the study of the radical reactions of arenediazonium salts.[191] The internal trap containing reagent, (*ortho*-allyloxyphenyl)lead triacetate (94), can be easily prepared from the corresponding boronic acid.[192] Reaction with various types of nucleophiles, such as ethyl 2-oxocyclopentanecarboxylate (86), mesitol (36), the sodium salt of nitropropane, iodide and azide always afforded the *C*-arylation products in high yield. No trace of the 3-substituted dihydrobenzofurans, expected in a mechanism involving the intermediacy of free radicals, could be detected.

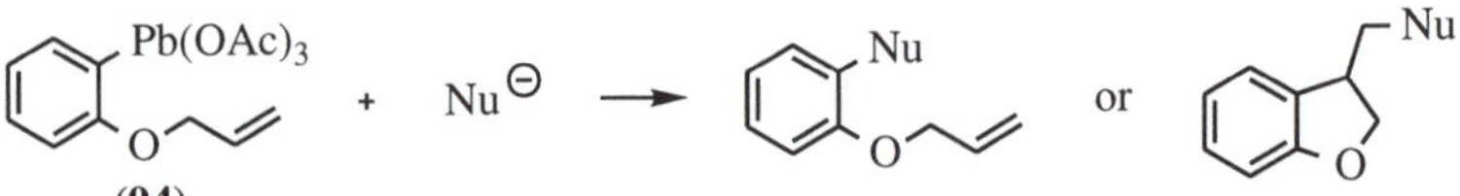

(94)

A ligand coupling mechanism has been proposed to explain the arylation reactions. For ambident substrates, two types of intermediates, resulting from ligand exchange in the first step, can be postulated. For phenols and β-dicarbonyl compounds, either an oxygen-lead or a carbon-lead intermediate are possible. In a second step, ligand coupling then affords the products.[12,109,126,166,193]

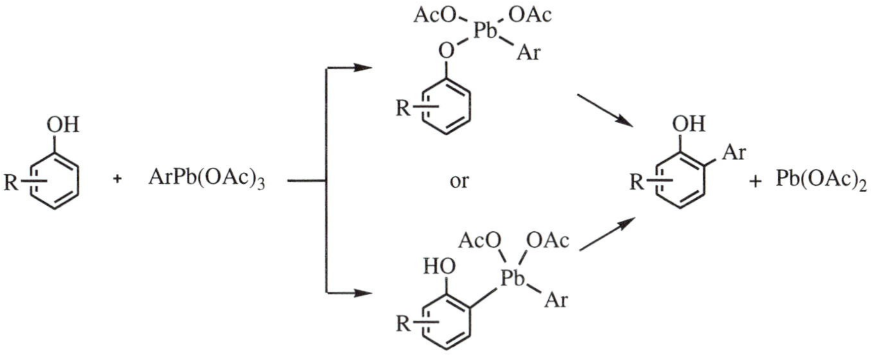

Scheme 7.15: Alternative intermediates in the phenol *C*-arylation reaction

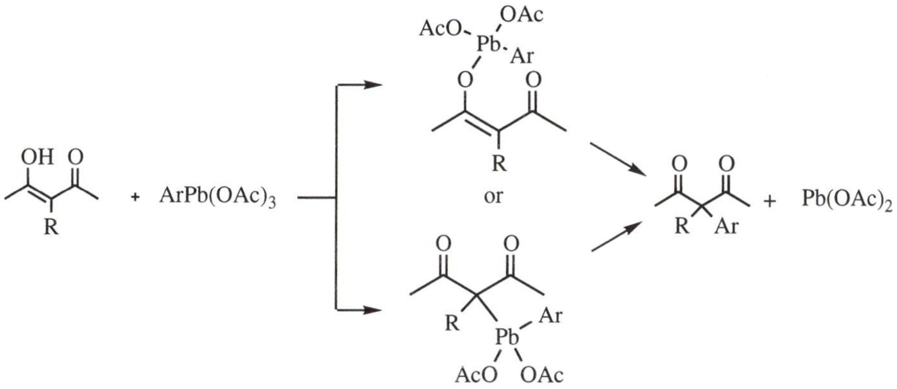

Scheme 7.16: Alternative intermediates in the β-diketone *C*-arylation reaction

A general feature of the arylation reactions with aryllead triacetate is the need for the presence of a complexing agent which can be either 2-3 equivalents of a coordinating base, such as pyridine or the stronger 2,2'-bipyridine or 1,10-phenanthroline, or a coordinating solvent such as DMSO. Therefore, in the case of the coordinating base, the intermediate involved in the arylation step must have two nitrogen ligands on the lead atom. For the detection of an intermediate, either spectroscopic or synthetic approaches can be used. Unfortunately, all attempts to detect such an intermediate by ^{1}H or [207]Pb NMR failed in the case of phenols as well as in the case of β-dicarbonyl compounds.[12,109,126] Pinhey *et al.* have prepared stable unsymmetrical diaryllead diacetates by boron-lead exchange.[194] They used this boron-lead exchange reaction to prepare a diaryllead-type intermediate such as the *O*-methyl ether analogue (**95**) of a possible intermediate bearing a carbon-lead bond, which was supposed to occur in the easy arylation of 3,4,5-trimethylphenol (**45**). However, the compound appeared to be quite stable as well in CHCl$_3$ / pyridine at 60°C even after 6 days as in DMSO under reflux.[193] The intermediate involved in the phenol arylation is therefore more likely to be the oxygen-bound (aryloxy)aryllead diacetate analogous to the intermediate involved in the Wessely α-acetoxylation of phenols.

(95)

In the arylation of methyl-substituted phenols, the rate of *ortho*-arylation increases with the of methyl groups. Moreover, *ortho*-methyl groups show a much greater effect than *meta*- o methyl groups. ^{13}C NMR studies have shown that the *ipso*-carbon of the aryllead reagent pres electrophilic character.[126] Therefore it reacts with the more electron-rich site, an overlap of π-systems of the phenolate ligand and of the Pb-bound aryl group facilitating the arylation.[12,126,193]

In the case of the arylation of β-dicarbonyl compounds, a lead enolate was suggeste intermediate, by analogy with the phenol arylation.[109] However in this case, indirect ev pointing to an intermediate with a covalent carbon-lead bond intervening in the ligand c process was obtained. Treatment of silyl enol ethers of alkylphenylketones with *p*-methoxyphe triacetate in the presence of a catalytic amount of BF$_3$.Et$_2$O afford diorganolead diacetates (see 7.2.3.1). In the case of the unsubstituted acetophenone ($R^1 = R^2 = H$), none of the α-arylketc produced after 3 days in CHCl$_3$ under reflux. But more substituted products (R^1 and/or $R^2 = al$ thermally unstable, and at 60°C they yield the α-arylated ketone, sometimes accompanied corresponding α-acetoxyketone. No symmetrical products, either biaryls or 1,4-diketon produced in these reactions.[166]

(96), 55-99%   OMe   20-50%

a - BF$_3$-Et$_2$O / CHCl$_3$ b - CHCl$_3$ / 60°C

In agreement with the hypothesis that similar intermediates are involved in β-dicarbonyl ary in which the steric decompression resulting from the ligand coupling step favours the arylatioi more substituted β-dicarbonyl substrates, the thermal instability of the (*p*-methoxyphenyl)ph lead diacetates increased in the series:

$$R^1 = R^2 = H \ < \ R^1 = Me, R^2 = H \ < \ R^1 = R^2 = Me$$

Attempts to produce the analogous carbon-lead compounds by reaction of silyl enol et β-dicarbonyl compounds with aryllead triacetates under similar conditions were unsuccessf only products which were detected were the α-arylated β-dicarbonyl derivatives.

An X-ray study of the (4-methoxyphenyl) phenacyllead diacetate (96) with $R^1 = Me$ and showed that the lead atom is heptacoordinate and that the compound presents a pen bipyramidal geometry.[166] The groups bound to the central lead atom by a carbon atom oc nearly axial position. This indicates that, in the first step, the organolead triacetate undergoes a exchange with the enolic compound to afford a species in which the two carbon-bound ligand *trans* diaxial configuration. Thermal pseudorotation of the phenacyl group leads to the *cis*-con which then undergoes the ligand coupling process, with the change of oxidation state from Pb(IV) to Pb(II) as the driving force.[12,166]

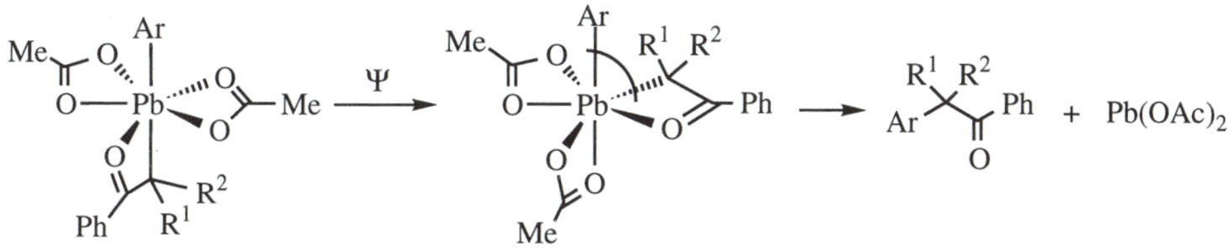

Scheme 7.17

Similarly, in the reaction of aryllead triacetate with phenols, an initial ligand exchange affords an aryl-(aryloxy)lead diacetate intermediate which can have the aryloxy ligand either in the axial position (**97**) or in the equatorial position (**98**). If the aryloxy group is in the axial position, pseudorotation can easily interconvert this conformer with the conformer possessing the aryloxy ligand in the equatorial position (**98**). In this spatial arrangement, the favourable overlap between the π-systems of the two aryl groups makes the ligand coupling process possible.[12,193]

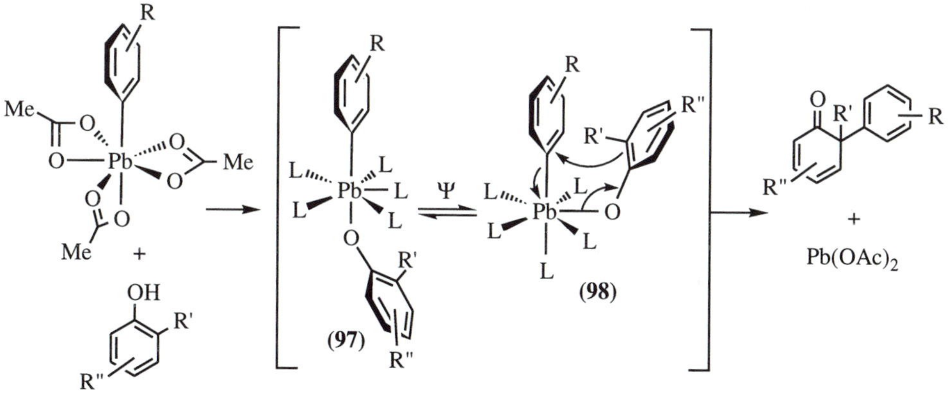

In the course of their studies towards the synthesis of 1,3,5-triphenyl-2,4,9-trithia-1,3,5-triplumba-adamantane, West *et al.* studied the mechanism of the decomposition of organolead iodides.[195] Iodinolysis of alkyltriphenyllead affords the corresponding diorganolead diiodide, which can produce alkyliodo(diphenyl)lead and an unstable alkyllead triiodide by redistribution reactions. The alkyllead triiodide decomposes eventually to give lead diiodide and alkyl iodide. As iodinolysis of (cyclopropyl-methyl)triphenylplumbane (**99**) yielded mainly iodomethylcyclopropane (**100**) instead of the ring-opened products derived from cyclopropylmethyl cations or radicals, these authors suggested that the reductive elimination of RI and lead diiodide from RPbI$_3$ is concerted, that is, in other words, a ligand coupling process.

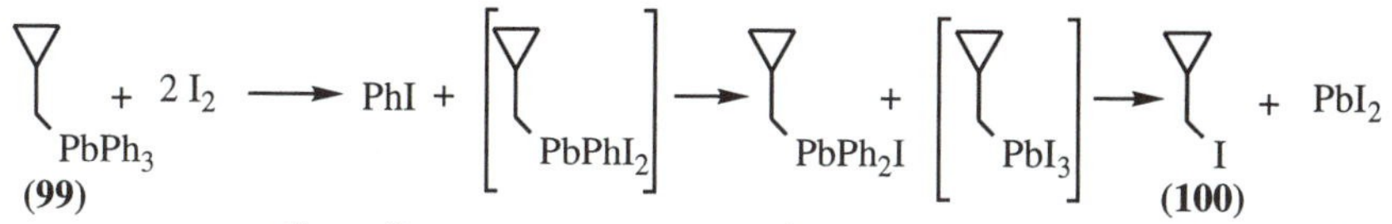

7.6 COPPER-CATALYSED ARYLATION REACTIONS

A different type of ligand-coupling reactions is involved in the copper-catalysed arylation of amines. As anilines react with lead tetraacetate to afford the corresponding α-acetoxyimino derivative (see section 7.1.2.1), their reaction with aryllead triacetates could have been expected to lead to *ortho*-arylanilines. However, such a reaction was never observed. In fact, in the presence of a catalytic amount of a copper salt, a completely different outcome was observed. Instead of the α-*C*-arylation, *N*-monoarylation of anilines was observed.[196-198] Moderate to good yields are generally obtained, for a variety of anilines, even for relatively hindered anilines such as mesitylamine (**101**), which gave high yields of the *N*-aryl products.

The efficiency of the arylation is very dependent upon the basicity of the amines. Electron-poor anilines do not react, while electron-rich anilines give high yields of *N*-arylation products. However, in the case of easily oxidised anilines, oxido-reduction of the aryllead reagent can compete with the *N*-arylation when the steric compression becomes too important. For example, in the case of mesitylamine (**101**), the copper-catalysed reaction with a variety of substituted phenyllead derivatives led to generally high yields of the diarylamines.

However, in its reaction with the sterically hindered 2,4,6-trimethoxyphenyllead triacetate, the predominant product was the product of oxido-reduction of the aryllead reagent.

A variety of azole derivatives reacted with *para*-tolyllead triacetate (**67**) in the presence of copper diacetate to afford the *N*-monoaryl derivatives in good to excellent yields.[199-203] For 1,2,4-triazole, 1,2,3-benzotriazole and indole derivatives, good yields of the *N*-aryl were only obtained upon treatment of the sodium salt of the substrate with *para*-tolyllead triacetate in the presence of copper diacetate at 60-80°C. In the case of aminobenzimidazole, chemoselectivity was observed with the exclusive formation of the *N*-arylaminobenzimidazole in 50% yield.[201]

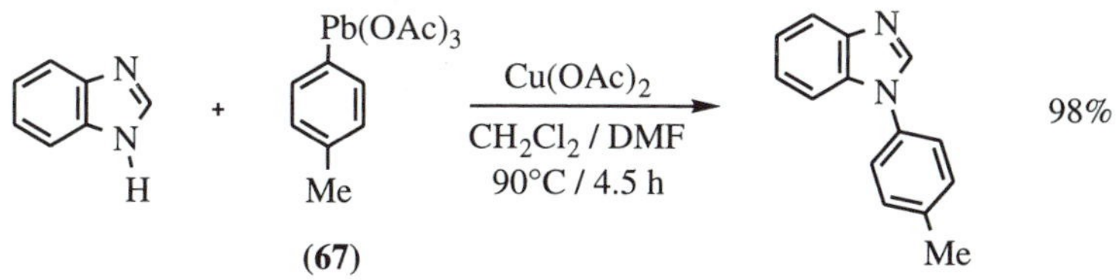

(67)

98%

Copper-catalysed arylation of heterocyclic amines such as piperidine,[198] tetrahydroisoquinoline[198] or 1,6-diazacyclodecane[204] with aryllead triacetates gave only modest to moderately good yields of the *N*-aryl derivative, and the reaction with aliphatic amines led to particularly poor yields of the derived anilines.

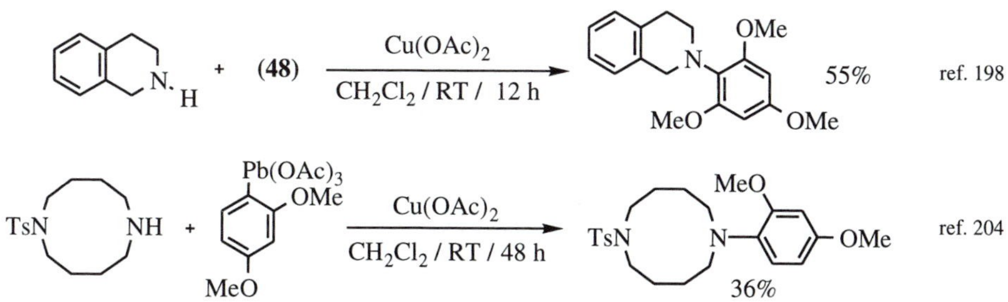

55%   ref. 198

36%   ref. 204

As addition of 1,1-diphenylethylene, a free radical trap, had no effect on the reaction, the involvement of a free-radical mechanism was excluded. A transmetallation between the organolead and a copper (I) species was suggested to be operating. That step is followed by a ligand-coupling type process in which the amine is either covalently bound to the arylcopper intermediate or in an arylcopper-amine complex.

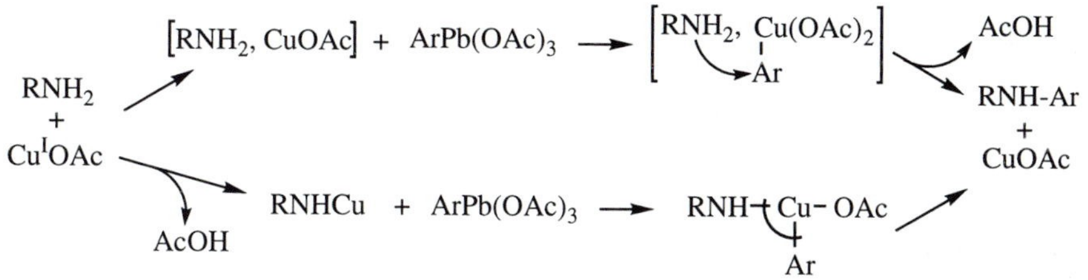

Amidic nitrogen atoms can also be *N*-arylated when their sodium salts are treated with *para*-tolyllead triacetate in CH_2Cl_2-DMF at 60-80°C under mild conditions. When a competition with a more reactive group is possible, chemoselectivity is observed, so that amino or β-diketone groups are selectively arylated.[153,205]

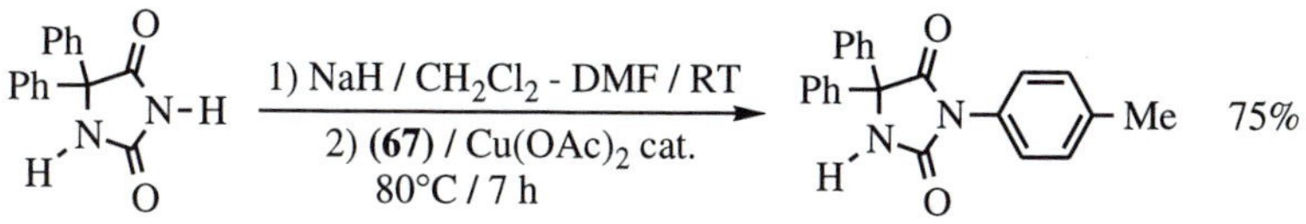

75%

A third type of copper-catalysed reactions involves diorganolead diacetates.[194] Diaryllead, arylvinyllead or divinyllead diacetates can be prepared by transmetallation between the appropriate organoboronic acid and the organolead triacetate or lead tetraacetate (for the synthesis of the divinyl compounds). They undergo a copper(I)-catalysed coupling reaction to afford biaryls, vinylaromatics or buta-1,3-dienes in high yields. With unsymmetrical diorganolead reagents, the three possible coupling products are obtained, resulting either from ligand exchange in the lead compounds or from a non-exclusively bimolecular process in the transmetallation step leading to statistical mixtures of organocopper intermediates.

7.7 REFERENCES

1. Hutchinson, A.; Pollard, W. *J. Chem. Soc.* **1896**, *69*, 212-226.
2. Dimroth, O.; Kämmerer, H. *Ber. Dtsch. Chem. Ges.* **1920**, *53B*, 471-480; Dimroth, O.; Friedemann, O.; Kämmerer, H. *Ber. Dtsch. Chem. Ges.* **1920**, *53B*, 481-487.
3. Rawlinson, D.J.; Sosnovsky, G. *Synthesis* **1973**, 567-603.
4. Rotermund, G.W. Blei-Verbindungen als Oxidationsmittel. In Oxidation Teil 2. In *Methoden der Organischen Chemie*; Müller, E., Ed.; Houben-Weyl; Thieme Verlag: Stuttgart, **1975**; Vol. IV/1b, pp. 204-413.
5. Butler, R.N. *Chem. Rev.* **1984**, *84*, 249-276.
6. Mihailović, M.Lj.; Čeković, Ž.; Lorenc, Lj. Oxidations with Lead Tetraacetate. In *Organic Syntheses by Oxidation with Metal Compounds*; Mijs, W.J.; De Jonge, C.H.R.I., Eds.; Plenum Press: New York, **1986**; pp. 741-816.
7. Leeper, R.W.; Summers, L.; Gilman, H. *Chem. Rev.* **1954**, *54*, 101-167.
8. Shapiro, H.; Frey, F.W. *The Organic Compounds of Lead*; J. Wiley & Sons: New York, **1968**.
9. Abramovitch, R.A.; Barton, D.H.R.; Finet, J.-P. *Tetrahedron* **1988**, *44*, 3039-3071.
10. Pinhey, J.T. *Aust. J. Chem.* **1991**, *44*, 1353-1382.
11. Pinhey, J.T. Lead. In *Comprehensive Organometallic Chemistry II*; Abel, E.W.; Stone, F.G.A.; Wilkinson, G., Eds.; Pergamon Press: Oxford, **1995**; Vol. 11, pp. 461-485.
12. Pinhey, J.T. *Pure Appl. Chem.* **1996**, *68*, 819-824.
13. Corey, E.J.; Schaefer, J.P. *J. Am. Chem. Soc.* **1960**, *82*, 918-929.
14. Henbest, H.B.; Jones, D.N.; Slater, G.P. *J. Chem. Soc.* **1961**, 4472-4478.
15. Marshall, J.A.; Bundy, G.L. *Chem. Commun.* **1966**, 500-501.
16. Wessely, F.; Lauterbach-Keil, G.; Schmid, F. *Monatsh. Chem.* **1950**, *81*, 811-818.

17. Wessely, F.; Sinwell, F. *Monatsh. Chem.* **1950**, *81*, 1055-1070.
18. Wessely, F.; Kotlan, J.; Sinwell, F. *Monatsh. Chem.* **1952**, *83*, 902-914.
19. Wessely, F.; Kotlan, J. *Monatsh. Chem.* **1953**, *84*, 291-297.
20. Wessely, F.; Kotlan, J.; Metlesics, W. *Monatsh. Chem.* **1954**, *85*, 69-79.
21. Wessely, F.; Zbiral, E.; Sturm, H. *Chem. Ber.* **1960**, *93*, 2840-2851.
22. Harrison, M.J.; Norman, R.O.C. *J. Chem. Soc. C* **1970**, 728-730.
23. Begley, M.J.; Gill, G.B.; Pattenden, G.; Stapleton, A.; Raphael, R.A. *J. Chem. Soc., Perkin Trans. 1* **1988**, 1677-1683.
24. Bichan, D.J.; Yates, P. *Can. J. Chem.* **1975**, *53*, 2054-2063.
25. Yates, P.; Auksi, H. *Can. J. Chem.* **1979**, *57*, 2853-2863.
26. Feldman, K.S.; Ensel, S.M. *J. Am. Chem. Soc.* **1993**, *115*, 1162-1163.
27. Feldman, K.S.; Ensel, S.M. *J. Am. Chem. Soc.* **1994**, *116*, 3357-3366.
28. Feldman, K.S.; Ensel, S.M. *J. Org. Chem.* **1995**, *60*, 8171-8178.
29. Quideau, S.; Feldman, K.S. *Chem. Rev.* **1996**, *96*, 475-503.
30. Feldman, K.S.; Smith, R.S. *J. Org. Chem.* **1996**, *61*, 2606-2612.
31. Cavill, G.W.K.; Solomon, D.H. *J. Chem. Soc.* **1955**, 4426-4429.
32. Fuson, R.C.; Maynert, E.W.; Tan, T.-L.; Trumbull, E.R.; Wassmundt, F.W. *J. Am. Chem. Soc.* **1957**, *79*, 1938-1941.
33. Cocker, J.D.; Henbest, H.B.; Phillipps, G.H.; Slater, G.P.; Thomas, D.A. *J. Chem. Soc.* **1965**, 6-11.
34. Ellis, J.W. *Chem. Commun.* **1970**, 406.
35. Lansbury, P.T.; Vacca, J.P. *Tetrahedron Lett.* **1982**, *23*, 2623-2626.
36. Moon, S.; Bohm, M. *J. Org. Chem.* **1972**, *37*, 4338-4340.
37. Johnson, W.S.; Gastambide, B.; Pappo, R. *J. Am. Chem. Soc.* **1957**, *79*, 1991-1994.
38. Oka, K.; Hara, S. *J. Am. Chem. Soc.* **1977**, *99*, 3859-3860.
39. Rubottom, G.M.; Gruber, J.M.; Kincaid, K. *Synth. Commun.* **1976**, *6*, 59-62.
40. Moriarty, R.M.; Penmasta, R.; Prakash, I. *Tetrahedron Lett.* **1987**, *28*, 873-876.
41. Rubottom, G.M.; Gruber, J.M.; Mong, G.M. *J. Org. Chem.* **1976**, *41*, 1673-1674.
42. Kashin, A.N.; Tul'chinskii, M.L.; Bumagin, N.A.; Beletskaya, I.P.; Reutov, O.A. *Zh. Org. Khim.* **1982**, *18*, 1588-1595.
43. Moriarty, R.M.; Penmasta, R.; Prakash, I.; Awasthi, A.K. *J. Org. Chem.* **1988**, *53*, 1022-1025.
44. Bell, H.C.; Pinhey, J.T.; Sternhell, S. *Aust. J. Chem.* **1982**, *35*, 2237-2245.
45. Pausacker, K.H.; Scroggie, J.G. *J. Chem. Soc.* **1954**, 4003-4006.
46. Dimroth, K.; Kalk, F.; Neubauer, G. *Chem. Ber.* **1957**, *90*, 2058-2071.
47. Richter, H.J.; Dressler, R.L. *J. Org. Chem.* **1962**, *27*, 4066-4068.
48. Baer, E.; Tosoni, A.L. *J. Am. Chem. Soc.* **1956**, *78*, 2857-2858.
49. Okazaki, R.; Hosogai, T.; Hashimoto, M.; Inamoto, N. *Bull. Chem. Soc. Jpn.* **1969**, *42*, 3559-3564.
50. Yoshida, K.; Nakajima, S.; Ohnuma, T.; Ban, Y.; Shibasaki, M.; Aoe, K.; Date, T. *J. Org. Chem.* **1988**, *53*, 5355-5359.
51. Adams, R.; Agnello, E.J.; Colgrove, R.S. *J. Am. Chem. Soc.* **1955**, *77*, 5617-5625.
52. Adams, R.; Brower, K.R. *J. Am. Chem. Soc.* **1956**, *78*, 4770-4773.
53. Adams, R.; Werbel, L.M. *J. Am. Chem. Soc.* **1958**, *80*, 5799-5803.
54. Corbani, F.; Rindone, B.; Scolastico, C. *Tetrahedron* **1973**, *29*, 3253-3257.
55. Kuehne, M.E.; Giacobbe, T.J. *J. Org. Chem.* **1968**, *33*, 3359-3369.
56. McKillop, A.; Taylor, E.C. *Chem. Br.* **1973**, *9*, 4-11.
57. Ahlbrecht, H.; Hagena, D. *Chem. Ber.* **1976**, *109*, 2345-2350.
58. Boar, R.B.; McGhie, J.F.; Robinson, M.; Barton, D.H.R.; Horwell, D.C.; Stick, R.V. *J. Chem. Soc., Perkin Trans. 1* **1975**, 1237-1241.

59.   Boar, R.B.; McGhie, J.F.; Robinson, M.; Barton, D.H.R. *J. Chem. Soc., Perkin Trans. 1* **1975**, 1242-1244.
60.   Corbani, F.; Rindone, B.; Scolastico, C. *Tetrahedron* **1975**, *31*, 455-457.
61.   Khetan, S.K. *J. Chem. Soc., Chem. Commun.* **1972**, 917.
62.   Carr, R.M.; Norman, R.O.C.; Vernon, J.M. *J. Chem. Soc., Perkin Trans. 1* **1980**, 156-162.
63.   Vernon, J.M.; Carr, R.M.; Sukari, M.A. *J. Chem. Res. (S)* **1982**, 115.
64.   Sukari, M.A.; Vernon, J.M. *Tetrahedron* **1983**, *39*, 793-796.
65.   Thomas, A.; Vishwakarma, J.N.; Apparao, S.; Ila, H.; Junjappa, H. *Tetrahedron* **1988**, *44*, 1667-1672.
66.   Riehl, J.J.; Lamy, F. *Chem. Commun.* **1969**, 406-407.
67.   Iffland, D.C.; Criner, G.X. *Chem. Ind. (London)* **1956**, 176-177.
68.   Iffland, D.C.; Salisbury, L.; Schafer, W.R. *J. Am. Chem. Soc.* **1961**, *83*, 747-749.
69.   Harrison, M.J.; Norman, R.O.C.; Gladstone, W.A.F. *J. Chem. Soc. C* **1967**, 735-739.
70.   Butler, R.N. *Chem. Ind. (London)* **1968**, 437-440.
71.   Stojiljković, A.; Orbović, N.; Sredojević, S.; Mihailović, M.Lj. *Tetrahedron* **1970**, *26*, 1101-1107.
72.   Barton, D.H.R.; McGhie, J.F.; Batten, P.L. *J. Chem. Soc. C* **1970**, 1033-1042.
73.   Gale, D.M.; Middleton, W.J.; Krespan, C.G. *J. Am. Chem. Soc.* **1966**, *88*, 3617-3623.
74.   Ciganek, E. *J. Org. Chem.* **1965**, *30*, 4193-4204.
75.   Butler, R.N.; Scott, F.L. *J. Chem. Soc. C* **1966**, 1202-1206.
76.   Gladstone, W.A.F.; Aylward, J.B.; Norman, R.O.C. *J. Chem. Soc. C* **1969**, 2587-2598.
77.   Butler, R.N.; King, W.B. *J. Chem. Soc., Perkin Trans. 1* **1975**, 61-65.
78.   Buckingham, J.; Guthrie, R.D. *J. Chem. Soc. C* **1968**, 1445-1451.
79.   Hoffmann, R.W.; Luthardt, H.J. *Chem. Ber.* **1968**, *101*, 3851-3860.
80.   Kotali, A.; Tsoungas, P.G. *Tetrahedron Lett.* **1987**, *28*, 4321-4322.
81.   Katritzky, A.R.; Harris, P.A.; Kotali, A. *J. Org. Chem.* **1991**, *56*, 5049-5051.
82.   Kotali, A. *Tetrahedron Lett.* **1994**, *35*, 6753-6754.
83.   Kropf, H.; Lambeck, R. *Liebigs Ann. Chem.* **1966**, *700*, 1-17.
84.   Kropf, H.; Lambeck, R. *Liebigs Ann. Chem.* **1966**, *700*, 18-28.
85.   Kaufmann, S.; Tökes, L.; Murphy, J.W.; Crabbé, P. *J. Org. Chem.* **1969**, *34*, 1618-1621.
86.   Shafiullah, D.; Ali, H. *Synthesis* **1979**, 124-126.
87.   Gilbert, B.C.; Norman, R.O.C. *J. Chem. Soc. B* **1966**, 86-91.
88.   Just, G.; Dahl, K. *Tetrahedron Lett.* **1966**, 2441-2448; *Tetrahedron* **1968**, *24*, 5251-5269.
89.   Kropf, H. *Angew. Chem., Int. Ed. Engl.* **1965**, *4*, 983 and 1091.
90.   Criegee, R. In *Oxidation in Organic Chemistry*; Wiberg, K.B., Ed.; Academic Press: New York, **1965**; Part A, pp. 277-366.
91.   Rubottom, G.M. In *Oxidation in Organic Chemistry*; Trahanovsky, W.S., Ed.; Academic Press: New York, **1982**; Part D, pp. 1-145.
92.   Lethbridge, A.; Norman, R.O.C.; Thomas, C.B. *J. Chem. Soc., Perkin Trans. 1* **1974**, 1929-1938.
93.   Lethbridge, A.; Norman, R.O.C.; Thomas, C.B.; Parr, W.J.E. *J. Chem. Soc., Perkin Trans. 1* **1975**, 231-241.
94.   Ephritikine, M.; Levisalles, J. *J. Chem. Soc., Chem. Commun.* **1974**, 429-430.
95.   Harvey, D.R.; Norman, R.O.C. *J. Chem. Soc.* **1964**, 4860-4868.
96.   Campbell, J.R.; Kalman, J.R.; Pinhey, J.T.; Sternhell, S. *Tetrahedron Lett.* **1972**, 1763-1766.
97.   Kalman, J.R.; Pinhey, J.T.; Sternhell, S. *Tetrahedron Lett.* **1972**, 5369-5372.
98.   Bell, H.C.; Kalman, J.R.; Pinhey, J.T.; Sternhell, S. *Tetrahedron Lett.* **1974**, 853-856.
99.   Lewis, J.R.; Tele, C.G. *Chem. Ind. (London)* **1987**, 858-859.
100. Bell, H.C.; Kalman, J.R.; May, G.L.; Pinhey, J.T.; Sternhell, S. *Aust. J. Chem.* **1979**, *32*, 1531-1550.

101. Norman, R.O.C.; Thomas, C.B.; Willson, J.S. *J. Chem. Soc., Perkin Trans. 1* **1973**, 325-332.
102. Uemura, S.; Ikeda, T.; Tanaka, S.; Okano, M. *J. Chem. Soc., Perkin Trans. 1* **1979**, 2574-2576.
103. McKillop, A.; Turrell, A.G.; Young, D.W.; Taylor, E.C. *J. Am. Chem. Soc.* **1980**, *102*, 6504-6512.
104. Funk, R.L.; Vollhardt, K.P.C. *J. Am. Chem. Soc.* **1979**, *101*, 215-217.
105. Bell, H.C.; Kalman, J.R.; Pinhey, J.T.; Sternhell, S. *Aust. J. Chem.* **1979**, *32*, 1521-1530.
106. Aylward, J.B. *J. Chem. Soc. B* **1967**, 1268-1270.
107. Kozyrod, R.P.; Pinhey, J.T. *Tetrahedron Lett.* **1983**, *24,* 1301-1302.
108. Kozyrod, R.P.; Morgan, J.; Pinhey, J.T. *Aust. J. Chem.* **1985**, *38*, 1147-1153.
109. Barton, D.H.R.; Donnelly, D.M.X.; Finet, J.-P.; Guiry, P.J. *J. Chem. Soc., Perkin Trans. 1* **1992**, 1365-1375.
110. Morgan, J.; Pinhey, J.T. *J. Chem. Soc., Perkin Trans. 1* **1990**, 715-720.
111. Suginome, H.; Orito, K.; Yorita, K.; Ishikawa, M.; Shimoyama, N.; Sasaki, T. *J. Org. Chem.* **1995**, *60*, 3052-3064.
112. Donnelly, D.M.X.; Fitzpatrick, B.M.; Finet, J.-P. *J. Chem. Soc., Perkin Trans. 1* **1994**, 1791-1795.
113. Pinhey, J.T.; Roche, E.G. *J. Chem. Soc., Perkin Trans. 1* **1988**, 2415-2421.
114. Yamamoto, M.; Izukawa, H.; Saiki, M.; Yamada, K. *J. Chem. Soc., Chem. Commun.* **1988**, 560-561.
115. Yamamoto, M.; Munakata, H.; Kishikawa, K.; Kohmoto, S.; Yamada, K. *Bull. Chem. Soc. Jpn.* **1992**, *65*, 2366-2370.
116. Moloney, M.G.; Pinhey, J.T. *J. Chem. Soc., Chem. Commun.* **1984**, 965-966.
117. Parkinson, C.J.; Pinhey, J.T.; Stoermer, M.J. *J. Chem. Soc., Perkin Trans. 1* **1992**, 1911-1915.
118. Parkinson, C.J.; Stoermer, M.J. *J. Organomet. Chem.* **1996**, *507*, 207-214.
119. Hashimoto, S.-i.; Miyazaki, Y.; Shinoda, T.; Ikegami, S. *Tetrahedron Lett.* **1989**, *30,* 7195-7198.
120. Bell, H.C.; Kalman, J.R.; Pinhey, J.T.; Sternhell, S. *Tetrahedron Lett.* **1974**, 857-860.
121. Briquet, A.A.S.; Hansen, H.-J. *Helv. Chim. Acta* **1994**, *77,* 1577-1584.
122. Bell, H.C.; May, G.L.; Pinhey, J.T.; Sternhell, S. *Tetrahedron Lett.* **1976**, 4303-4306.
123. Bell, H.C.; Pinhey, J.T.; Sternhell, S. *Aust. J. Chem.* **1979**, *32*, 1551-1560.
124. Greenland, H.; Kozyrod, R.P.; Pinhey, J.T. *J. Chem. Soc., Perkin Trans. 1* **1986**, 2011-2015.
125. Barton, D.H.R.; Donnelly, D.M.X.; Guiry, P.J.; Reibenspies, J.H. *J. Chem. Soc., Chem. Commun.* **1990**, 1110-1111.
126. Barton, D.H.R.; Donnelly, D.M.X.; Guiry, P.J.; Finet, J.-P. *J. Chem. Soc., Perkin Trans. 1* **1994**, 2921-2926.
127. Kovtonyuk, V.N.; Kobrina, L.S. *J. Fluorine Chem.* **1993**, *63*, 243-251.
128. Pinhey, J.T.; Rowe, B.A. *Aust. J. Chem.* **1979**, *32*, 1561-1566.
129. Emsley, J.; Freeman, N.J.; Bates, P.A.; Hursthouse, M.B. *J. Chem. Soc., Perkin Trans. 1* **1988**, 297-299.
130. Emsley, J.; Ma, L.Y.Y.; Bates, P.A.; Motevalli, M.; Hursthouse, M.B. *J. Chem. Soc., Perkin Trans. 2* **1989**, 527-533.
131. Emsley, J.; Ma, L.Y.Y.; Karaulov, S.A.; Motevalli, M.; Hursthouse, M.B. *J. Mol. Struc.* **1990**, *216,* 143-152.
132. Emsley, J.; Ma, L.Y.Y.; Nyburg, S.C.; Parkins, A.W. *J. Mol. Struc.* **1990**, *240*, 59-67.
133. Pinhey, J.T.; Rowe, B.A. *Aust. J. Chem.* **1983**, *36*, 789-794.
134. Collins, D.J.; Cullen, J.D.; Fallon, G.D.; Gatehouse, B.M. *Aust. J. Chem.* **1984**, *37*, 2279-2294.
135. Pinhey, J.T.; Rowe, B.A. *Aust. J. Chem.* **1980**, *33*, 113-120.
136. Kozyrod, R.P.; Pinhey, J.T. *Organic Syntheses* **1984**, *62,* 24-29.
137. Kozyrod, R.P.; Pinhey, J.T. *Tetrahedron Lett.* **1982**, *23,* 5365-5366; Kozyrod, R.P.; Pinhey, J.T. *Aust. J. Chem.* **1985**, *38*, 1155-1161.

138. Krapcho, A.P.; Lovey, A.J. *Tetrahedron Lett.* **1973**, 957-960.
139. Bonjoch, J.; Casamitjana, N.; Quirante, J.; Garriga, C.; Bosch, J. *Tetrahedron* **1992**, *48*, 3131-3138.
140. Barton, D.H.R.; Donnelly, D.M.X.; Finet, J.-P.; Guiry, P.J.; Kielty, J.M. *Tetrahedron Lett.* **1990**, *31*, 6637-6640.
141. Orito, K.; Yorita, K.; Suginome, H. *Tetrahedron Lett.* **1991**, *32*, 5999-6002.
142. Orito, K.; Sasaki, T.; Suginome, H. *J. Org. Chem.* **1995**, *60*, 6208-6210.
143. Donnelly, D.M.X.; Finet, J.-P.; Kielty, J.M. *Tetrahedron Lett.* **1991**, *32*, 3835-3836.
144. Donnelly, D.M.X.; Finet, J.-P.; Rattigan, B.A. *J. Chem. Soc., Perkin Trans. 1* **1993**, 1729-1735.
145. Donnelly, D.M.X.; Finet, J.-P.; Rattigan, B.A. *J. Chem. Soc., Perkin Trans. 1* **1995**, 1679-1683.
146. Mérour, J.Y.; Chichereau, L.; Finet, J.-P. *Tetrahedron Lett.* **1992**, *33*, 3867-3870.
147. Shimizu, I.; Tsuji, J. *J. Am. Chem. Soc.* **1982**, *104*, 5844-5846; Tsuji, J.; Nisar, M.; Shimizu, I. *J. Org. Chem.* **1985**, *50*, 3416-3417.
148. Kopinski, R.P.; Pinhey, J.T. *Aust. J. Chem.* **1983**, *36*, 311-316.
149. Barton, D.H.R.; Donnelly, D.M.X.; Finet, J.-P.; Guiry, P.J. *Tetrahedron Lett.* **1989**, *30*, 1539-1542.
150. Barton, D.H.R.; Donnelly, D.M.X.; Finet, J.-P.; Guiry, P.J. *Tetrahedron Lett.* **1990**, *31*, 7449-7452.
151. Donnelly, D.M.X.; Molloy, D.J.; Reilly, J.P.; Finet, J.-P. *J. Chem. Soc., Perkin Trans. 1* **1995**, 2531-2534.
152. Donnelly, D.M.X.; Finet, J.-P.; Guiry, P.J.; Hutchinson, R.M. *J. Chem. Soc., Perkin Trans. 1* **1990**, 2851-2852.
153. López-Alvarado, P.; Avendaño, C.; Menéndez, J.C. *J. Org. Chem.* **1996**, *61*, 5865-5870.
154. Ackland, D.J.; Pinhey, J.T. *Tetrahedron Lett.* **1985**, *26*, 5331-5334.
155. Ackland, D.J.; Pinhey, J.T. *J. Chem. Soc., Perkin Trans. 1* **1987**, 2689-2694.
156. Ackland, D.J.; Pinhey, J.T. *J. Chem. Soc., Perkin Trans. 1* **1987**, 2695-2700.
157. Parkinson, C.J.; Pinhey, J.T. *J. Chem. Soc., Perkin Trans. 1* **1991**, 1053-1057.
158. Pinhey, J.T.; Rowe, B.A. *Tetrahedron Lett.* **1980**, *21*, 965-968.
159. Kopinski, R.P.; Pinhey, J.T.; Rowe, B.A. *Aust. J. Chem.* **1984**, *37*, 1245-1254.
160. Pearson, R.G.; Dillon, R.L. *J. Am. Chem. Soc.* **1953**, *75*, 2439-2443.
161. Kozyrod, R.P.; Morgan, J.; Pinhey, J.T. *Aust. J. Chem.* **1991**, *44*, 369-376.
162. Ref. 8, p. 298.
163. Koen, M.J.; Morgan, J.; Pinhey, J.T. *J. Chem. Soc., Perkin Trans. 1* **1993**, 2383-2384.
164. Morgan, J.; Pinhey, J.T. *Tetrahedron Lett.* **1994**, *35*, 9625-9628.
165. Konopelski, J.P.; Hottenroth, J.M.; Oltra, H.M.; Véliz, E.A.; Yang, Z.-C. *Synlett* **1996**, 609-611.
166. Morgan, J.; Buys, I.; Hambley, T.; Pinhey, J.T. *J. Chem. Soc., Perkin Trans. 1* **1993**, 1677-1681.
167. Donnelly, D.M.X.; Kielty, J.M.; Cormons, A.; Finet, J.-P. *J. Chem. Soc., Perkin Trans. 1* **1993**, 2069-2073.
168. Donnelly, D.M.X.; Fitzpatrick, B.M.; O'Reilly, B.A.; Finet, J.-P. *Tetrahedron* **1993**, *49*, 7967-7976.
169. Donnelly, D.M.X.; Fitzpatrick, B.M.; Ryan, S.M.; Finet, J.-P. *J. Chem. Soc., Perkin Trans. 1* **1994**, 1797-1801.
170. Santosh, K.C.; Balasubramanian, K.K. *14th International Congress of Heterocyclic Chemistry* Antwerp, Belgium, **1993**.
171. May, G.L.; Pinhey, J.T. *Aust. J. Chem.* **1982**, *35*, 1859-1871.
172. Kozyrod, R.P.; Pinhey, J.T. *Tetrahedron Lett.* **1981**, *22*, 783-784.
173. Kozyrod, R.P.; Pinhey, J.T. *Aust. J. Chem.* **1985**, *38*, 713-721.
174. Ref. 8, p. 297.

175. Huber, M.L.; Pinhey, J.T. *J. Chem. Soc., Perkin Trans. 1* **1990**, 721-722.
176. De Meio, G.V.; Pinhey, J.T. *J. Chem. Soc., Chem. Commun.* **1990**, 1065-1066.
177. De Meio, G.V.; Morgan, J.; Pinhey, J.T. *Tetrahedron* **1993**, *49*, 8129-8138.
178. Corey, E.J.; Wollenberg, J. *J. Am. Chem. Soc.* **1974**, *96*, 5581-5583.
179. Larock, R.C.; Oertle, K.; Beatty, K.M. *J. Am. Chem. Soc.* **1980**, *102*, 1966-1974.
180. Moloney, M.G.; Pinhey, J.T.; Stoermer, M.J. *J. Chem. Soc., Perkin Trans. 1* **1990**, 2645-2655.
181. Pinhey, J.T.; Stoermer, M.J. *J. Chem. Soc., Perkin Trans. 1* **1991**, 2455-2460.
182. Moloney, M.G.; Pinhey, J.T. *J. Chem. Soc., Perkin Trans. 1* **1988**, 2847-2854.
183. Shibasaki, M.; Torisawa, Y.; Ikegami, S. *Tetrahedron Lett.* **1982**, *23*, 4607-4610.
184. Hashimoto, S.-i.; Shinoda, T.; Ikegami, S. *J. Chem. Soc., Chem. Commun.* **1988**, 1137-1139.
185. Hambley, T.W.; Holmes, R.J.; Parkinson, C.J.; Pinhey, J.T. *J. Chem. Soc., Perkin Trans. 1* **1992**, 1917-1922.
186. Moloney, M.G.; Pinhey, J.T.; Roche, E.G. *Tetrahedron Lett.* **1986**, *27*, 5025-5028.
187. Moloney, M.G.; Pinhey, J.T.; Roche, E.G. *J. Chem. Soc., Perkin Trans. 1* **1989**, 333-341.
188. see ref. 10, p.1379-1380 and ref. 68 therein.
189. Hashimoto, S.-i.; Miyazaki, Y.; Shinoda, T.; Ikegami, S. *J. Chem. Soc., Chem. Commun.* **1990**, 1100-1102.
190. Barton, D.H.R.; Finet, J.-P.; Giannotti, C.; Halley, F. *J. Chem. Soc., Perkin Trans. 1* **1987**, 241-249.
191. Beckwith, A.L.J.; Meijs, G.F. *J. Org. Chem.* **1987**, *52*, 1922-1930; Beckwith, A.L.J.; Palacios, S.M. *J. Phys. Org. Chem.* **1991**, *4*, 404-412.
192. Morgan, J.; Pinhey, J.T. *J. Chem. Soc., Perkin Trans. 1* **1993**, 1673-1676.
193. Morgan, J.; Hambley, T.W.; Pinhey, J.T. *J. Chem. Soc., Perkin Trans. 1* **1996**, 2173-2177.
194. Morgan, J.; Parkinson, C.J.; Pinhey, J.T. *J. Chem. Soc., Perkin Trans. 1* **1994**, 3361-3365.
195. Kobayashi, M.; Latour, S.; Wuest, J.D. *Organometallics* **1991**, *10*, 2908-2913.
196. Barton, D.H.R.; Yadav-Bhatnagar, N.; Finet, J.-P.; Khamsi, J. *Tetrahedron Lett.* **1987**, *28*, 3111-3114.
197. Barton, D.H.R.; Donnelly, D.M.X.; Finet, J.-P.; Guiry, P.J. *Tetrahedron Lett.* **1989**, *30*, 1377-1380.
198. Barton, D.H.R.; Donnelly, D.M.X.; Finet, J.-P.; Guiry, P.J. *J. Chem. Soc., Perkin Trans. 1* **1991**, 2095-2102.
199. López-Alvarado, P.; Avendaño, C.; Menéndez, J.C. *Heterocycles* **1991**, *32*, 1003-1012.
200. López-Alvarado, P.; Avendaño, C.; Menéndez, J.C. *Tetrahedron Lett.* **1992**, *33*, 659-662.
201. López-Alvarado, P.; Avendaño, C.; Menéndez, J.C. *J. Org. Chem.* **1995**, *60*, 5678-5682.
202. Boyer, G.; Galy, J.-P.; Barbe, J. *Heterocycles* **1995**, *41*, 487-496.
203. Morel, S.; Boyer, G.; Coullet, F.; Galy, J.-P. *Synth. Commun.* **1996**, *26*, 2443-2447.
204. Schneider, R.; Hosseini, M.W.; Planeix, J.-M. *Tetrahedron Lett.* **1996**, *37*, 4721-4724.
205. López-Alvarado, P.; Avendaño, C.; Menéndez, J.C. *Tetrahedron Lett.* **1992**, *33*, 6875-6878.

Chapter 8

Ligand Coupling Involving Other Heteroatoms

In the previous chapters, the reactions involving the five most important elements of the main group p-block, which show a behaviour typical of ligand coupling, have been reviewed. However, in the p-block group, a number of other elements have seen scattered reports of ligand coupling reactions. From the point of view of an organic chemist, their significance goes from a purely mechanistic interest to the very specific functionalizations, which are effected with the help of some organic derivatives of these elements. In this chapter, we describe a number of these reactions which are ligand coupling reactions as defined in chapter 1. However, in his reviews, Oae has considered as ligand coupling some reactions which do not fit entirely with our criteria. This is the case for example of the following reactions involving silicon and tin. Although the reaction of the silicon derivative involves the formation of a bond between two ligands bonded to the central atom, it is rather explained by intramolecular S_N2 displacement on an *in situ* generated transient hypervalent species. The reaction of alkyltrifluorosilanes with hydrogen peroxide or *m*-chloroperbenzoic acid, which is catalysed by a small amount of potassium fluoride, leads after work-up to the corresponding alcohol.[1] In view of the predominant retention of the stereochemistry, this oxidative cleavage of alkyltrifluorosilanes was considered by Oae to be a ligand coupling reaction.[2] (Scheme 8.1)

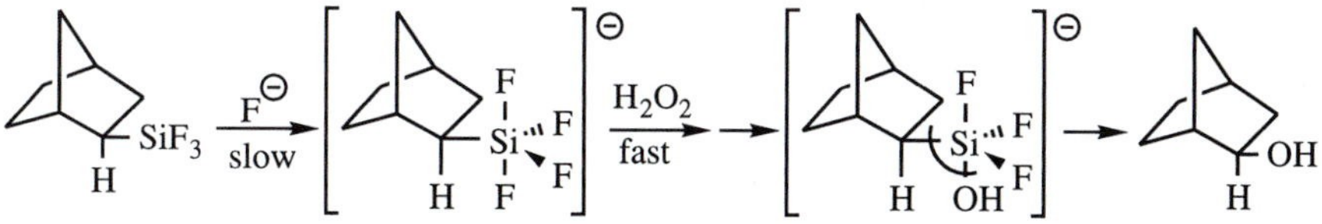

Scheme 8.1: Ligand coupling mechanism in the oxidation of alkyltrifluorosilanes

However, the generally accepted mechanism involves the formation of an alkoxysilane after the ligand transfer. Therefore, the mechanism cannot be considered to be a ligand coupling and is better described as an intramolecular migration.[1,3] (Scheme 8.2)

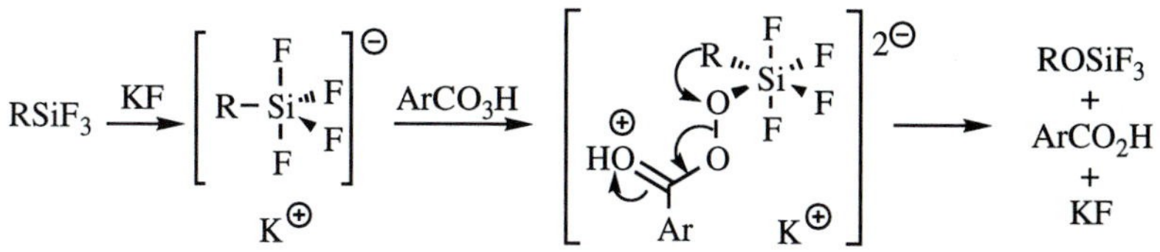

Scheme 8.2: Intramolecular migration mechanism in the oxidation of alkyltrifluorosilanes

Although some other reactions may be viewed as typical ligand coupling processes, we will not consider them in this chapter, as they result from the classical sequence in transition organometallic chemistry: oxidative addition - reductive elimination. For example, the reaction of organotin compounds (1) described by Reich is in accordance with a ligand coupling mechanism.[4-6]

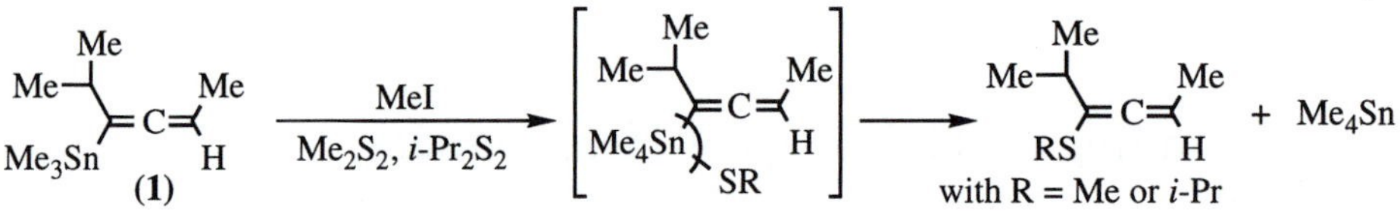

A second type of ligand coupling reaction of vinylstannane was reported by Tius and Kawakami.[7] The reaction of vinylstannanes with *in situ* generated trifluoromethanesulfonyl hypofluorite leads to vinyl fluoride according to the mechanism depicted in scheme 8.3.

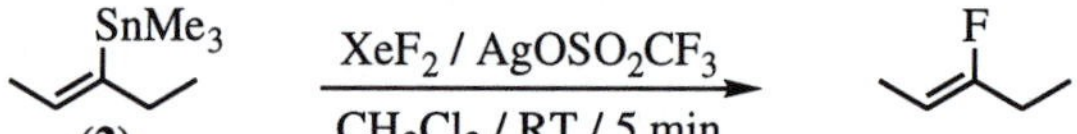

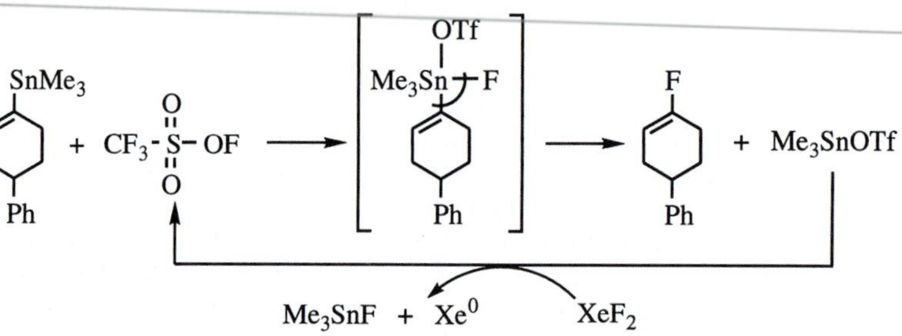

Scheme 8.3: Ligand coupling mechanism in the substitution of organotin derivatives

In this chapter, we will focus on the chemistry of organoantimony, organoselenium, organotellurium and organothallium compounds, for which some typical ligand coupling reactions, although in a relatively limited number, have been reported over the years.

8.1 ORGANOANTIMONY

The application of organoantimony compounds to synthetic organic chemistry has been increasing recently.[8,9] However, the scope of synthetically useful reactions remains still rather narrow. In spite of the relatively weak energy of the antimony-element bond, compared to the analogous phosphorus compounds for example, the reactions of ligand transfer with organoantimony are fairly scarce. Among these reactions, only a limited number of ligand coupling reactions have been reported with different pentavalent antimony compounds.

The thermal decomposition of pentaphenylantimony (3) is very dependent upon the nature of the solvent and the reaction conditions. This can lead to a variety of products formed along different pathways. In hot benzene, pentaphenylantimony (3) is relatively stable. Heated in a sealed ampoule in an autoclave, a solution of [1-^{14}C]-pentaphenylantimony in benzene required 3 hours at 220°C for the decomposition of pentaphenylantimony into triphenylstibane (4) and biphenyl (5) to be observed. No reaction took place at 175°C. As the radioactivity was detected only in the products but not in the solvent, an intramolecular ligand coupling process was considered to take place.[10] The decomposition of pentaphenylantimony can be catalysed by copper diacetate, and the reaction takes place at room temperature.[11]

$$Ph_5Sb \xrightarrow[\text{3 h / sealed tube}]{225°C \ / \ C_6H_6} Ph_3Sb \ + \ Ph-Ph$$

(3)   (4), 89%   (5), 90%

Using mixed pentaarylantimony compounds (6), Akiba *et al.* studied the influence of the electronic nature of the aryl groups on the relative ratios of ligand coupling products. With Ar_nSbTol_{5-n} compounds [with Ar = 4-(CF_3)-C_6H_4], a greater selectivity to afford Ar-Ar and Ar-Tol compared to Tol-Tol was generally observed. (Table 8.1) The reaction can also be catalysed by copper acetylacetonate or by $[3,5-(CF_3)_2C_6H_3]_4BLi$.[12]

Table 8.1: Selectivity of the ligand coupling reactions of pentaarylantimony[12]

$$Tol_nSbAr_{5-n} \xrightarrow{220°C \ / \ C_6H_6} Tol-Tol \ + \ Tol-Ar \ + \ Ar-Ar$$

(6)

	Tol-Tol	Tol-Ar	Ar-Ar
$TolSbAr_4$	0	100	-
Tol_2SbAr_3	10	2	88
Tol_3SbAr_2	2	98	0
Tol_4SbAr	-	21	79

Tol = 4-MeC_6H_4 and Ar = 4-$CF_3C_6H_4$

In chloroform, the decomposition of pentaphenylantimony (3) led to tetraphenylstibonium chloride (7), benzene and a small amount of biphenyl (5). In carbon tetrachloride, the reaction of decomposition of pentaphenylantimony led to tetraphenylstibonium chloride, chlorobenzene and a small amount of biphenyl. These reactions were first reported by Razuvaev *et al.* in 1960 who considered them to be free radical processes.[13]

in $CHCl_3$ $$Ph_5Sb \xrightarrow[\text{25 h / sealed tube}]{CHCl_3 \ / \ 100°C} Ph_4SbCl \ + \ PhH \ + \ Ph-Ph$$
(3)   (7)   (5)

in CCl_4 $$Ph_5Sb \xrightarrow[\text{25 h / sealed tube}]{CCl_4 \ / \ 100°C} Ph_4SbCl \ + \ PhCl \ + \ Ph-Ph$$
(3)   (7)   (5)

Later in 1974, Mc Ewen and Lin reinvestigated more extensively the reaction of pentaarylantimony compounds with carbon tetrachloride. They found that the reaction of pentaphenylantimony (3) led to a more complex mixture than originally described.[14]

$$Ph_5Sb \xrightarrow[\text{1 month / dark}]{CCl_4 \ / \ 52.3°C} Ph_4SbCl \ + \ PhCl \ + \ PhH \ + \ CCl_3-CCl_3 \ + \ CHCl_3 \ + \ Ph-Ph$$

(3)   (7), 67.3%   33.3%   8.5%   21.9%   1.7%   (5), 21.4%

They concluded that three types of behaviour were in fact competing: 1) a ligand coupling, which gives biaryl without intervention of any free aromatic species, such as free radical or anion; 2) a homolytic fission of the antimony-carbon bond, and the two radical species evolve by complicated radical chain reactions; 3) an ionic fission of the antimony-carbon bond forming an aryl anion, the source of the arenes which are eventually obtained.[14] This latter type of behaviour has been used in the reaction of pentaphenylantimony with allylic halides, acid chlorides and carbonyl compounds.[15,16]

$$\text{ligand coupling} \qquad Ar_5Sb \longrightarrow Ar_3Sb + Ar\text{-}Ar$$

$$\text{free radical} \qquad Ar_5Sb \longrightarrow Ar_4Sb^{\bullet} + Ar^{\bullet}$$

$$\text{ionic} \qquad Ar_5Sb \longrightarrow Ar_4Sb^{\oplus} + Ar^{\ominus}$$

Pentaarylantimony compounds undergo solvolysis fairly rapidly in hydroxylic solvents in which they are soluble.[13,17-24] Otherwise, the reaction requires a co-solvent. The hydrolysis can be easily performed in a mixture of dioxane-water.[22,23] Under the reaction conditions, the products are generally stable. However, under more forcing conditions, a second substitution can happen to lead to a disubstituted triarylantimony derivative such as (**10**).[13,17,18]

$$\underset{(\mathbf{3})}{Ph_5Sb} + ROH \xrightarrow[\text{some hours}]{\text{reflux}} \underset{(\mathbf{8})}{Ph_4Sb\text{-}OR} \qquad R = H,^{22,23} \text{ alkyl},^{13,17,18,20,21}$$
$$\text{triphenylsilyl}^{21}$$

$$\underset{(\mathbf{3})}{Ph_5Sb} + ArOH \xrightarrow{RT} \underset{(\mathbf{9})}{Ph_4Sb\text{-}OAr} \qquad \text{ref. 19,24}$$

$$\underset{(\mathbf{3})}{Ph_5Sb} + MeOH \xrightarrow[\text{5 weeks}]{\text{reflux}} \underset{(\mathbf{10})}{Ph_3Sb{<}^{OMe}_{OMe}} \qquad \text{ref. 13,17,18}$$

A very easy and high-yielding solvolysis was realized by treatment of pentaphenylantimony with a range of carboxylic acids, which led to acyloxytetraphenylantimony (**11**) in good to high yields.[25]

$$\underset{(\mathbf{3})}{Ph_5Sb} + RCOOH \xrightarrow{RT} \underset{(\mathbf{11}),\ 69\text{-}98\%}{Ph_4Sb\text{-}OCOR}$$

The thermal decomposition of the various types of oxy derivatives of tetraarylantimony compounds takes place by various pathways, depending on the nature of the oxygen bound substituent. A free radical chain mechanism was shown by Mc Ewen *et al.* to operate in the decomposition of tetraarylantimony hydroxide.[22,23] However, this mechanism was later questioned as the reaction is not affected by the presence of a free radical trap, 1,1-diphenylethylene.[26]

$$Ar_4SbO^{\bullet} \longrightarrow Ar_3SbO + Ar^{\bullet}$$
$$Ar^{\bullet} + Ar_4SbOH \longrightarrow ArH + Ar_4SbO^{\bullet}$$

In the case of alkoxytetraarylantimony compounds, the thermal decomposition occurs intramolecularly to afford either oxidation products or the *O*-aryl ethers.[19,21] In the case of the (methoxy)tetraphenylantimony compound (**12**), anisole was obtained in a poor yield (10%). The *O*-aryl ether was completely absent in the thermolysis of the analogous (benzyloxy)tetraphenylantimony compound which led exclusively to the oxidation product, benzaldehyde.[21]

Ph₄Sb-OMe → HCHO + PhH + Ph₃Sb   Oxidation
                                    (4)

(12) ← Ph-OMe + Ph₃Sb   Arylation
                (4)

In the presence of copper diacetate, the reaction of pentaphenylantimony with methanol and other simple aliphatic alcohols took place under milder conditions to afford only the *O*-phenyl ethers. Under these conditions, the products of oxidation were no longer formed.[11]

$$Ph_5Sb + ROH \xrightarrow{Cu(OAc)_2} Ph\text{-}OR$$
(3)

In the case of *O*-silyl ethers, thermolysis of (triphenylsilyloxy)tetraphenylantimony (13) did not afford the ligand coupling product, phenyl(triphenylsilyl)ether. Instead, a disproportionation reaction took place to give the bis(silyloxy) ether (14) and pentaphenylantimony (3), which decomposed into triphenylantimony (4) and biphenyl (5) under the conditions of the reaction.[21]

$$2\ Ph_4Sb\text{-}O\text{-}SiPh_3 \longrightarrow Ph_3Sb(O\text{-}SiPh_3)_2 + Ph_5Sb \longrightarrow Ph_3Sb + Ph\text{-}Ph$$
(13) (14) (3) (4) (5)

Intramolecular decomposition of (aryloxy)tetraphenylantimony (9) by thermolysis leads to the corresponding diarylethers (15) in modest to high yields.[19,24]

$$Ph_4Sb\text{-}O\text{-}Ar \xrightarrow[2\text{-}3\text{ h}]{200\text{-}220°C} Ph\text{-}O\text{-}Ar + Ph_3Sb$$
(9)   (15), 30-90%   (4)

The thermal decomposition of (acyloxy)tetraphenylantimony (11) required also relatively harsh conditions and led to the acyloxybenzene derivatives (16) in good yields.[25]

$$Ph_4Sb\text{-}O\text{-}COR \xrightarrow[1\text{ h}]{200°C} Ph\text{-}O\text{-}COR + Ph_3Sb$$
(11)   (16), 76-96%   (4)

By contrast with the reaction of lithium *tert*-butoxide with triphenylbismuth diacetate, which gave *tert*-butyl phenyl ether (*t*-BuOPh) in 66% yield, the same reaction with triphenylantimony diacetate appeared to stop at the stage of the formation of (*tert*-butoxy)acetoxytriphenylantimony. The ligand coupling did not proceed, and *tert*-butanol was recovered after hydrolysis of the mixture.[27]

Ph₃Sb⟨O-t-Bu / OAc ⟵ M = Sb ⟵ *t*-BuOK + Ph₃M(OAc)₂ ⟶ M = Bi ⟶ *t*-BuOPh 66%

(Arylthio)tetraphenylantimony compounds like (17) are conveniently prepared by the reaction of tetraphenylantimony chloride with thiols in the presence of triethylamine. These compounds are relatively stable at room temperature. However, they decompose upon heating to afford a mixture of products in which the arylphenylmercaptide (18) is usually largely predominant. Although a free radical pathway was considered to constitute the major part of the reaction, a direct ligand coupling pathway was also viewed as playing a relatively significant role.[28] Similarly, tetramethylantimony mercaptides (19), prepared from pentamethylantimony and thiols, decomposed at room temperature to afford ligand coupling type products. No mechanistic studies have been reported on this decomposition reaction which can be regarded as a ligand coupling process although a free radical pathway can also be operative.[29]

$$Ph_4Sb\text{-}SAr \xrightarrow{140°C \, / \, 3 \, h} Ph_3Sb + ArS\text{-}SAr + Ph\text{-}S\text{-}Ar + Ph\text{-}Ph + PhH$$

(17)   (4), 93-99%   3-40%   (18), 50-97%   3-21%   3-13%

$$Me_5Sb + RSH \xrightarrow{0\text{-}15°C} Me_4Sb\text{-}SR \xrightarrow[48 \, h]{RT} Me_3Sb + RS\text{-}Me$$

(19), 73-95%

Triaryl(phenylethynyl)antimony (20) are easily prepared by the action of lithium phenylacetylide with triarylantimony dibromide.[12,30] Their thermal decomposition took place at 110°C to lead to mixtures of arylphenylalkyne (21) and diphenylbutadiyne (22). The amount of the arylphenylalkyne (21), one of the ligand coupling products, increased as the aryl group became more electronegative.[12,30]

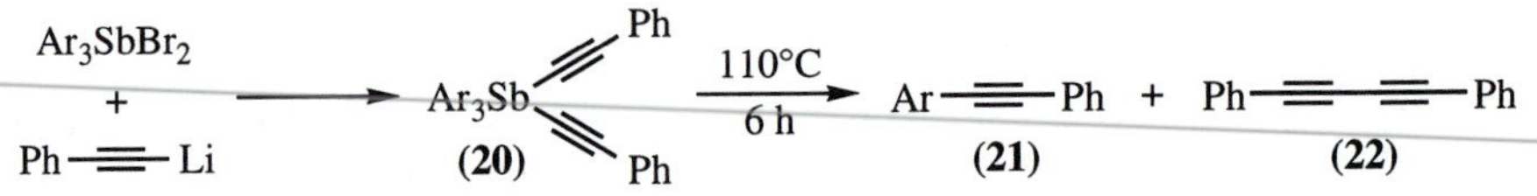

$$
\begin{array}{c}
Ar_3SbBr_2 \\
+ \\
Ph\text{---}\equiv\text{---}Li
\end{array}
\longrightarrow
Ar_3Sb
\begin{array}{c}
\equiv\text{---} Ph \\
\\
\equiv\text{---} Ph
\end{array}
\xrightarrow[6 \, h]{110°C}
Ar\text{---}\equiv\text{---}Ph + Ph\text{---}\equiv\text{---}\equiv\text{---}Ph
$$

(20)   (21)   (22)

Scheme 8.4: Synthesis and ligand coupling of triaryl(phenylethynyl)antimony (20)

8.2 ORGANOSELENIUM

The chemistry of organoselenium compounds shows a number of striking similarities with the reactions of the analogous sulfur compounds. The nucleophilicity and ease of oxidation of divalent selenium compounds are greater than that of the corresponding sulfur compounds. The selenium group is therefore introduced and manipulated more easily than the sulfur analogs. On the other hand, the stability of the hypervalent selenium compounds is lower than that of the sulfur compounds. The selenoxide elimination takes place under very mild conditions. Therefore, the ligand coupling reactions of alkylpyridylsulfoxides have not been extended to the analogous selenoxides. The chemistry of organoselenium which takes place by ligand coupling mechanisms has essentially been observed in the reactions of selenonium compounds.

8.2.1 Reactions of arylselenium compounds

Similarly to tetraarylsulfuranes which are decomposed to biaryls and diarylsulfides, the analogous reaction has been observed in the selenium series.[31-37] In 1915, Strecker and Willing observed only the formation of diphenylselenide (23) as a result of the reaction of selenium oxychloride and phenylmagnesium bromide.[31] In 1952, Wittig and Fritz showed that the reaction of triphenylselenonium chloride with phenyllithium led to diphenylselenide (23) and biphenyl (5).[32] The formation of biphenyl (5) was explained as resulting from the *in situ* decomposition of tetraphenylselenurane (24).

$$SeOCl_2 + Ph\text{-}MgBr \longrightarrow Ph_2Se \qquad\qquad \text{ref. 31}$$

(23)

$$Ph_3SeCl + Ph\text{-}Li \xrightarrow{RT} [\, Ph_4Se \,] \longrightarrow Ph_2Se + Ph\text{-}Ph \qquad \text{ref. 32}$$

(24)   (23)   (5)

The formation of triphenylselenonium bromide (25) can be observed when milder reaction conditions are used. However, it is accompanied by major amounts of diphenylselenide (23) and biphenyl (5). The more important formation of diphenylselenide (23) compared to biphenyl (5) in the reactions involving a selenium-oxygen derivative could only be explained by a side reaction going through a direct reduction of diphenylselenoxide by the Grignard reagent. By contrast, the reaction of phenylmagnesium bromide with dibromodiphenylselenurane (26) led to equivalent amounts of these two products (23) and (5).[33,34]

$$SeOCl_2 + Ph\text{-}MgCl \xrightarrow{MgBr_2} Ph_3Se^{\oplus} Br^{\ominus} + Ph_2Se + Ph\text{-}Ph$$

(25), 37%   (23), 50%   (5), 32%

$$Ph_2SeO + Ph\text{-}MgBr \longrightarrow Ph_3Se^{\oplus} Br^{\ominus} + Ph_2Se + Ph\text{-}Ph$$

(25), 25%   (23), 64%   (5), 10%

$$Ph_2SeBr_2 + Ph\text{-}MgBr \longrightarrow Ph_2SeBr_2 + Ph_2Se + Ph\text{-}Ph$$
(26)

(26), 40%   (23), 48%   (5), 42%

Tetraphenylselenurane (24) was later detected by low temperature (- 100°C) NMR experiments, during the reactions of either diphenylselenoxide or triphenylselenonium bromide (25) with phenyllithium at - 78°C.[35-37] Raising the temperature resulted in high yields of the ligand coupling products. The activation parameters for the ligand coupling were in agreement with an unimolecular rate determining step.[37] (Scheme 8.5)

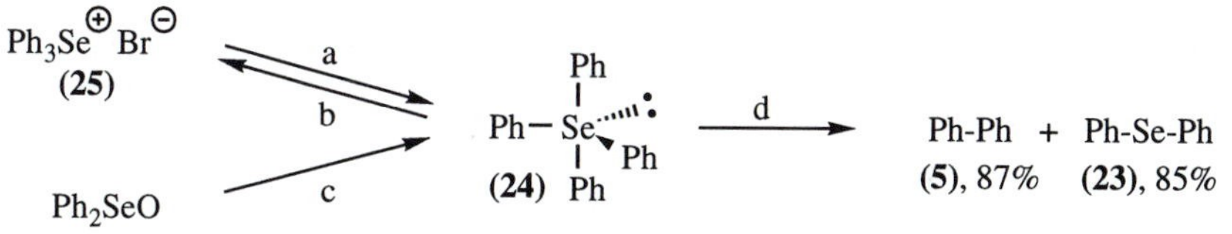

a) PhLi (1 equiv.) / - 78°C b) aq. HBr c) PhLi (2 equiv.) / - 78°C d) RT

Scheme 8.5: Synthesis and ligand coupling of tetraphenylselenurane (24)

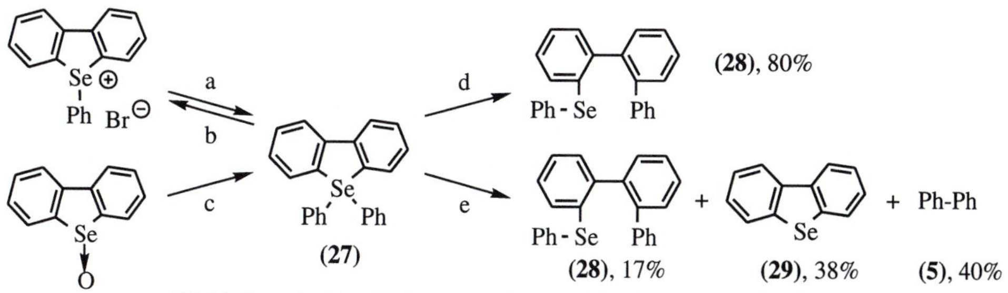

a) PhLi (1 equiv.) / - 78°C b) aq. HBr c) PhLi (2 equiv.) / - 78°C
d) RT e) HMPA, RT

Scheme 8.6: Synthesis and ligand coupling of 2,2'-biphenylylenediphenylselenurane (27)

The behaviour of the 2,2'-biphenylylenediphenylselenurane (**27**) was closely similar to the tetraphenyl compound (**24**). It can be prepared by action of phenyllithium either on the biphenylyl phenylselenonium salt or on the selenoxide.[35,36,38] However, the decomposition appeared to be solvent-dependent. Due to the non-symmetry of the molecule, the ligand coupling can occur by two ways. (See section 2.2.2.2) The favoured pathway takes place at room temperature to afford the phenyl *ortho*-terphenyl selenide (**28**). In the presence of HMPA (hexamethylphosphoramide), a more complex mixture was formed. The usually favoured pathway gave only a minor amount of the product (**28**), whereas the major pathway led to the formation of dibenzoselenophene (**29**) and biphenyl (**5**) in equivalent amounts.[35] (Scheme 8.6)

Similarly to the sulfur analog, the bis(2,2'-biphenylylene)selenurane (**30**) is more stable than compounds (**24**) and (**27**). It was even sufficiently stable to be isolated as a solid, from the reaction of biphenylylene *N*-tosylselenimide with 2,2'-dilithiobiphenyl.[39] The activation parameters for the pseudorotation of bis(2,2'-biphenylylene)selenurane were measured: the activation energy $\Delta G^{\neq}$ is 13.1 kcal/mole.[40]

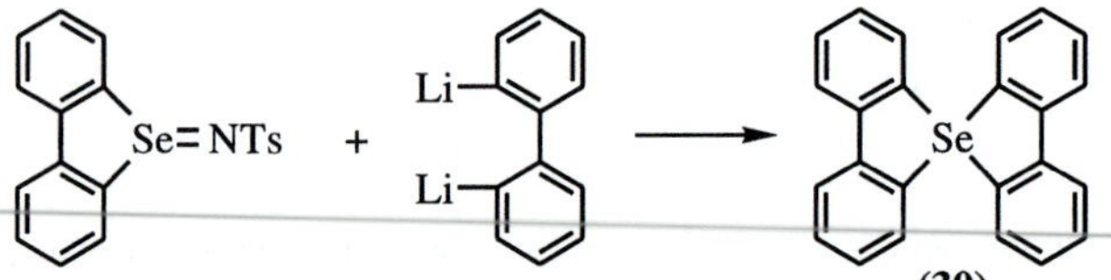

The bis(2,2'-biphenylylene)selenurane (**30**) reacted with alcohols, phenol, thiols and phenylselenol to afford a coloured solution at - 78°C, resulting from the proton-initiated ring opening followed by formation of an intermediate. This intermediate can be either a selenurane (**31**) or a selenonium salt (**32**). Although not clearly demonstrated, the selenurane structure (**31**) was nevertheless considered as more likely. Upon warming, a ligand coupling or an *ipso*-substitution took place to lead to the coupled products (**33**) - (**35**).[41] (Scheme 8.7)

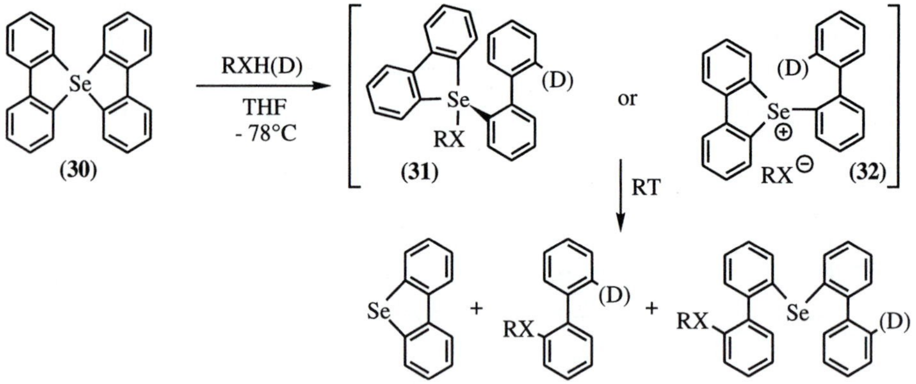

	(33)	(34)	(35)
X = O	71-99	65-89	-
X = Se	80	61	14
X = S	75-80	60-74	19-25

Scheme 8.7: Reaction of bis(2,2'-biphenylylene)selenurane (**30**) with acidic reagents and ligand coupling of the ring opened products (**31**), (**32**)[41]

The decomposition of triarylselenonium halides was reported to produce aryl halide.[42] This reaction is likely to proceed through a ligand coupling mechanism.

$$Ar_3SeX \longrightarrow Ar-X + Ar_2Se$$

However, in the course of a recent mechanistic study of the electrophilic chlorination of toluene in the presence of a mixture of bis(4-chlorophenyl)selenide-Lewis acid as catalyst, Graham *et al.* studied the decomposition of bis(4-chlorophenyl) (4-methylphenyl)selenonium chloride (**36**). They found that the ligand coupling pathway occured only in the presence of aluminum (III) chloride. Moreover, only chlorotoluene derivatives were obtained: dichlorobenzene was not detected. The reductive elimination appears to be obviously mostly intramolecular. However, the important amount of *ortho*-chlorotoluene is not consistent with a single ligand coupling mechanism.[43]

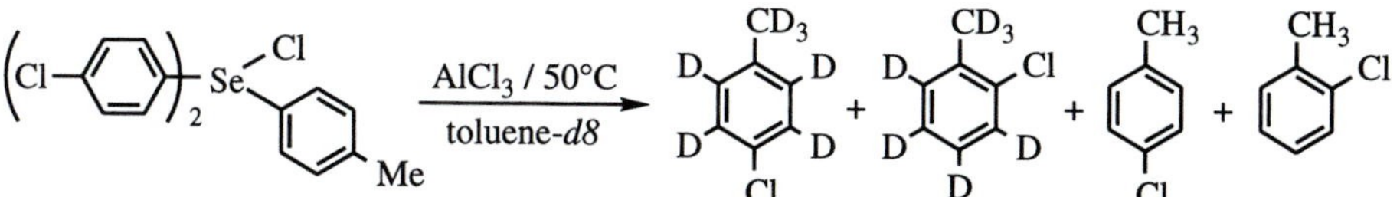

(**36**) 2.42% 2.51% 48.60% 16.10%

The reaction of diarylselenoxides with organolithium has been scarcely studied. Treatment of bis(2-pyridyl)selenoxide (**37**) with phenyllithium gave a moderate yield of bipyridyl (**38**), the ligand coupling product, together with a range of other products. With phenyl(2-pyridyl)selenoxide (**39**), the reaction evolved through a series of ligand exchange and ligand coupling pathways.[44,45] But the reaction of phenyl *p*-tolylselenoxide (**40**) with *tert*-butyllithium afforded only the products of rapid ligand exchange and disproportionation.[46]

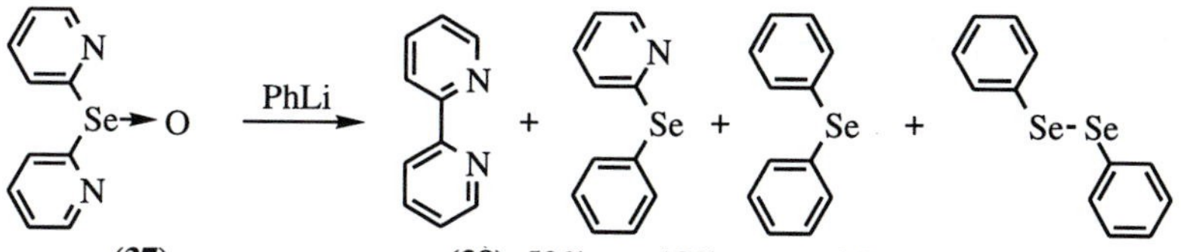

(**37**) (**38**), 53% 19% 25% 30%

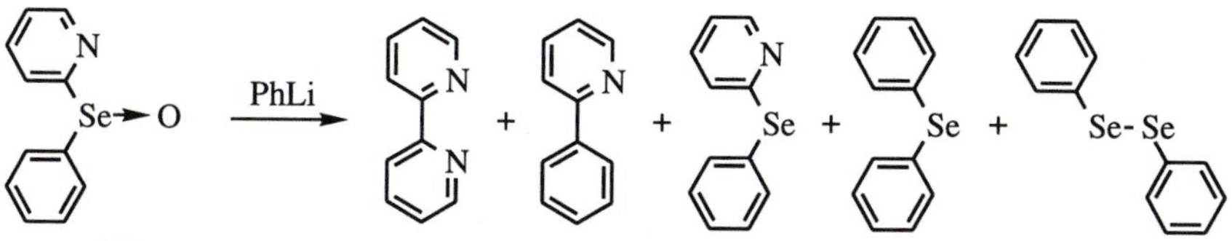

(**39**) (**38**), 13% 27% 19% 37% 6%

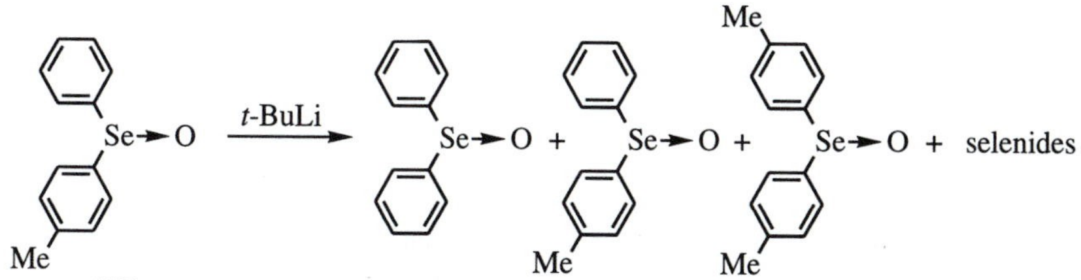

(**40**) 10% 20% 16% 30%

8.2.2 Reactions of alkylselenium compounds

The reaction of the 2-methylisoselenochromane derivatives (**41**) or (**42**) with nucleophiles led to substituted methyl(phenylethyl)selenides (**43**), the product of ring opening, as well as to styrene derivatives (**44**), the product of base catalysed β-elimination. Depending on the nature of the counteranion, the heterocyclic selenophene derivatives exist either as selenonium salts (**41**) or as selenurane (**42**). The reaction with various types of nucleophiles leading to ring opened products was considered to go through a common transition state resulting from the cleavage of the selenurane to a selenonium salt. However, it cannot be excluded that the reaction occurs through a common tetracoordinate selenurane intermediate, which would undergo a ligand coupling reaction.[47] Grignard reagents reacted with the isoselenochromanium salts to lead exclusively to a different outcome: only SET reduction was observed to afford the reductive ring opened products.[48]

With oxygen nucleophiles:

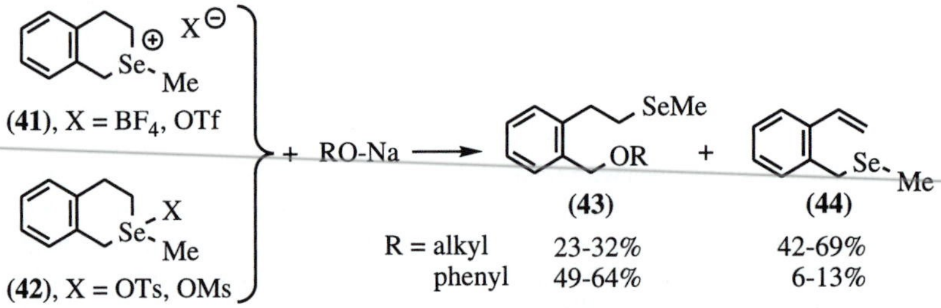

With carbon nucleophiles:

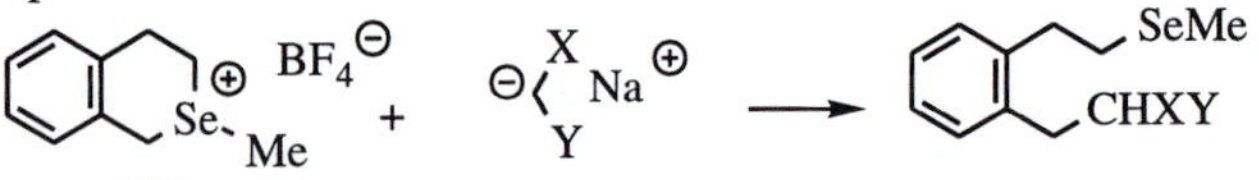

$$X = Y = COMe \qquad 42\%$$
$$X = COMe, Y = CO_2Me \quad 52\%$$
$$X = Y = CO_2Me \qquad 44\%$$

The reaction of dibenzylselenide with difluorotriphenylbismuth led to benzyl fluoride. The mechanism may be viewed as a ligand transfer giving difluorodibenzylselenurane (**45**), followed by ligand coupling.[49]

$$Ph_3BiF_2 + (PhCH_2)_2Se \xrightarrow{120°C} \left[(PhCH_2)_2SeF_2 \right] \longrightarrow PhCH_2F \quad 29\%$$
$$(45)$$

8.2.3 Reactions of perfluoroselenonium salts

Electrophilic perfluoroalkylation of nucleophiles is not as common as the electrophilic alkylation of nucleophiles.[50] However, a variety of *S*-trifluoromethyl dibenzothiophenium salts have been used for a range of substrates. (See section 3.1.5) Among the other chalcogenium systems which have been studied, the selenium analogs, the *Se*-trifluoromethyl dibenzoselenophenium salts, behave as milder electrophilic trifluoromethylating agents towards nucleophiles. Usually less reactive than the sulfur analogues, they gave good yields of the trifluoromethylated products, sometimes more efficiently than the sulfur reagent. The range of selenium compounds is more limited than in the sulfur series. The reactivity comparisons were made with the sulfur reagents (46), (48) and the selenium reagents (47) and (49).[51,52]

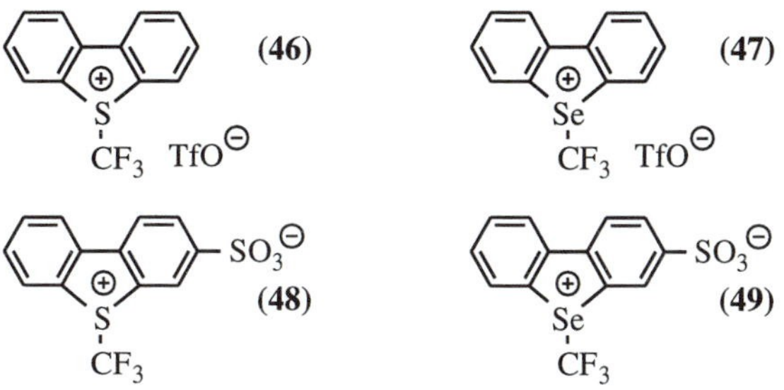

Table 8.2: Comparative reactivity of the sulfur reagents (46), (48) and the selenium reagents (47), (49) in the trifluoromethylation reaction of various types of nucleophiles [51,52]

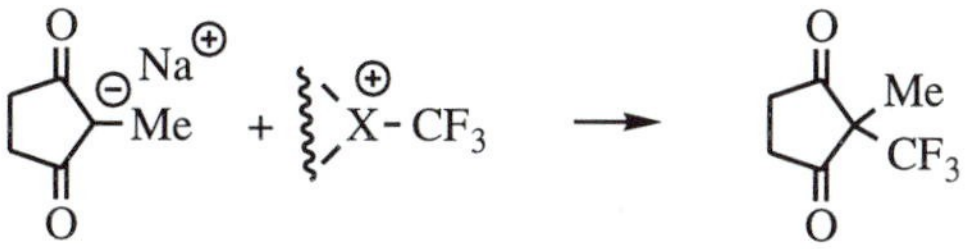

S reagents:	(46)	6 h	84%	ref. 51
	(48)	2 h	86%	ref. 52
Se reagents:	(47)	24 h	84%	ref. 51
	(49)	3.5 h	81%	ref. 52

| S reagent: | (46) | 1.25 h | 58% | ref. 51 |
| Se reagent: | (47) | 1.25 h | 89% | ref. 51 |

| S reagent: | (46) | 0.5 h | 47% | ref. 51 |
| Se reagent: | (47) | 0.5 h | 87% | ref. 51 |

In the case of the reaction of cyclohexanone potassium enolate in the presence of the boron complexing agent (50), the selenium reagent (47) appeared to be significantly less efficient as a regioselective trifluoromethylation agent than the sulfur reagent (46).[53]

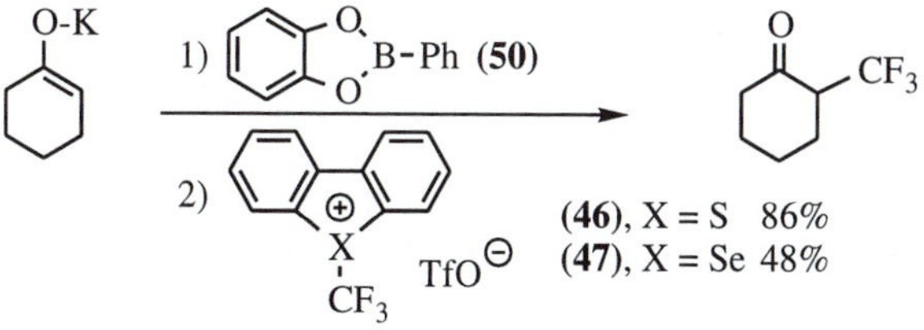

(46), X = S 86%
(47), X = Se 48%

8.2.4 Reactions of benzeneseleninic anhydride

Benzeneseleninic anhydride (**51**), (PhSeO)$_2$O, reacts with phenols to lead preferentially to the products of *ortho*-hydroxylation. Depending on the structure of the phenol, minor by-products may be observed resulting from *para*-oxidation or from *ortho*-phenylselenylation.[54-56] The formation of these different types of products has been explained by the intervention of two competing mechanisms. The major one, leading to *ortho*-oxidation involves a series of steps, the key one being a [2,3]-sigmatropic rearrangement of the intermediately formed aryl benzeneseleninate.[54-56] This sequence can be viewed as an example of ligand coupling involving the creation of a σ bond between two sp^2 atoms. The minor ones, leading to *para*-oxidation or *ortho*-phenylselenylation follow ene routes.[56,57]

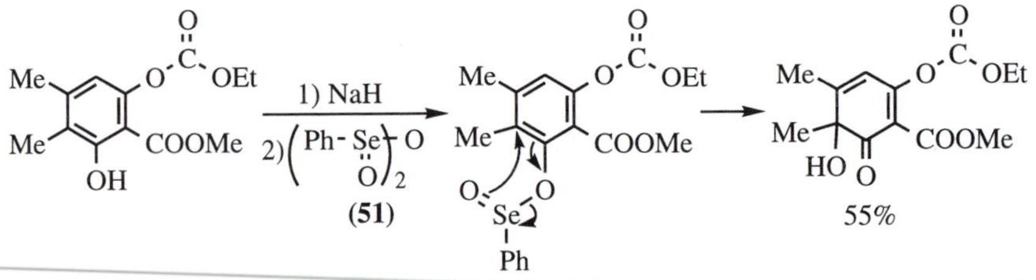

Scheme 8.8: *ortho*-Hydroxylation of a phenol with benzeneseleninic anhydride (**51**)[55,56]

When the reaction of a phenol with benzeneseleninic anhydride (**51**) is performed in the presence of hexamethyldisilazane, the corresponding *N*-phenylselenoimine is generally obtained, again with high *ortho*-selectivity.[58] The aminating reagent was shown to be a polymeric form of the intermediate RSeN (**52**).[59]

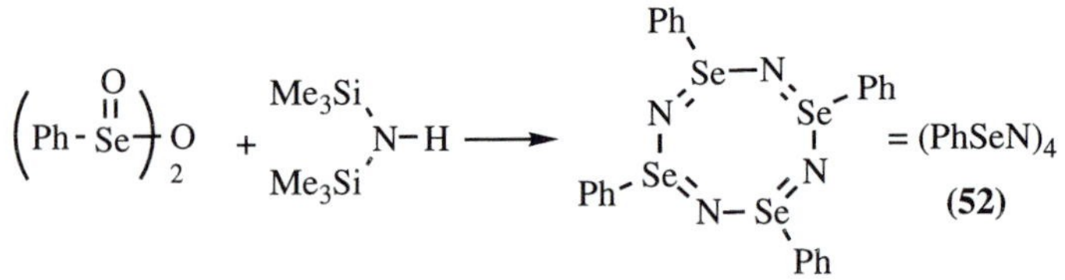

Scheme 8.9: Mechanism of the reaction of phenols with benzeneseleninic anhydride and hexamethyldisilazane: generation of RSeN (**52**)[59]

This cyclotetraselenazatetraene (**52**) reacted with the phenolic substrate to afford the *N*-phenylselenoimine (**53**). The key step, the [2,3]-sigmatropic rearrangement, has also been viewed as an example of ligand coupling involving the creation of a σ bond between two sp^2 atoms.[59]

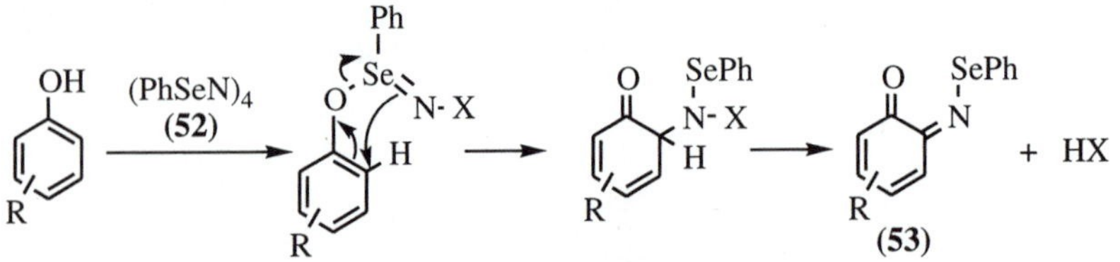

Scheme 8.10: Mechanism of the reaction of phenols with benzeneseleninic anhydride and hexamethyldisilazane: reaction of the reagent RSeN (**52**) with the phenol[59]

8.3 ORGANOTELLURIUM

The chemistry of organotellurium derivatives has only recently been developed into a range of reagents useful for mild selective transformations in organic synthesis.[60-64] Although the first studies were mostly extrapolations of sulfur and selenium chemistry, organotellurium reagents have now grown into a full-fledged group of selective reagents for specific transformations. They involve various types of mechanistic pathways, such as ionic, radical or intramolecular rearrangements. The ligand coupling mechanism has also been invoked to explain a series of transformations.[65]

8.3.1 Aryl-aryl coupling

The synthesis of tetraphenyltellurane (54) was first reported by Wittig and Fritz in 1952.[32] This compound, easily prepared by the reaction of phenyllithium with tellurium tetrachloride, is a relatively stable substance which decomposes above 100°C to give diphenyltelluride (55) and biphenyl (5). Later, other tetraaryltelluranes have been prepared by the same method.[66,67]

$$TeCl_4 + 4\ PhLi \longrightarrow Ph_4Te \xrightarrow{\ T > 100°C\ } Ph_2Te + Ph\text{-}Ph$$

$$\text{(54)} \qquad\qquad \text{(55)} \qquad \text{(5)}$$

Extensive mechanistic studies have shown that the thermal decomposition of (54) occurs through a concerted pathway, without intervention of either free radical or ionic species.[66,67] The activation parameters revealed a higher energy barrier to carbon-chalcogen bond scission in the tellurane compared to the selenurane and are in support of the unimolecularity of the rate determining step. The activation energy for the ligand coupling reaction was measured to be $E_{act} = 29$ kcal/mole, compared with 21.3 for Ph_4Se and 10.9 for Ph_4S.[37]

The reaction of pentafluorophenyllithium (56) with tellurium tetrachloride afforded a mixture of tetrakis(pentafluorophenyl)tellurane (57), bis(pentafluorophenyl)telluride (58) and perfluorobiphenyl (59). Moreover, the tetraaryltellurane compound (57) decomposes into perfluorobiphenyl (59) and bis(pentafluorophenyl)telluride (58) upon heating in a sealed tube to 200-220°C.[68]

$$\text{(56)} \qquad\qquad\qquad \text{(57)} \qquad \text{(58), 30\%} \qquad \text{(59)}$$

Treatment of arylmagnesium bromides (4 mol. equiv.) with tellurium tetrahalides gave rise directly to diaryltelluride in high yields. As by-product, the biaryl compounds were also formed. Therefore, it is likely that the reaction goes through the intermediacy of tetraaryltelluranes which undergo ligand coupling under the reaction conditions.[69,70]

$$TeX_4 + ArMgBr \xrightarrow[\text{reflux}]{\text{ether / benzene}} Ar_2Te + Ar\text{-}Ar$$

Ar = Ph; X = Cl, Br, I
Ar = 1-Naphthyl, *o*-MeC₆H₄, *p*-MeC₆H₄; X = Br

The 2,2'-biphenylylenediphenyltellurane (**60**) can be conveniently prepared through the reaction of 2,2'-dilithiobiphenyl with dichlorodiphenyltellurane. It is a relatively stable compound, which can be isolated at room temperature. However, at higher temperatures (~ 140°C), it decomposes to give the ligand coupling products (**61**), (**62**) and (**5**).[38] However, the mechanistically more favoured ligand coupling pathway, leading to the terphenyl compound (**61**), appeared to play a minor role in the tellurium series. The decomposition pathways are in fact more likely to involve phenyl free radicals, which can either couple to afford biphenyl or further react with the solvent.

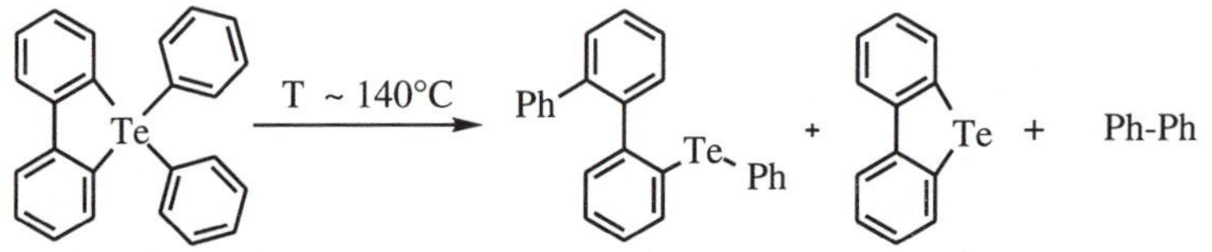

| (**60**) | (**61**), 8% | (**62**), 69% | (**5**), 36% |

The spiro compound, bis(2,2'-biphenylylene)tellurane (**63**), has been prepared and isolated in good yields.[71,72] It is a relatively stable compound, which requires fairly high temperatures to decompose into dibenzotellurophene (**64**) and other products derived from the diradical species, the dibenzocyclobutane (**65**) being the most important one.[72] The activation parameter for the pseudorotation was determined by variable temperature NMR to be $\Delta G^{\neq}$ = 9.2 kcal/mole.[40]

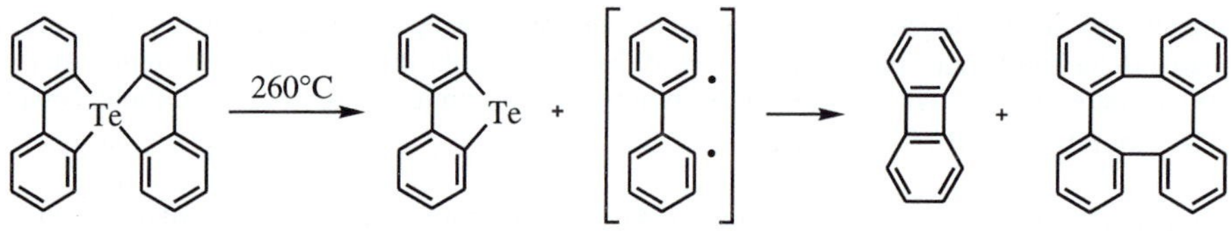

| (**63**) | (**64**), 76% | (**65**), 51% | traces |

Scheme 8.11

8.3.2 Alkynyl-alkynyl coupling

Dialkynyltellurides are easily formed by the reaction of alkynyl Grignard reagents or alkynyl lithium reagents with tellurium tetrachloride.[73,74] The reaction mixtures contain significant amounts of the butadiyne, a typical ligand coupling reaction product.

With Grignard reagents:[73]
The reaction of (phenylethynyl)magnesium bromide with tellurium tetrachloride afforded modest yields of the dialkynyltelluride (**66**) together with an equimolecular amount of 1,4-diphenylbutadiyne (**22**). Because of the difficulties for the purification, the dialkynyltelluride (**66**) was isolated as the diiodide in a 33% yield.

$$4\ Ph-C\equiv C-MgBr\ +\ TeCl_4 \longrightarrow Ph-C\equiv C-Te-C\equiv C-Ph\ +\ Ph-C\equiv C-C\equiv C-Ph$$
$$\qquad\qquad\qquad\qquad\qquad\qquad\qquad\qquad (\mathbf{66})\qquad\qquad\qquad\qquad (\mathbf{22})$$

With lithium reagents:[74]
The reaction of an excess of 1-alkynyllithium reagents with tellurium tetrachloride afforded good yields of the dialkynyltellurides (**67**) (42-69%). Moreover, diynes (**68**) were isolated as major by-

products, when R = *t*-Bu, Me$_3$Si and Ph. In the case of the *tert*-butyl derivative, equivalent amounts of the telluride (**67**) and the diyne (**68**) were obtained in an 85% overall yield.

$$4 \ \text{R-C}\equiv\text{C-Li} + \text{TeCl}_4 \longrightarrow \text{R-C}\equiv\text{C-Te-C}\equiv\text{C-R} + \text{R-C}\equiv\text{C-C}\equiv\text{C-R}$$

R = Me, Et, *n*-Pr, *t*-Bu, Me$_3$Si and Ph   (**67**), 42-69%   (**68**)

8.3.3 Aryl-heteroatomic nucleophile coupling

The bis(2,2'-biphenylylene)tellurane (**63**) reacted with phenol to afford the product of protonolysis (**69**) in high yield. The X-ray crystallographic analysis of (**69**) revealed a tellurane structure, and not the alternative onium structure.[75]

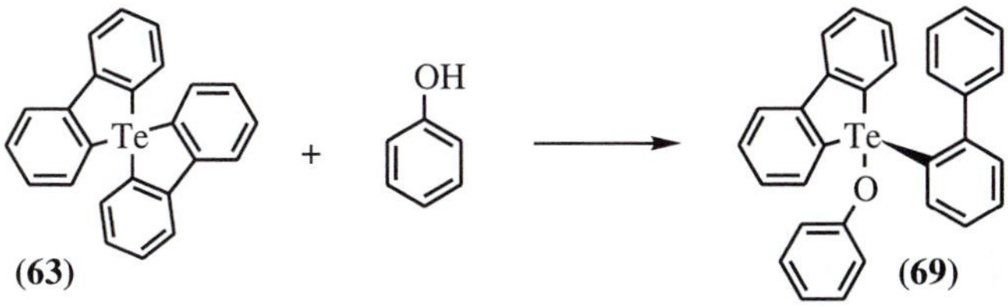

In marked contrast with the selenium analog (**41**), this compound (**69**) was thermally stable, even at more than 100°C for 10 hours. Only at temperature above 190°C, it decomposed to afford the ligand coupling products (**64**), (**70**) and (**71**).

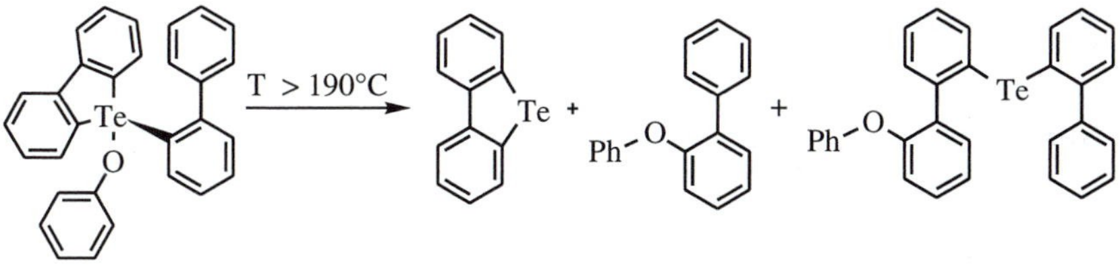

With other substituted phenols, the tellurane structure was not significantly altered by the presence of strongly electron-withdrawing substituents in the *para* position. Only strong steric hindrance led to a change in the structure and the length of the tellurium-oxygen bond. (Table 8.3)

Table 8.3: Length of the tellurium-oxygen bond in aryloxytelluranes [75,76]

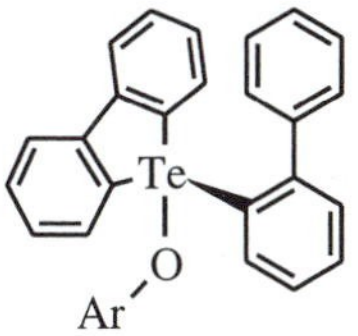

(**69**), (**72**), (**73**)

Ar	(**69**), C$_6$H$_5$	(**72**), 4-NO$_2$C$_6$H$_4$	(**73**), 2,4,6-Cl$_3$C$_6$H$_2$
Te-O (Å)	2.294	2.461	2.76 and 2.787

For example, the 4-nitrophenoxytellurane (**72**) presented a distorted tellurane structure as in (**69**), but the 2,4,6-trichlorophenoxytellurane (**73**) appeared to be a centrosymmetric *O*-bridged dimer, with a distorted square pyramidal onium-type structure. The pyrolysis of these products in the solid state led to the ligand coupling products which showed the same general selectivity patterns as those observed in the case of (**69**).[76]

Triphenyltelluronium salts react with sodium aryltellurolate to afford the ligand coupling products. Metastable crystals of the ditellurium intermediate (**74**) have even be isolated. Their thermal decomposition led to the ligand coupling products.[77]

$$Ph_3TeCl + ArTeNa \xrightarrow{THF} Ar\text{-}Te\text{-}Ph + Ph_2Te + NaCl$$

$$\begin{matrix} Ph_3TeCl \\ + \\ ArTeNa \end{matrix} \xrightarrow[-60°C]{MeOH / EtOH} \underset{(74)}{Ph_3Te\text{-}TeAr} \longrightarrow Ar\text{-}Te\text{-}Ph + Ph_2Te + Ph\text{-}Ph + Ar_2Te_2$$

Similarly, triphenyltelluronium salts react with sodium arylselenolate to afford the intermediate selenium-tellurium intermediate (**75**), which decomposed relatively easily. In this reaction, the ratio diselenide : selenide is approximatively 2:1, whereas only minute amounts of the ditelluride were observed in the reaction involving the ditelluride intermediate (**74**).[77]

$$\begin{matrix} Ph_3TeCl \\ + \\ ArSeNa \end{matrix} \xrightarrow[-60°C]{MeOH / EtOH} \underset{(75)}{Ph_3Te\text{-}SeAr} \longrightarrow Ar\text{-}Se\text{-}Ph + Ph_2Te + Ph\text{-}Ph + Ar_2Se_2$$

8.3.4 Halodetelluration reactions

The transformation of a carbon-tellurium bond into a carbon-halogen bond can be performed with several types of organotellurium trihalide or diorganotellurium dihalide compounds, in which the organic group is an alkyl, an alkenyl or an aryl substituent. A variety of reaction conditions have been described to realize these transformations, which belong to three main types:

1 - oxidative procedures:[78,79]
2 - photolytic procedures:[79,80]
3 - thermolytic procedures:[79,81]

In the oxidative procedure, the substrate is treated with an oxidant, such as *tert*-butyl hydroperoxide, in refluxing dioxane, acetic acid or acetonitrile.[78,79] The yields are good to high for cyclohexanone derivatives, moderately good for aryl derivatives and rather modest for vinylic substrates. In all the reactions involving cyclohexyl and alkyl derivatives, retention of configuration was observed. Diaryltellurium dichlorides did not react, and, in the case of arylcyclohexyl derivatives, the cyclohexyl group is more reactive than the aryl group towards ligand coupling.

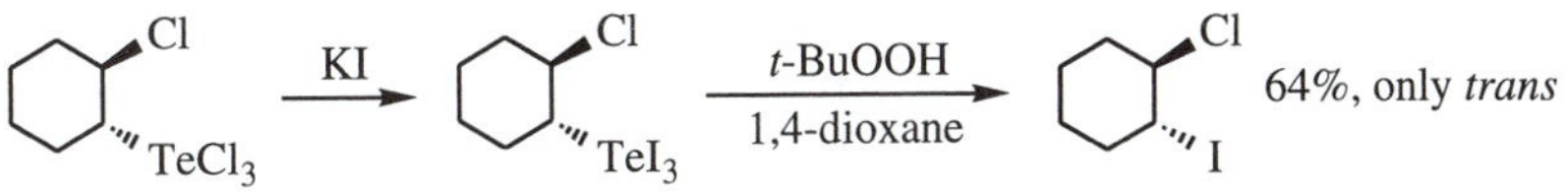

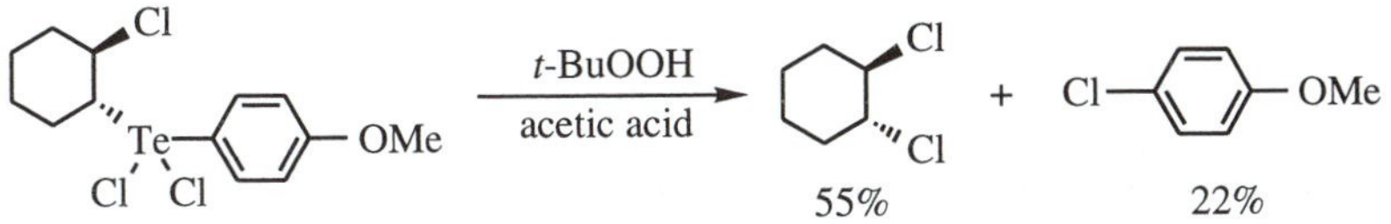

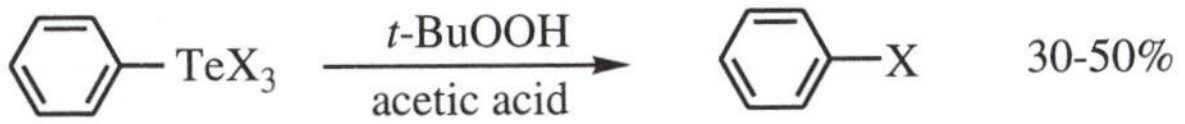

The mechanism of the oxidative α-elimination has been suggested to proceed through an hypervalent tellurone(VI) intermediate (**76**), which undergoes ligand coupling realizing an intramolecular 1,2-tellurium halogen shift.[78,79] (Scheme 8.12)

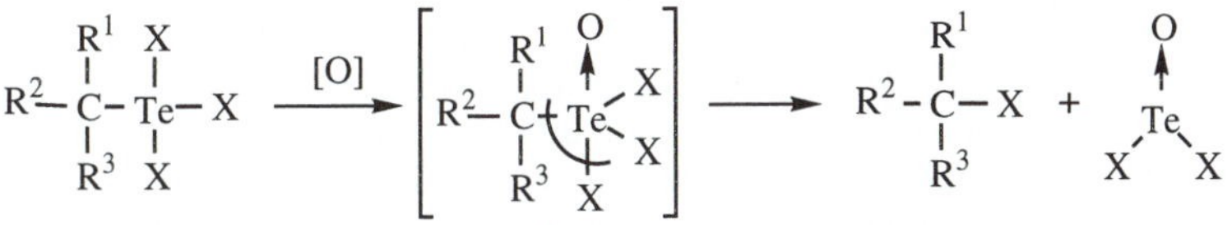

Scheme 8.12: Oxidative halodetelluration with a peroxidic oxidant

The role of oxidant can be played by a molecule of halogen which oxidizes the diorganyl-tellurium(IV) dihalide to a tellurone analogue, a diorganyltellurium(VI) tetrahalide, which undergoes the intramolecular 1,2-tellurium halogen shift to afford similarly the alkyl halide and a monoorganyl-tellurium(IV) trihalide.[64] (Scheme 8.13)

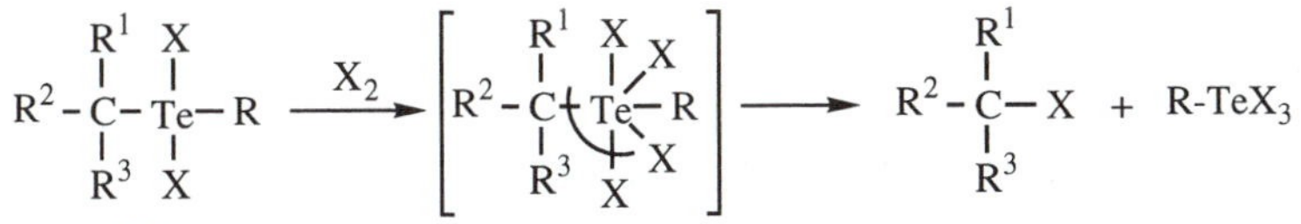

Scheme 8.13: Oxidative halodetelluration with halogens as oxidant

This mechanism may explain a number of oxidative halodetelluration reactions which have been reported. This is the case of the following examples:

1) the halogenodetelluration of α-dichloroaryltelluroketones:

Treatment of α-dichloroaryltelluroketones (**77**) with one molecular equivalent of chlorine in CH_2Cl_2 leads to the corresponding α-chloroketones (**78**), at room temperature, in good yields.[82]

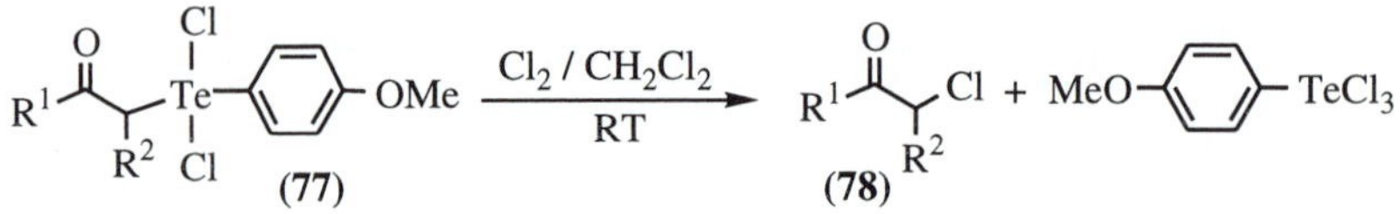

$$R^1 = Ph, t\text{-Bu}; \quad R^2 = H \qquad\qquad 75\text{-}78\%$$
$$R^1, R^2 = (CH_2)_3, (CH_2)_4, (CH_2)_5 \qquad 65\text{-}71\%$$

2) the halogenodetelluration of acetylenic tellurides:

Treatment of acetylenic tellurides (79) with three molecular equivalents of bromine or iodine leads to trihalovinylic derivatives (80). The reactions involve the oxidative halogenolysis of the carbon-tellurium bond followed by addition of the halide to the acetylenic bond.[83]

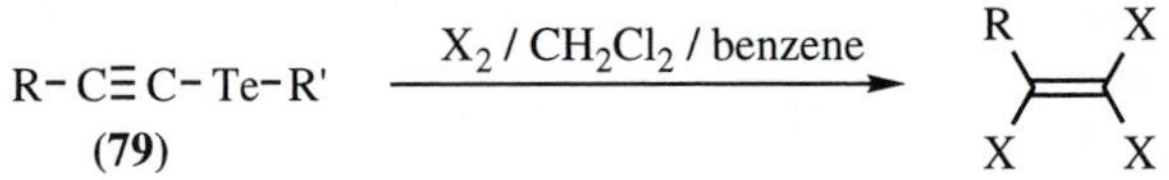

$$X = Br \quad R = n\text{-}C_5H_{11}, C_6H_5, 4\text{-}BrC_6H_4; \quad R' = n\text{-Bu} \qquad 53\text{-}76\%$$
$$R = C_6H_5; \quad R' = 4\text{-MeOC}_6H_4 \qquad\qquad 51\%$$
$$X = I \quad R = n\text{-}C_5H_{11}; \quad R' = n\text{-Bu} \qquad\qquad 74\%$$

3) the halogenodetelluration of benzylic tellurium derivatives:

Treatment of dibenzyltellurides or dibenzylditellurides with an excess of bromine leads to benzyl bromide.[84,85] The intermediate dibenzyltellurium dibromide is too unstable to be isolated and it decomposes directly to benzyl bromide. A similar carbon-tellurium cleavage was observed in the treatment of benzyltellurocyanide with an excess of bromine. Two consecutive oxidation-ligand coupling reactions sequence take place to afford eventually cyanogen bromide and benzyl bromide.[86]

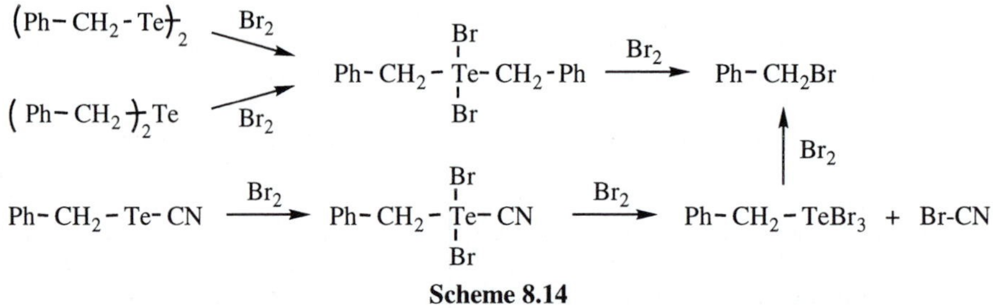

Scheme 8.14

Pyrolysis of organotellurium dibromides at 200-250°C without solvent under reduced pressure afforded the organyl bromides in good to high yields. When applied to alkyl- or aryl-tellurium trichlorides, no halogen-containing products were produced. As this pyrolysis reaction gives the same products as those obtained in the previous oxidative halodetelluration reaction, it is likely that the mechanism of the pyrolytic reaction is also an intramolecular ligand coupling process.[79]

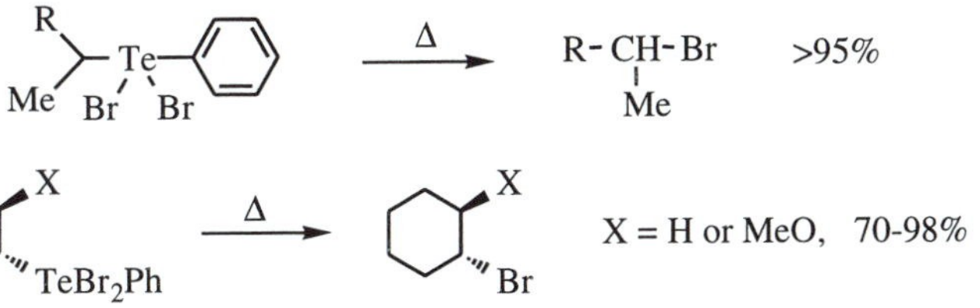

$$R\text{-}CH\text{-}Br \quad >95\%$$
$$|$$
$$Me$$

X = H or MeO, 70-98%

8.3.5 Trifluoromethylation reactions

Among a series of trifluoromethyl dibenzoheterocyclic onium salts derived from chalcogens, a small number of tellurium compounds have been prepared. As electrophilic trifluoromethylating agents, the tellurium compounds appeared as the least reactive.[50] With a very reactive carbanion, the unsubstituted dibenzotellurophenium salt (81) gave only 9% of the trifluoromethylated product. With the β-diketone enolate, (81) did not react. However, the 3,7-dinitrotellurophenium salt (82) showed a relative activity similar to that of the unsubstituted selenium compound (47).[51]

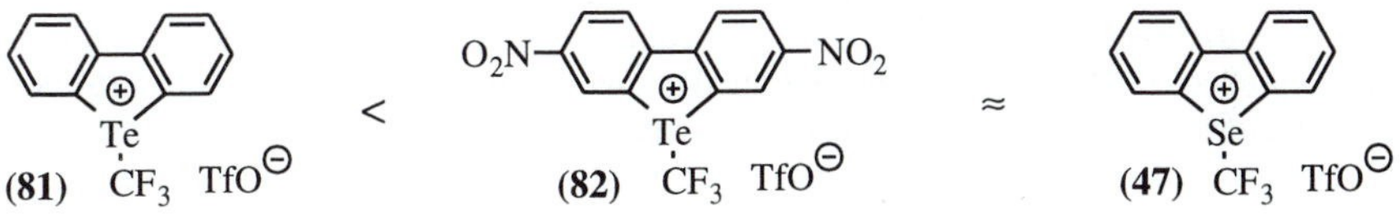

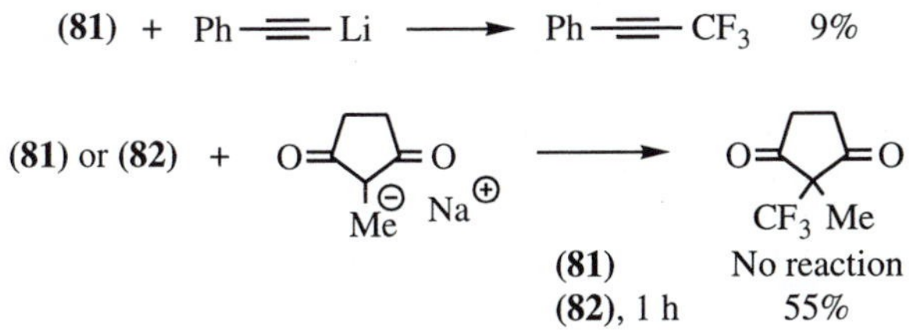

(81) + Ph≡Li ⟶ Ph≡CF₃   9%

(81) or (82) +

	(81)	(82), 1 h
	No reaction	55%

8.4 ORGANOTHALLIUM

Long considered "dull and predictable",[87] the chemistry of organothallium compounds has seen an extensive growth since the seminal works of McKillop and Taylor and of Uemura, who have independently discovered a number of selective transformations involving organothallium intermediates or reagents. Most of these transformations were explained by conventional mechanisms such as electrophilic additions, nucleophilic substitutions or radical mechanisms.[87-91] However, some reactions have been explained by intramolecular decompositions, which can be classified as ligand coupling reactions. Two major types of organothallium compounds evolve into the reaction end-products by ligand coupling: the arylthallium(III) compounds and the alkyl or alkenylthallium(III) compounds.

8.4.1 Arylthallium(III) compounds

Arylthallium(III) compounds (83) are generally prepared by electrophilic aromatic thallation with thallium tris(trifluoroacetate) (84).

$$Ar\text{-}H + Tl(OCOCF_3)_3 \xrightarrow{CF_3COOH} Ar\text{-}Tl(OCOCF_3)_2$$
$$(84) \qquad\qquad\qquad\qquad\qquad\qquad (83)$$

This method allows a good control of the regioselectivity of the substitution in the case of substituted aromatic nuclei. The electrophilic character of thallium tris(trifluoroacetate) (84) can be increased by addition of Lewis acids, such as boron trifluoride or antimony pentafluoride.[92,93] Side reactions are observed with electron-rich substrates which undergo only oxidative coupling.[94] Thallation is obtained for these substrates by performing the reaction in the presence of a co-solvent which inhibits the solvation of the intermediate radical ion pair by CF_3CO_2H. In a 1:1 mixture of CF_3CO_2H and ether, the thallated products derived from electron-rich arenes were isolated in good yields.[95] In the case of acid-sensitive electron-rich substrates, these systems are not suitable but milder conditions avoid these drawbacks, by using boron trifluoride as a catalyst and 1,2-dichloroethane as the solvent.[96]

$$\text{Ar-H} \ + \ \text{Tl(OCOCF}_3)_3 \ \xrightarrow[\text{BF}_3\text{-Et}_2\text{O}]{\text{ClCH}_2\text{-CH}_2\text{Cl}} \ \text{Ar- Tl(OCOCF}_3)_2$$
$$\qquad\qquad\qquad\quad (84) \qquad\qquad\qquad\qquad\qquad\qquad\qquad\qquad\qquad (83)$$

The thallation orientation can be controlled by proper choice of the reaction conditions.[97] Aromatic thallation is a reversible reaction with a moderately large energy of activation (27 kcal/mole) and an extremely large steric requirement.[98] In the case of cumene, the reaction under kinetic conditions leads to the *para*-product (85), whereas under thermodynamic control, the *meta*-product (86) is largely predominant. However, the outcome of the reaction may be dramatically influenced by the presence of a complexing group. Indeed, thallation of benzoic acid leads mostly to the *ortho*-thallated product (87), and benzyl alcohol affords only the *ortho*-product. A number of heteroatom-containing groups are able to orient the thallation reaction selectively on the *ortho*-position. This is the case of CO_2R, CO_2H, CH_2OH, OR, $NHCOR$.

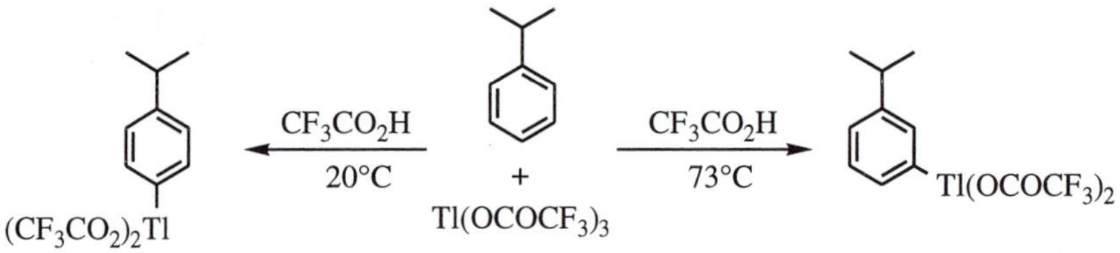

para-isomer (85): 94%   meta-isomer (86): 85%

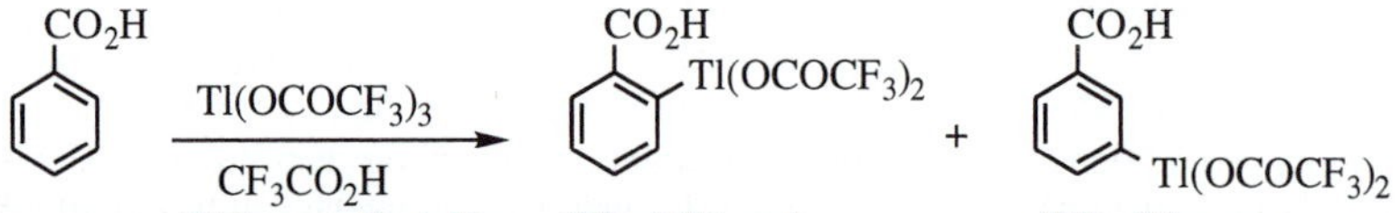

76% overall yield   (87), 95% ortho   (88), 5% meta

The arylthallium(III) compounds can be easily converted into a variety of substituted aromatic derivatives. This conversion can be performed by three major reaction types, two of them being either typical ligand coupling or ligand exchange reactions involving the presence of a copper species, either as a reagent or as a catalyst.

a - Reaction with a nucleophile:

Reaction of the arylthallium(III) compound with a nucleophile affords a new intermediate arylthallium derivative (89), which then undergoes ligand coupling.[99,100]

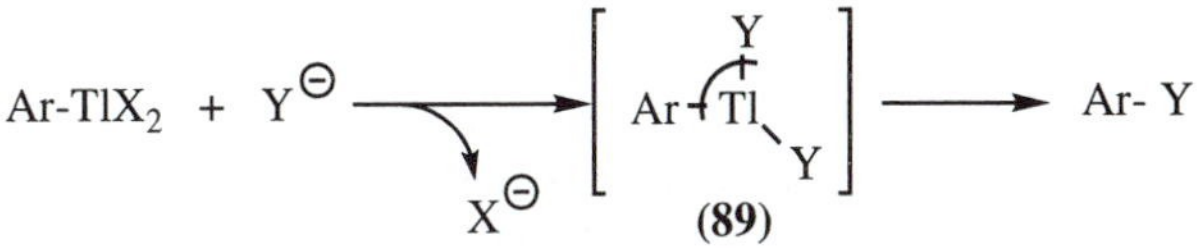

$$\text{Ar-TlX}_2 \ + \ Y^{\ominus} \ \longrightarrow \ \left[\text{Ar} \overset{Y}{\underset{Y}{\overset{+}{-}}}\text{Tl} \right] \ \longrightarrow \ \text{Ar-Y}$$

$$X^{\ominus} \qquad \textbf{(89)}$$

Scheme 8.15

b - Copper salt mediated reaction:

When the reaction of the arylthallium compound $ArTlX_2$ with the nucleophile Nu^- leads to a too stable intermediate $ArTlNu_2$, competing protonolysis may become predominant. This is circumvented by using copper salts, which favours the removal of a ligand from the thallium metal. According to Uemura *et al.*, this reaction leads to substitution in a concerted process by reductive displacement of the group TlX.[101-103]

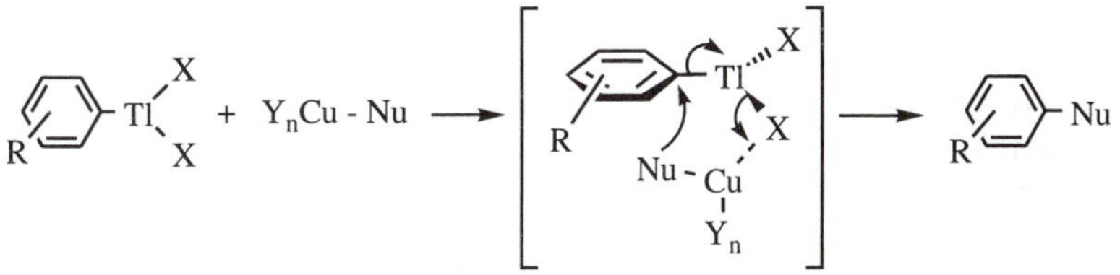

Scheme 8.16

Recently, Somei *et al.* invoked a closely related mechanism which involves the formation of a π-complex between the arylthallium and the copper species, favouring the attack of the *ipso*-carbon by a coordinated nucleophilic ligand.[104]

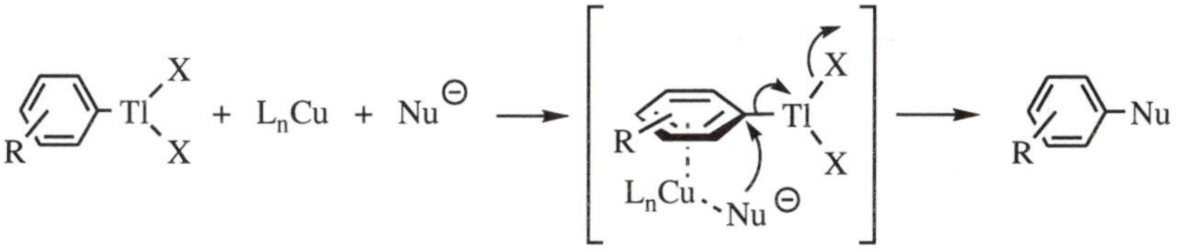

Scheme 8.17

A different mechanism was suggested by McKillop and Taylor, who considered that the copper mediated reactions involve single electron transfer from the copper Cu(I) species to the thallium substrate. The arylthallium(II) species loses a thallium(I) salt and the aryl radical. This aryl radical secondarily reacts with the copper(II) species formed in the first step by electron transfer.[105,106]

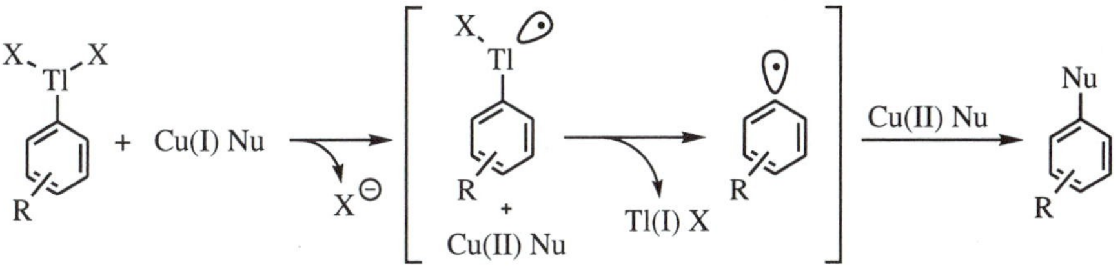

Scheme 8.18

c - Electrophilic substitution:

Reaction of the arylthallium(III) compound with an electrophile affords an intermediate cationic arylthallium derivative (90), which then undergoes elimination of the thallium moiety. Examples of this type of reactivity are the reactions of arylthallium(III) compounds with bromine affording aryl bromide,[107] with iodine affording aryl iodide,[108-110] or with nitrosyl chloride giving nitroso-benzene.[111]

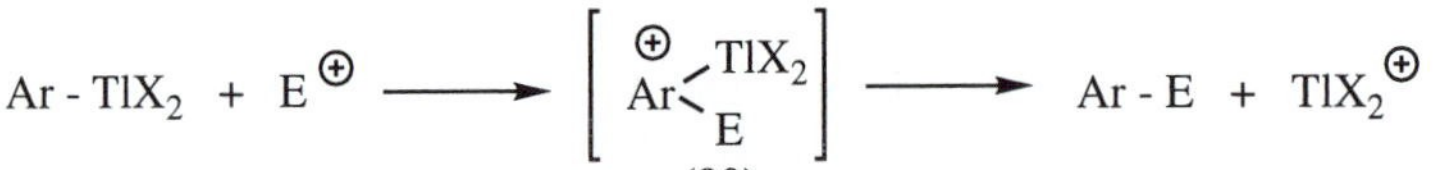

$$ \text{Ar - TlX}_2 \; + \; \text{E}^{\oplus} \longrightarrow \left[\overset{\oplus}{\underset{\text{E}}{\text{Ar}}}\diagdown \text{TlX}_2 \right] \longrightarrow \text{Ar - E} \; + \; \text{TlX}_2^{\oplus} $$

(90)

Scheme 8.19

The ligand coupling and ligand transfer mechanisms explain a number of transformations of organothallium compounds. The best studied reaction is the carbon-iodine conversion for the synthesis of aryl iodides, but other efficient conversions have been also described.

8.4.1.1 C-Tl to C-I

The formation of iodobenzene by treatment of phenylthallium(III) compounds with potassium iodide was reported, without experimental details, by Challenger *et al.* in the 1930's.[112,113] The potential and synthetic interest of this iododethallation reaction was extensively studied by McKillop and Taylor in the early 1970's.[87-89] Although arylthallium(III) compounds prepared by reaction of the arenes with thallium tris(trifluoroacetate) (84) can be isolated, they can also be directly converted into aryl iodides by addition of aqueous potassium iodide to the thallation reaction mixture. An intermediate arylthallium(III) diiodide (91) was suggested to be formed and to decompose intramolecularly to lead to the aryl iodide.[99]

$$ \underset{(83)}{\text{Ar-Tl(OCOCF}_3)_2} \; + \; \text{KI} \longrightarrow \left[\underset{(91)}{\text{Ar} - \text{Tl}} \overset{I}{\underset{I}{\diagup}} \right] \longrightarrow \text{Ar- I} $$

Scheme 8.20

The thallation-iodination sequence has been used by McKillop and Taylor in their mechanistic studies of the aromatic thallation reaction, as the aromatic iodides are easier to isolate and analyse than the thallated products.[98-100,114] The thallation-iodination sequence has also been used for the preparation of a variety of aromatic as well as heteroaromatic iodides. Among the aromatic iodides, a number of simple substituted benzene derivatives have been prepared as well as more functionalised derivatives, such as 1-iododibenzosuberone,[115] *o*-iodotoluic acid,[116] *p*-iodophenylpentadecanoic acid derivatives,[117] iodopentafluorobenzene,[93] 1-iodo-4-methoxytetrafluorobenzene,[109] 5-iodo-2-methoxy-6-(8-hydroxyoctanyl)benzoic acid lactone (92)[96] or *p*-iodocalix[4]arenes (93).[118] Sometimes, diiodo compounds are obtained selectively by using an excess of trifluoroacetic acid[119] or by adding antimony pentafluoride as a catalyst.[92] Heteroaromatic iodinated compounds have been prepared with various heterocycles such as thiophene,[99,100] pyrrole,[120,121] 1,3,5,2,4-trithiadiazepine (94)[125,126] indole,[120,122] and indoline.[123,124] In the indolic systems, the regioselectivity of the thallation can be controlled. Thallation of 3-carbonylindoles affords the 4-thallated products which can be treated with a nucleophile such as iodide to afford selectively the 4-iodo 3-carbonylindole [(95), X = I]. Thallation-iodination of *N*-acetylindoline leads to the 7-iodo derivative [(96), X = I] which can then be hydrolysed and oxidised to give the corresponding 7-iodoindole derivative.[124] This reaction of indolic compounds has been applied to the selective functionalisation of the complex system of bipolaramide analogues [(**97**), X = I].[127,128] For the selected examples reported in scheme 8.21, the yields are given for the two steps sequence: 1 - thallation, 2- iodination.

$$\text{Ar-H} \quad \xrightarrow[\text{2) KI}]{\text{1) Tl(OCOCF}_3)_3} \quad \text{Ar-I}$$

$$\text{X = H} \quad \longrightarrow \quad \text{X = I}$$

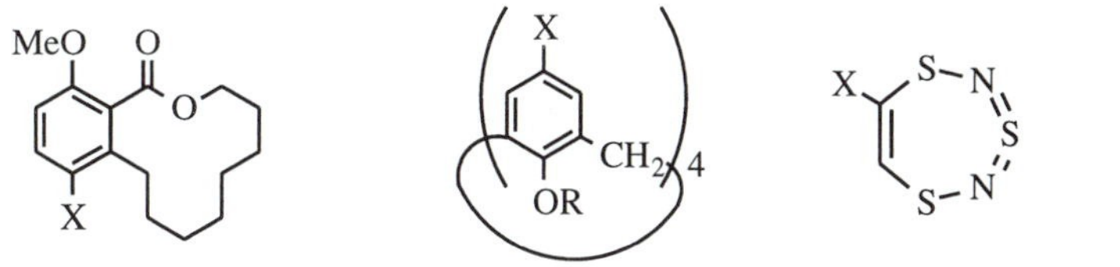

(**92**), 92%[96]

(**93**), 35%[118]

(**94**), 80%[125,126]

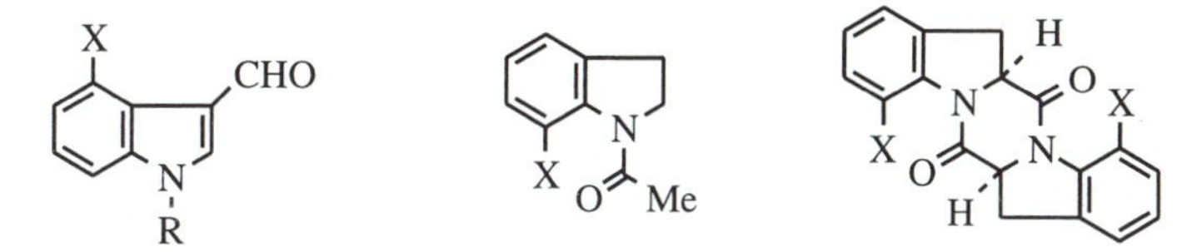

(**95**), R = H   50%[120]
R = OMe   91%[122]

(**96**), 74%[123,124]

(**97**), 29%[127]

Scheme 8.21

Improved yields were observed for the 4-iodination reaction of [(**95**), R = H, X = Tl(OCOCF$_3$)$_2$] by using a combination of copper(I) iodide and iodine, as a iodinating reagent, as the 4-iodo compound was obtained in 94% yield instead of 71% with potassium iodide.[129] However, this improvement was not always observed: in the case of the bipolaramide derivative [(**97**), X = Tl(OCOCF$_3$)$_2$] it gave a lower yield [16% with copper(I) iodide and iodine instead of 20% with potassium iodide].[127]

The monoorganothallium(III) iodide derivative can also be prepared from the diarylthallium compound (**98**), which is easily made from the Grignard reagent. Treatment of the diarylthallium compound with iodine gives two molecules of the corresponding aryl iodide, one directly by electrophilic substitution, and the second one derived from the ligand coupling decomposition of the intermediately formed aryliodothallium(III) trifluoroacetate (**99**). This reaction afforded the aryl iodides in yields ranging from 74 to 94%.[130]

$$\text{Ar}_2\text{TlOCOCF}_3 + \text{I}_2 \xrightarrow{\text{CHCl}_3} \text{ArI} + \text{Ar-Tl}\overset{\text{OCOCF}_3}{\underset{\text{I}}{\diagdown}} \longrightarrow \text{Ar-I} + \text{TlOCOCF}_3$$

(**98**)   (**99**)

8.4.1.2 C-Tl to C-X

The reaction of arylthallium(III) compounds (**83**) with potassium bromide gave the dibromoaryl thallium derivatives (**100**), which decomposed rapidly either on storage or on gentle heating into aryl bromides and thallium(I) bromide.[131] With other metal halide salts (NaCl and KF), only the stable arylthallium dihalides were isolated.[131] However, when arylthallium difluorides **(101)** were treated with boron trifluoride in non-polar solvents, the corresponding aryl fluorides were obtained in 50-70% yields, presumably through formation of arylthallium(III) bis(tetrafluoroborates).[132]

$$ArTl(OCOCF_3)_2 + 2\ KBr \longrightarrow ArTlBr_2 \xrightarrow{\Delta} Ar\text{-}Br + TlBr$$
$$\quad\quad (83) \quad\quad\quad\quad\quad\quad\quad\quad\quad (100)$$

$$ArTl(OCOCF_3)_2 + 2\ KF \longrightarrow ArTlF_2 \xrightarrow{BF_3} Ar\text{-}F$$
$$\quad\quad (83) \quad\quad\quad\quad\quad\quad\quad (101)$$

The aryl chlorides and aryl bromides are easily prepared by treatment of the arylthallium(III) bis(trifluoroacetates) with the corresponding copper(I) or copper(II) halides. The best yields were obtained with the copper(II) halides in dioxane under reflux.[102] These reactions have been successfully applied by Somei *et al.* to the synthesis of indole derivatives by reaction of the arylthallium compounds with copper(II) salts in DMF.[127,133,134]

$$X = Tl(OCOCF_3)_2 \longrightarrow X = Hal$$

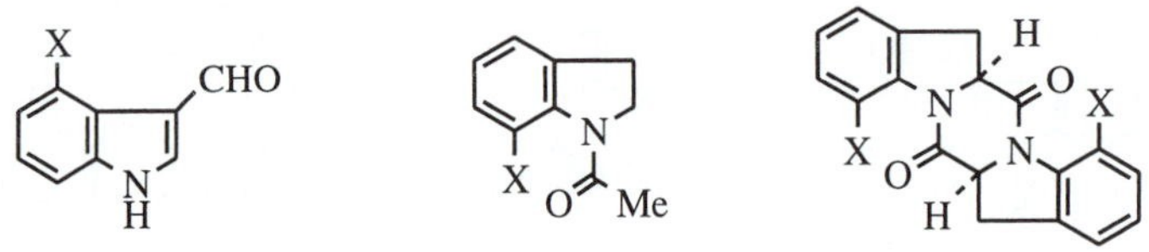

| Hal = Cl | 48%[133] | 42%[134] | 78%[127] |
| Hal = Br | 58%[133] | 62%[134] | - |

8.4.1.3 C-Tl to C-OH

The direct synthesis of phenols by treatment of the intermediate arylthallium(III) compounds with hydroxyde anions does not take place as the intermediate arylthallium oxides are stable.[135] However, the conversion of arylthallium(III) compounds to aryloxide derivatives is efficiently performed by different indirect procedures, such as:
- reaction with lead tetraacetate, followed by treatment with triphenylphosphine and eventually hydrolysis to the phenol.[135]
- reaction with diborane, followed by oxidation of the arylboron derivative with hydrogen peroxide to the phenol.[136,137] If the the product of transmetallation, the arylboron derivative, is hydrolysed under non-oxidizing conditions, the arylboronic acid derivative can be isolated.
- reaction with water in the presence of copper(II) sulfate in DMF.[124,138] This reaction was used for the synthesis of bipolaramide **(102)**.[127,128]

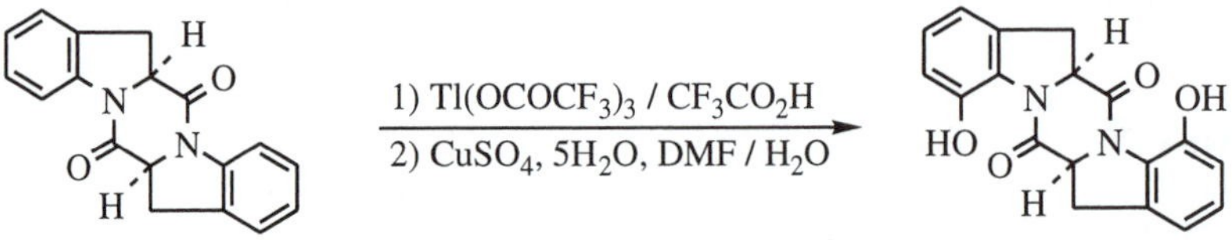

1) Tl(OCOCF_3)_3 / CF_3CO_2H
2) CuSO_4, 5H_2O, DMF / H_2O

(102), 35%

8.4.1.4 C-Tl to C-N

Treatment of arylthallium(III) compounds with metal nitrites in trifluoroacetic acid affords the corresponding nitrosoarenes, which are oxidised to the nitroarenes. The reaction was suggested to occur by electrophilic attack of NO^+ on the carbon-thallium bond.[139,140]

$$ArTlX_2 \;+\; NaNO_2 \xrightarrow{\;CF_3CO_2H\;} ArNO \xrightarrow{\;[Ox]\;} Ar\text{-}NO_2$$

Direct synthesis of the nitro derivatives (103) - (105) was reported to occur in the treatment of indolic derivatives with copper(II) sulfate and sodium nitrite in DMF.[104]

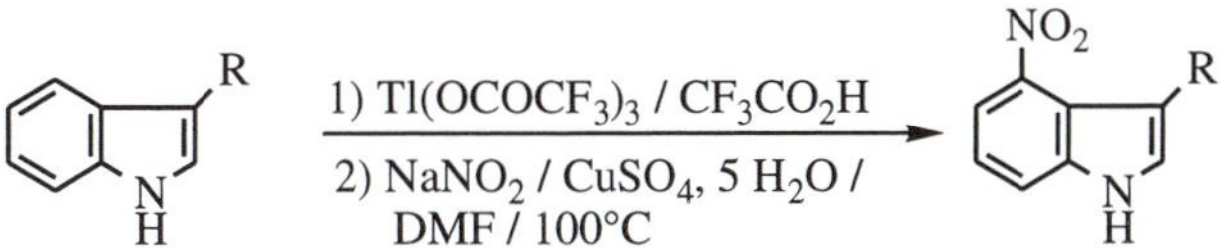

(103), R = CHO   65%
(104), R = COOMe 54%

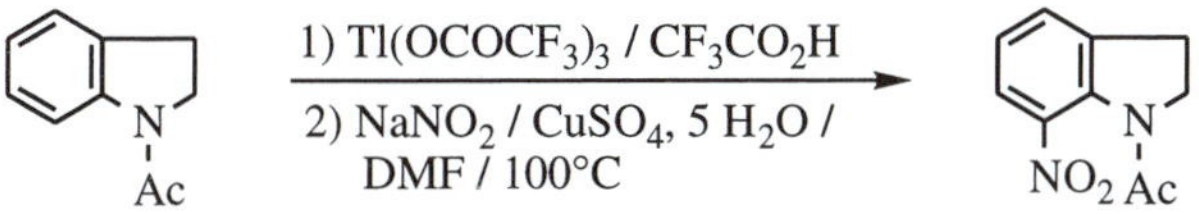

(105), 58%

Azides (106) and (107) can be obtained by a similar reaction system, using sodium azide instead of sodium nitrite.[104] In the treatment of indolic derivatives with copper(II) sulfate and sodium azide in DMF, very poor yields were sometimes observed. Replacing copper(II) sulfate by copper(I) iodide resulted in a considerable increase of the yields.[141]

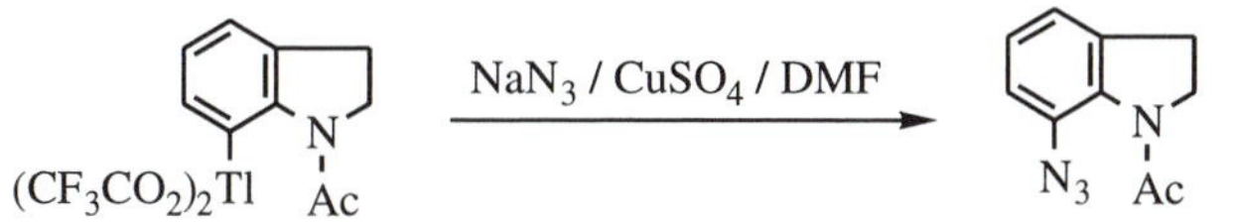

(106), 90%   ref. 104

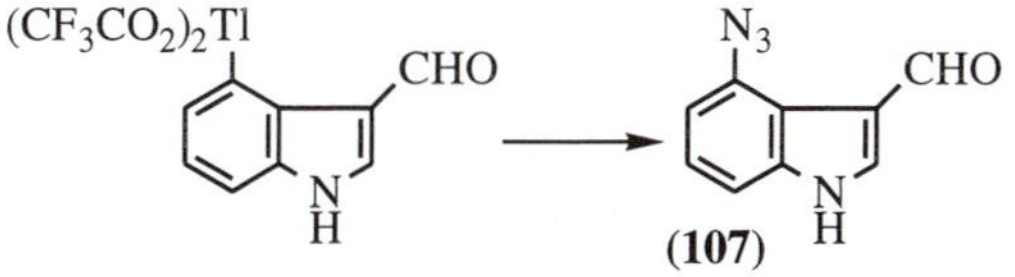

NaN$_3$ / CuSO$_4$ / DMF: 10-30%   ref. 104
NaN$_3$ / CuI / DMF: 80-87%   ref. 141

These reactions were applied to the synthesis of nitro and azido derivatives of bipolaramide. In this case, the nitration reaction afforded only a modest yield of the product (108).[128]

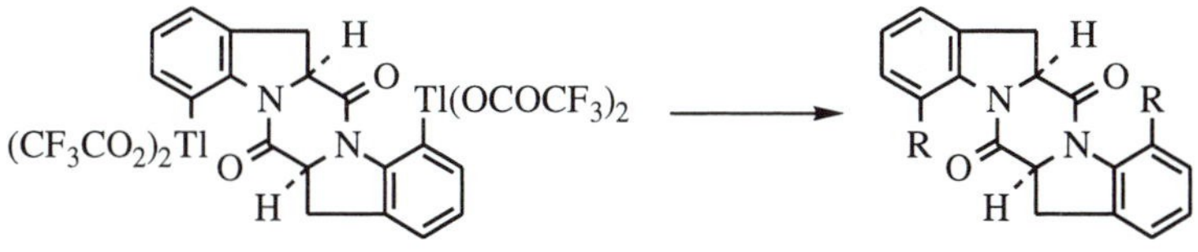

NaNO$_2$ / CuSO$_4$, 5 H$_2$O / DMF $\longrightarrow$ (108), R = NO$_2$ 21%
NaN$_3$ / CuI / DMF $\longrightarrow$ (109), R = N$_3$ 64%

8.4.1.5 C-Tl to C-S

Thiophenols are obtained by photolysis of the appropriate arylthallium dithiocarbamate.[88]

$$ArTl\left(S-\overset{\overset{\displaystyle S}{\|}}{C}-NMe_2\right)_2 \xrightarrow{\;h\nu\;} \begin{array}{c} ArS\text{-}SAr \\ + \\ Ar-S-\overset{\overset{\displaystyle S}{\|}}{C}-NMe_2 \end{array} \xrightarrow{\;H_2O\;} Ar\text{-}SH$$

The reaction of phenylthallium diacetate with the sodium salt of thiophenol did not afford diphenylsulfide. Instead, the product of disproportionation of the thallium reagent was observed. Although not isolated, this product was suggested to be diphenylthallium phenylsulfide (110).[142]

$$PhTl(OAc)_2 \;+\; PhSNa \xrightarrow{\;MeOH\;} Ph_2Tl\text{-}SPh \;+\; Ph_2S_2$$
$$(110)$$

Arylthiocyanates (111),[103] as well as the related arylselenocyanates (112),[143] are obtained by the reaction of the arylthallium(III) compound with Cu(SCN)$_2$ or Cu(SeCN)$_2$ respectively.

$$ArS\text{-}CN \xleftarrow{\;Cu(SCN)_2\;} ArTlX_2 \xrightarrow{\;Cu(SeCN)_2\;} ArSe\text{-}CN$$
$$(111) \hspace{6cm} (112)$$

Arylthiocyanates can also be prepared by treatment of the thallium compound with potassium thiocyanate. However, it requires photochemical activation to proceed, the bis(thiocyanato)thallium compound being too stable to decompose directly by ligand coupling.[144,145]

$$ArTlX_2 \xrightarrow[\;2)\ h\nu\;]{\;1)\ KSCN\;} ArS\text{-}CN \qquad (111),\ 36\text{-}58\%$$

Diarylsulfones are easily prepared by treatment of the thallium compound with copper(II) benzenesulfinate generated *in situ* by the combination of copper(II) sulfate with an excess of sodium benzenesulfinate.[146]

$$ArTlX_2 \;+\; Cu(SO_2Ph)_2 \xrightarrow{\;dioxane\ /\ H_2O\;} Ar\text{-}SO_2\text{-}Ph$$

8.4.1.6 C-Tl to C-C

The conversion of a carbon-thallium bond into a carbon-carbon bond has been performed directly only in a relatively limited number of cases: formation of arylcyanides, reaction with nitroalkane salts as well as the reaction with copper acetylides. Other carbon-carbon bond formation reactions have been reported with a carbon-thallium substrate. However, they all involve palladium catalysis, and this is beyond the scope of this book.[147]

The reaction of arylthallium(III) compounds with potassium cyanide affords stable arylthallium(IV) cyanide complexes,[113,135] which can only decompose upon photolytic activation, indicative of a radical mechanism.[135] Decomposition of these complexes by heating does not afford the aryl cyanide in efficient yields. Pyrolysis afforded only trace amounts of the cyanide.[135] Heating a water solution of the potassium phenylthallium tricyanide complex (113) led to the formation of diphenylthallium cyanide.[113]

$$\text{ArTlX}_2 \ + \ \text{KCN} \ \longrightarrow \ \text{ArTl(CN)}_3^{\ominus} \ \text{K}^{\oplus} \ \xrightarrow{h\nu} \ \text{Ar-CN} \quad 27\text{-}80\%$$

(113)

However, aryl cyanides are more directly obtained by reaction of the arylthallium compound with various copper cyanides. Different reaction systems have been reported, such as CuCN or Cu(CN)$_2$ in acetonitrile or pyridine with arylthallium acetate perchlorate monohydrates.[101] Other systems involve the arylthallium bistrifluoroacetate with CuCN in acetonitrile[105] or in DMF.[148] With these systems, a number of aryl cyanides (114),[105] heteroaryl cyanides (115)[125,126] and indolyl cyanides (116) and (117)[124,148] have been prepared.

$$X = \text{Tl(OCOCF}_3)_2 \ \longrightarrow \ X = \text{CN}$$

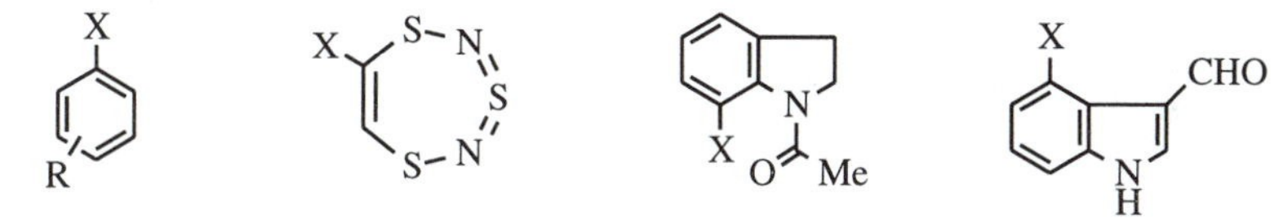

(114), 62-85%[105]   (115), 85%[125,126]   (116), 69%[124]   (117), 53%[148]

The reaction of arylthallium(III) diacetates with nitroalkane anions under photolysis gave the arylated nitroalkanes (118) in 60-70% yields. As photolytic activation was required, radical intermediates were postulated to be generated by electron transfer activation of the carbon-thallium bond. However, a relatively good yield of the *C*-aryl product was obtained as well at ambient light as in the dark, indicating the significant contribution of a different pathway, likely to be a ligand coupling process.[149,150]

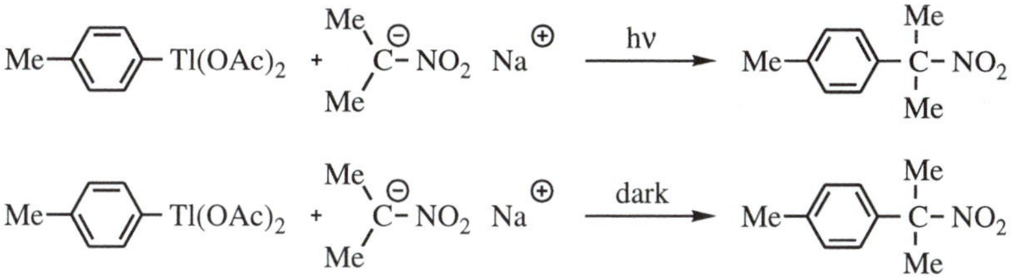

Thallation of benzanilides with thallium tris(trifluoroacetate) in a mixture of trifluoroacetic acid and ether gave the *ortho*-thallated products. Reaction of these products with copper(I) acetylide in acetonitrile led to the 2-benzamidotolanes (119), which were eventually elaborated into the 2-phenyl-indole derivatives.[106]

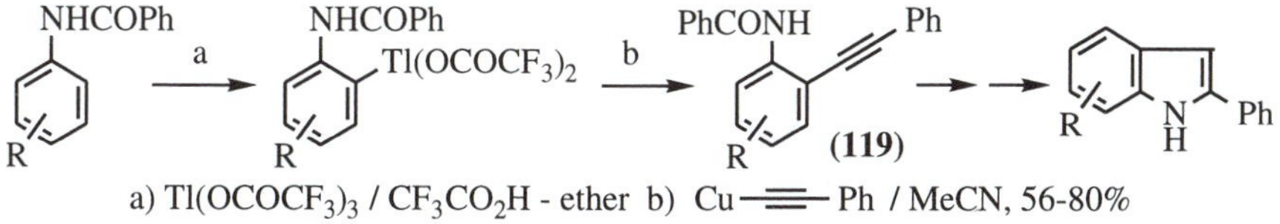

a) Tl(OCOCF$_3$)$_3$ / CF$_3$CO$_2$H - ether b) Cu$\equiv$Ph / MeCN, 56-80%

Scheme 8.22

The attempt to substitute the thallium moiety of phenylthallium diacetate with acetylacetone failed to give the 3-phenylpentan-2,4-dione. Instead, the presence of the disproportionation product (120) of the thallium reagent was detected by NMR.[142]

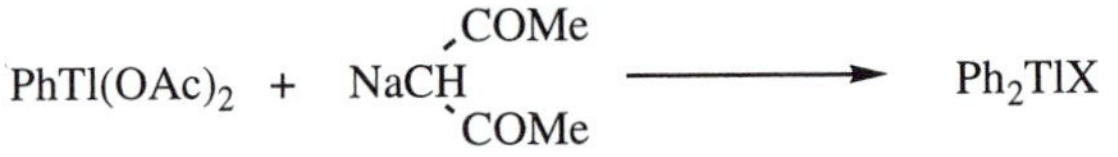

$$\text{PhTl(OAc)}_2 \ + \ \text{NaCH}\underset{\text{COMe}}{\overset{\text{COMe}}{<}} \longrightarrow \text{Ph}_2\text{TlX} \quad \textbf{(120)}, \textit{ca.}\ 90\% \text{ by NMR}$$

8.4.2 Alkyl and vinylthallium(III) compounds

Alkyl and vinylthallium(III) compounds are mostly obtained by oxythallation of alkenes or alkynes. They can also be formed by transmetallation between, for example, an organomercury compound and thallium(III) salts. Various types of reactivity have been observed for these compounds. The evolution of the carbon-thallium bond into functionalised products usually takes place homolytically or heterolytically. The ligand coupling process has also been evoked to rationalize a small number of transformations.

Similarly to the reaction of arylthallium(III) compounds, the reaction of potassium iodide with (121), the product of oxythallation of alkenes, afforded the derived oxy-iodination product (123) in good yields through the alkylthallium diiodide intermediate (122).[151]

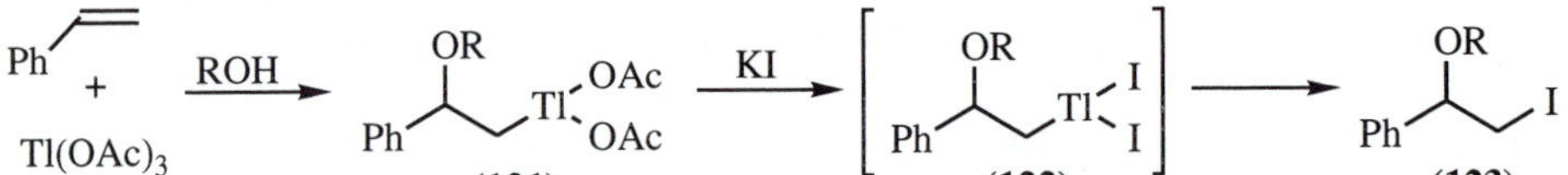

Scheme 8.23

With other halides, the reaction is conveniently performed by treatment of the alkylthallium compound with the appropriate copper(I) halide. The product yield is increased by the addition of potassium halide.[151] The thallium moiety can also be replaced by other groups, similarly to the arylthallium compounds. The cyano group is introduced by reaction with copper(I) cyanide,[151] the thiocyano group by reaction with potassium and/or copper thiocyanate[151,152] and the selenocyano group by treatment with potassium selenocyanate.[143]

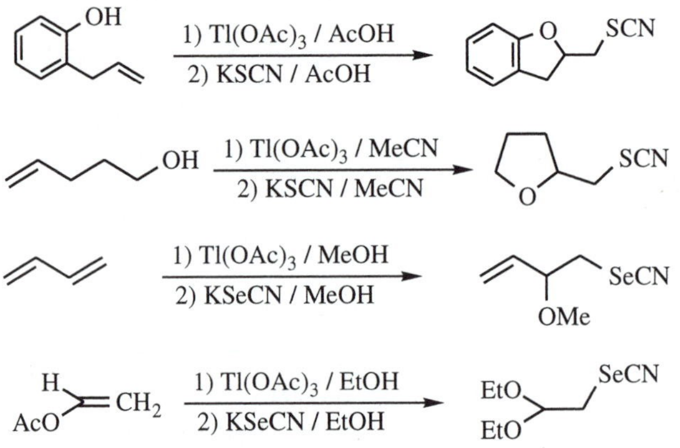

In all these reactions, the ligand transfer or ligand coupling takes place on the carbon to which the thallium group was originally bound. By contrast with a number of oxidation reactions with thallium salts, phenyl migration was not observed. In the reaction of oxythallation followed by conversion to SCN or to SeCN, transient organothallium(III) dithiocyanates [(124), X=S] or diselenocyanates [(124), X=Se] were postulated to be formed and to decompose by ligand coupling.[152]

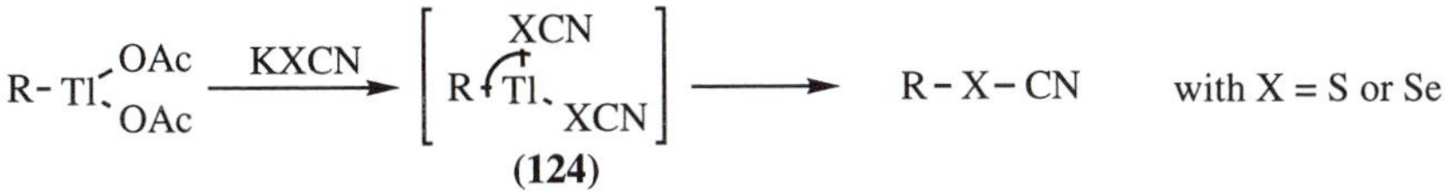

Scheme 8.24

The bromodethalliation of methoxyalkylthallium diacetate (125) by potassium bromide in the presence of 2,6-dimethyl-18-crown-6 affords 1-bromo-2-methoxyalkanes. A S_N2 displacement of the thallium group after or before anion exchange was suggested to be the most likely, although the possibility of a ligand coupling involving transfer of the bromine from thallium to the C-1 atom was not ruled out.[153] (Scheme 8.25)

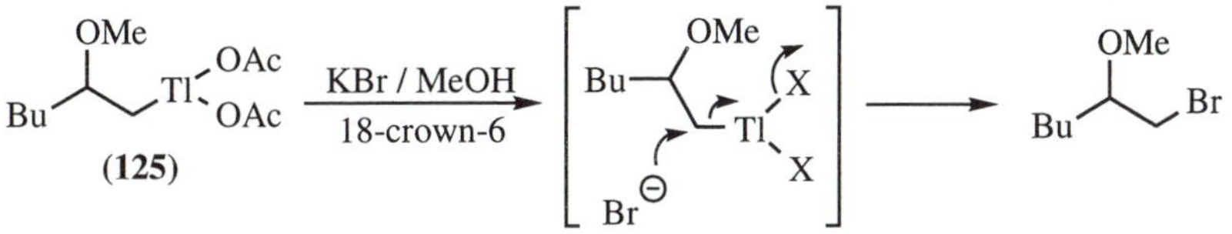

Scheme 8.25

By contrast, the methoxythallation adduct (126) obtained through reaction of *n*-hex-1-ene with thallium tris(trifluoroacetate) (84) undergoes rapidly dethallation to lead to 1,2-dimethoxyhexane and 2-methoxyhexan-1-ol. These compounds arose from a first ligand exchange giving rise to the new intermediates (127) with R = H or Me, followed by transfer of the methoxy or hydroxy groups from thallium to the C-1 atom in a ligand coupling process.[153] (Scheme 8.26)

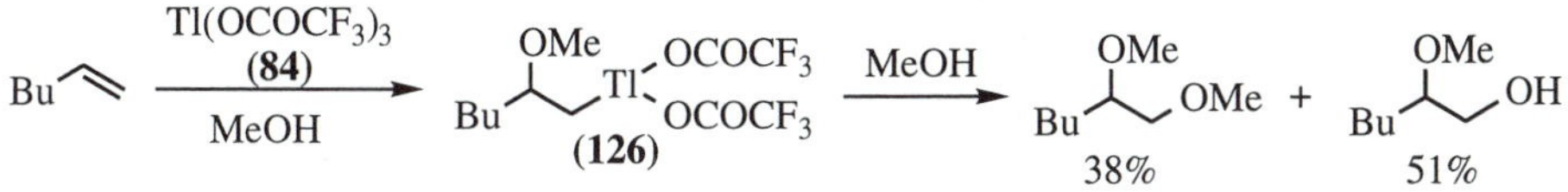

Mechanism:

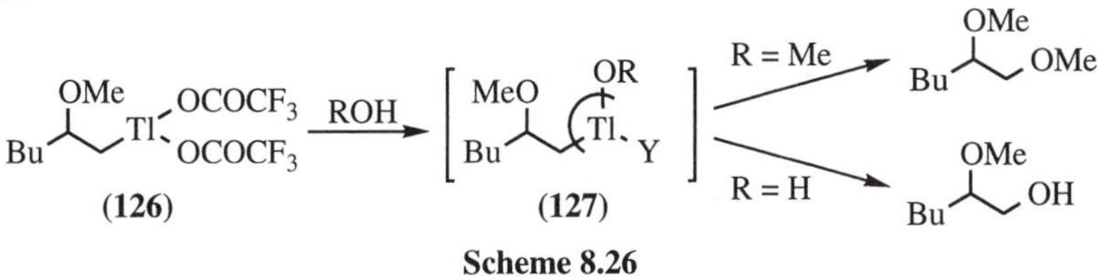

Scheme 8.26

The mechanism of the copper(I) halide-mediated halogenodethallation of the 2-methoxyalkyl-thallium diacetate (128) appeared to depend on the reaction temperature. In acetonitrile at 80°C, a free radical pathway was predominant.[154] (Scheme 8.27) However at 60°C, the ionic pathway via a S_N2 displacement of the thallium group became more important.[155] (Scheme 8.28)

Radical pathway:

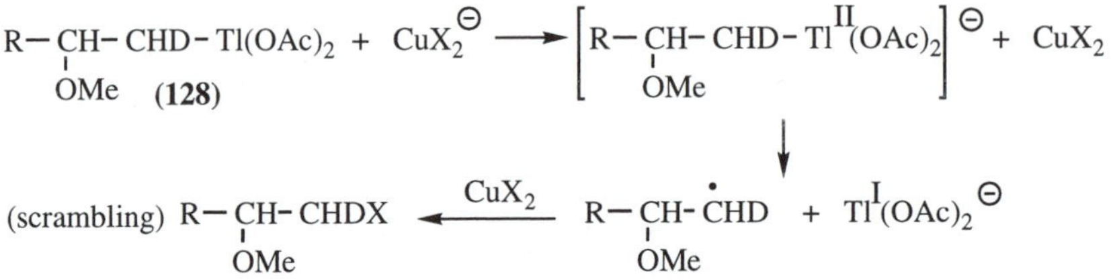

$$R-\underset{\underset{OMe}{|}}{CH}-CHD-Tl(OAc)_2 \ + \ CuX_2^{\ominus} \ \longrightarrow \ \left[R-\underset{\underset{OMe}{|}}{CH}-CHD-Tl^{II}(OAc)_2\right]^{\ominus} + \ CuX_2$$

$$(128)$$

$$(\text{scrambling}) \ R-\underset{\underset{OMe}{|}}{CH}-CHDX \ \xleftarrow{\ CuX_2\ } \ R-\underset{\underset{OMe}{|}}{CH}-\overset{\bullet}{C}HD \ + \ Tl^I(OAc)_2^{\ \ominus}$$

Scheme 8.27

Ionic pathway:

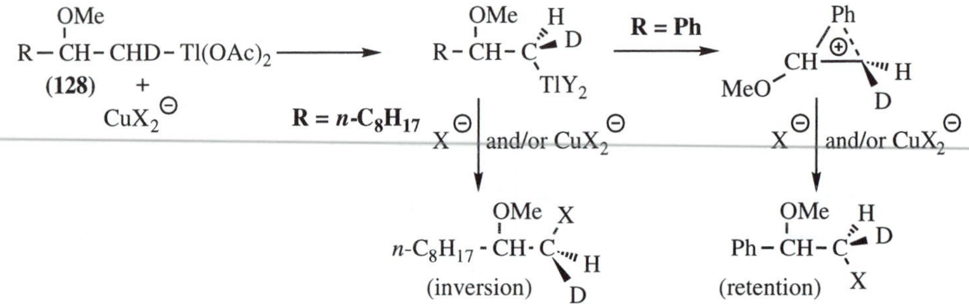

$$\underset{\underset{(128)}{}}{\overset{OMe}{\underset{|}{R-CH-CHD-Tl(OAc)_2}}} \ \longrightarrow \ R-\underset{\underset{|}{OMe}}{CH}-\overset{H}{\underset{TlY_2}{C{\blacktriangleleft}D}} \ \xrightarrow{\ R=Ph\ } \ \underset{MeO}{\overset{Ph}{\underset{}{\oplus}}}CH\!-\!\!\overset{H}{\underset{D}{\ldots}}$$

$$\begin{array}{cc} (128) & \\ + & R = n\text{-}C_8H_{17} \\ CuX_2^{\ominus} & X^{\ominus}\, \text{and/or}\, CuX_2^{\ominus} \end{array} \qquad\qquad X^{\ominus}\,\text{and/or}\, CuX_2^{\ominus}$$

$$n\text{-}C_8H_{17}-\underset{\underset{}{|}}{\overset{OMe}{CH}}\cdot\overset{X}{\underset{D}{C\cdots_{H}}} \qquad\qquad Ph-\underset{\underset{}{|}}{\overset{OMe}{CH}}-\overset{H}{\underset{X}{C\blacktriangleleft D}}$$

$$\text{(inversion)} \qquad\qquad\qquad \text{(retention)}$$

Scheme 8.28

The formation of a carbon-carbon bond by reaction of alkylthallium(III) compounds has been described in the α-nitroalkylation of alkanes. The reaction of alkylthallium **(129)** with nitroalkane anions leads to the nitroalkyl derivatives **(130)**. In this case, radical intermediates generated by electron transfer activation of the carbon-thallium bond are involved in a non-chain substitution process.[149,150]

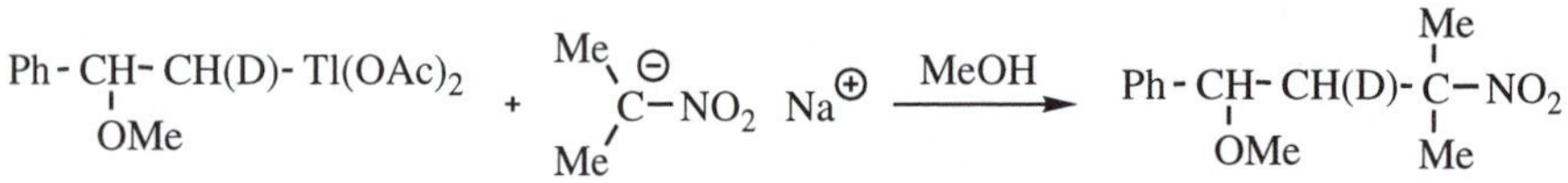

$$Ph-\underset{\underset{OMe}{|}}{CH}-CH(D)-Tl(OAc)_2 \ + \ \underset{Me}{\overset{Me}{>}}\!\!C\!\!-\!\!NO_2^{\ \ominus}\ Na^{\oplus} \ \xrightarrow{\ MeOH\ } \ Ph-\underset{\underset{OMe}{|}}{CH}-CH(D)-\underset{\underset{Me}{|}}{\overset{Me}{\underset{|}{C}}}\!\!-\!\!NO_2$$

$$\textbf{(129)},\ erythro \qquad\qquad\qquad\qquad\qquad \textbf{(130)},\ 90\%$$
$$(threo:erythro\ =\ 1{:}1)$$

Vinylthallium(III) diacetates **(131)** react also with nitronate anions. However, the mechanism is now completely different, as the reaction proceeds with retention of stereochemistry and is not affected by irradiation or by oxygen.[149,150] It was suggested that the reaction proceeds by a multi-step vinylic nucleophilic substitution. However, a ligand coupling mechanism is also fully compatible with the experimental observations.

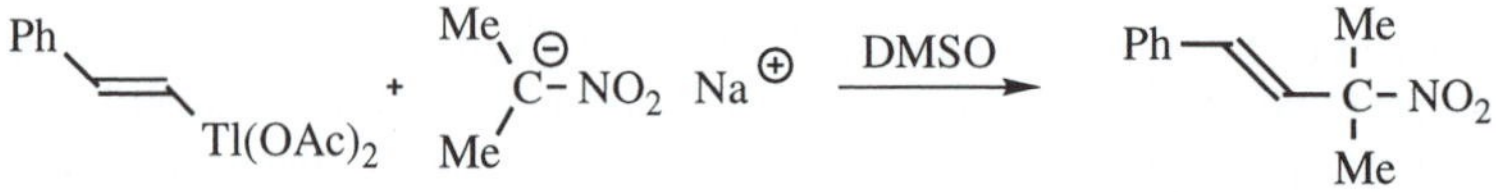

$$\underset{\underset{Tl(OAc)_2}{}}{\overset{Ph}{\diagdown}}\!\!\diagup \quad + \quad \underset{Me}{\overset{Me}{>}}\!\!C\!\!-\!\!NO_2^{\ \ominus}\ Na^{\oplus} \ \xrightarrow{\ DMSO\ } \ \overset{Ph}{\diagdown}\!\!\diagup\!\!\overset{Me}{\underset{\underset{Me}{|}}{\underset{|}{C}}}\!\!-\!\!NO_2$$

$$\textbf{(131)},\ (Z:E = 0{:}100) \qquad\qquad\qquad\qquad 99\%\ (Z:E = 0{:}100)$$

$$Ph-\!\!\!=\!\!\!-Tl(OAc)_2 \quad + \quad \underset{Me}{\overset{Me}{\underset{}{}}}\!\!\!\overset{\ominus}{C}\!\!-NO_2 \;\; Na^{\oplus} \quad \xrightarrow{\;DMSO\;} \quad Ph-\!\!\!=\!\!\!-\underset{Me}{\overset{Me}{\underset{|}{\overset{|}{C}}}}\!\!-NO_2$$

(131), *(Z : E = 70:30)*   93% *(Z : E = 65:35)*

Other monovinylthallium(III) compounds such as **(132)**, the product of oxythallation of acetylenes, react analogously to the arylthallium compounds. Halogenodethallation and pseudohalogeno-dethallation reactions result from the interaction of the oxyalkenylthallium compound with either potassium iodide or with the appropriate copper derivatives.[156]

$$\underset{AcO}{\overset{Ph}{}}\!\!\!\diagdown\!\!=\!\!\!\diagup\!\!\!\underset{\underset{Me}{(132)}}{\overset{Tl(OAc)_2}{}} \quad + \quad KI \quad \xrightarrow{\;MeCN\;} \quad \underset{AcO}{\overset{Ph}{}}\!\!\!\diagdown\!\!=\!\!\!\diagup\!\!\!\underset{Me}{\overset{I}{}}$$

$$\underset{\underset{(132)}{AcO}}{\overset{Ph}{}}\!\!\!\diagdown\!\!=\!\!\!\diagup\!\!\!\underset{Me}{\overset{Tl(OAc)_2}{}} \quad + \quad CuX_2 \quad \xrightarrow{\;MeCN\;} \quad \underset{AcO}{\overset{Ph}{}}\!\!\!\diagdown\!\!=\!\!\!\diagup\!\!\!\underset{Me}{\overset{X}{}} \qquad X = Br, Cl \text{ or } CN$$

8.5 REFERENCES

1. Tamao, K.; Akita, T.; Iwahara, T.; Kanatani, R.; Yoshida, J.; Kumada, M. *Tetrahedron* **1983**, *39*, 983-990.
2. Oae, S.; Uchida, Y. *Acc. Chem. Res.* **1991**, *24*, 202-208.
3. Chuit, C.; Corriu, R.J.P.; Reye, C.; Young, J.C. *Chem. Rev.* **1993**, *93*, 1371-1448.
4. Reich, H.J., Lecture at Université Louis Pasteur, Strasbourg, France, April 1987, cited in references 5 and 6.
5. Oae, S. *Reviews on Heteroatom Chemistry* **1988**, *1*, 304-335.
6. Oae, S. *Main Group Chemistry News* **1996**, *4*, 10-17.
7. Tius, M.A.; Kawakami, J.K. *Tetrahedron* **1995**, *51*, 3997-4010.
8. Freedman, L.D.; Doak, G.O. in *The Chemistry of the Metal-Carbon Bond*; Hartley, F.R., Ed.; J. Wiley & Sons: New York, **1989**; Vol. 5, Ch. 9, pp. 397-433.
9. Huang, Y.-Z. *Acc. Chem. Res.* **1992**, *25*, 182-187.
10. Shen, K.-w.; McEwen, W.E.; Wolf, A.P. *J. Am. Chem. Soc.* **1969**, *91*, 1283-1288.
11. Dodonov, V.A.; Bolatova, O.P.; Gushchin, A.V. *Zh. Obshch. Khim.* **1988**, *58*, 711.
12. Akiba, K-y. *Pure Appl. Chem.* **1996**, *68*, 837-842.
13. Razuvaev, G.A.; Osanova, N.A.; Shulaev, N.P.; Tsigin, B.M. *Zh. Obshch. Khim.* **1960**, *30*, 3234-3237.
14. McEwen, W.E.; Lin, C.-T. *Phosphorus* **1974**, *4*, 91-96.
15. Fujiwara, M.; Tanaka, M.; Baba, A.; Ando, H.; Souma, Y. *J. Organomet. Chem.* **1996**, *508*, 49-52.
16. Fujiwara, M.; Tanaka, M.; Baba, A.; Ando, H.; Souma, Y. *J. Organomet. Chem.* **1996**, *525*, 39-42.
17. McEwen, W.E.; Briles, G.H. *Tetrahedron Lett.* **1966**, 5191-5196.
18. McEwen, W.E.; Briles, G.H.; Giddings, B.E. *J. Am. Chem. Soc.* **1969**, *91*, 7079-7084.
19. Razuvaev, G.A.; Osanova, N.A. *J. Organomet. Chem.* **1972**, *38*, 77-82.
20. Lanneau, G.F.; Wikholm, R.J.; Lin, C.T.; McEwen, W.E. *J. Organomet. Chem.* **1975**, *85*, 179-191.
21. Razuvaev, G.A.; Osanova, N.A.; Brilkina, T.G.; Zinovjeva, T.I.; Sharutin, V.V. *J. Organomet. Chem.* **1975**, *99*, 93-106.

22. Chupka, F.L.; Knapczyk, J.W.; McEwen, W.E. *J. Org. Chem.* **1977**, *42*, 1399-1402.
23. McEwen, W.E.; Chupka, F.L. *Phosphorus* **1972**, *1*, 277-282.
24. Sharutin, V.V.; Zhidkov, V.V.; Muslin, D.V.; Liapina, N.Sh.; Fukin, G.K.; Zakharov, L.N.; Yanovsky, A.I.; Struchkov, Yu.T. *Izv. Akad. Nauk, Ser. Khim.* **1995**, 958-963.
25. Sharutina, O.K.; Sharutin, V.V.; Senchurin, V.A.; Fukin, G.K.; Zakharov, L.N.; Yanovsky, A.I.; Struchkov, Yu.T. *Izv. Akad. Nauk, Ser. Khim.* **1996**, 194-198.
26. Barton, D.H.R.; Finet, J.-P.; Giannotti, C.; Halley, F. *J. Chem. Soc., Perkin Trans. 1* **1987**, 241-249.
27. Gushchin, A.V.; Dyomina, O.P.; Dodonov, V.A. *Izv. Akad. Nauk, Ser. Khim.* **1995**, 964-967.
28. Wardell, J.L.; Grant, D.W. *J. Organomet. Chem.* **1980**, *188*, 345-351.
29. Schmidbaur, H.; Mitschke, K.-H. *Chem. Ber.* **1971**, *104*, 1837-1841.
30. Akiba, K.-y.; Okinaka, T.; Nakatani, M.; Yamamoto, Y. *Tetrahedron Lett.* **1987**, *28*, 3367-3368.
31. Strecker, W.; Willing, A. *Ber. Dtsch. Chem. Ges.* **1915**, *48*, 196-206.
32. Wittig, G.; Fritz, H. *Liebigs Ann. Chem.* **1952**, *577*, 39-46.
33. Ishii, Y.; Iwama, Y.; Ogawa, M. *Synth. Commun.* **1978**, *8*, 93-97.
34. Iwama, Y.; Aragi, M.; Sugiyama, M.; Matsui, K.; Ishii, Y.; Ogawa, M. *Bull. Chem. Soc. Jpn.* **1981**, *54*, 2065-2067.
35. Ogawa, S.; Sato, S.; Erata, T.; Furukawa, N. *Tetrahedron Lett.* **1991**, *32*, 3179-3182.
36. Ogawa, S.; Sato, S.; Masutomi, Y.; Furukawa, N. *Phosphorus, Sulfur, and Silicon* **1992**, *67*, 99-102.
37. Ogawa, S.; Sato, S.; Furukawa, N. *Tetrahedron Lett.* **1992**, *33*, 7925-7928.
38. Sato, S.; Furukawa, N. *Tetrahedron Lett.* **1995**, *36*, 2803-2806.
39. Hellwinkel, D.; Fahrbach, S.I. *Liebigs Ann. Chem.* **1968**, *715*, 68-73.
40. Ogawa, S.; Sato, S.; Erata, T.; Furukawa, N. *Tetrahedron Lett.* **1992**, *33*, 1915-1918.
41. Sato, S.; Furukawa, N. *Chem. Lett.* **1994**, 889-892.
42. Leicester, H.M.; Bergstrom, F.W. *J. Am. Chem. Soc.* **1931**, *53*, 4428-4436.
43. Graham, J.C.; Feng, C.-H.; Orticochea, M.; Ahmed, G. *J. Org. Chem.* **1990**, *55*, 4102-4109.
44. Furukawa, N.; Ogawa, S.; Matsumura, K.; Fujihara, H. *J. Org. Chem.* **1991**, *56*, 6341-6348.
45. Oae, S. *Reviews on Heteroatom Chemistry* **1991**, *4*, 195-225.
46. Furukawa, N.; Ogawa, S.; Matsumura, K.; Fujihara, H. *J. Org. Chem.* **1991**, *56*, 6341-6348.
47. Hori, M.; Kataoka, T.; Shimizu, H.; Tsutsumi, K. *Chem. Pharm. Bull.* **1990**, *38*, 779-782.
48. Hori, M.; Kataoka, T.; Shimizu, H.; Miyagaki, M. *Tetrahedron Lett.* **1989**, *30*, 981-984.
49. Lermontov, S.A.; Rakov, I.M.; Zefirov, N.S.; Stang, P.J. *Phosphorus, Sulfur, and Silicon* **1994**, *92*, 225-229.
50. Umemoto, T. *Chem. Rev.* **1996**, *96*, 1757-1777.
51. Umemoto, T.; Ishihara, S. *J. Am. Chem. Soc.* **1993**, *115*, 2156-2164.
52. Ono, T.; Umemoto, T. *J. Fluorine Chem.* **1995**, *74*, 77-82.
53. Umemoto, T.; Adachi, K. *J. Org. Chem.* **1994**, *59*, 5692-5699.
54. Barton, D.H.R.; Brewster, A.G.; Ley, S.V.; Rosenfeld, M.N. *J. Chem. Soc., Chem. Commun.* **1976**, 985-986.
55. Barton, D.H.R.; Ley, S.V.; Magnus, P.D.; Rosenfeld, M.N. *J. Chem. Soc., Perkin Trans. 1* **1977**, 567-572.
56. Barton, D.H.R.; Finet, J.-P.; Thomas, M. *Tetrahedron* **1988**, *44*, 6397-6406.
57. Henriksen, L. *Tetrahedron Lett.* **1994**, *35*, 7057-7060.
58. Barton, D.H.R.; Brewster, A.G.; Ley, S.V.; Rosenfeld, M.N. *J. Chem. Soc., Chem. Commun.* **1977**, 147-148.
59. Barton, D.H.R.; Parekh, S.I. *Pure Appl. Chem.* **1993**, *65*, 603-610.
60. Engman, L. *Acc. Chem. Res.* **1985**, *18*, 274-279.
61. Petragnani, N.; Comasseto, J.V. *Synthesis* **1986**, 1-30.

62. Sadekov, I.D.; Rivkin, B.B.; Minkin, V.I. *Russ. Chem. Rev.* **1987**, *56,* 343-354.
63. Petragnani, N.; Comasseto, J.V. *Synthesis* **1991**, 793-817.
64. Petragnani, N.; Comasseto, J.V. *Synthesis* **1991**, 897-919.
65. For a recent general survey of organotellurium compounds in organic synthesis: Petragnani, N., *Tellurium in Organic Synthesis*; Academic Press: London, **1994**.
66. Barton, D.H.R.; Glover, S.A.; Ley, S.V. *J. Chem. Soc., Chem. Commun.* **1977**, 266-267.
67. Glover, S.A. *J. Chem. Soc., Perkin Trans. 1* **1980**, 1338-1344.
68. Cohen, S.C.; Reddy, M.L.N.; Massey, A.G. *J. Organomet. Chem.* **1968**, *11*, 563-566.
69. Rheinboldt, H.; Petragnani, N. *Chem. Ber.* **1956**, *89,* 1270-1276.
70. McWhinnie, W.R.; Patel, M.G. *J. Chem. Soc., Dalton Trans.* **1972**, 199-202.
71. Hellwinkel, D.; Fahrbach, S.I. *Tetrahedron Lett.* **1965**, 1823-1827.
72. Hellwinkel, D.; Fahrbach, S.I. *Liebigs Ann. Chem.* **1968**, *712,* 1-20.
73. Campos, M.M.; Petragnani, N. *Tetrahedron* **1962**, *18,* 527-530.
74. Gedridge, R.W. Jr.; Brandsma, L.; Nissan, R.A.; Verkruijsse, H.D.; Harder, S.; de Jong, R.L.P.; O'Connor, C.J. *Organometallics* **1992**, *11*, 418-422.
75. Sato, S.; Kondo, N.; Furukawa, N. *Organometallics* **1994**, *13*, 3393-3395.
76. Sato, S.; Kondo, N.; Furukawa, N. *Organometallics* **1995**, *14*, 5393-5398.
77. Jeske, J.; du Mont, W.-W.; Jones, P.G. *Angew. Chem., Int. Ed. Engl.* **1996**, *35*, 2653-2655.
78. Uemura, S.; Fukuzawa, S.-i. *J. Chem. Soc., Chem. Commun.* **1980**, 1033-1034.
79. Uemura, S.; Fukuzawa, S.-i. *J. Organomet. Chem.* **1984**, *268,* 223-234.
80. Uemura, S.; Fukuzawa, S.-i. *Chem. Lett.* **1980**, 943-946.
81. Chikamatsu, K.; Otsubo, T.; Ogura, F.; Yamaguchi, H. *Chem. Lett.* **1982**, 1081-1084.
82. Stefani, H.A.; Chieffi, A.; Comasseto, J.V. *Organometallics* **1991**, *10*, 1178-1182.
83. Dabdoub, M.J.; Comasseto, J.V.; Barros, S.M.; Moussa, F. *Synth. Commun.* **1990**, *20*, 2181-2183.
84. Vicentini, G. *Chem. Ber.* **1958**, *91,* 801-805.
85. Spencer, H.K.; Cava, M.P. *J. Org. Chem.* **1977**, *42,* 2937-2939.
86. Spencer, H.K.; Lakshmikantham, M.V.; Cava, M.P. *J. Am. Chem. Soc.* **1977**, *99,* 1470-1473.
87. McKillop, A.; Taylor, E.C. *Chem. Brit.* **1973**, *(9),* 4-11.
88. McKillop, A.; Taylor, E.C. *Acc. Chem. Res.* **1970**, *3,* 338-346.
89. McKillop, A.; Taylor, E.C. Thallium(III) salts as Oxidants in Organic Synthesis, in *Organic Syntheses by Oxidation with Metal Compounds*; Mijs, W.J.; De Jonge, C.H.R.I., Eds; Plenum Press: New York, **1986**; pp. 695-407.
90. Uemura, S. Preparation and use of organothallium(III) compounds in organic synthesis, in *The Chemistry of the Metal-Carbon Bond*; Hartley, F.R., Ed.; J. Wiley & Sons: New York, **1987**; Vol.4, Ch. 5, pp. 473-538.
91. Marko, I.E.; Leung, C.W. Thallium, in *Comprehensive Organometallic Chemistry II*; Abel, E.W.; Stone, F.G.A.; Wilkinson, G., Eds.; Pergamon Press: Oxford, **1995**; Vol. 11, pp. 437-459.
92. Deacon, G.B.; Smith, R.N.M. *Aust. J. Chem.* **1982**, *35,* 1587-1597.
93. Deacon, G.B.; Smith, R.N.M. *J. Fluorine Chem.* **1980**, *15,* 85-88.
94. McKillop, A.; Turrell, A.G.; Young, D.W.; Taylor, E.C. *J. Am. Chem. Soc.* **1980**, *102,* 6504-6512.
95. Taylor, E.C.; Katz, A.H.; Alvarado, S.I.; McKillop, A. *J. Organomet. Chem.* **1985**, *285,* C9-C12.
96. dos Santos, M.L.; de Magalhães, G.C.; Filho, R.B. *J. Organomet. Chem.* **1996**, *526,* 15-19.
97. McKillop, A.; Fowler, J.S.; Zelesko, M.J.; Hunt, J.D.; Taylor, E.C.; McGillivray, G. *Tetrahedron Lett.* **1969**, 2423-2426.
98. Taylor, E.C.; Kienzle, F.; Robey, R.L.; McKillop, A.; Hunt, J.D. *J. Am. Chem. Soc.* **1971**, *93,* 4845-4850.

99. McKillop, A.; Fowler, J.S.; Zelesko, M.J.; Hunt, J.D.; Taylor, E.C.; McGillivray, G. *Tetrahedron Lett.* **1969**, 2427-2430.
100. McKillop, A.; Hunt, J.D.; Zelesko, M.J.; Fowler, J.S.; Taylor, E.C.; McGillivray, G.; Kienzle, F. *J. Am. Chem. Soc.* **1971**, *93*, 4841-4844.
101. Uemura, S.; Ikeda, Y.; Ichikawa, K. *Tetrahedron* **1972**, *28*, 3025-3030.
102. Uemura, S.; Ikeda, Y.; Ichikawa, K. *Tetrahedron* **1972**, *28*, 5499-5504.
103. Uemura, S.; Uchida, S.; Okano, M.; Ichikawa, K. *Bull. Chem. Soc. Jpn.* **1973**, *46*, 3254-3257.
104. Somei, M.; Yamada, F.; Hamada, H.; Kawasaki, T. *Heterocycles* **1989**, *28*, 643-648.
105. Taylor, E.C.; Katz, A.H.; McKillop, A. *Tetrahedron Lett.* **1984**, *25*, 5473-5476.
106. Taylor, E.C.; Katz, A.H.; Salgado-Zamora, H.; McKillop, A. *Tetrahedron Lett.* **1985**, *26*, 5963-5966.
107. McKillop, A.; Bromley, D.; Taylor, E.C. *J. Org. Chem.* **1972**, *37*, 88-92.
108. Taylor, E.C.; Altland, H.W.; McKillop, A. *J. Org. Chem.* **1975**, *40*, 3441-3443.
109. Deacon, G.B.; Tunaley, D. *Aust. J. Chem.* **1979**, *32*, 737-753.
110. Ishikawa, N.; Sekiya, A. *Bull. Chem. Soc. Jpn.* **1974**, *47*, 1680-1682.
111. Taylor, E.C.; Danforth, R.H.; McKillop, A. *J. Org. Chem.* **1973**, *38*, 2088-2089.
112. Challenger, F.; Parker, B. *J. Chem. Soc.* **1931**, 1462-1467.
113. Challenger, F.; Richards, O.V. *J. Chem. Soc.* **1934**, 405-411.
114. Taylor, E.C.; Kienzle, F.; Robey, R.L.; McKillop, A. *J. Am. Chem. Soc.* **1970**, *92*, 2175-2177.
115. Hollins, R.A.; Salim, V.M. *Tetrahedron Lett.* **1979**, 591-592.
116. Carruthers, W.; Pooranamoorthy, R. *J. Chem. Soc., Perkin Trans.1* **1974**, 2405-2409.
117. Goodman, M.M.; Kirsch, G.; Knapp, F.F. Jr *J. Med. Chem.* **1984**, *27*, 390-397.
118. Arduini, A.; Pochini, A.; Rizzi, A.; Sicuri, A.R.; Ungaro, R. *Tetrahedron Lett.* **1990**, *31*, 4653-4656.
119. Deacon, G.B.; Tunaley, D.; Smith, R.N.M. *J. Organomet. Chem.* **1978**, *144*, 111-123.
120. Hollins, R.A.; Colnago, L.A.; Salim, V.M.; Seidl, M.C. *J. Heterocycl. Chem.* **1979**, *16*, 993-996.
121. Monti, D.; Sleiter, G. *Gazz. Chim. Ital.* **1990**, *120*, 771-774.
122. Somei, M.; Ohnishi, H.; Shoken, Y. *Chem. Pharm. Bull.* **1986**, *34*, 677-681.
123. Somei, M.; Saida, Y. *Heterocycles* **1985**, *23*, 3113-3114.
124. Somei, M.; Kawasaki, T.; Ohta, T. *Heterocycles* **1988**, *27*, 2363-2365.
125. Morris, J.L.; Rees, C.W.; Rigg, D.J. *J. Chem. Soc., Chem. Commun.* **1985**, 396-397.
126. Rees, C.W.; Surtees, J.R.J. *J. Chem. Soc., Perkin Trans.1* **1991**, 2945-2953.
127. Somei, M.; Kawasaki, T. *Chem. Pharm. Bull.* **1989**, *37*, 3426-3428.
128. Somei, M.; Kawasaki, T. *Heterocycles* **1996**, *42*, 281-287.
129. Somei, M.; Yamada, F.; Kunimoto, M.; Kaneko, C. *Heterocycles* **1984**, *22*, 797-801.
130. Taylor, E.C.; Altland, H.W.; McKillop, A. *J. Org. Chem.* **1975**, *40*, 2351-2355.
131. McKillop, A.; Hunt, J.D.; Taylor, E.C. *J. Organomet. Chem.* **1970**, *24*, 77-88.
132. Taylor, E.C.; Bigham, E.C.; Johnson, D.K.; McKillop, A. *J. Org. Chem.* **1977**, *42*, 362-363.
133. Somei, M.; Kizu, K.; Kunimoto, M.; Yamada, F. *Chem. Pharm. Bull.* **1985**, *33*, 3696-3708.
134. Somei, M.; Saida, Y.; Funamoto, T.; Ohta, T. *Chem. Pharm. Bull.* **1987**, *35*, 3146-3154.
135. Taylor, E.C.; Altland, H.W.; Danforth, R.H.; McGillivray, G.; McKillop, A. *J. Am. Chem. Soc.* **1970**, *92*, 3520-3522.
136. Breuer, S.W.; Pickles, G.M.; Podesta, J.C.; Thorpe, F.G. *J. Chem. Soc., Chem. Commun.* **1975**, 36-37.
137. Pickles, G.M.; Spencer, T.; Thorpe, F.G.; Ayala, A.D.; Podesta, J.C. *J. Chem. Soc., Perkin Trans.1* **1982**, 2949-2951.
138. Somei, M.; Iwasa, E.; Yamada, F. *Heterocycles* **1986**, *24*, 3065-3069.
139. Uemura, S.; Toshimitsu, A.; Okano, M. *Bull. Chem. Soc. Jpn.* **1976**, *49*, 2582-2584.
140. Uemura, S.; Toshimitsu, A.; Okano, M. *J. Chem. Soc., Perkin Trans.1* **1978**, 1076-1079.

141. Hamabuchi, S.; Hamada, H.; Hironaka, A.; Somei, M. *Heterocycles* **1991**, *32*, 443-448.
142. Kurosawa, H.; Sato, M. *Organometallics* **1982**, *1*, 440-443.
143. Uemura, S.; Toshimitsu, A.; Okano, M.; Ichikawa, K. *Bull. Chem. Soc. Jpn.* **1975**, *48*, 1925-1928.
144. Taylor, E.C.; Kienzle, F.; McKillop, A. *Synthesis* **1972**, 38.
145. Mitani, H.; Ando, T.; Yukawa, Y. *Mem. Inst. Sci. Ind. Res., Osaka Univ.* **1973**, *30*, 81-82; *Chem. Abstr.* **1973**, *79*, 31613n.
146. Hancock, R.A.; Orszulik, S.T. *Tetrahedron Lett.* **1979**, 3789-3790.
147. For examples of palladium-catalysed reactions involving arylthallium compounds, see ref. 90 and 91.
148. Yamada, F.; Somei, M. *Heterocycles* **1987**, *26*, 1173-1176.
149. Kurosawa, H.; Sato, M.; Okada, H. *Tetrahedron Lett.* **1982**, *23*, 2965-2968.
150. Kurosawa, H.; Okada, H.; Sato, M.; Hattori, T. *J. Organomet. Chem.* **1983**, *250*, 83-97.
151. Uemura, S.; Zushi, K.; Tabata, A.; Toshimitsu, A.; Okano, M. *Bull. Chem. Soc. Jpn.* **1974**, *47*, 920-927.
152. Mitani, H.; Ando, T.; Yukawa, Y. *Chem. Lett.* **1972**, 455-458.
153. Bloodworth, A.J.; Lapham, D.J. *J. Chem. Soc., Perkin Trans.1* **1981**, 3265-3271.
154. Uemura, S.; Toshimitsu, A.; Okano, M.; Kawamura, T.; Yonezawa, T.; Ichikawa, K. *J. Chem. Soc., Chem. Commun.* **1978**, 65-66.
155. Bäckvall, J.E.; Ahmad, M.U.; Uemura, S.; Toshimitsu, A.; Kawamura, T. *Tetrahedron Lett.* **1980**, *21*, 2283-2286.
156. Uemura, S.; Tara, H.; Okano, M.; Ichikawa, K. *Bull. Chem. Soc. Jpn.* **1974**, *47*, 2663-2671.

Index